FRACTURE 1977

ADVANCES in RESEARCH on the STRENGTH and FRACTURE of MATERIALS

TITLES IN THIS SERIES

Vol. 1	–	An Overview
Vol. 2A	–	The Physical Metallurgy of Fracture
Vol. 2B	–	Fatigue
Vol. 3A	–	Analysis and Mechanics
Vol. 3B	–	Applications and Non-Metals
Vol. 4	–	Fracture and Society

FRACTURE 1977

ADVANCES in RESEARCH on the STRENGTH and FRACTURE of MATERIALS

D.M.R. TAPLIN
Editor

Vol. 3A—Analysis and Mechanics

Fourth International Conference on Fracture
June 1977
University of Waterloo, Canada

PERGAMON PRESS
New York / Oxford / Toronto / Sydney / Frankfurt / Paris

Pergamon Press Offices:

U.S.A.	Pergamon Press Inc., Maxwell House, Fairview Park, Elmsford, New York 10523, U.S.A.
U.K.	Pergamon Press Ltd., Headington Hill Hall, Oxford OX3, OBW, England
CANADA	Pergamon of Canada, Ltd., 75 The East Mall, Toronto, Ontario M8Z 5W3, Canada
AUSTRALIA	Pergamon Press (Aust) Pty. Ltd., 19a Boundary Street, Rushcutters Bay, N.S.W. 2011, Australia
FRANCE	Pergamon Press SARL, 24 rue des Ecoles, 75240 Paris, Cedex 05, France
WEST GERMANY	Pergamon Press GmbH, 6242 Kronberg/Taunus, Pferdstrasse 1, West Germany

Library of Congress Cataloging in Publication Data

International Conference on Fracture, 4th, University
of Waterloo, 1977.
Fracture 1977.

Includes indexes.
CONTENTS: v. 1. An overview.--v. 2A. The physical metallurgy of fracture.--v. 2B. Fatigue.--v. 3A. Analysis and mechanics. [etc.]
1. Fracture mechanics--Congresses. 2. Strength of materials--Congresses. I. Taplin, David M. R.
II. Title
TA409.I44 1977 620.1'126 77-15623

ISBN 0-08-022136-X Vol. 1
0-08-022138-6 Vol. 2A
0-08-022140-8 Vol. 2B
0-08-022142-4 Vol. 3A
0-08-022144-0 Vol. 3B
0-08-022146-7 Vol. 4
0-08-022130-0 6 -vol-set

Printed in the United States of America

To Diana and Justin

Contents

Session 2 – Dynamic Processes and Stress Intensity Factors

Session 3 – Fracture Analysis

Foreword

The Fourth International Conference on Fracture, or ICF4 as it came to be known, was planned over a period of about four years. The Conference was intended as a state-of-the-art summary of our understanding of fracture in a wide variety of materials. In this respect ICF4 was very successful, and the presence of approximately 750 participants from 38 countries attested to the drawing power of the subject.

If we compare the present Conference with those preceding, several long-range trends may be deduced. There is now less concern with micromechanisms of the cleavage of iron, but there is more emphasis on effects of the environment. There is a growing realization that fracture of real materials may be dominated by the presence of inclusions and chemical segregates. We now have more work on polymers and ceramics and the beginnings of some efforts on biological materials. There is a vast range of subject matter in these papers, which will become a primary reference for workers in the field.

ICF4 had features which were absent in the earlier conferences. First was the emphasis on fracture in large structures. This has become exceedingly important with the proliferation of big ships, big aircraft, big nuclear reactors, big pipelines, big bridges and big buildings, where fractures can become major catastrophes. It was interesting to observe the different concerns of those who deal with large structures and those who are accustomed to working on a laboratory scale. The interaction was useful, and we must find ways of making it better.

Other innovations were the sessions on *Fracture, Education and Society* and on *Fracture, Politics and Society*. Public attention is now being directed towards questions of safety and of the environmental consequences of failures in large structures. These public concerns are being translated into legislation, regulation and lawsuits. We are thus being propelled, willy nilly, into one side or the other of questions of public policy, and few of us are adequately prepared to cope with this situation. If we shirk this responsibility, other more legal minds will assume this role, and we will lose the opportunity to make an important contribution to society. Our discussions in this area probe only the outer bounds of the problem and we must now learn how to become more effective in matters where the public is directly involved.

All ICF meetings are sponsored by the parent organization, The International Congress on Fracture, and the growth of these Conferences is a tribute to the vision of the Founder-President, Professor Takeo Yokobori of Tohoku University. ICF4 was organized by Professor D.M.R. Taplin of the University of Waterloo with the assistance of a Canadian Organizing Committee. Professor Taplin was also the Editor-in-Chief of the Editorial Board. The success of the Conference and the excellence of the Proceedings are the consequence of their hard work, and I would like to express my personal appreciation as well as the thanks of the International Congress on Fracture.

B. L. Averbach
President (1973-77)
International Congress on Fracture

June 24, 1977

Foreword

My first duty as the new President of the International Congress on Fracture turns out to be one of the most pleasant – to write this short message for the permanent record of the 1977 Waterloo Conference, ICF4. A pleasant task because ICF4 was such a pleasant and successful conference, due in the main to the dedication and hard work shown by Professor David Taplin and all his co-workers. The pattern of the meeting, with its effective plenary lectures each morning and the several workshop sessions running in parallel later in the day, allowed one at the same conference to obtain both the detailed discussion of a particular interest and the general overview of many fracture disciplines which is so much a feature of the concept of ICF.

The siting together of almost all of the delegates on the beautiful and comfortable campus, together with the alternating afternoon and evening free period, actively encouraged free and informal technical discussions. Anyone who looked into the Village Bar or the Faculty Club any evening will have seen the strange paradox of the bringing together of many people whose main technical interest is that of separation.

It would be invidious, and indeed virtually impossible, for me to single out particular technical contributions for praise, but important technical contributions there were, and I am sure these volumes will be the reference works on Fracture for some years to come. In addition to the more usual form of technical papers, I commend the two panel discussions on education and on the relationship of fracture to politics and society, both of which contain much hard sense and emphasise the important role that this Congress can play in the improvement of the standards of life. In this respect it is worth adding that Waterloo also provided the opportunity for a number of meetings of the ICF Council and Executive, in which discussions took place which will encourage various activities in these areas. If any readers have ideas, please be sure that your Council and I would be pleased to hear from you.

But to return to the pleasures of ICF4, one of the features that contributed so greatly to this will not be found explicitly in these volumes. The friendliness of our Canadian hosts, the effectiveness of their social arrangements, the high standard of the catering and domestic arrangements and the major contributions made by the wives of the organising team, were all factors that made this meeting one that will be long remembered as a happy occasion.

So I will close by thanking all those concerned with making ICF4 such a success that we all look forward with pleasant anticipation to meeting in France at ICF5 in 1981.

Roy W. Nichols
President (1977-1981)
International Congress on Fracture

June 28, 1977

Preface

Preface to the Conference Edition

The International Congress on Fracture was founded by Professor Takeo Yokobori at the First International Conference on Fracture held at Sendai, Japan in 1965. This was followed by the Second Conference in Brighton, England, 1969, and the Third Conference in Munich, West Germany, 1973. The purpose of the Congress is to foster research in the mechanics and mechanisms of fracture, fatigue and strength of materials; to promote international cooperation among scientists and engineers covering the many disciplines involved in fracture research; and to assist in making available the results of research and development. To this end ICF decided to hold an international conference on fracture at least once every four years and the ICF Executive suggested that the Fourth Conference in the series be held in Canada.

The Canadian Fracture Committee was thus established with industrial, government and university representatives and it was decided to hold ICF4 under the auspices of the Faculty of Engineering of the University of Waterloo. Waterloo was chosen as the location for ICF4, being a convenient and compact setting which would provide a forum for the formal proceedings and an intimate campus environment to promote extensive and informal discussions between delegates. Waterloo is also one of the principal centres of fracture research in Canada. The date was chosen to ensure pleasant weather and to mesh with other major conferences in North America. The Canadian Fracture Committee has been incorporated and after ICF4 will continue to be the national arm of ICF in Canada. A National Fracture Conference is envisaged for the periods between international conferences.

The structure of the conference programme consists of: Plenary Sessions each morning where 40 invited papers are to be presented (published in Volume 1 of the Proceedings); Workshop Sessions alternately during afternoons and evenings where 325 contributed research papers are to be presented (published in Volumes 2 and 3 of the Proceedings); and Panel Discussions. An International Editorial Board was established to review the contributed papers. This was based mainly on scientists resident in North America. Refereeing has been rigorous and extended over the period from April 1976 when the first paper was submitted to March 1977 when the last paper was received and later accepted. The workshop papers were classified into 7 parts:

Part		No. of Papers
I	Physical Metallurgy	43
II	Voids, Cavities, Forming	42
III	Fatigue: Micromechanisms	40
IV	Fatigue: Mechanics	45
V	Analysis and Mechanics	64
VI	Applications	42
VII	Non-Metals	49

Parts I-IV make up Volume 2 of the Proceedings and Parts V-VII, Volume 3. More than 700 scientists were involved in the writing and refereeing of these papers. In all, some 38 countries were represented in this process including:

Argentina	Brazil	Denmark
Australia	Canada	England
Austria	Cuba	Finland
Belgium	Czechoslovakia	France

East Germany	Luxembourg	South Africa
West Germany	Mexico	Spain
Hong Kong	Netherlands	Sweden
Hungary	Norway	Switzerland
India	People's Republic of China	Turkey
Israel		U.A.R.
Italy	Poland	U.S.A.
Jamaica	Romania	U.S.S.R.
Japan	Scotland	Wales

The Conference is certainly international in character and representatives and delegates are surely welcome at ICF4 from any country in the world.

To ensure a high level of effective *spoken* interaction and communication at the conference at least half the time has been allotted in each Workshop Session for an extended period of vigorous, directed discussion led by a Workshop Foreman. Workshop Papers themselves will be presented in outline only in groups of 5-7 related papers. Authors will be strictly limited to a maximum of seven minutes and about five slides for their address. Speakers should assume that those present have studied the written text of the papers since a major feature of ICF4 is that the whole proceedings will be prepublished one month prior to the conference. Focus at the conference itself can therefore be upon the latest developments, outstanding problems and spoken communication. The aim is to integrate written and spoken communication, minimizing some of the problems associated with such a large-scale conference.

On the Wednesday afternoon a panel discussion on the *Teaching of Fracture* has been organized to study the task of the educator in developing an understanding of the failure of materials under stress. On the final afternoon of the conference, following the broad review paper by Professor Bruce Bilby, a closing panel discussion has been scheduled to examine the topic *Fracture, Politics and Society*. Professor Max Saltsman MP will give the introductory address, followed by a commentary by Dr. John Knott on the interview with Sir Alan Cottrell FRS entitled *Fracture and Society*. This final plenary session is designed to provide a vigorous and integrated conclusion to the conference giving a basis for a full appreciation of fracture problems in relation to wider social issues.

Notwithstanding the rather overwhelming number and range of papers, the overall objective has been to search for strong unifying themes and connections. Conceptual links between the different *types* of fracture have been especially sought along with the development of the interface between the mechanics and the micromechanistic approaches. An important focus for the conference has been the application of fracture research to large scale structures, covering, for example, nuclear reactors, ships, pipelines, aircraft, and risk analysis.

It is hoped that these proceedings will provide a sound basis for further progress and spoken interchange at the conference itself. The proceedings are destined to become the essential primary archive reference for fracture research, at least until ICF5.

April 8, 1977

Preface to the General Edition

This General Edition of the Proceedings of the Fourth International Conference on Fracture differs in content and format from the Conference Edition, *Fracture 1977*, which was published prior to the Conference for the registered delegates. The expanded title, *Fracture 1977 – Advances in Research on the Strength and Fracture of Materials*, was used to ensure a clear distinction between this edition and its antecedent. The General Edition incorporates a full Subject Index, in addition to the Author Index, plus corrections of textual and typographical errors. Where possible, *Overviews* of the individual Parts of the Workshop Programme have been incorporated and these appear in Volume 4. Messages from the incoming and outgoing Presidents of ICF are also included in a Foreword, plus certain crucial papers and documents received after publication of the Conference Edition.

In order to produce books of more manageable size, the General Edition of the Proceedings appears in six volumes, the original page numbering being retained; thus the content of these six volumes as follows:

Volume 1, *An Overview*, comprises all the invited Plenary papers received when the Conference Edition went to press, and is thus similar in content in the two editions. The same page numbering and citation index for the plenary papers is retained and the full Author Index has been added.

Volume 2A, *The Physical Metallurgy of Fracture*, consists of the papers presented in Parts I and II of the Workshop Programme, which appeared in the first half of *Fracture 1977*, Volume 2; hence it contains pages 1 through 678 of this Volume. The full Author Index is also included.

Volume 2B, *Fatigue*, consists of the papers presented in Parts III and IV of the Workshop Programme, which appeared in the second half of *Fracture 1977*, Volume 2; hence it contains pages 679 through 1392 of this Volume, which includes the full Author Index.

Volume 3A, *Analysis and Mechanics*, consists of the papers presented in Part V of the Workshop Programme, which appeared in the first half of *Fracture 1977*, Volume 3; hence it contains pages 1 through 522 of this Volume. This volume also includes the full Author Index.

Volume 3B, *Applications and Non-Metals*, consists of the papers presented in Parts VI and VII of the Workshop Programme, which appeared in the second half of *Fracture 1977*; hence it contains pages 523 through 1232 of this Volume. Volume 3B contains a full Subject Index to the Proceedings in addition to the Author Index and Citation Index.

Volume 4, *Fracture and Society*, contains the papers issued in a softbound supplementary volume, published a few hours before the Conference began, plus the edited transcript of the two Plenary Panel Discussions *Fracture, Education and Society* and *Fracture, Politics and Society* held under this general title. Included are the ICF4 Interview with Sir Alan Cottrell FRS, the paper *Political and Social Decision Making in Relation to Fracture, Failure, Risk Analysis and Safe Design*, by Max Saltsman MP and the full text of the general survey paper *Fracture*, presented at the conclusion of the Plenary Programme by Professor Bruce Bilby FRS. Also included are the crucial plenary and workshop papers received too late for publication in the earlier volumes. Volume 4 will contain both the full Subject Index and a Citation Index to the complete Proceedings.

A further book, *Conference Theory and Practice,* will also be published. This will be a full report of the Waterloo Conference, also providing some general guidelines for the planning of large-scale Technical Conferences.

It is recommended that references to papers in the volumes be cited in the following way:

Reference to the General Edition –

King, J.E., Smith, R.F. and Knott, J.F., "Fracture 1977 – Advances in Research on the Strength and Fracture of Materials," ed. D.M.R. Taplin, Vol. 2A, Pergamon Press, New York, 1977, page 279 (Conference Edition, University of Waterloo Press).

Reference to the Conference Edition –

Rabotnov, Yu. N. and Polilov, A.N., "Fracture 1977," ed. D.M.R. Taplin, Vol. 3, University of Waterloo Press, 1977, page 1059 (General Edition, Pergamon Press, New York).

These Proceedings will serve as a very substantial physical reminder of the large and significant technical content of ICF4. It is hoped that the memory of other aspects of the Conference, the friends and acquaintances made and renewed, the formal and informal technical discussions, the planned and impromptu social activities, will prove equally enduring and valuable to the 750 participants from some 40 countries who assembled in Waterloo.

A final innovation at ICF4 was the distribution of a detailed questionnaire inviting criticism and comment on the organisation of the Conference, to aid the planning of ICF5 and other similar conferences. Responses were generally complimentary about the technical programme and many kind comments were received on the quality of the hospitality and accommodation offered and on the beauty and compactness of the facilities available on the Waterloo Campus. The structure of the Workshop Sessions came under some criticism, perhaps not surprisingly in view of its innovative nature. Some authors seemed unable or unwilling to describe the main points of their work brought up to date (June 1977) in the eight minutes allotted, preferring to attempt a full formal presentation delivered at a gallop. Problems seemed to arise only where speakers did not study the very full instructions provided. This is a common failing of us all. However, this aspect of the Conference also gained many very positive comments and most speakers came extremely well prepared. Certainly this approach merits repetition in a similar form at ICF5.

The Plenary Sessions were positively received – indeed plenary speakers came extremely well prepared and chairmen were strict in control of the sessions. The *essence* of each paper was presented as required under the instructions, with full up-dating of the work to June 1977, such that it was possible to cover virtually the whole field of fracture in an up-to-date way at the highest possible level. To have had fewer Plenary papers with more time for each presentation, as suggested by some respondents to the questionnaire, would have left significant gaps and failed in this purpose. Plenary speakers are surely to be highly complimented on the unusually commanding quality of the presentations. The fact that little time was available for immediate discussion in Plenary Sessions, a criticism of others, is, frankly, hardly avoidable. With an audience of about 750, controlled and effective discussion is impossible. Discussion of Plenary papers, in fact, occurred in the appropriate Workshops. It should be recorded that these Workshop Discussions were often extremely lively and effective and many positive comments to this effect were received. Two other points seem worth mentioning. It would certainly have been beneficial to have scheduled the Plenary Panel Discussions earlier in the Conference – perhaps even on the first two days. Also, earlier and stronger measures could perhaps have been taken to involve the national and international media, and thereby the public at large, in the problems of fracture and failure in our advanced technological society. This suggestion in fact came from Mr. Robert Maxwell, the Publisher of this General Edition and it was also emphasised in the comments of Max Saltsman MP. These items are worthy of consideration for ICF5.

A point mentioned in Dr. Nichols' Foreword is worth re-emphasis here. The names of all the new Executive Officers of ICF (1977-1981 term) are listed in each Volume of these Proceedings. Any of these would certainly welcome suggestions on the organisation of further Conferences and any other activities which ICF might usefully initiate or co-ordinate, particularly in regard to Publications. The success of ICF4 derived from the whole-hearted participation of many people. ICF wishes to serve all those around the world working on Fracture Problems. This purpose can be achieved effectively only by the further active involvement of us all and the continuing recognition of ICF as the appropriate world organisation and "umbrella" for coordinating work on fracture.

July 19, 1977

Acknowledgements

On behalf of the Canadian Fracture Committee I would like to thank all those who contributed to the preparations of the Fourth International Conference on Fracture. We are particularly grateful to the University of Waterloo and the International Congress on Fracture under whose joint auspices the Conference was organized. Financial assistance from the following organizations is gratefully acknowledged:

National Research Council of Canada
Ontario Hydro
Babcock and Wilcox
University of Waterloo
Atomic Energy of Canada
ALCAN
Westinghouse Canada Limited
International Nickel Company Limited
Canadian Welding Development Institute
Consumers' Gas Company
Trans Canada Pipelines
Noranda Mines Limited
Alberta Gas Trunk Line Company
Canadian Vickers Limited
Atlas Steels Company
Dominion Foundries and Steel Company
STELCO
International Congress on Fracture
Dominion Bridge Company Limited
General Electric
Algoma Steel
American Society for Metals
Union Carbide of Canada
Pratt and Whitney of Canada
MTS Corporation
Instron of Canada

On behalf of the Editorial Board I wish to record our gratitude to the authors of the workshop papers published in these proceedings (Volumes 2 and 3). The standard was high and yet authors were forbearing with critical editorial comments. All papers were extensively reviewed by the Board through the assistance and cooperation of a large body of expert referees. The technical quality of the publication is also directly related to these efforts, which are greatly appreciated but, according to custom, remain anonymous.

I am pleased to record a special thanks in these proceedings to the Publications Group of the Solid Mechanics Division of the University of Waterloo – Professor D.E. Grierson, Technical Editor; Miss Pam Umbach and Miss Linda Heit, Editorial Staff; and Mrs. Cynthia Jones – to the secretarial staff, Miss Elizabeth Krakana, Mrs. Daniela Michiels and Mrs. Jana Karger – and to Mr. David Bartholomew, Graphic Designer. I am also pleased to acknowledge my appreciation to Mr. Robert N. Miranda, Senior Vice President and Mrs. Sylvia M. Halpern, Chief Manuscript Editor of Pergamon Press, Inc., Elmsford, New York, and to Mrs. Patty Patrick of LithoCrafters, Inc.

International Congress on Fracture

EXECUTIVE COMMITTEE 1973 - 1977

Founder-President	Takeo Yokobori	Japan
President	B. L. Averbach	U.S.A.
Vice-Presidents	A. Kochendörfer	West Germany
	R. V. Salkin	Belgium
	S. N. Zhurkov	U.S.S.R.
Directors	R. W. Nichols	United Kingdom
	C. J. Osborn	Australia
	Yu. N. Rabotnov	U.S.S.R.
	D. M. R. Taplin	Canada
	H. C. van Elst	Netherlands
Treasurer	J. Nemec	Czechoslovakia
Secretary-General	T. Kawasaki	Japan
Members	P. Haasen	West Germany
	A. K. Head	Australia
	N. J. Petch	United Kingdom
	J. L. Swedlow	U.S.A.
	M. L. Williams	U.S.A.

MEMBERS OF COUNCIL 1973 - 1977

Australia

F. P. Bullen
A. K. Head*
C. J. Osborn

Belgium

R. V. Salkin
W. Soete*
A. Vinckier

Canada

M. R. Piggott
L. A. Simpson
D. M. R. Taplin*

Czechoslovakia

M. Klesnil
V. Linhart
J. Nemec*

Denmark

F. Niordson*

France

D. Francois*
R. Labbens
G. Sanz

Hungary

L. G. Gillemot*

Israel

A. Buch*
J. Glucklich
R. Rotem

Italy

F. Gatto*
L. Lazzarino
F. Manna

Japan

T. Kawasaki
T. Yokobori*

Luxembourg

E. A. Hampe*

The Netherlands

D. Broek
H. C. van Elst*
C. A. Verbraak

Norway

H. Wintermark*

Poland

Z. Pawlowski*
T. Pelczynski

South Africa

Z. T. Bieniawski*
F. R. N. Nabarro
L. O. Nicolaysen

Spain

C. Nuñez*

Sweden

B. Broberg
J. Carlsson*
J. Hult

Switzerland

E. Amstutz
M. J. Briner*

United Kingdom

M. J. May
K. J. Miller
P. L. Pratt
R. W. Nichols*
N. J. Petch

U.S.A.

B. L. Averbach*
J. L. Swedlow
M. L. Williams

U.S.S.R.

Yu. N. Rabotnov
S. N. Zhurkov*

West Germany

W. Dahl
H. H. Kausch
F. Kerkof
P. Haasen
A. Kochendörfer*

*voting member

International Congress on Fracture

Executive Officers Elected at ICF4 for 1977-1981

Founder-President	T. Yokobori	Japan
President	R. W. Nichols	United Kingdom
Vice-Presidents	Yu. N. Rabotnov	U.S.S.R.
	D.M.R. Taplin	Canada
	H.C. van Elst	Netherlands
Directors	A. J. Carlsson	Sweden
	W. Dahl	West Germany
	D. Francois	France
	C. J. McMahon, Jr.	U.S.A.
	J. Pelczynski	Poland
Treasurer	H. H. Kausch	Switzerland
Secretary-General	T. Kawasaki	Japan

Honorary Fellows of ICF (Elected at ICF4)

T. Yokobori
R. W. Nichols
A. Kochendörfer
D.M.R. Taplin
B. L. Averbach

ICF Nominating Committee 1977-1981

Chairman	R. V. Salkin	Belgium
	Z. T. Bieniawski	South Africa
	K. B. Broberg	Sweden
	A. G. Evans	U.S.A.
	A. K. Head	Australia
	T. Kunio	Japan
	Liu Chun Tu	People's Republic of China
	J. Nemec	Czechoslovakia
	P. Rama Rao	India
	E. Smith	United Kingdom
	V. I. Vladimirov	U.S.S.R.

ICF Publications and Finance Committee (1977-1981)

Chairman	D.M.R. Taplin
	B. L. Averbach
	D. Francois
	H. H. Kausch
	J. F. Knott
	J. L. Swedlow
	T. Yokobori

ICF Committees on Regional Liaison and a Standing Committee on Policy (Chairman R.W. Nichols) are also being commissioned by the President. Through the work of these various committees we look forward to a new phase in the development of the International Congress on Fracture as an influential world body.

Fourth International Conference on Fracture

Waterloo, June 19 - 24, 1977

CANADIAN FRACTURE COMMITTEE

Chairman	D. M. R. Taplin	(Waterloo)
	C. M. Bishop	(de Havilland)
	D. J. Burns	(Waterloo)
	J. Dunsby	(NRC)
	W. H. Erickson	(Defence Dept)
	J. D. Embury	(McMaster)
	J. Hood	(Stelco)
	R. R. Hosbons	(AECL)
	H. H. E. Leipholz	(Waterloo)
	I. Le May	(Saskatchewan)
	J. T. McGrath	(CWDI)
	D. Mills	(Ontario Hydro)
	K. R. Piekarski	(Waterloo)
	M. R. Piggott	(Toronto)
	L. A. Simpson	(AECL)
	T. A. C. Stock	(Alcan)
	T. H. Topper	(Waterloo)
	G. C. Weatherly	(Toronto)
Conference Secretary	R. F. Smith	(Waterloo)

LOCAL ARRANGEMENTS COMMITTEE

Chairman:	D. M. R. Taplin
	J. R. Cook
	K. D. Fearnall
	H. W. Kerr
	H. H. E. Leipholz
	D. Mills
	R. J. Pick
	Sherry Pick
	K. R. Piekarski
	M. R. Piggott
	A. Plumtree
	Betty Statham
	Diana Theodores Taplin
	T. H. Topper
Conference Manager:	R. F. Smith

Standard Nomenclature List

In order to minimize unnecessary confusion, a standard nomenclature for commonly used quantities has been adopted for ICF4. This coincides closely with other developing nomenclatures in the field of fracture and it is hoped that this notation will become widely used. SI units have been used throughout the Proceedings with fracture toughness reported as $MPam^{1/2}$. It was originally thought that this quantity might be designated the "griffith". Whilst we surely wished to honour the father of the science of fracture in this way, we thought better of taking any unilateral action at this time. Thus only informal use of the griffith is recommended at ICF4.

A	Area of Cross-Section of a Specimen
A_o	Area of Cross-Section of a Specimen at the Start of Testing
A_f	Area of Cross-Section of a Specimen at Fracture
a	Crack Length - One-Half the Total Length of an Internal Crack or Depth of a Surface Crack
a_o	Original Crack Length - One-Half of Total Length of an Internal Crack at the Start of a Fracture Toughness Test, or Depth of a Surface Crack at the Start of a Fracture Toughness Test
a_p	Measured Crack Length - One-Half the Total Length of an Internal Crack or Depth of a Surface Crack as Measured by Physical Methods
a_e	Effective Crack Length - One-Half the Effective Total Length of an Internal Crack **or** Effective Depth of a Surface Crack (Adjusted for the Influences of a Crack-Tip Plastic Zone)
Δa, Δa_p, ...	Crack Growth Increment
da/dN	Rate of Fatigue Crack Propagation
B	Test Piece Thickness
b	Atomic Interval (Burgers Vectors Magnitude)
d	Average Grain Diameter
D_L	Lattice Diffusion Rate
D_B	Grain Boundary Diffusion Rate
D_S	Surface Diffusion Rate
E	Young's Modulus of Elasticity
exp	Exponential Base of Natural Logarithms

G	Strain Energy Release Rate with Crack Extension per unit length of Crack Border of Crack Extension Force
G_I G_{II} G_{III}	Crack Extension Forces for Various Modes of Crack Opening
h	Planck's Constant
I	Moment of Inertia
J	Path-Independent Integral Characterizing Elastic/Plastic Deformation Field Intensity at Crack Tip; also, Energy Release Rate for Non-Linear Elastic Material
K	Stress Intensity Factor - A Measure of the Stress-Field Intensity near the Tip of a Perfect Crack in a Linear-Elastic Solid
K_c	Fracture Toughness - The Largest Value of the Stress-Intensity Factor that exists prior to the Onset of Rapid Fracture
K_{max}	Maximum Stress-Intensity Factor
K_{min}	Minimum Stress-Intensity Factor
K_{th}	Threshold Stress Intensity Factor Below which Fatigue Crack Growth Will Not Occur
K_I	Opening Mode Stress Intensity Factor
K_{IC}	Plane-Strain Fracture Toughness as Defined by ASTM Standard Designation E 399-74
K_{Ii}	Elastic Stress-Intensity Factor at the Start of a Sustained-Load Flaw-Growth Test
K_{ISCC}	Plane-Strain K_I Threshold Above Which Sustained-Load Flaw-Growth Occurs
K_{II}	Edge-Sliding Mode Stress Intensity Factor
K_{III}	Tearing Mode Stress Intensity Factor
$\dot{K}$	Rate of Change of Stress-Intensity Factor with Time
ΔK	Stress Intensity Range
k	Boltzmann Constant
k_y	Parameter that Determines Grain-Size Dependence of Yield Strength
l_o	Gauge Length
ln	Natural Logarithm
log	Common Logarithm

m	Strain-Rate Sensitivity Exponent
N_f	Number of Cycles to Failure
n	Strain Hardening Exponent
P	Force
P_{max}	Maximum Force
P	Pressure
Q	Activation Energy
Q_a	Activation Energy for Crack Growth
Q_c	Activation Energy for Creep
Q_d	Activation Energy for Self Diffusion
T	Temperature
T_M	Absolute Melting Temperature
T_D	Brittleness Transition Temperature
t	Time
t_o	Time at the Onset of a Test
t_f	Fracture Time
U	Potential Energy
δz	Thickness of Grain Boundary Layer
γ_s	True Surface Energy
γ_B	Grain Boundary Surface Energy
γ_p	Effective Surface Energy of Plastic Layer
δ	Value of Crack Opening Displacement
δ_c	Critical Crack Opening Displacement, Being One of the Following: (1) Crack Opening Displacement at Fracture (2) Crack Opening Displacement at First Instability or Discontinuity (3) Crack Opening Displacement at Which an Amount of Crack Growth Commences
δ_m	Crack Opening Displacement at First Attainment of Maximum Force
ε	Normal Strain

Nomenclature

ε_e	Normal Strain, Elastic
ε_p	Normal Strain, Plastic
ε_T	Normal Strain, Total
ε_{max}	Normal Strain, at Maximum Tensile Load
ε_E	Engineering Normal Strain
ε_f	Normal Strain, Critical Value at Fracture
ε_i	Principal Strains (i = 1, 2, 3)
ε_{pi}	Principal Strains, Plastic
ε_x ε_y ε_z	Cartesian Strain Components
ε_{ij}	Strain Tensor
$\dot{\varepsilon}$	Strain Rate
$\dot{\varepsilon}_e$	Strain Rate, Elastic
$\dot{\varepsilon}_p$	Strain Rate, Plastic
$\dot{\varepsilon}_o$	Strain Rate, Initial Value
$\Delta\varepsilon$	Strain Range
$\Delta\varepsilon_p$	Plastic Strain Range
ν	Poisson's Ratio
σ	Normal Stress
σ_y	Yield Stress Under Uniaxial Tension
σ_1 σ_2 σ_3	Principal Normal Stresses
σ_e	Fatigue Strength, Endurance Limit
σ_f	Fracture Stress
σ_{max}	Maximum Stress
σ_x σ_y σ_z	Cartesian Components of Normal Stress
$\dot{\sigma}$	Stress Rate
τ	Shear Stress
τ_o	Critical Shear Stress
τ_1 τ_2 τ_3	Principal Shear Stresses
τ_{max}	Shear Stresses, Maximum Value
Ω	Atomic Volume

Conversion Units

To Convert From	To	Multiply By
inch	meter (m)	2.54×10^{-2}
pound force	newton (N)	4.448
kilogram force	newton (N)	9.807
kilogram force/meter^2	pascal (Pa)	9.807
pound mass	kilogram mass(kg)	4.536×10^{-1}
ksi	pascal (Pa)	6.895×10^{6}
ksi $\sqrt{in}$	MPam$^{1/2}$ **(Gr)**	1.099
ton	pascal (Pa)	1.333×10^{2}
torr	pascal (Pa)	1×10^{5}
angstrom	meter (m)	1×10^{-10}
calorie	joule (J)	4.184
foot-pound	joule (J)	1.356
degree Celsius	kelvin (K)	$T_K = T_C + 273.15$

Important Multiples

Multiplication Factor	Prefix	Symbol
10^{-12}	pico	p
10^{-9}	nano	n
10^{-6}	micro	µ
10^{-3}	milli	m
10^{3}	kilo	k
10^{6}	mega	M
10^{9}	giga	G

Part V
Analysis and Mechanics

Chairmen:
J. L. Swedlow
H. C. van Elst

Fracture 1977, Volume 3, ICF4, Waterloo, Canada, June 19 - 24, 1977

A NUMERICAL APPROACH FOR STABLE CRACK-GROWTH AND FRACTURE CRITERIA

G. Rousselier*

I. INTRODUCTION

The behaviour of a cracked body in large-scale yielding conditions has been intensively studied in the past years. The well-known crack-tip parameters like J-integral or C.O.D. are generally computed and the influence of plasticity studied. However the computations are made for a stationary crack and give no information about stable crack-growth and corresponding fracture criteria.

Stable crack-growth has been studied by Andersson [1] by performing successive relaxation of crack-tip nodal forces in a finite-element programme. In this paper we attempt to refine this approach by introducing on the extension of the crack-line special finite-elements modelling the behaviour of the end-region and allowing the elimination of stress and strain singularities. Stable and unstable crack-growth will be connected to the fracture properties of the material submitted to complex loading.

II. FRACTURE CRITERION FOR AN ELASTIC BODY

The usual boundary conditions on the crack-line, mode I: $u_2 = 0$, $\sigma_{12} = 0$, ($\sigma_2 = \sigma_{12} = 0$ on the crack faces), lead to infinite stresses and strains at the crack-tip. Finite stresses and strains are obtained with the boundary condition $\sigma_2 = f(u_2)$ instead of $u_2 = 0$, as in the Barenblatt's model [2] and also in the Dugdale's model [3]. The main difficulty lies in the interpretation of the normal displacement u_2 in a continuum model. In this paper u_2 is interpreted from the strain $\varepsilon_2 = u_2/h$ of a strip with height 2 h located on the extension of the crack line, in a way similar to the rigid-plastic strip model introduced by Rice [4]. Dugdale's model is based on the Tresca criterion, which gives $\sigma_2 = f(u_2) = \sigma_y$ for an elastic-perfectly plastic material in plane stress conditions. In this paper a state of plane deformation is assumed; the relation $\sigma_2 = f(u_2) = g(\varepsilon_2)$ represents the local stress-strain curve in the strip and is related to the flow rule of the material.

The geometry of the crack-tip is modified by the insertion of the strip. But such a model is perhaps more realistic in this highly strained region than the usual reference to the initial geometry of a cut with zero crack-tip radius.

In this way, for an infinite linear-elastic medium (plane strain), the problem is reduced to the one-dimensional integral equation:

* Laboratoire de Mécanique des Solides, Ecole Polytechnique, 91120 - PALAISEAU, France and E.d.F., Etudes Matériaux, Les Renardières, 77250 - MORET SUR LOING, France

$$u_2\ (x) = [2\ (1-\nu^2)/\pi E] \int_{-\infty}^{+\infty} \sigma_2\ (t)\ \ln\ |t-x|\ dt\ . \qquad (1)$$

Andersson and Bergkvist [5] have resolved numerically this equation with a law $\sigma_2 = f\ (u_2)$ linearly increasing than decreasing; in this paper we consider a more general law. The non-linear part of f (u_2) will be used to define the length of a "plastic zone" limited to the strip.

The model gives an interesting possibility to connect the global criterion of fracture with a local criterion at the crack-tip, defined by $\sigma_2 = \sigma_f$ or $\varepsilon_2 = \varepsilon_f$. We show numerically that in small-scale yielding the local criterion defined above yields the global criterion $K_I = K_{IC}$ with a good accuracy. We obtain $K_{IC}^2 = k\ S_f$, where S_f is the area under the curve $\sigma = g\ (\varepsilon)$ up to the limiting strain ε_f for which the stress vanishes.

The J-integral is found to be path-independent outside the "plastic zone". For a remote path, at the onset of fracture, its value is $J_c = [(1-\nu^2)/E]\ K_{IC}^2$, while for a path along the boundary of the "plastic zone" $J_c = 2h\ S_f$. This latter result is the same as given by Rice [6] but is based on a different model. So the constant k should be equal to $2hE/(1-\nu^2)$. This is verified numerically with a good accuracy.

III. ELASTIC-PLASTIC BODY. STABLE CRACK-GROWTH

A finite-element approach is convenient in the case of an elastic plastic body. The incremental plastic deformation, in plane strain conditions, is taken according to the Prandtl-Reuss flow-rule along with the von Mises criterion. An "implicit" algorithm recently given by Nguyen, Q. S. [7] is used. The implicit algorithm eliminates all the numerical and systematic errors usually found in the "explicit" method. Furthermore loading, unloading and reloading can be easily done.

The elastic-plastic constants are: E = 200,000 MPa, $\nu = 0.3$, $\sigma_y = 700$ MPa, linear hardening with a 1,000 MPa modulus. The law $\sigma = g\ (\varepsilon)$ in the strip is not the conventional curve obtained in the tension test, but a deduced curve corresponding to uniaxial strain. It is chosen to represent the complex stress state at the crack-tip. The fracture of each element occurs at a critical stress $\sigma_2 = \sigma_f = g\ (\varepsilon_f)$; for $\varepsilon > \varepsilon_f$ the curve g (ε) drops to zero. During crack growth, the crack-tip nodal force is relaxed in five equal steps.

We study a three-point bend specimen (width W, span S = 4 W, thickness B, initial crack-length a_o, $b = W-a_o$). In order to avoid the effect of element size the elements adjacent to the uncracked ligament near the crack-tip have the same dimension, and remain unchanged for specimens of different sizes. The side s of these triangular elements is taken equal to 0.1 mm for a width W from 5 to 200 mm. The strip height h is no longer the characteristic length of the process of fracture as it was the case for an elastic body. In fact h may be related to the crack-tip radius, and the results are independent of h if it is sufficiently small (here for h < 0.01 mm). It is s which is the characteristic length: fracture occurs when σ_2 (mean) > σ_f over a distance s from the crack-tip.

The load-deflection curves P (d) for σ_f = 2,300 MPa are given in Figure 1 along with the crack-growth curves P (Δa) for W = 20 and 5 mm. The points of crack-growth initiation and unstable crack-growth depend on the size of the specimen. Moreover the second one depends on the loading conditions. With load-control the instability occurs after a few steps of crack-growth. With displacement-control the crack-growth is stable; it goes on under a quasi-constant, than decreasing load (curve 5a); the maximum load can be somewhat greater than that obtained with load-control (curve 3a).

IV. FRACTURE CRITERIA

Figure 2 shows different critical values of the stress-intensity factor K_I as function of specimen size. These values are deduced from Figure 1 as follows: K_{max} is the value of K_I at maximum load; K^d_{max} at the load obtained by extrapolation of the linear part of the P - d curve up to the displacement at maximum load; K_Q at the load defined by the intersection with "5%-secant"; K_B is computed according to the equivalent energy concept introduced by Witt [8]; K_{Ji} is deduced from the J-integral at the onset of stable crack-growth. K_Q and K_{max} decrease for b < 1 to 1.5 $(K_{IC}/\sigma_y)^2$ as it is verified with medium-strength steels if the thickness is sufficient. K^d_{max} and K_B are more constant and bracket the value of K_{IC}; these two values give a good estimate of the fracture toughness K_{IC} with "medium-size" specimens [between 0.25 and 1.5 $(K_{IC}/\sigma_y)^2$]. With smaller specimens K^d_{max} and K_B are no longer well-defined for the computed value of the displacement at maximum load is not accurate.

The J-integral is computed for two cases: σ_f = 2,300 MPa and 3,200 MPa. $V = -(1/B)[dU/da]_d$ and $V^* = -(2/B)(U^*/b)$ are also computed[1]. The obtained values are converted into K_J, K_V and K^*_V by the usual plane strain formula $J = [(1-\nu^2)/E]K^2$. They fit well together, at least until full plasticity, which justifies the experimental determination of J (see Figure 3).

K_J at the initiation of crack-growth, i.e., K_{Ji}, is practically independent of the specimen size and depends only of the material considered: $K_{Ji} \simeq 54$ MPa•m$^{1/2}$ for σ_f = 2,300 MPa (see Figure 2), $K_{Ji} \simeq 100$ MPa•m$^{1/2}$ for σ_f = 3,200 MPa (for W = 20 mm, a_o/W = 0.5, 0.6, 0.7 - for W = 50, 100 and 200 mm, a_o/W = 0.5). The J-integral gives a good criterion for the initiation of stable crack-growth. However, since K_J deviates very little from the linear curve, K^d_I (Figure 3) is a more simple, but approximate, criterion for the initiation.

For the two cases investigated K_{Ji} is notably smaller than K_{IC}, about 30% for the weaker material (σ_f = 2,300 MPa) and 70% for the tougher one (σ_f = 3,200 MPa: in that case it was not possible to reach the point of instability, because of computer limitations; K_{IC} is greater than 300 MPa•m$^{1/2}$; this shows the high dependence of K_{IC} with σ_f). It does not seem that the value of the J-integral at the onset of stable crack-growth allows the direct determination of fracture toughness on small specimens.

V. FURTHER DEVELOPMENTS

The magnitude orders of the computed values K_{IC} and K_{Ji} are quite good. However the model will be refined in the following ways.

---

[1] These two values are given by the well-known relations used for the experimental determination of J, the former with a few specimens, the latter with a single deep-cracked specimen ($a/W \geq 0.6$) [9].

First, the uniaxial strain hypothesis will no longer be imposed to the special crack-elements. The algorithm for the finite-elements in plasticity will be used also for the special crack-elements, that is to say the increments of stresses will be given as functions of actual stresses, hardening parameter and increments of strains $\Delta\varepsilon_1 = \Delta u_1/\Delta x_1$, $\Delta\varepsilon_2 = \Delta u_2/h$, $\Delta\varepsilon_{12} = 0$.

Second, instead of a critical stress σ_f, a local criterion $F(\sigma_1, \sigma_2, \sigma_3) = 0$ will be used. It will be related to tests on notched - but not cracked-specimens of a given material.

With these improvements an agreement is hoped between numerical and experimental values of K_{IC} for the given material. Moreover theoretical and experimental results in large-scale yielding conditions will be compared.

REFERENCES

1. ANDERSSON, H., J. Mech. Phys. Solids, 21, 1973, 337.
2. BARENBLATT, G. I., Advances in Applied Mechanics, VIII, 1962, Academic Press, New York.
3. DUGDALE, D. S., J. Mech. Phys. Solids, 8, 1960, 100.
4. RICE, J. R., First Int. Conf. on Fracture, Sendai, Japan, I, 1965, 283.
5. ANDERSSON, H. and BERGKVIST, H., J. Mech. Phys. Solids., 18, 1970, 1.
6. RICE, J. R., Fracture II, 1968, 234.
7. NGUYEN, Q. S., Contribution à la théorie macroscopique de l'élasto-plasticité avec écrouissage, Thesis, CNRS AO 9317, Paris, 1973.
8. WITT, F. J., Nuclear Engineering and Design, 20, 1972.
9. PELISSIER-TANON, A., Application of the J-Integral for Fracture Toughness Measurements, Advanced Seminar on Fracture Mechanics, October, 1975, ISPRA, Italy

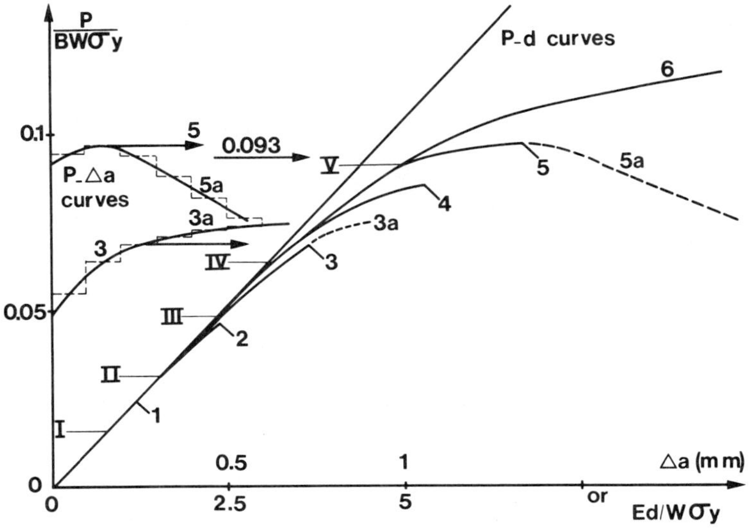

Figure 1 Computed Dimensionless P versus d Curves with σ_f = 2,300 MPa for Different Specimen Sizes (1 : W = 200 mm, 2 : 50 mm, 3 : 20 mm, 4 : 10 mm, 5 : 5 mm; a_o/W = 0.5) with Load Control (Curves 1 to 5) or Displacement Control (Curves 3a and 5a). Points of Crack-Growth Initiation are Shown on Each Curve by Roman Numerals (I to V). Curve 6 is for a Stationary Crack. Load P versus Crack-Growth Δa Curves are Shown for W = 20 mm and 5 mm. P/BW σ_y = 0.093 is the Limit Load from the Slip-Line Field Theory

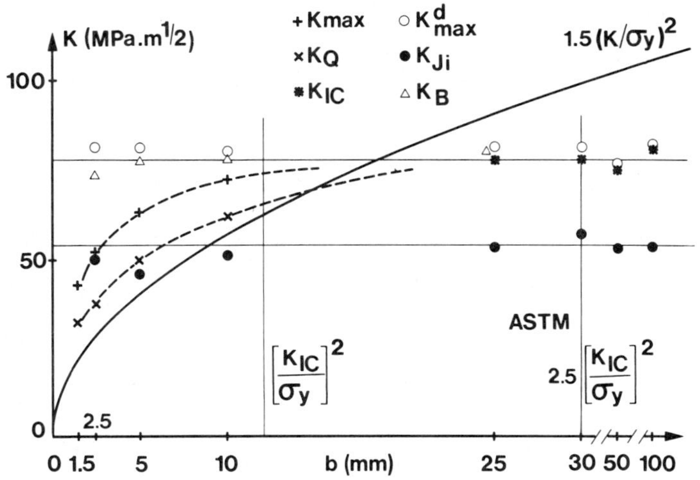

Figure 2 Computed Critical Values of K_I and K_J for σ_f = 2,300 MPa as Function of Specimen Size (W = 200, 100, 50, 20, 10 and 5 mm; a_o/W = 0.5 and 0.7). Symbols are Defined in the Main Text

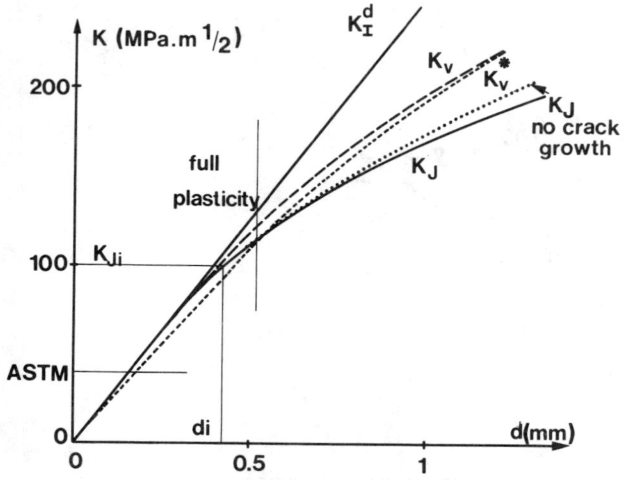

Figure 3 Computed K versus d Curves with σ_f = 3,200 MPa for a W = 20 mm, a_o/W = 0.6 Specimen (——— K_J, ... K_J for a Stationary Crack, —— —— —— K_V, ----- K_V*). Symbols are Defined in the Main Text

ON THE APPLICATION OF CTD TO THE FRACTURE MECHANICS

Y. Mitani* and H. Miyamoto**

INTRODUCTION

The purpose of this paper is to combine the continuum theory of dislocations and conventional fracture mechanics, in order to clarify the physical aspects of deformation to seek a criterion of fracture from a microscopical point of view. These ideas were firstly realized by the BCS-Model [1], where one dimensional distribution of dislocations is assumed and the elastic-plastic -- problem is replaced by the problem of finding out an equilibrium distribution of dislocation under some suitable boundary conditions. Further progress has been made by Yokobori et al [2] and Lardner [3]. These direct applications of the BCS type solution, however, complicate the mathematical treatment, and seem to be inconvenient to the general plane problems. In this analysis the concept of dislocation theory is incorporated with Finite Element Method (FEM) to find out the equilibrium distribution of dislocations which satisfies the boundary conditions. It is noted that the estimation of the internal stress field due to the dislocation density is made fictitiously after Eshelby by self-consistent method [4] with the assumption that uniform plastic strain is produced by dislocation migration, if the dislocation were not confined in the elastic domain. The numerical analysis based on these fundamental concepts, (termed CTD hereafter) is carried out for monotonic loading condition, and the compatibility of the plastic deformation is examined for a centrally cracked plate.

THEORETICAL FUNDAMENTALS AND NUMERICAL PROCEDURES

The equilibrium equation of BCS model may be rewritten, associated with a slip plane as:

$$\tau_a + \tau_{int} = \tau_c , \tag{1}$$

in the plastic zone whose dimension appears as the integral limit of internal stress term τ_{int} as an unknown as well as the dislocation density - function. Here, τ_a, τ_c denote the shear stress due to the external force and the frictional stress, respectively. The limitation of assuming the one-dimensional dislocation distribution will be overcome by properly estimating the internal stress field due to the dislocation and by examining the equation (1) in iterative fashion.

We assume the following mechanism for the calculation of the internal stress.

---

* National Institute of Technology, Mexico, D.F., Mexico.
** University of Tokyo, Tokyo, Japan.

(a) The plastic zone with uniform ε_p in the elastic domain is represented by the surface forces after Eshelby

$$T_i = n_j \cdot c_{ijkl} \cdot \varepsilon^p{}_{kl} , \tag{2}$$

which are expressed by the nodal forces in FEM.

(b) The uniform plastic strain associated with the slip plane γ^p is related with the excess dislocation density N with burgers vector b as

$$\gamma^p = (1/2)Nb , \tag{3}$$

where N is derived from the number of the equilibrium dislocation of BCS type solution for one-side pile-up [5] with a plane correction factor n*

$$N = \frac{\pi(1-\nu)n^*d}{\mu b} \cdot \tau_{eff} , \tag{4}$$

and

$$\tau_{eff} = (\tau_a + \tau_{int}) - \tau_c , \tag{5}$$

where τ_{int} vanishes for the initial stage.

Then we obtain the hardening ratio for uni-axial tension H' by considering $2\tau_c = \sigma_y$

$$H' = \frac{d\sigma}{d\varepsilon^p} = \frac{2}{\pi} \frac{\mu}{1-\nu} \frac{1}{n^* \cdot d} , \tag{6}$$

which shows the inverse proportional relation to the average grain diameter d. In this way we equivalently assumed the simple bilinear material characteristics by excess dislocation, and further discussions of dislocation distribution focus on the "geometrically necessary dislocation" after Ashby's terminology [6].

(c) Taking the irreversibility of plastic deformation into account, we obtain the internal stress field with free boundary condition for the specimenconfiguration. Hence the internal stress field is the sum of the stress due to the dislocation τ_d and its image stress τ_d^{im}

$$\tau_{int} = \tau_d + \tau_d^{im} . \tag{7}$$

Hence, for each external stress increment, equation (7) is calculated by using relations (4), (3) and (2). Then equation (1) is examined until it holds. In the case of isotropic material, the slip plane is determined equal to the maximum shear plane of each incremental step.

The material constants used in this analysis are $E = 2.058 \times 10^{11}$Pa, $\tau_c = 9.80$ MPa, $V = 0.333$, b = 1.0 A and $d = 10^{-6}$ m.

The specimen configuration is half width $W/2 = 100$, thickness $B = 1.0$ and half crack length $a = 10$ mm, respectively. The dislocation density VD and the total number of dislocation TND are defined in a yielded element as $VD = N/d$ and $TND = NA/d$, respectively, where A is the area of the element in which stress and strain are defined as constant by FEM.

RESULTS AND DISCUSSIONS

The typical feature obtained by this analysis is the bursting phenomena which depends highly on the plastic characteristics expressed by n^*, that is, in the course of examination of equation (1), such loading stage appears where no convergence is obtained. It is seen in the divergence of TND in Figure 1, which implies that the compatible state can not exist and, hence the assumed condition (a) doesn't hold any more. Then it can be inferred that the excess dislocation which violate the compatibility might be emitted from the crack surface at this loading point, in order to recover the compatibility by the geometric change of the crack surface [7]. Since the effect of the internal stress field is highly localized near the crack tip, crack opening displacement increases non-linearly as shown in Figure 2, corresponding to the tendency of TND, where COD is plotted for the nearest joint of the crack tip.

These localized effect of the internal stress field influences also the spread of the plastic zone, and the resultant effects with the dislocation mobility show clearly the differences between the materials which have different n^* as shown in Figures 3 and 4. It is noted that the bursting phenomena occur indifferent to the plastic zone size, instead, those materials which have n^* smaller than 1.5, so far as this analysis is concerned, recover the stress singularity characteristics at the crack tip element is yielded before bursting occurs. This result implies the transition of fracture conditions according to the plastic property of the material from brittle to ductile. Figure 6 shows an example of stress singularity recovery after its relaxation.

Comparison is made in Figure 5 on plastic zone size. The discrepancy between the small scale yielding solution and the present analysis after the loading level around 0.4 presumably stems from the ignorance of the internal response due to the plastic deformation in s.s.y. analysis [8]. The difference to the BCS solution is due to the difference of the plastic zone dimension. The change of the internal state in terms of dislocation density distribution is clearly seen in Figure 7 which shows the difference of the equilibrium distributions not only by the amount but also by the position of the peak, where numbers in the Figure 7 denote those of the nearest elements around the crack tip from the front to the back.

From these facts we can conclude that s.s.y. concept is applicable to those materials which recover the stress singularity without violating the compatibility conditions and does not always mean the geometrical amount ratio to the other dimensions. The greater the n^* is, the more easily the compatibility conditions is violated. Hence, for those materials which have smaller n^* the conventional fracture criterion as the stress intensity factor might be applicable. In Table 1 the results are summarized, where NELM denotes the element number which has the maximum dislocation density just before the bursting occurs, and from which the excess dislocation might be emitted to the crack surface.

CONCLUSIONS

The application of CTD to the elastic-plastic analysis is useful for discussing fracture phenomena for the following reasons:

(a) It analyses the compatibility of the plastic deformation which is responsible for the fracture in the case of nonhomogeneous deformation.

(b) The internal responses of the material or mechanisms can be discussed by the introduction of the concept of dislocation distribution which is ignored in the conventional fracture mechanics (especially in LFM).

The blunting phenomenon is explained by the emission of dislocations which violate the compatibility condition.

ACKNOWLEDGEMENTS

One of the authors (Y. Mitani) expresses his heartiest gratitude to Professor E. Kröner of the University of Stuttgart for his valuable discussions and advice, and also to DAAD for their scholarship which enabled him to study in the above mentioned University. Their thanks are due to Dr. M. Kikuchi whose offer of FEM techniques and valuable discussions enabled the prompt completion of this work which is based on the dissertation of Y. Mitani at the University of Tokyo.

REFERENCES

1. BILBY, B. A., COTTRELL, A. H. and SWINDEN, K. H., Proc. Roy. Soc. London, A272, 1963, 304.
2. YOKOBORI, T., YOSHIDA, M., KURODA, H., KAMEI, A. and KOUNOSU, S., Eng. Frac. Mech., 7, 1975, 377.
3. LARDNER, R. W., Proc. Roy. Soc. London, A317, 1970, 199.
4. ESHELBY, J. D., Solid State Phys., 3, Academic Press, New York, 1968, 79.
5. LOTHE, J. P. and HIRTH, J., "Theory of Dislocations", McGraw-Hill, New York, 1968, 704.
6. ASHBY, M. F., Phil. Mag., 21, 1970, 399.
7. KRÖNER, E., Private Communication.
8. RICE, J. R., Fracture 2, edited by H. Liebowitz, Academic Press, New York, 1969, 191.

Table 1 Bursting Stresses by Monotonic Loading for Different Correction Factor

n*	Burst. Stress	TND	VD max. x 10^4	NELM	δ x 10^{-4}
1.0	-	-	-	-	-
1.5	0.315	6767.83	74.76	1	0.9208
2.0	0.239	3634.25	24.68	1	0.8297
2.5	0.163	340.09	4.85	5	0.3914

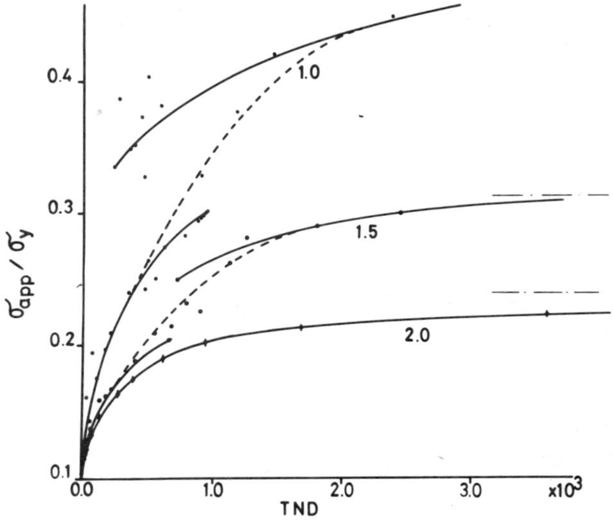

Figure 1 Variation of Total Number of Dislocation for Different Correction Factor

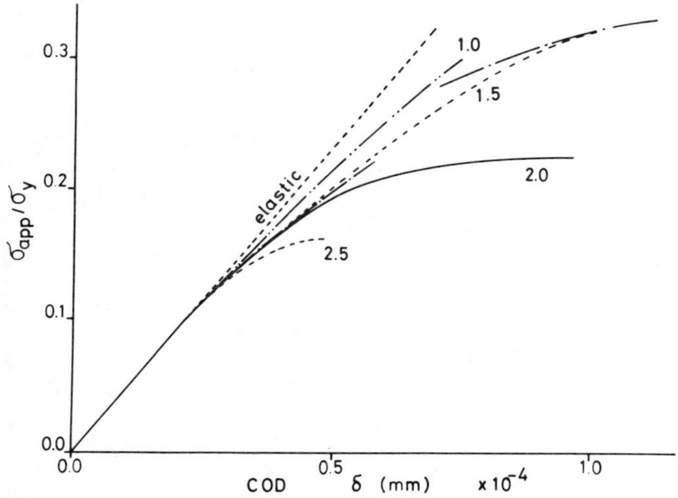

Figure 2 Variation of Crack Opening Displacement for Different Correction Factor

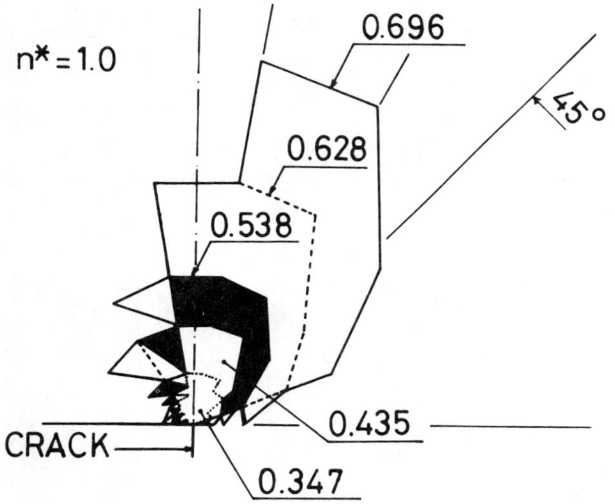

Figure 3 Spread of Plastic Zone for n* = 1.0

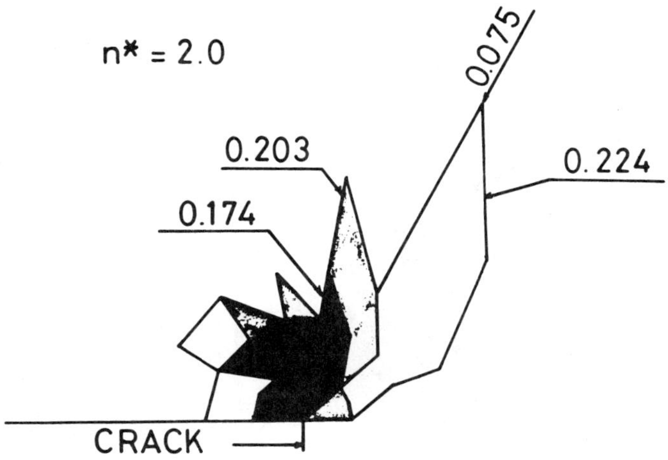

Figure 4 Spread of Plastic Zone for n* = 2.0

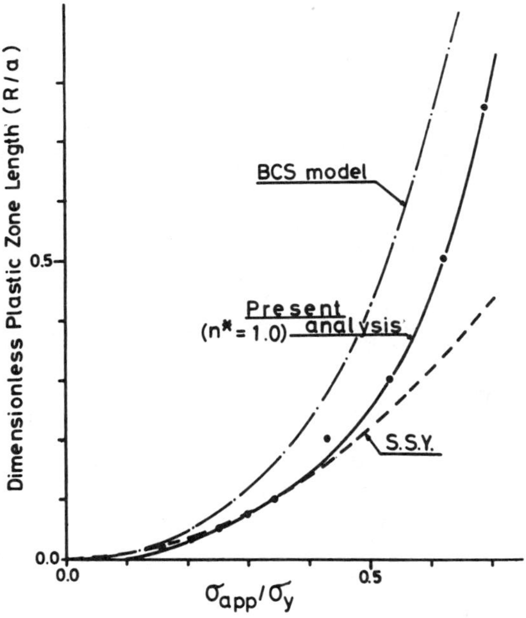

Figure 5 Comparison of Plastic Zone Length

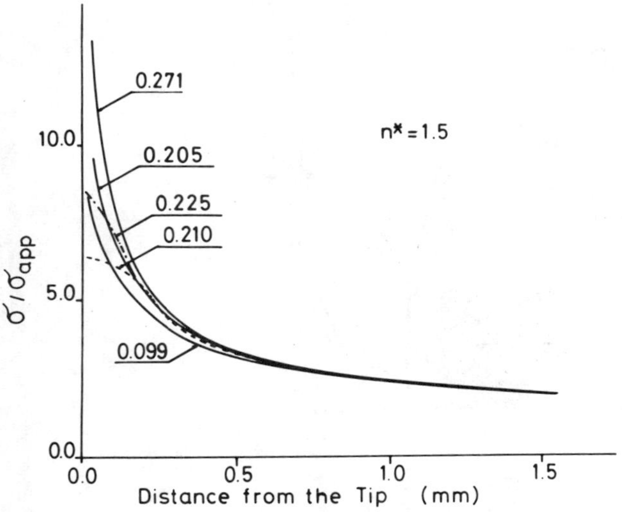

Figure 6 Variation of Stress Distribution Ahead of the Crack Tip for Different Loading Level

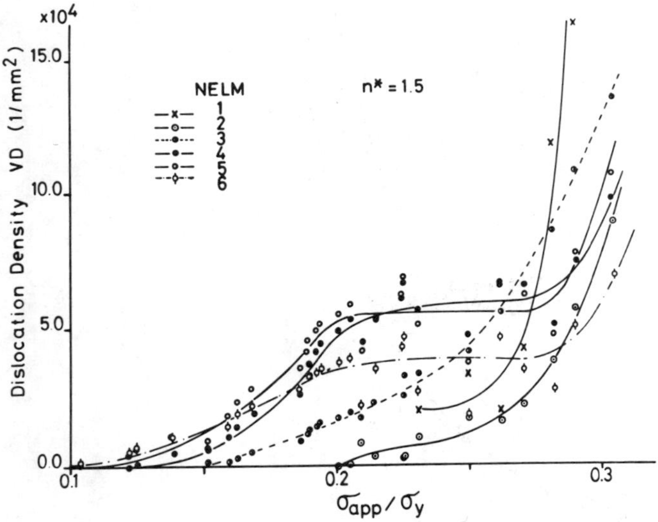

Figure 7 Variation of Dislocation Density Around the Crack Tip for $n^* = 1.5$

Fracture 1977, Volume 3, ICF4, Waterloo, Canada, June 19 - 24, 1977

DETERMINATION OF CRACK GROWTH IN A MIXED MODE LOADING SYSTEM

A. de S. Jayatilaka, I. J. Jenkins and S. V. Prasad*

INTRODUCTION

Studies on the propagation of cracks under an applied load are not readily analyzed when the cracks are not lying perpendicular to the applied stress. When a crack is orientated at some angle to the applied field, the mode of fracture is not a simple Mode 1 or Mode 2 but a combination of both.

Griffith [1] considered crack propagation to be primarily an energetic process and that the crack extends in a plane coincident to the plane containing the original crack. In reality the application of a load to a system containing a crack rarely leads to a discrete mode of fracture, usually two or three Modes act simultaneously at the crack front. Erdogan and Sih [2] considered the problem of mixed mode fracture by studying the initial direction of crack growth in a combined stress field. Further studies on mixed mode cracking were made by Sih [3] who showed that the condition for the direction of initial crack growth is given when strain energy density, S, attains a minimum value.

In this paper a new theory is proposed for the direction of crack initiation on the basis of the distortion strain energy. A 'failure curve' for a biaxial loading system is also suggested.

THEORY

To compute the strain energy around a crack tip, consider a sharp slit approximation and an element distance, r, away at an angle θ to the crack. The solutions of the stresses are given below for the combined Modes 1 and 2 in terms of the coordinate system given in Figure 1.

$$\begin{bmatrix}\sigma_x\\ \sigma_y\\ \tau_{xy}\end{bmatrix} = K_I/(2\pi r)^{1/2}\begin{bmatrix}\cos\theta/2\,(1-\sin\theta/2\sin3\theta/2)\\ \cos\theta/2\,(1+\sin\theta/2\sin3\theta/2)\\ \sin\theta/2\cos\theta/2\cos3\theta/2\end{bmatrix} + K_{II}/(2\pi r)^{1/2}\begin{bmatrix}-\sin\theta/2\,(2+\cos\theta/2\cos3\theta/2)\\ \sin\theta/2\cos\theta/2\cos3\theta/2\\ \cos\theta/2\,(1-\sin\theta/2\sin3\theta/2)\end{bmatrix} \quad (1)$$

* Materials Division, School of Engineering and Applied Sciences, University of Sussex, Brighton BN1 9QT, England

$$\sigma_z = \nu(\sigma_x + \sigma_y) \qquad \text{(plane strain)}$$

$$\tau_{xz} = \tau_{zy} = 0$$

$$E = 2(1 + \nu)\mu \,, \qquad \text{where } \mu \text{ is the shear modulus.}$$

The total elastic strain energy, dW_T, stored in an element $dV = dx.dy.dz$ under three dimensional stress system is given by

$$dW_T = \left[\frac{1}{2E}(\sigma_x{}^2+\sigma_y{}^2+\sigma_z{}^2) - \frac{\nu}{E}(\sigma_x\sigma_y+\sigma_y\sigma_z+\sigma_z\sigma_x) + \frac{1}{2\mu}(\tau_{xy}{}^2+\tau_{xz}{}^2+\tau_{zy}{}^2)\right]dV \quad .(2)$$

The total strain energy, dW_T, consists of the sum of the strain energy due to change in volume, dW_V, and the strain energy due to distortion, dW_d.

$$dW_T = dW_V + dW_d \; . \tag{3}$$

The strain energies due to volumetric change (hydrostatic component) and due to distortion are given below.

$$dW_V = \frac{1-2\nu}{6E}\left[\sigma_x{}^2+\sigma_y{}^2+\sigma_z{}^2+2(\sigma_x\sigma_y+\sigma_y\sigma_z+\sigma_z\sigma_x)\right] dV \tag{4}$$

$$dW_d = \frac{1+\nu}{3E}\left[\sigma_x{}^2+\sigma_y{}^2+\sigma_z{}^2-(\sigma_x\sigma_y+\sigma_y\sigma_z+\sigma_z\sigma_x)+3(\tau_{xy}{}^2+\tau_{xz}{}^2+\tau_{zy}{}^2)\right]dV. \tag{5}$$

Upon the substitution of equation (1) into equations (2, 4, 5), the total, hydrostatic and distortion strain energy densities may be written in the form given below.

$$\frac{dW_T}{dV} = \frac{1}{r}S, \qquad \frac{dW_V}{dV} = \frac{1}{r}S_V, \qquad \frac{dW_d}{dV} = \frac{1}{r}S_d \; . \tag{6}$$

Here S, S_V and S_d are independent of r but depend on the elastic constants, K_I, K_{II} and the angle, θ.

It should be noted that the strain energy densities tend to infinity as r tends to zero. As a result, a cut off point of $r > 0$ has to be introduced to allow for the discontinuity of strain energy density at the crack tip.

THE DIRECTION OF CRACK INITIATION AND PROPAGATION

In a mixed mode system, the strain energy density comprises of that due to a volume change and that due to a distortion effect as shown in Figure 2. A crack in an elastic material in pure hydrostatic tension will extend along a line coincident with the axis, A A', of the crack as shown in Figure 3. It appears that the direction of crack growth takes place along the direction where the distortion strain energy is minimum.

$$\frac{dS_d}{d\theta} = 0 \quad \text{at} \quad \theta = \theta_o \text{ and } \frac{d^2S_d}{d\theta^2} > 0 \; . \tag{7}$$

The resistance to crack growth is determined by the total strain energy density which reaches a critical value, S'_{crit}, at $\theta = \theta_o$,

$$S'_{crit} = S_V + S_d, \quad \text{at } \theta = \theta_o . \tag{8}$$

UNIAXIAL TENSION

For a crack extending in a mixed mode system, K_I and K_{II} are given by the following expressions.

$$K_I = \sigma\sqrt{\pi a}\,\sin^2\beta \quad , \quad K_{II} = \sigma\sqrt{\pi a}\,\sin\beta\,\cos\beta . \tag{9}$$

Figure 4 shows the variation of θ_o as a function of β. Table 1 is a comparison of theoretical results of θ_o due to Sih [3], the authors and the experimental results of Erdogan and Sih [2]. The experimental work was performed on plexiglass of dimensions 9" x 18" x 3/16" with a central crack of 2 in. The values of θ_o at $\nu = 0.0$ and 0.33 have been considered to give a meaningful comparison of experimental data with the theoretical values. The reason for this is that the dimensions of the samples tested would not give rise to a pure plane strain behaviour but a combination of plane strain and plane stress. When the normalized strain energy density term is plotted for different values of β, the resulting curves are similar to that of Sih. Hence these curves are omitted in this paper.

UNIAXIAL COMPRESSION

In this case, the values of θ_o for varying angles of β are shown in Figure 5. To determine the failure strength as a function of β, in compression, experiments have been carried out by the authors on "perspex" specimens of 100 x 100 x 6 mm dimensions. Cracks of 35 mm long were introduced using a circular saw of 0.2 mm thickness. These preslotted specimens were tested in uniaxial compression using an Instron machine at a cross head speed of 10 mm/min. Buckling of the samples was avoided by using a special jig. The results given in Figure 6 show an excellent agreement with the theory.

PURE SHEAR

Figure 7 shows the case of plane shear. K_I and K_{II} for this system are given by the following expressions.

$$K_I = 0 , \quad K_{II} = \tau(\pi a)^{1/2} . \tag{10}$$

The values of θ_o are computed as for previous cases and are given in Table 2 along with those due to Sih [3]. The classical theory due to Griffith postulates that a crack subjected to a pure shear stress is assumed to extend along the plane of the axis of the crack. As evident from the values of Sih and those due to authors, the crack extends in a line which is not coincident with the axis of the initial crack.

BIAXIAL LOADING SYSTEM

The theory described for the determination of crack growth in a uniaxial loading system can also be extended to biaxial loading systems. Such a system is shown in Figure 8 where the applied stress can be either both compressive or tensile or a combination of both. For this system K_I and K_{II} are given by the following expressions.

$$K_I = \sigma_2(\pi a)^{0.5}\left[\frac{\sigma_1}{\sigma_2}\sin^2\beta + \cos^2\beta\right] \tag{11}$$

$$K_{II} = \sigma_2(\pi a)^{0.5}\left[\frac{\sigma_1}{\sigma_2} - 1\right]\sin\beta\cos\beta\ . \tag{12}$$

For a known ratio of σ_1/σ_2, the minimum value of $\sigma_{2,crit}$ may be obtained similarly to the previous cases. Figure 9 shows a 'failure diagram' which is drawn on the assumption that the critical flaw size in a material lies in the direction in which $\sigma_{2,crit}$ is minimum.

DISCUSSION

A method for the determination of crack initiation has been put forward on the basis that the direction of initiation is given when S_d reaches a minimum value. It is evident from the results that the proposed theory is different to that of Sih. His work showed that θ_0 takes lower values for lower values of ν when compared with the authors. However, the resistance to crack growth which determines the failure stress shows a good agreement with the work of Sih.

An important outcome of this work and also that due to Sih's studies is that it confirms that the use of the expression, given below, for the strain energy release rate, G, for mixed mode systems, needs to be revised.

$$G = K_I{}^2(1-\nu^2)/E + K_{II}{}^2(1-\nu^2)/E + K_{III}{}^2(1+\nu)/E \tag{13}$$

(plane strain)

It is important to note that the ratio of compression to tensile strength is not easily evaluated for a real material, due to the random distribution of flaw sizes and their orientations within a material. This argument may also be applied to the biaxial system. The application of statistical methods to determine the strength of brittle materials for both uniaxial and biaxial systems is under progress.

ACKNOWLEDGEMENTS

One of the authors (I. J. Jenkins) acknowledges the support by the Science Research Council in the form of a studentship.

REFERENCES

1. GRIFFITH, A. A., "The Theory of Rupture", Proceedings of the 1st Int. Congress on Applied Mechanics, Delft, 1924, 55.
2. ERDOGAN, F. and SIH, G. C., J. Basic Eng., 6, 1970, 519.
3. SIH, G. C., Int. J. of Fracture, 10, 3, 1974, 305.

Table 1 A Comparison of Theoretical and Experimental Values of θ_0

β	30	40	50	60	70	80	ν
3	-48.0	-39.0	-32	-23	-15	-10	0.0
	-63.5	-56.7	-49.5	-41.5	-31.8	-18.5	0.333
2	-62.4±2.1	-55.6±1.2	-51.6±0.8	-43.1±1.0	-30.7±0.8	-17.3±0.7	0.333
Authors	-64	-56	-49	-42	-32	-19	0.00
	-69.5	-62	-53.5	-45.0	-33	-19	0.333

Table 2 Theoretical Values of $|\theta_0|$ for Pure Shear

ν	0	0.1	0.2	0.3	0.4	0.5	
θ_0	84°	86°	88°	89°	90°	-	Authors
θ_0	70.5	75.6	79.3	83.3	87.2	90.0	3

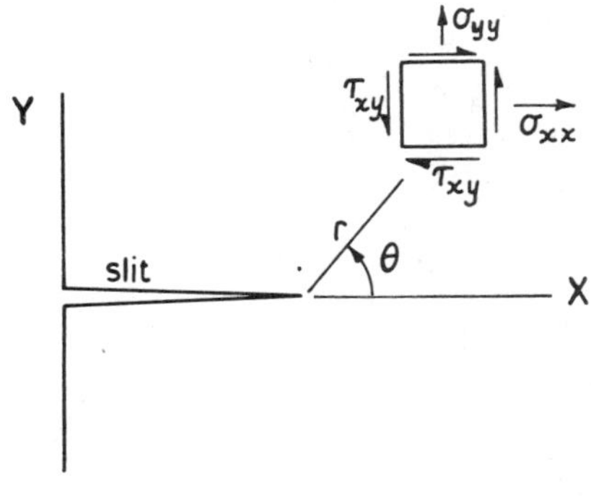

Figure 1 Crack Tip Stresses, Showing Rectangular Components

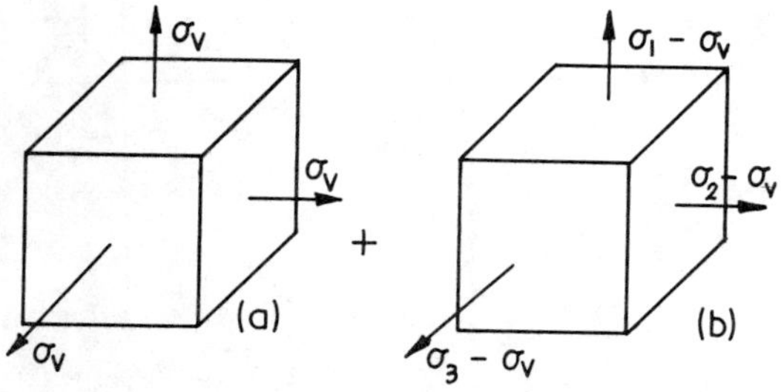

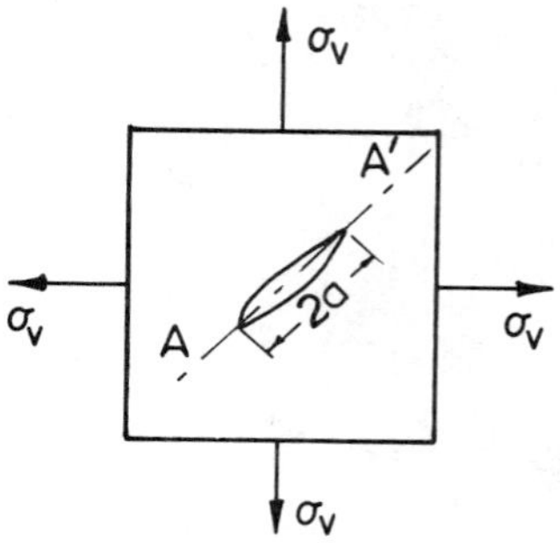

Figure 2 The Stress System for (a) Hydrostatic, (b) Distortion Effects. $\sigma_V = 1/3(\sigma_1+\sigma_2+\sigma_3)$ where σ_1, σ_2, σ_3, are the Principal Stresses

Figure 3 The Direction of Crack Extension in Hydrostatic Loading

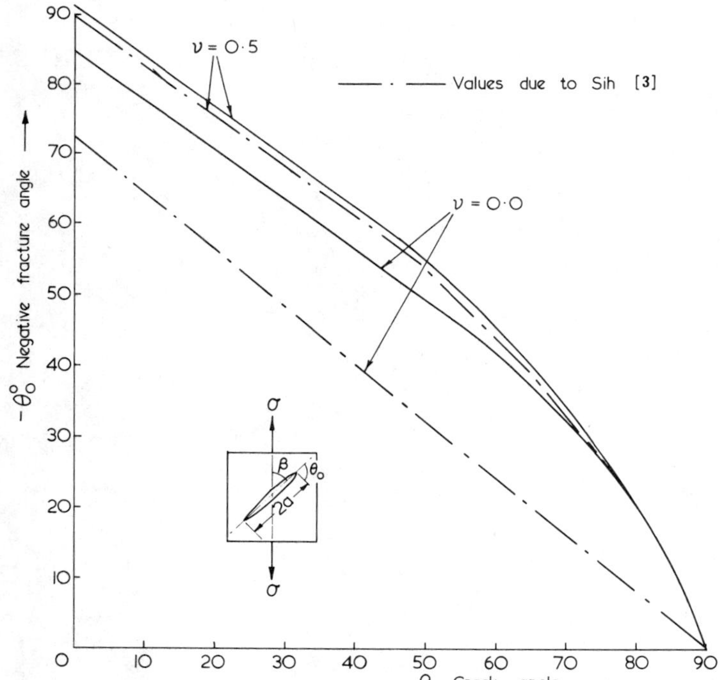

Figure 4 Crack Angle Versus Fracture Angle for the Tension Case

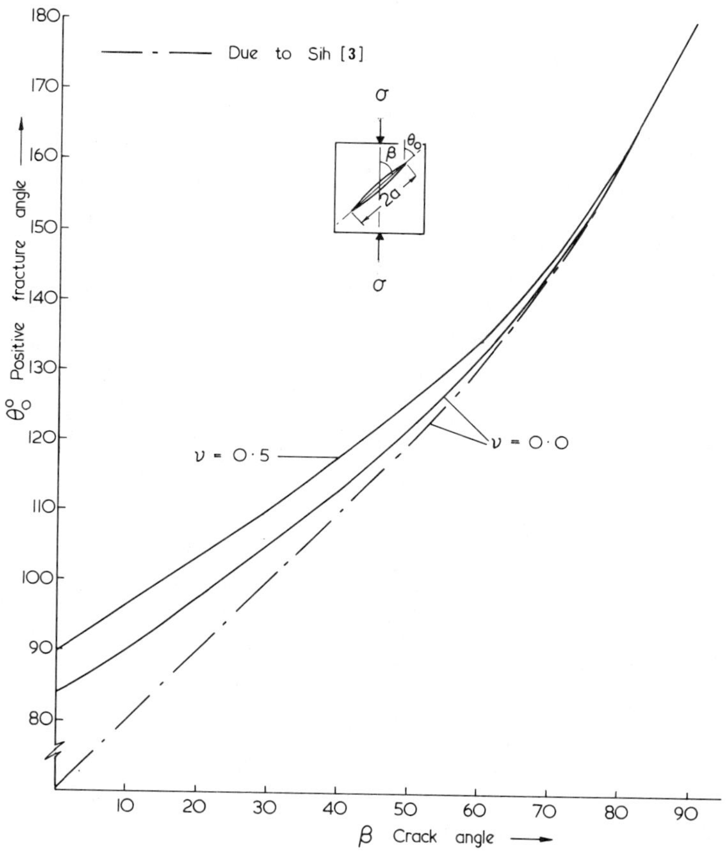

Figure 5 Crack Angle Versus Fracture Angle for the Compression Case

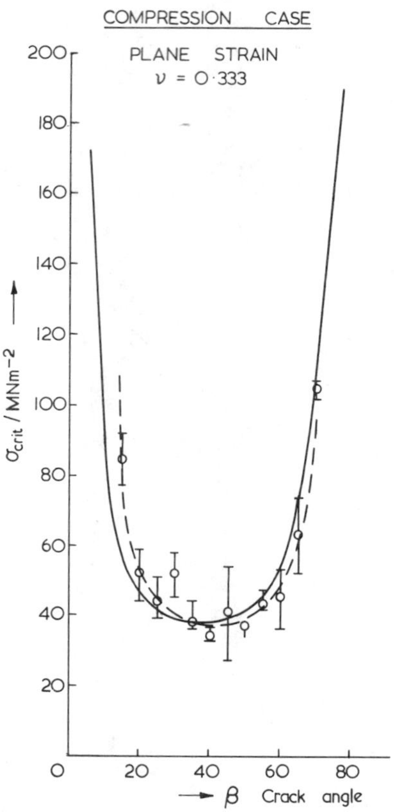

Figure 6 Stress to Failure Versus Crack Angle for Plane Strain Conditions in Compressive Loading. The Theoretical Curve was Drawn with its Minimum Point Having the Same Value of σ_{crit} of the Experimental Curve at the Same Angle. Solid Line Indicates the Theoretical Curve

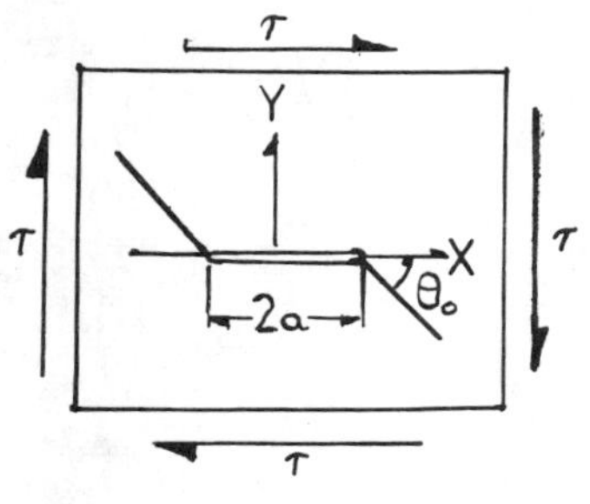

Figure 7 A Line Crack in Shear

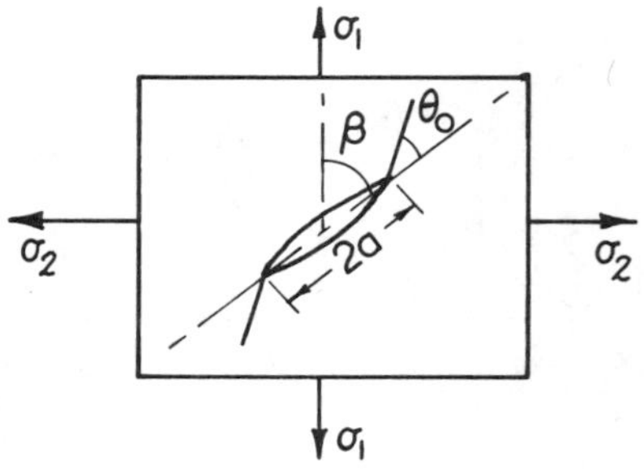

Figure 8 An Inclined Crack in a Biaxial Stress System

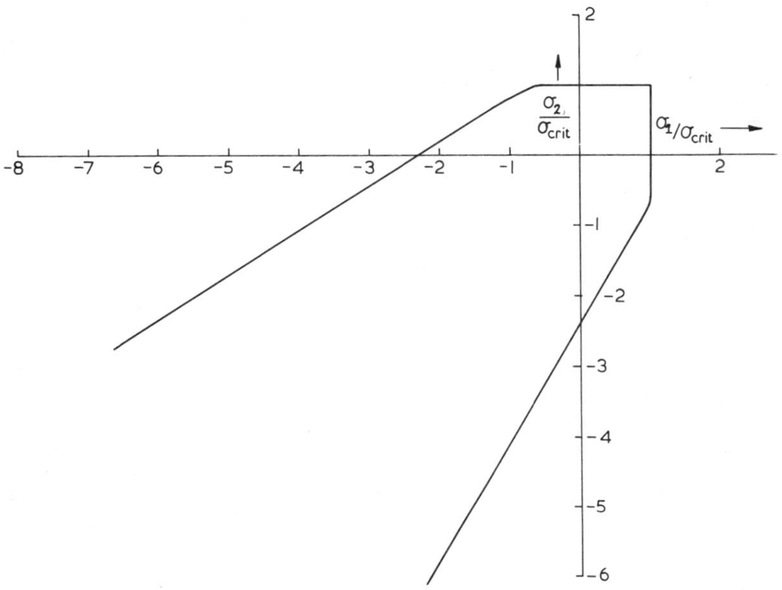

Figure 9 A Biaxial Failure Diagram, Where σ_{crit} is the Stress to Failure for Mode 1. '$\nu = 0.3$'

Fracture 1977, Volume 3, ICF4, Waterloo, Canada, June 19 - 24, 1977

A CONTRIBUTION TO THE ANALYSIS OF QUASI STATIC CRACK GROWTH IN SHEET MATERIALS

A. U. de Koning*

INTRODUCTION

In general, aircraft structures have to satisfy conditions with respect to strength, stiffness and weight. These requirements have led to the application of thin walled stiffened structures made of high strength aluminium alloys. The linear elastic behaviour of these structures can be described accurately by the simple mathematical models from the linear theory of elasticity, and in combination with the finite element technique the stiffness of complex aircraft structures can be analysed accurately for general loading cases.

With respect to the analysis of strength the situation is much more complicated. Non-linear effects such as buckling and plastic deformation have a pronounced influence on the load level that can be applied to the structure. Another important aspect that frequently has to be considered is the sensitivity of structural strength to fatigue damage. As a result of increasing lifetimes and the character of the loading conditions, initiation of fatigue cracks cannot always be prevented. In this situation it is important that the rate of crack growth is limited, and also that the strength of the weakened structure is not reduced in such a way that its reliability is endangered.

In the present investigation the behaviour of cracked structures subjected to high peak loading conditions is considered. Using the Finite Element Method the phenomenon of slow stable crack growth can be analysed. In this way detailed information is obtained about the behaviour during crack growth of well known crack growth parameters such as the value of the path-independent J integral, the crack extension force G and the crack opening displacement. It is thought that for an analysis of residual strength of structures such information can be very useful.

Of course, the occurrence of a peak loading situation also affects the subsequent crack growth behaviour under fatigue loading conditions. Nowadays, it is known that the crack growth rate is strongly affected by peak loads. In the present paper some aspects of crack growth retardation and acceleration are considered.

For both studies the Finite Element Method is applied intensively. The results obtained are verified by a comparison with experimental data.

SCOPE OF THE PRESENT STUDY

Plasticity effects and the complex changes of boundary conditions associated with crack growth, crack opening and crack closure can be treated by

*National Aerospace Laboratory NLR, Amsterdam, The Netherlands.

the Finite Element Method. At the NLR laboratory a FEM computer program was developed specifically to deal with these situations. In the present investigation crack growth is realised by disconnecting finite elements. In view of the incremental description of plasticity effects this procedure has to be executed in small steps. In section 3 it is shown that application of the disconnecting procedure has consequences with regard to the energy dissipation in the crack tip zone. These consequences are analysed.

In the study of slow stable crack growth the decision whether elements must be disconnected or not is taken in such a way that the load versus crack length curve that was measured for the specimen under consideration is realised also in the FEM analysis of that particular specimen. Using this procedure experimental results can be analysed in detail. The method is applied in an analysis of the crack growth behaviour in two different sheet specimens weakened by central fatigue cracks of lengths respectively 61 and 123 mm. The results are presented in section 4. In the same section it is concluded that soon after initiation of crack growth the angle between the crack surfaces near the crack tip seems to be constant. This result provides a ground for the assumption that after initiation of crack growth the severity of the crack tip situation can be characterised by the crack tip opening angle CTOA. The critical value of the CTOA can also be measured accurately using an optical microscope, a fact that is an important advantage with regard to other crack growth parameters.

In the analysis of peak loading effects on fatigue crack growth, three aspects are considered. Firstly, a peak load is applied to a structure weakened by a flat central initial fatigue crack. At the maximum load level the elastic-plastic deformations and the crack opening displacements are computed. Then the loading forces are decreased. During this part of the loading trajectory the crack closure behaviour and the reversed plastic flow near the crack tip are studied. It will be shown in section 4 that both effects depend strongly on the geometry of the deformed crack surfaces in the maximum peak load situation. The results show the important influences of the peak load level and the geometry of the deformed crack on the stress distribution in the crack tip zone. Finally, subsequent fatigue crack growth through the plastic zone is analysed. In the finite element model the crack length can be increased by a simple procedure. In this way the effect of a relatively large plastic zone, located at a fixed position near the initial crack tip, on the crack extension force was computed for different actual crack lengths. From this computation it can be concluded that in terms of the distance between initial and actual crack tip, the effect of a simple overload is limited. Nevertheless, it is thought that the results obtained justify further effort to improve the methods used here and to arrive at a more complete model of fracture.

DISCRETIZATION OF CRACK GROWTH

In an application of the Finite Element Method crack growth can be realised in different ways. In this investigation a nodal force relaxation technique was applied. Adopting this method the forces connecting two finite elements are relaxed in a stepwise manner. For reasons of simplicity, in this analysis linear displacement finite elements (trim 3) were used. During relaxation of a nodal force the crack length cannot be defined in an unambiguous manner, but after complete relaxation the crack length has increased by the length of one element side.

It is noted that, in an analysis of energy dissipation near the crack tip, the work rate associated with the relaxation of nodal forces has to be considered. To estimate the significance of this complication, during relaxation a linear relation is assumed between the connecting force p and the displacement u at a node. Denoting the work rate by $\Delta R/\Delta a$, it follows that

$$\Delta R/\Delta a = \frac{1}{2}\sum_i p_i^a \, u_i^{a+\Delta a}/\Delta a \tag{1}$$

where Δa is identified with the discrete increment of crack growth (= the element size). Further, p_i^a and $u_i^{a+\Delta a}$ denote respectively the connecting forces before and the displacements after completion of the relaxation procedure. In the present 2D-analysis the crack line is the axis of symmetry. Thus, in p_i^a two components of equal magnitude are significant. Therefore, in equation (1) the subscript i will be dropped. In the case of purely elastic material behaviour it is clear that $\Delta R/\Delta a$ equals $-G$, where G denotes the crack extension force. In the case of ideal plastic material behaviour the value of $\Delta R/\Delta a$ is estimated. It will be assumed that

$$p^a = -\,\beta\,\sigma_y \Delta a\ ,\quad u^{a+\Delta a} = \alpha\,\Delta a/2 \tag{2}$$

In these formulas β is a constraint fractor and σ_y is the uniaxial yield limit. The scalar α stands for the crack tip opening angle CTOA. In section 4 the value of α is identified as a material constant. Then after substitution of the formulas from (2) into equation (1), for one crack tip, it follows that

$$\Delta R/\Delta a = -\,\alpha\,\beta\,\sigma_y \Delta a/2 \tag{3}$$

From this consideration it appears that the work rate $\Delta R/\Delta a$ varies in proportion to the element size selected at the crack tip, and approaches zero if Δa tends to zero. As shown in Figure 1 for the real material behaviour a similar relation was found from the result of the FEM computations. This, in fact, invalidates a computation of the energy dissipation in the crack tip zone. However, the apparent error can be made arbitrarily small by proper choice of the element size. In the present investigation the element size was chosen in such a way that value of $\Delta R/\Delta a$ is small when compared with the plastic energy dissipation.

ANALYSES AND RESULTS

In the present analysis of slow stable crack growth the behaviour of two different sheet specimens weakened by central fatigue cracks of lengths respectively 61 and 123 mm is studied. The specimens were cut from 2024-T3 aluminium. The sheet thickness was 2 mm and the width of the specimens 600 mm.

From the FEB solutions obtained, the value of the J-integral was computed along two different contours through the purely elastic part of the specimen. Path independence was verified. The values of J are presented in Figure 2. The initial crack lengths are indicated. It is seen that the value of J depends on the amount of crack growth $a-a_o$. Further, the significant terms in the balance between rates of deformation energy quantities and the work rate associated with the external loading forces were computed. The balance was verified. From the results it follows that the elastic deformation energy is not strongly affected by plasticity effects: at the onset of unstable crack growth a difference of about 5 per cent was found between the FEM result and the value computed using an analytical formula that is based on purely elastic material behaviour. However, the rate of plastic energy dissipation and thus also the work rate associated with the external loading forces was much larger than predicted by certain analytical formulas. Moreover, the rate of energy dissipation seems to depend on $(a-a_o)/a_o$, instead of a-a as frequently suggested in the R-curve approach. In Figure 3 the values of $\Delta W/\Delta a$ are plotted in relation to $(a-a_o)/a_o$. The present results raise some questions regarding the validity of the R-curve approach in its current formulation.

At different load levels the crack opening displacements were computed. These displacements determine the actual shape of the crack. The results are given in Figure 4. It is concluded that soon after initiation of crack growth the angle between the crack surfaces near the tip is constant. This angle will be called the Crack Tip Opening Angle CTOA. It was also observed that this angle is nearly independent of the element size applied in the crack tip region. The computed values of the CTOA are given in Figure 5. The crack opening displacements were also measured using an optical microscope. In Figure 6 the measuring result is indicated. It is seen that the CTOA value obtained from the tangent modulus of this curve is in agreement with the computational result. These results provide grounds for the assumption that after initiation of crack growth the severity of the crack tip situation can be characterised by the CTOA. In this circumstance it seems logical to assume that the crack grows when the value of the CTOA exceeds some critical value. Thus, a deformation type of crack growth criterion is obtained. It is to be noted that the critical value of the CTOA can be obtained by direct measurement. Hence, adopting the CTOA as crack growth parameter, some important sources of inaccuracy associated with the determination of values for the usual crack growth parameters can be eliminated.

From the displacement pattern shown in Figure 4 an important conclusion can be drawn: for load levels below the level for initiation of stable crack growth the crack tip is blunted, while for higher peak load levels a sharp crack tip is observed. To analyse the effect of the geometry of the crack on the subsequent crack closure behaviour the loading forces are reduced. From the FEM solutions obtained at different load levels during this unloading procedure, it follows that the sharp crack tip immediately starts closing. Subtracting the "elliptic" elastic solution from the linear pattern from Figure 4 this behaviour can be verified. Further, it is concluded that the compressive stresses acting at the closed crack surfaces considerably delay reversed plastic flow and, thus preserve the plastic strain field produced as result of the peak load.

The FEM solutions obtained for a crack with blunted tips show a completely different behaviour. It was found that reversed plastic flow is immediately initiated, while crack closure is first observed at about zero load level midway between the crack tips. From there on crack closure proceeds towards the tips. However, all FEM results obtained for a blunted crack tip show that the tip itself remains open, even after application of compressive loading forces of the same order of magnitude as the original tensile peak load. From these observations it is concluded that stable crack growth has a tremendous effect on the deformation fields in the crack tip zone. Although, in general, the load level for initiation of stable crack growth is not sharply defined, it is thought that insight into the extreme situations discussed here will help to explain the effect of peak loads on the subsequent fatigue crack growth behaviour.

Analysing subsequent fatigue crack growth, it is recognised that the crack grows through the relatively large plastic zone associated with the peak load. Then, after a certain amount of fatigue crack growth, the plastic zone is situated behind the crack tip and the distance between plastic zone and crack tip is increasing. As a result of this increasing distance the effect of the peak load on the actual crack tip situation will diminish. Simulating fatigue crack growth by the nodal force relaxation model applied at a relatively low load level this effect can be analysed by the FEM. The method was applied and after completion of each nodal force relaxation the elastic deformation energy in the structure was computed. From this array values of the crack extension force G were obtained for different distances between plastic zone and actual crack tip. From the results it follows that after an amount of crack growth of the same order of magnitude as the plastic zone size the effect of the peak load on G has damped out. So, in terms of crack length the effect of a single peak load on G is limited.

CONCLUSIONS

1) The value of the J integral seems to be a useful parameter to predict initiation of stable crack growth in centrally cracked sheets.
2) During stable crack growth the crack tip opening angle CTOA is nearly constant and independent of the element size applied to discretize the crack tip zone.
3) The R curve concept needs revision.
4) After initiation of stable crack growth the CTOA seems to be an attractive parameter to describe stable crack growth.
5) The effect of peak loads on subsequent fatigue crack growth can be analysed by the Finite Element Method.

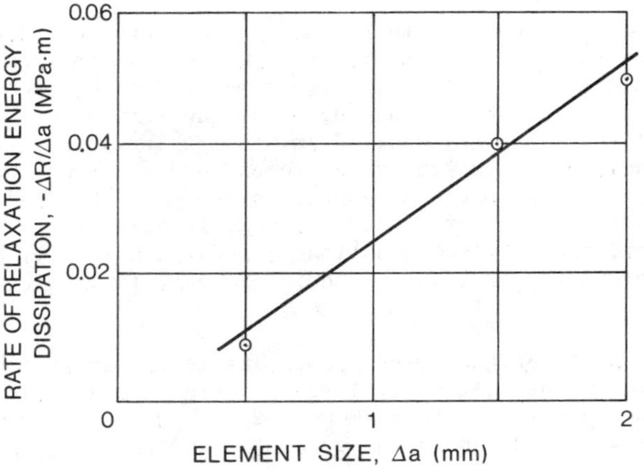

Figure 1 The relation between the work rate associated with the relaxation of nodal forces and the element size applied in the crack tip region

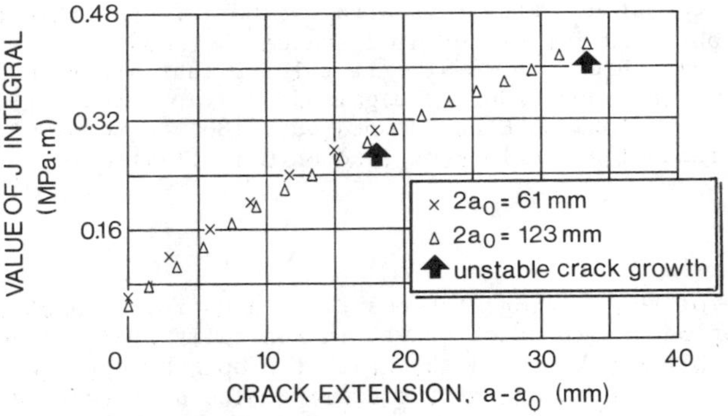

Figure 2 The relation between the value of J and the amount of crack growth

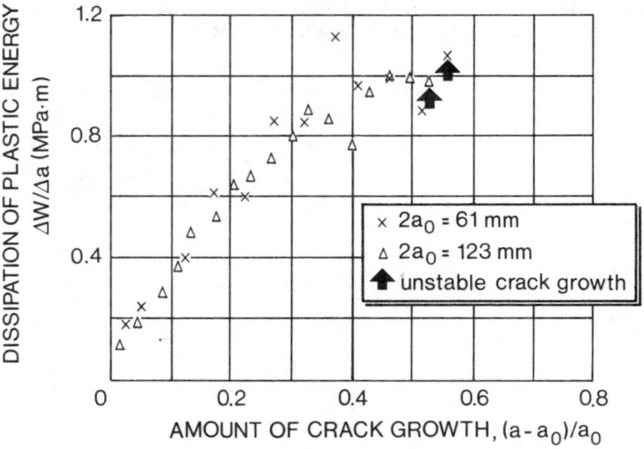

Figure 3 The relation between the rate of energy dissipation in the plastic zone and the amount of crack growth

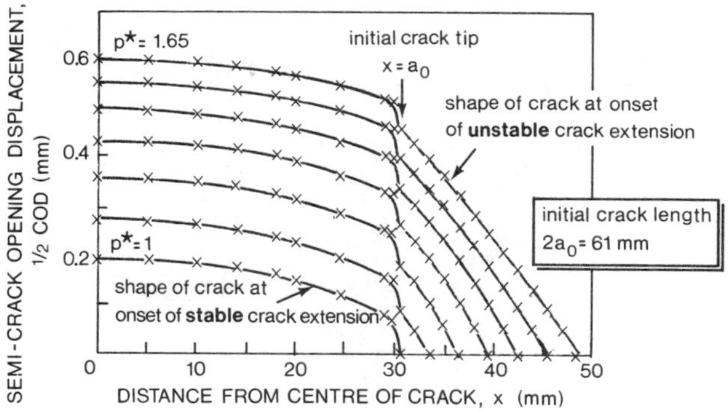

Figure 4   Shape of the crack as computed at different stages of loading

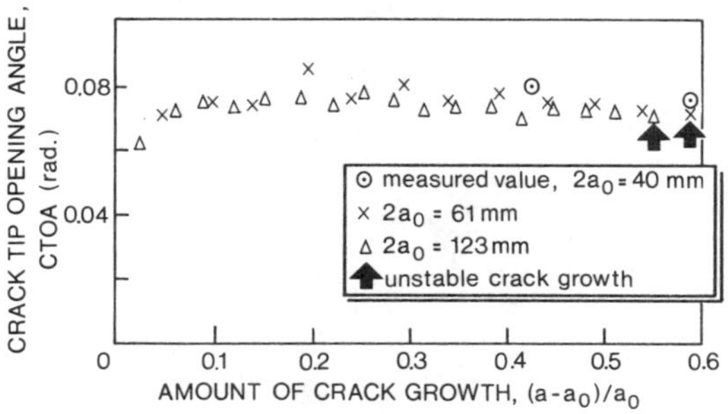

Figure 5   Computed and measured values of the crack tip opening angle in relation to the amount of crack growth

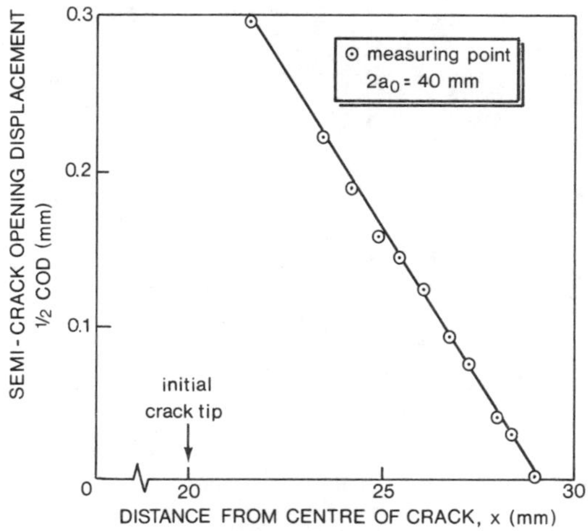

Figure 6   Shape of the crack as measured after 40 percent crack growth

Fracture 1977, Volume 3, ICF4, Waterloo, Canada, June 19 - 24, 1977

SELF-SIMILAR PRESSURE PROFILES IN THE SYMMETRICALLY EXTENDING PLANE CRACK

Finn Ouchterlony*

INTRODUCTION

Much of the author's recent work has involved Fracture Mechanics analysis of rock destruction. In one case, the different stages of the breaking process in rock blasting were modelled by various static radial crack systems in [1]. Another case, which prompted the work in [2], is the splitting of plane rectangular slabs of rock or concrete by wedging tools. A static analysis of mechanically induced rock fractures may be adequate but in rock blasting it is sure to leave out vital parts such as stress wave effects, crack branching phenomena, and the crack propagation speed. This note concerns the latter part. It is based on the author's report [3] and illustrates the effects of gas penetration on the energy release rates at the tips of a symmetrically extending plane crack.

Since the advent of Broberg's analysis [4] many papers on self-similarly extending cracks in elastic media have appeared, see for example [5 - 12] for in-plane situations. The papers by Willis [7] and Norwood [9 - 10] outline general solution procedures but we prefer to base our analysis on the papers by Craggs [5] and Cherepanov and Afanas'ev [8]. It is restricted to symmetric crack extension even though Norwood [10] and Brock [11 - 12] recently have treated some non-symmetric cases and it leads to a Green's function for the stress intensity factor which is analogous to the static one derived by Sih et al [13]. A numerical integration gives results for any symmetric self-similar pressure profile.

PROBLEM FORMULATION AND SOLUTION

Consider the following conditions in plane linear elastodynamics: At $t = 0$ a crack is somehow initiated at the origin of a polar coordinate system (r, θ) in an undisturbed homogeneous medium described by the Young's modulus E and the Poisson'a ratio ν. It's tips propagate in opposite directions along the x-axis with the same constant velocity v which is less than the Rayleigh wave velocity c_R and thus they lie inside the shear wave front at $r = c_2t$, see Figure 1. The crack faces are opened by two pairs of concentrated line loads of magnitude $F = pvt$ which are moving in opposite directions with the velocity $v_F < v$. The disturbed medium is encompassed by the longitudinal wavefront at $r = c_1t$. The boundary conditions along the x-axis are

$$\left.\begin{aligned} &v_\theta = 0 \text{ when } |x| > vt \text{ and} \\ &\sigma_\theta = pvt[\delta(x+v_Ft) + \delta(x-v_Ft)]H(t) \text{ when } |x| < vt\ . \end{aligned}\right\} \quad (1)$$

---

* Swedish Detonic Research Foundation, P. O. Box 32058, S-12611 Stockholm, Sweden

Here $\delta(t)$ denotes the delta function and $H(t)$ the unit step function.

The problem consists of solving the governing wave equations with the appropriate initial and boundary conditions. The method of solution follows in the vein of Craggs [14, 5] and Cherepanov and Afanas'ev [8]. The self-similarity of the problem reduces the independent variables to r/t and θ. The semi-circular regions above the x-axis and inside the respective wave fronts $c_1 t$ and $c_2 t$ are mapped on the upper halves of the ζ_1 and ζ_2 planes respectively with

$$\zeta_j = \frac{1}{r}\left[c_j t \cos\theta + i\sqrt{(c_j t)^2 - r^2}\sin\theta\right] \text{ and } j = 1,\ 2\ . \tag{2}$$

The whole problem then reduces to a Keldysh-Sedov problem for one analytic function in the half-plane $\mathrm{Im}\zeta_1 > 0$.

The details of the solution are found in [3]. For both tips, the resulting expression for the stress intensity factor becomes

$$K_I(v,t,s) = \frac{2pvt}{\sqrt{\pi vt}} \cdot \frac{f(sv)}{f(v)} \cdot \frac{F(v,s)}{\sqrt{1-s^2}}\ . \tag{3}$$

Here $s = v_F/v$ denotes the relative speed of the loading points, $f(v)$ denotes a universal function of the crack speed given by

$$f(v) = \sqrt{1-M_1^2}/(1-\nu)R(v) \text{ where} \tag{4}$$

$$R(v) = -[(2-M_2^2)^2 - 4\sqrt{(1-M_1^2)(1-M_2^2)}]/M_2^2 \tag{5}$$

denotes the Rayleigh function with $M_1 = v/c_1$, and $M_2 = v/c_2$ being Mach numbers and m the ratio c_1/c_2, and $F(v,s)$ denotes the expression

$$F(v,s) = \left\{-\left[\frac{4P(v,s)}{m^2R(sv)} + \frac{s^2}{1-s^2}\right] \cdot [M(v) - L(v) + 4J(v,s)] + \right.$$
$$\left. + 8\ N(v,s)\right\}(1-s^2)/L(v)\ . \tag{6}$$

The function $F(v,s)$ contains in turn the following functional expressions: First there is $P(v,s)$ given by

$$P(v,s) = \frac{(2-s^2M_2^2)\ (6-8s^2M_1^2+s^2M_2^2)}{4(1-s^2M_1^2)} - (3-2s^2M_2^2)\sqrt{\frac{1-s^2M_1^2}{1-s^2M_2^2}}\ . \tag{7}$$

Then we have the well known Broberg function [4]

$$L(v) = \left[\frac{(M_2^2-2M_1^2)^2}{1-M_1^2} + 4M_1^2\right]K_1 - \left[\frac{(2-M_2^2)^2}{1-M_1^2} + 4\right]E_1 - 4M_2^2K_2 + 8E_2\ , \tag{8}$$

where K_1 and E_1 denote complete elliptic integrals of the first and second kind respectively with the modulus $\sqrt{1-M_1^2}$ and the definition of index 2 follows in analogy. For details see Byrd and Friedmann [15]. Next,

$$M(v) = M_1^2\left[\frac{4}{3} + \frac{(2-M_2^2)^2}{1-M_1^2}\right]K_1 - \left[\frac{8}{3}(1+M_1^2)+4(1-M_2^2) + \frac{(2-M_2^2)^2}{1-M_1^2}\right]E_1$$

$$- \frac{16}{3}M_2^2K_2 + \frac{4}{3}(8-M_2^2)E_2 \,. \quad (9)$$

The functions J(v,s) and N(v,s), finally, both contain complete elliptic integrals of the third kind. π_1 has the same modulus as K_1 and E_1 and the parameter $n_1^2 = (1-M_1^2)/(1-s^2M_1^2)$. π_2 is found by switching the indici. The expressions read

$$J(v,s) = -\left[\frac{(2-s^2M_2^2)^2}{4(1-s^2M_1^2)}M_1^2\pi_1-E_1\right]/s^2 + \left[(2+2M_1^2-3M_2^2)E_1-M_1^2K_1\right]/3$$

$$+ (M_2^2\pi_2-E_2)/s^2 - \left[(2-M_2^2)E_2-M_2^2K_2\right]/3 \text{ and} \quad (10)$$

$$s^2N(v,s) = \frac{(2-s^2M_2^2)(2+s^2M_2^2-4s^2M_1^2)}{4(1-s^2M_1^2)}M_1^2\pi_1-E_1-(M_2^2\pi_2-E_2)$$

$$+ \frac{(2-s^2M_2^2)^2}{(1-s^2M_1^2)}\left[(1-s^2M_1^2-n_1^2)\pi_1-E_1+s^2M_1^2K_1\right]/8(1-s^2)$$

$$- \left[(1-s^2M_2^2-n_2^2)\pi_2-E_2+s^2M_2^2K_2\right]/2(1-s^2) \,. \quad (11)$$

In the quasi-static limit equation (3) yields the same result as would Sih et al [13] for a symmetric loading.

The case when the forces act at the origin is contained in equation (3). Setting $v_F = s = 0$ we obtain

$$K_I(v,t,0) = \frac{2F}{\sqrt{\pi vt}} \cdot [M(v)/L(v)-1]/f(v). \quad (12)$$

Somewhat surprisingly, this implies that the stress intensity factor, and hence the energy release rate, will become zero at a crack speed v_{max} which is less than the Rayleigh wave speed. See Figure 2 for details.

ENERGY RELEASE RATES CAUSED BY SELF-SIMILAR PRESSURE PROFILES

Our purpose is to study the effects of gas penetration on the energy release rate at various crack speeds. More specifically, we choose a one parameter family of binomial type pressure profiles given by

$$p(s,q) = p_o(1+q)(1-|s|)^q \quad \text{when} \quad -1 \leq s \leq 1 \,. \quad (13)$$

Here p_o is some reference pressure and the parameter q is a measure of the peakedness or, equivalently, the gas penetration. All pressure profiles exert the same opening force on the crack faces due to the factor (1+q). The stress intensity factor is obtained from an integration of equation (3) with p(s,q) as a weight functions as

$$K_I(v,t,q) = \int_0^1 K_I(v,t,s)\, p(s,q)ds \ . \qquad (14)$$

The energy release rate in turn is given by the universal relation

$$G_I(v,t,q) = (1-\nu^2)K_I^{\,2}(v,t,q)f(v)/E \ , \qquad (15)$$

see Freund and Clifton [16] for example. Plane strain conditions are assumed.

We prefer to present normalized values g of the energy release rates with the static Broberg crack taken as the reference state. Thus

$$g(v,q) = G_I(v,t,q)/G_{Io} \quad \text{where } G_{Io} = \pi(1-\nu^2)p_0{}^2a/E \ , \qquad (16)$$

and a = vt is the instantaneous crack length. Hereby g(v,q) becomes independent of a. With the aid of equations (3), and (13 - 16), we may write

$$g(v,q) = g(v,0) \cdot \left[\frac{2(1-\nu)(1+q)L(v)}{\pi m} \int_0^1 (1-s)^q f(sv)\, \frac{F(v,s)}{\sqrt{1-s^2}}\, ds\right]^2 , \qquad (17)$$

where g(v,0) is the normalized energy release rate of the Broberg crack which is given by

$$g(v,0) = m^2M_1{}^4/L^2(v)f(v) \ . \qquad (18)$$

The squared factor in equation (17) is denoted I(v,q) and must be evaluated numerically. For the Broberg crack a weighted Gauss-Chebyshev formula with ten sampling points yields $I^2(v,0) = 1.000$ and that confirms equation (3).

It is also possible to deduce other limiting forms for g(v,q). From equation (12) there follows that for an infinitely peaked pressure distribution the normalized energy release rate is given by

$$\lim_{q\to\infty} g(v,q) = \frac{4}{\pi^2}\, [M(v)/L(v)-1]^2/f(v) \ . \qquad (19)$$

For a static crack the results of Sih et al [13] yield the formula

$$g(0,q) = \frac{4}{\pi^2} \left[(q+1) \sum_{k=0}^{\infty} (-1)^k (1/2)_k/(q+1/2)_{k+1}\right]^2 , \qquad (20)$$

where the expression $(\ )_k$ denotes the Pochhammer's symbol.

The results of the calculations are plotted in Figure 3 where g(v,q) is shown as a function of crack speed for $\nu = 1/3$ and with q as a parameter.

The exact limiting forms for $q = 0$ and $q = \infty$, given by equations (18) and (19) respectively, are represented by full lines. For intermediate values of q, where a numerical evaluation of $I(v,q)$ is necessary, the results are indicated by broken lines. The static limit given by equation (20) is drawn as circles on the vertical axis and the limiting speed on the horizontal one.

We observe that the extent of gas penetration has a profound effect on the energy release rate. At one extreme, total gas penetration is modelled by $q = 0$ and the curve $g(v,0)$ may be regarded as an upper limit to possible values of the energy release rate. Even a slight decrease in the gas penetration will lower it markedly. For a linearly decreasing gas pressure for example, when $q = 1$, the available energy is reduced to about half or less. For a more realistic pressure profile we would expect it to drop even more and to approximately correspond to the other extreme value which is no gas penetration and modelled by $q \to \infty$. Based on these curves we also expect that the theoretical limiting velocity may be lower than the Rayleigh wave velocity in a situation where the gas doesn't entirely fill a propagating crack.

The results presented above can only form a qualitative basis for the equation of motion of a pressurized crack in a blasting situation. If they are supplemented by an expansion law for the gaseous detonation products, an expression for the crack surface displacement, and measurements of the effective fracture surface energy γ_F as a function of the crack velocity then one may perhaps hope to use $G_I = 2\gamma_F$ as an equation of motion for crack growth, much in the same way as Bergkvist [17] has done for a central crack in an infinite sheet of PMMA.

ACKNOWLEDGEMENT

This research has been supported by the Atlas Copco and Nitro Nobel companies.

REFERENCES

1. OUCHTERLONY, F., Proc. Third Congress, ISRM IIB, 1974, 1377.
2. OUCHTERLONY, F., Int. J. Eng. Sci., in press.
3. OUCHTERLONY, F., Swedish Detonic Res. Found. Rept. 1976:13.
4. BROBERG, K. B., Arkiv för Fysik, 18, 1960, 159.
5. CRAGGS, J. W., Fracture of Solids, Wiley Interscience, New York, 1962, 51.
6. ATKINSON, C., Arkiv för Fysik, 35, 1967, 469.
7. WILLIS, J. R., Phil. Trans. Roy. Soc., A274, 1973, 435.
8. CHEREPANOV, G. P. and AFANAS'EV, E. F., Int. J. Eng. Sci., 12, 1974, 665.
9. NORWOOD, F. R., Int. J. Solids and Structures, 9, 1973, 789.
10. NORWOOD, F. R., Int. J. Eng. Sci., 14, 1976, 477.
11. BROCK, L. M., Int. J. Eng. Sci., 13, 1975, 951.
12. BROCK, L. M., Int. J. Eng. Sci., 14, 1976, 181.
13. SIH, G. C., PARIS, P. C. and ERDOGAN, F., J. Appl. Mech., 29, 1962, 306.
14. CRAGGS, J. W., Proc. Cambridge Phil. Soc., 56, 1960, 269.
15. BYRD, P. F. and FRIEDMAN, M. D., Handbook of Elliptic Integrals for Engineers and Scientists, Second Edition, Springer Berlin, 1971.
16. FREUND, L. B. and CLIFTON, R. J., J. Elasticity, 4, 1974, 293.
17. BERGKVIST, H., J. Mech. Phys. Solids, 21, 1973, 229.

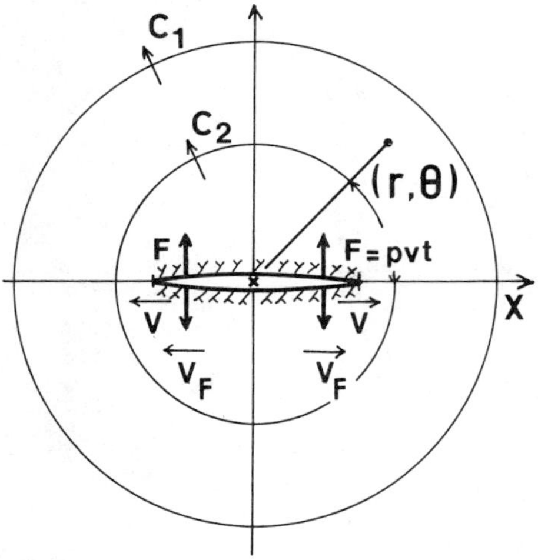

Figure 1 Crack Geometry

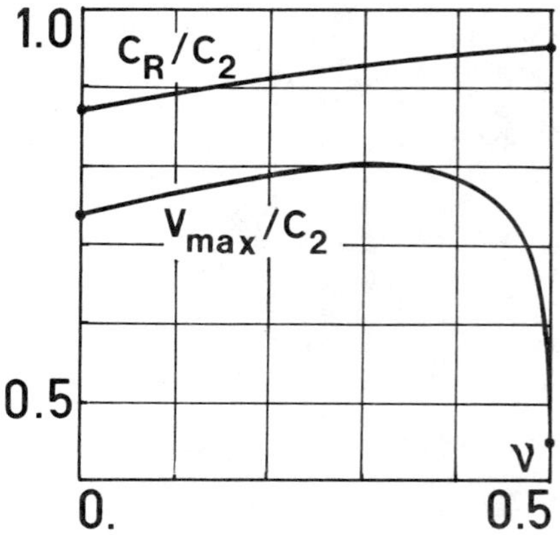

Figure 2 Limiting Crack Speeds

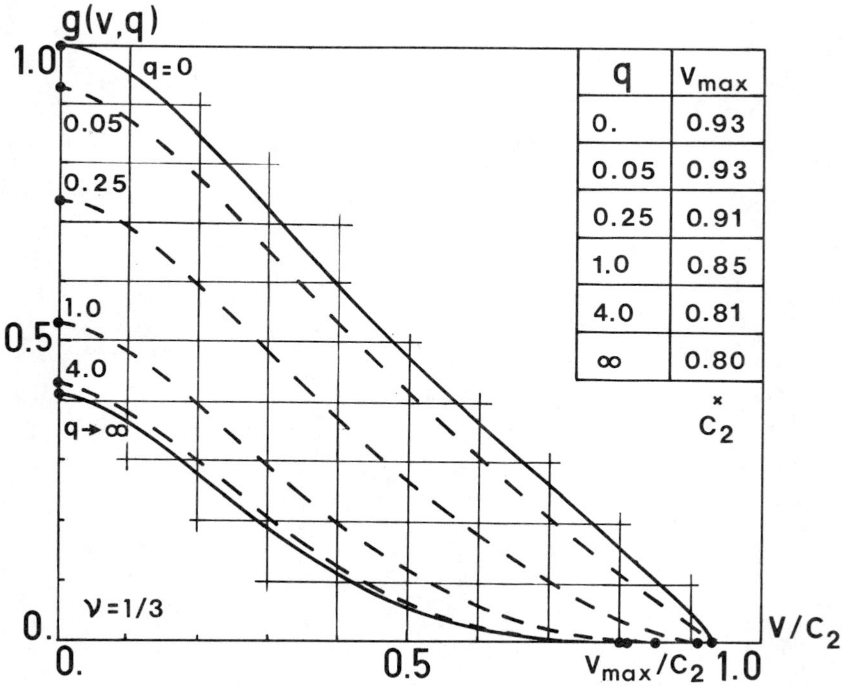

Figure 3 Normalized Energy Release Rates Versus Crack Speed

Fracture 1977, Volume 3, ICF4, Waterloo, Canada, June 19 - 24, 1977

A CRACK GROWTH GAGE FOR ASSESSING FLAW GROWTH POTENTIAL IN STRUCTURAL COMPONENTS

A. F. Grandt, Jr.*, R. L. Crane* and J. P. Gallagher**

INTRODUCTION

The objective of this paper is to describe an approach for monitoring operational service of individual structures or components for potential crack extension. Although much prior effort has gone into development of techniques and procedures for estimating remaining useful structural life [1 - 5], the current emphasis on increased requirements for tracking crack growth potential during service (e.g. references [6 and 7]) remains a formidable challenge.

The approach suggested here consists of mounting a precracked specimen or "gage" onto a load bearing member as shown schematically in Figure 1. Linear elastic fracture mechanics analyses are employed to relate crack growth in the gage with extension of a real or assumed initial flaw located in the structure. Crack growth in the gage can then be monitored during service for an indication of corresponding extension of the assumed structural defect. Moreover, as shown schematically in Figure 2, this relationship permits allowable maximums for the structural crack size (based on safety criteria or repair economics) to specify corresponding gage limits.

Derivation of the analytical relation between the two crack lengths is described below and demonstrated with sample calculations for various flaw geometries. Experimental verification of a portion of the mathematical model is also presented, followed by a summary discussion of the crack gage concept for tracking structural damage.

ANALYSIS

Considering Figure 1, assume that a small precracked coupon (crack length = a_g) is fixed along its ends to a large structural component containing a crack of length a_s. The problem here is to correlate growth of a_s with extension of a_g. In all subsequent discussion, it will be assumed that linear elastic fracture mechanics conditions are satisfied in both the gage and structure during service loading. In addition, the gage is sufficiently small that its attachment does not change the stresses in the structure.

Relation Between Gage and Structural Loads

The objective here is to determine the gage load P caused by application of the uniform structural stress σ_s shown in Figure 1. Since the gage endpoints are fixed to the structure, the total displacement δ along the

---

*Air Force Materials Laboratory, WPAFB, Ohio, U.S.A.
**Air Force Flight Dynamics Laboratory, WPAFB, Ohio, U.S.A.

gage length L equals that of the attached structure and is given by

$$\delta = \int_0^L \varepsilon_s dL = \frac{\sigma_s L}{E_s} \quad (1)$$

Here ε_s is the uniform strain over L, and E_s is the modulus of elasticity for the structure. Similarly, the gage has a component of displacement δ' given by

$$\delta' = \frac{PL}{BWE_g} \quad (2)$$

where B, W, and E_g are, respectively, the thickness, width, and elastic modulus of the gage.

The gage also has another component of displacement δ'' due to the presence of the flaw. Using the compliance concept outlined in reference [8], this additional deflection is given by

$$\delta'' = P\,\lambda \quad (3)$$

where λ is the crack compliance related to the strain energy release rate G, and the stress intensity factor K_g of the gage by the plane strain relationship

$$G = \frac{P^2}{2B}\frac{\partial\lambda}{\partial a_g} = \frac{1-\nu^2}{E_g} K_g^{\,2} \quad (4)$$

Here ν is Poisson's ratio for the gage. For plane stress equation (4) is given by $G = K_g^{\,2}/E_g$. Expressing the stress intensity factor in the form

$$K = \frac{P}{BW}\sqrt{\pi a}\;\beta \quad (5)$$

where β is a dimensionless geometry factor which can depend on crack length, equations (4) and (5) reduce to

$$\lambda = \frac{2(1-\nu^2)\pi}{E_g BW^2}\int_0^{a_g} a\,\beta_g^{\,2}\,da \quad (6)$$

Thus, the displacement of the gage is given by

$$\delta = \delta' + \delta'' = \frac{PL}{E_g BW} + P\lambda = \frac{\sigma_s L}{E_s} \quad (7)$$

which when solved for gage load with equation (6) becomes

$$P = \sigma_s BW \left\{ \frac{E_g}{E_s}\left[\frac{L}{L + \frac{2(1-\nu^2)}{W}\int_0^{a_g} a\beta_g^{\,2} da}\right]\right\} = \sigma_s BW\, f \quad (8)$$

Thus the load in the cracked gage is directly related to the uniform gross stress in the structure. This uniform stress is the same stress that influences the crack growth behaviour at the structural detail of interest. It now remains to describe how the crack growth behaviour of the detail is related to that in the gage; i.e. to provide the transfer function.

Gage and Structural Crack Relation

The structural crack length a_s will now be found as a function of gage crack length a_g. Assume that the fatigue crack growth rate in each member can be expressed in the form [9]

$$\frac{da}{dF} = C\,\overline{K}^{\,m} \tag{9}$$

where da/dF is the average crack extension per block of service usage (e.g. an aircraft flight) and C and m are empirical constants. The parameter $\overline{K}$ is a stress intensity factor that senses the influence of stress history on the crack growth process. As such, $\overline{K}$ is normally obtained by multiplying a stress history characterizing parameter (e.g. $\overline{\sigma}$ = rms stress range) by the stress intensity factor coefficient for the geometry of interest. For the structure, $\overline{K}$ would be

$$\overline{K} = \overline{\sigma}\left(\frac{K}{\sigma}\right) = \overline{\sigma}_s\sqrt{\pi a_s}\;\beta_s \tag{10}$$

For brevity, assume that the gage and structure are made from the same material and have the same C and m in equation (9). Using the fact that both gage and structure see the same number of loading blocks F, integrating equation (9) for a fixed number of flights with equations (5) and (10) yields

$$F = \int_{a_{os}}^{a_s} \frac{da}{C\left\{\overline{\sigma}_s\sqrt{\pi a}\;\beta_s\right\}^m} = \int_{a_{og}}^{a_g} \frac{da}{C\left\{\frac{P}{BW}\sqrt{\pi a}\;\beta_g\right\}^m} \tag{11}$$

which reduces by employing equation (8) and cancelling like quantities to

$$\int_{a_{os}}^{a_s} \frac{da}{\left\{\beta_s\sqrt{\pi a}\right\}^m} = \int_{a_{og}}^{a_g} \frac{da}{\left\{f\beta_g\sqrt{\pi a}\right\}^m} \tag{12}$$

Note that equation (12) is independent of stress history (explicitly), so the a_s versus a_g response is also anticipated to be independent of stress history. This is a first order assumption that might have to be modified when sufficient data become available.

A numerical integration scheme was employed here to solve equation (12) for a_s as a function of a_g. First, the integration of the right-hand side of equation (12) was carried out with the trapezoidal rule together with Romberg's extrapolation method. The upper bound of the absolute error for this procedure was specified to be less than 1×10^{-5}. Next, an upper limit for the left hand side of equation (12) was chosen and the integration performed as before. Depending on the agreement of the left hand value with the previously determined right hand side, an adjustment was made in the upper limit (a_s) of the left integral and the process repeated until the values of the two integrals agreed to within 0.02%.

EXAMPLE RESULTS

Solving equation (12) by the numerical procedure described above, the relation between structural and gage flaw lengths was found for several geometric configurations. Results from two sample cases are briefly described below. In both examples, the structure and gage had the same C, E, and m, while Poisson's ratio for the gage was 0.333.

Consider an edge cracked coupon (50 mm long by 25 mm wide) attached to a large plate containing a 6.4 mm diameter radially cracked hole (length = 1.3 mm) as shown in Figure 3. Numerical results from equation (12) for m = 4 (a constant amplitude fatigue crack growth rate exponent typical of many structural materials) are shown in Figure 3 for various initial gage flaw sizes (a_{og} = 1.3, 1.9, 2.5, and 3.8 mm). Note that the results show a strong dependence on initial crack size, varying from a fast growing structural crack ($a_{og} = a_{os}$ = 1.3 mm) to a response where gage crack growth significantly amplifies corresponding extension of the structural flaw (a_{os} = 1.3 mm and a_{og} = 3.8 mm). Thus, varying the initial crack size provides one means for designing a gage for various degrees of amplification of structural crack growth.

If the gage flaw is located in the structural component rather than in an attached coupon and sees the same remote stress, f = 1 in equation (12). Experimental data [10] for this special case, provided a means of checking equations (9) and (12) of the model. Briefly, the experimental set-up was as follows. Long specimens of 7075-T651 aluminum (width = 150 mm, thickness = 12.7 mm) containing both a radially cracked hole and a centre crack in series as shown in Figure 4, were subjected to complex variable amplitude loading representative of an aircraft stress history. Since the crack growth exponent m would not be known a priori, computations were made for m = 3, 4, and 5, a range encompassing expected values for this material. Note in Figure 4 that these analytical predictions closely bound the test data. Thus, this excellent agreement between experiment and analysis lends credence to equation (9), and subsequent development of equation (12). Again it should be emphasized here that the numerical calculations required no knowledge of the actual load history applied to the test specimen.

CONCLUDING DISCUSSION

A concept for monitoring potential flaw growth in structural components with a flawed gage has been presented, along with a mathematical model for the relation between the structural and gage flaw sizes. This relationship, given by equation (12), and demonstrated in Figures 3 - 4, provides the means for designing a simple crack growth gage capable of "tail number" tracking a fleet of structures for extension of potential or known flaws. The proposed gage would have no moving or electronic components to malfunction, need only minimum instrumentation, and could be designed for various degrees of amplification between structural and gage crack lengths (i.e. see Figure 3).

Since equation (12) relates the gage crack length with the assumed structural flaw size and depends only on geometric factors and material properties, extensive records of service loads would not be required to estimate flaw growth. Indeed, the gage provides a direct measure of crack growth, acting as an analog computer which collects, stores, and analyzes the severity of the input loads, and then responds with the appropriate flaw extension. Thus, the loading conditions which affect flaw growth (i.e. stress level, overloads, temperature, environment, etc.) should be integrated in the gage prediction of structural crack length on a real time basis. Although extensive experimental testing of this gage capability remains for future work, it is encouraging that the data shown in Figure 4 provide a preliminary verification of the transfer function model described in equation (12).

The authors believe that the crack gage described here can be used by logistics management for maintenance decisions in both of the following two ways: (1) as a simple "go/no go" measure of the necessity for inspecting or modifying any given structure, or (2) in conjunction with a Normalized Crack Growth Curve [9]. The crack gage-structural crack transfer function (equation (12)) and the Normalized Crack Growth Curve associated with the structural crack would collectively provide a direct indication of structural crack size and an estimation of remaining useful service life. This second decision making concept is explored more fully in reference [10].

REFERENCES

1. SHEWMAKER, A. P. and WAGNER, J. A., Proc. Air Force Conference on Fatigue and Fracture of Aircraft Structures and Materials, AFFDL-TR-70-144, WPAFB, Ohio, 1970, 833.
2. WHITFORD, D. H. and DOMINIC, R. J., Proc. Air Force Conference on Fatigue and Fracture of Aircraft Structures and Materials, AFFDL-TR-70-144, WPAFB, Ohio, 1970, 847.
3. HAGLAGE, T. L. and WOOD, H. A., Proc. Air Force Conference on Fatigue and Fracture of Aircraft Structures and Materials, AFFDL-TR-70-144, WPAFB, Ohio, 1970, 137.
4. HARTING, D. R., Experimental Mechanics, 6, 1966, 19A.
5. SPANNER, J. C. and McELROY, ed., Monitoring Structural Integrity by Acoustic Emission, ASTM STP 571, 1975.
6. COFFIN, M. D. and TIFFANY, C. F., Journal of Aircraft, 13, 1976, 93.
7. Aircraft Structural Integrity Program, Airplane Requirements, Military Standard MIL-STD-1530A (USAF), 11 December 1975.
8. OKAMURA, H., WATAMBE, K. and TAKANO, T., Progress in Flaw Growth and Fracture Toughness Testing, ASTM STP 536, 1973, 423.
9. GALLAGHER, J. P. and STALNAKER, H. D., Proc. AIAA/ASME/SAE 17th Structures, Structural Dynamics and Materials Conference, Valley Forge, Pa., 1976, 486.
10. GALLAGHER, J. P., GRANDT, A. F. and CRANE, R. L., "Tracking Crack Growth Damage at Control Points", paper in preparation for presentation at AIAA/ASME/SAE 18th Annual Structures, Structural Dynamics, and Materials Conference, March 1977.

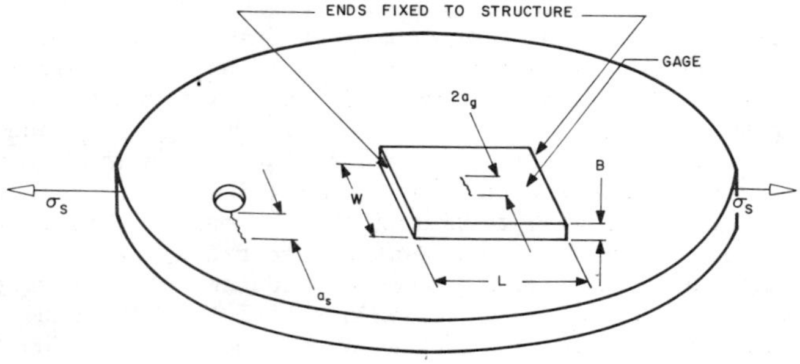

Figure 1 Schematic View of Crack Growth Gage Attached to Flawed Structural Component

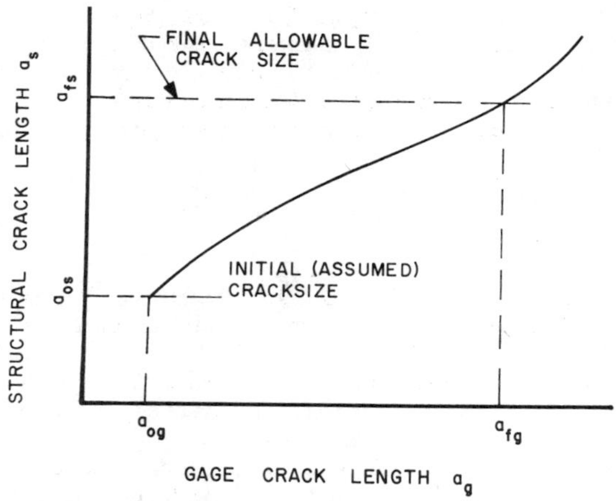

Figure 2 Schematic Representation of Relationship Between Crack Length in Gage (a_g) and Corresponding Structural Flaw Size (a_s)

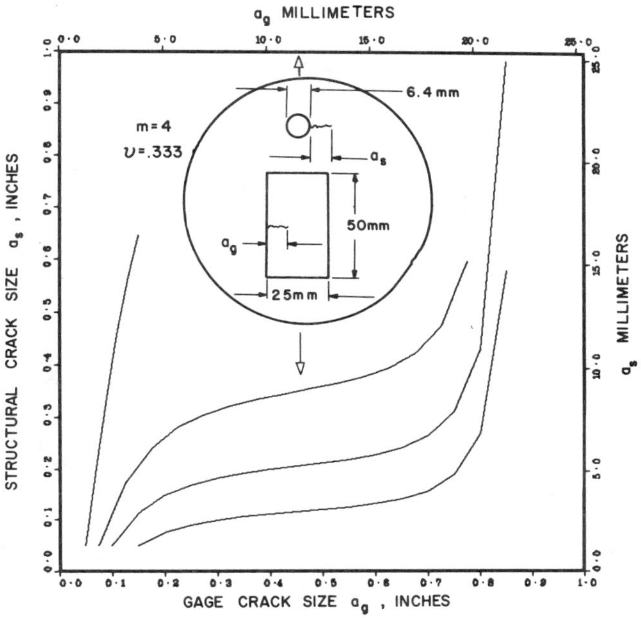

Figure 3 Analytical Results Showing Effect of Initial Gage Flaw Size on a Typical Gage/Structural Crack Relationship

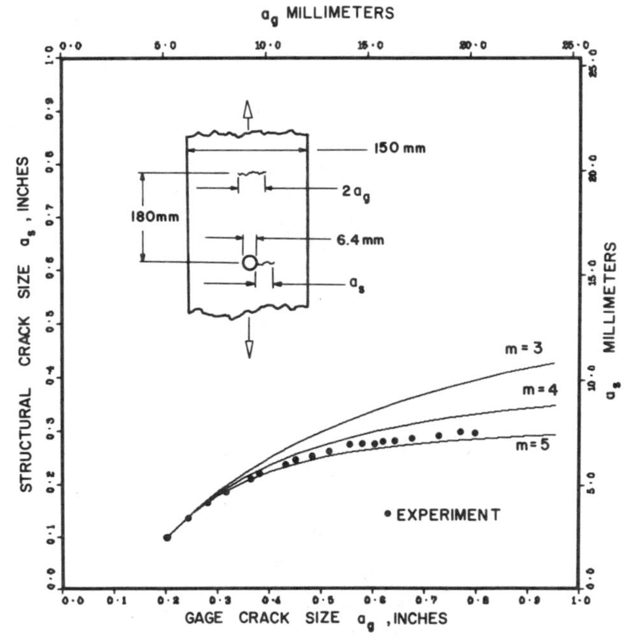

Figure 4 Comparison of Experimental Data with Analytical Prediction for the Relationship Between Two Different Flaw Geometries in the Same Specimen (f = 1 in equation (12))

Fracture 1977, Volume 3, ICF4, Waterloo, Canada, June 19 - 24, 1977

CRACK PROPAGATION INITIATION IN DUCTILE STRUCTURES

J. Lebey and R. Roche*

INTRODUCTION

Linear Elastic Fracture Mechanics is now well established. The determination of K_I is a simple mathematical problem, and the methods for determining its critical value are well known. LEFM can only be applied in cases where inelastic strain (plasticity, creep) is strictly localized. In actual practice, fractures of mechanical parts are preceded by significant plastic strains. In such cases, LEFM is incapable of clarifying the fracture conditions [1] [2] [3]. Initiation of the propagation of an existing crack occurs at loads lower than those specified by LEFM. This is especially true for ductile metallic materials such as standard structural steels.

This points to a pressing need for the development of Post Yield Fracture Mechanics, for a better knowledge and prediction of fracture conditions governing a large number of structures. In this area, a number of criteria have already been proposed. The best known are the Crack Opening Displacement and the J integral [4] [5] [6]. However, it is always difficult to substantiate the validity of a criterion, and the latter, like many others, have been subject to debate. Hence it appears indispensable to increase the number of experimental results which can help to define the field of application of any specific criterion. It is with this in mind that the Research Centre at Saclay undertook a programme concerned with thin structures of structural elements, in which crack propagation initiation occurs with substantial plastic strain. This paper gives the results obtained with two types of structures:

(a) centre cracked plates from a single steel previously subjected to various degrees of strain hardening,

(b) spheres of different dimensions.

CENTRE CRACKED PLATES

The plates, the dimensions of which are given in Figure 1, were machined from XC10 steel† 8 mm thick. After measurement of the mechanical properties of the metal as received, a number of rough test pieces were cold worked before final machining.

The cold working process involved elongation of the metal by longitudinal tension (previous elongation PE).

The initial elongation obtained is expressed as a percentage of the proportional elongation A_p at maximum load of the metal as received.

*Centre d'Etudes Nucléaires de Saclay, France.
†Composition : C 0.10; Mn 0.40; Si 0.25; S + P < 0.035

Three previous elongations were adopted : 20%, 50% and 80% of A_p.

With the non cold worked metal (as received), this provided four different groups of mechanical properties, as shown in Table 1.

The initial length $2a_o$ of the notches ranged from 5 to 50 mm; the end radius of the notch, obtained by electro-erosion, was 0.5 mm, with fatigue pre-cracks.

During the tests, the stress and displacement at the centre of the notch were recorded on an XY recorder. The crack initiation at the notch end was observed visually by means of a binocular microscope.

The results given in Table 2 are concerned exclusively with crack initiation conditions, in which :

$$\sigma_i \text{ net} = \frac{\text{load at initiation}}{(W-2a_o) \text{ x thickness}}$$

$$\sigma_i \text{ gross} = \frac{\text{load at initiation}}{W \text{ x thickness}}$$

Δ_{1i} = variation in central opening at initiation

J_{Ic} = critical value of the J integral, measured by the method indicated in [7].

The recordings obtained did not enable the measurement of J_{Ic} in cases of previous elongation 50% and 80%.

The different results are presented in graphic form in Figures 2, 3, 4 and 5.

Figure 2 shows that, in all cases, the yield stress must be reached on the remaining ligament for crack initiation to occur; no embrittlement occurs due to strain hardening. In all cases, the cracks were subsequently propagated in a stable manner.

Figure 4 shows that the critical value of the central opening depends on the state of strain hardening of the metal.

The determination of J_{Ic} is complicated by the need to derive the experimental results. The results obtained do not make it possible to confirm that the criterion is adequately substantiated for a ductile material in the probable case of plane stresses.

SPHERES

Tests were performed on manganese-molybdenum steel spheres of three different dimensions :

5 spheres diameter D 363 mm thick e 3 mm	theoretical dimensions
2 spheres diameter D 918 mm thick e 7 mm	
3 spheres diameter D 1800mm thick e 15mm	

The spheres featured thru notches terminated by radii of about 0.1 mm, with fatigue pre-crack.

The experimental method is described in previous publications [8] [9] [10], together with part of the results, obtained previously.

The mechanical properties and test results are given in Tables 3 and 4, in which the stresses indicated as σ_i and σ_p are respectively :

σ_i crack initiation stress

σ_p unstable crack propagation stress

Two crack propagation modes were observed, depending on the test; they are illustrated by the R curves in Figure 6 concerning 1800 mm diameter spheres.

(a) stable propagation from the initiation stress σ_i to the unstable propagation stress σ_p, at which the crack propagates rapidly at constant pressure. This type of fracture was observed in all tests on spheres 363 and 918 mm in diameter, and with the 1800 mm diameter No. 3, it corresponds to ductile tears.

(b) sudden fracture without stable propagation period; in this case, σ_i and σ_p coincide; this fracture mode was observed with 1800 mm diameter spheres, Nos. 1 and 2.

These results highlight the effect of the scale factor, already investigated elsewhere [11], on the strength of cracked vessels. The initiation stress values, related to the yield stress of the metal, are indicated in Figure 7 as a function of the relative length of the initial notch. Figure 8 shows the appearance of sphere No. 2 after sudden fracture.

In view of the thin dimensions, it proved impossible to take valid measurements of toughness (K_{Ic}) by standard methods [12]. However, an estimate of toughness can be made by using the method of equivalent energy (K_{Icd}) [13] or by measuring J_{Ic} experimentally by two different methods [7], [14], and by calculating K_{Ic} with the values thus obtained. This enables calculation of the theoretical crack initiation stresses, for comparison with the experimental values. Table 5 shows a number of these comparisons drawn up by calculating the initiation stresses from toughness measurements taken with CT specimens taken from spheres (and of substantially identical thickness to that of the spheres). The σ_i values calculated were obtained as follows :

column A : $$\sigma_i = \frac{K_{Icd}}{\alpha\sqrt{\pi a}}$$

column B : $$\sigma_i = \frac{\sqrt{J_{Ic} . E}}{\alpha\sqrt{\pi a}} \quad \text{with } J_I = \frac{2U}{A} \quad [14]$$

column C : $$\sigma_i = \frac{\sqrt{J_{Ic} . E}}{\alpha\sqrt{\pi a}} \quad \text{with } J_I = -\frac{dU}{da} \quad [7]$$

where : $\alpha = \sqrt{1 + \frac{1.9a^2}{R.\ e}}$ (Folias)

with R = radius of sphere
e = thickness of sphere

CONCLUSIONS

The results obtained are only applicable to a limited area; however, it nevertheless appears clear that it is not possible to exceed the generalized plasticity load without this resulting at least in stable propagation of existing cracks. In most cases stable propagation occurs before reaching generalized plasticity, but it seems possible, by using criteria of the equivalent energy or J_{Ic}, to predict with reasonable accuracy the load which causes propagation initiation [15]. Note that in all cases the thicknesses were too low for a valid measurement of K_{Ic}. While one cannot draw a general conclusion from the foregoing, it appears in the present case that propagation initiation occurs at the lower of the two following loads :

(a) limit loading

(b) loading calculated by means of a criterion of the J integral or equivalent energy type.

ACKNOWLEDGEMENTS

This study was sponsored by the Nuclear Safety Delegation of the CEA. The authors wish to thank those members of the Nuclear Safety Department who provided them with assistance.

REFERENCES

1. HAHN, G. T., SARRATE. M. and ROSENFIELD, R., Int. Journal of Fracture Mechanics S. 1969.
2. EIBER, R. J., MAXEY, W. A. and DUFFY, A. R., Rapport BMI 1883, 1970.
3. HEALD, P. T., SPINK, G. M. and WORTHINGTON, P. J., Materials Science Eng. 10, 1972.
4. NICHOLS, R. W., BURDEKIN, G. M., COWAN, A., ELLIOTT, D. and INGHAM, T., Symposium on Fracture Toughness Concepts for Weldable Structural Steels, I.I.W., IX 655-69 X-534-69.
5. RICE, J. R., in Fracture, Academic Press, 1968, 191.
6. RICE, J. R., Journal of Appl. Mech., June 1968.
7. LANDES, J. D. and BEGLEY, J. A., ASTM STP 514, 1972.
8. DORE, A., FOGLIA, S., LEBEY, J. and TOMACHEVSKY, E., Second Int. Conf. on Structural Mechanics in Reactor Technology, Berlin, 1973.
9. FOGLIA, S., LEBEY, J. and ROCHE, R., National Conf. on Technology in Pressure Vessels, Paris, 1975.
10. LEBEY, J. and ROCHE, R., Third Int. Conf. on Structural Mechanics in Reactor Technology, London, 1975.
11. DOWLING, A. R., Rapport CEGB RD/B/N 2796, October 1973.
12. ASTM Part 31, 1970.
13. WITT, F. J., 4th National Symposium on Fracture Mechanics, Pittsburg, 1971.

14. RICE, J. R., PARIS, P. C. and MERKLE, J. G., ASTM STP 536.
15. NEALE, B. K. and TOWNLEY, C. H., Rapport C.E.G.B. RD/B/N 3358, May 1975.

Table 1 Center Cracked Plates. Mechanical Characteristics

Previous Elongation (P.E.)	Yield Strength σ_y 0,2 (hb)	U.T.S. σ_R (hb)	% Elongation at Maximum Load	% Elongation at Failure	Strain Hardening Exponent
As received	28,5	42	23	40	0,20
20%	38	46	10	20	0,095
50%	48	51	5,5	16	0,053
80%	53	54	3	13	0,029

Table 2 Center Cracked Plates. Test Results at Initiation

Specimen n°	$2a_o$ (mm)	Thickness (mm)	Previous Elongation	σ_i Net (hb)	σ_i Gross (hb)	Δ_{li} (mm)	J_{Ic} (J/mm²)
16	5	3	As received	37,1	35,5	1,44	
9	10	3		31	28,4	1,60	
10	20	3		31,2	26	1,10	0,22
11	30	3		31,1	23,3	1,35	0,27
13	40	3		31,5	21	1,12	0,23
14	40	6		33,4	22,2	1,6	0,31
12	50	3		32,2	18,8	1,7	0,33
15	50	6		32,6	19	1,54	0,30
7	10	3	20%	41,5	38	1	0,17
8	50	3		43,4	25,3	1,18	0,206
5	10	3	50%	49,5	45,3	0,40	
6	50	3		51	29,8		
2	10	3	80%	54,4	49,8	0,16	
3	10	3		54,5	50		
4	50	3		54,5	31,9		

Table 3 Spheres. Mechanical Characteristics and Test Results

D (mm)	Sphere n°	Mechanical Characteristics σ_y 0,2 (hb)	Mechanical Characteristics σ_R (hb)	e (mm)	2a (mm)	P (bars) P_i	P (bars) P_p	(hb) σ_i	(hb) σ_p
	9			2,45	35	90	98	33	36,2
				2,25	70	51	63	20,5	25,5
				2,20	105	32	42	13,2	17,3
	10	36,2	51,8	2,70	40	95	99	31,8	33,2
				2,70	59	60	68	20	22,8
	13	36,2	52,5	2,15	35	80	84	33,6	36,2
				2,35	50	66	76	25,4	29,2
				2,50	65	60	72	21,7	26
				2,52	72	50	65	18	23,6
363				2,50	80	46	56	16,6	20,5
				2,20	95	32		13,2	
				2,10	102	29		12,5	
				2,20	110	28	37	11,5	15,4
				2,35	125	26	33	10	12,7
	15	39,7	52,5	2,17	15	88	86	36,8	41,5
				2,12	25	99	92	36,8	39,5
	19	36,4	52	2,45	56	55	76	20,3	28
				2,80	61	63	73	20,4	23,5
				2,55	75	52	63	18,5	22,5
				2,80	99	35	50	11,3	15,5

Table 4 Spheres. Mechanical Characteristics and Test Results

D (mm)	Sphere n°	Mechanical Characteristics σ_y 0,2 (hb)	Mechanical Characteristics σ_R (hb)	e (mm)	2a (mm)	P (bars) P_i	P (bars) P_p	(hb) σ_i	(hb) σ_p
	1	28	48,5	4,70	110	50	56	24	27
				4,75	140	38	50	18,3	23,8
				4,90	170	33	46	15,6	21,6
				4,70	205	25	37	12	18,1
				5	230	24	38	11	17,5
				4,75	260	22	33	10,2	16
				4,90	318	16	30	7,5	13,8
	2	27,1	48,8	4,65	173	32	41	15,5	20,3
	1	50,5	65,7	14	400	32	32	10,4	10,4
1800	2	50	68	14	300	48	48	15,2	15,2
	3	50	69,5	14	550	19	40	6	12,8

Table 5 Spheres. Calculated and Experimental Stress σ_i for Initiation of Crack

D (mm)	Sphere n°	$2a_o/D$	σ_i calculated (hb) (A)	(B)	(C)	σ_i Experimental (hb)	Propagation
363	13	from 0,090	54,6			33,6	Stable
		to 0,345	10,6			10	
1800	2	0,167	11	13,6	12,4	15,2	Unstable
	1	0,222	8,1			10,4	Unstable
	3	0,305	5,4	5,9	5,69	6	Stable

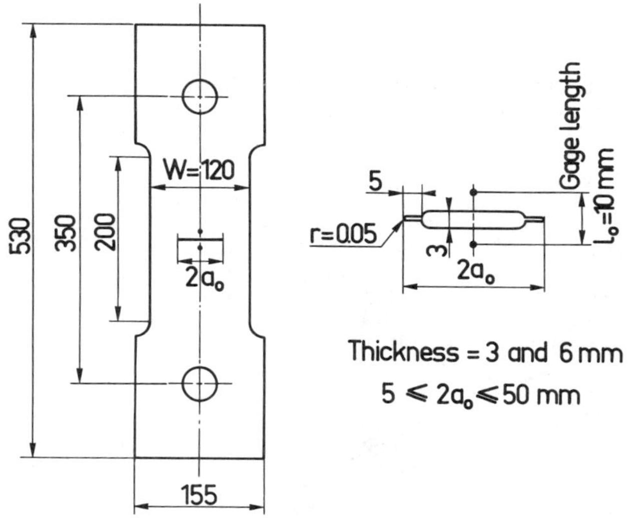

Figure 1 Center Cracked Plates Dimensions

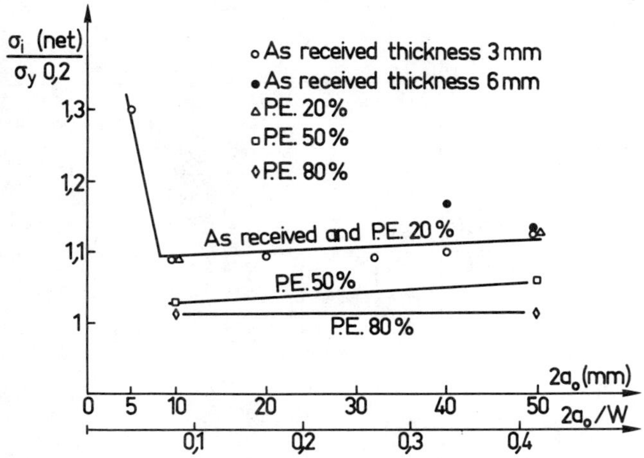

Figure 2 Center Cracked Plates. Relation Between the Net Initiation Stress and the Crack Length

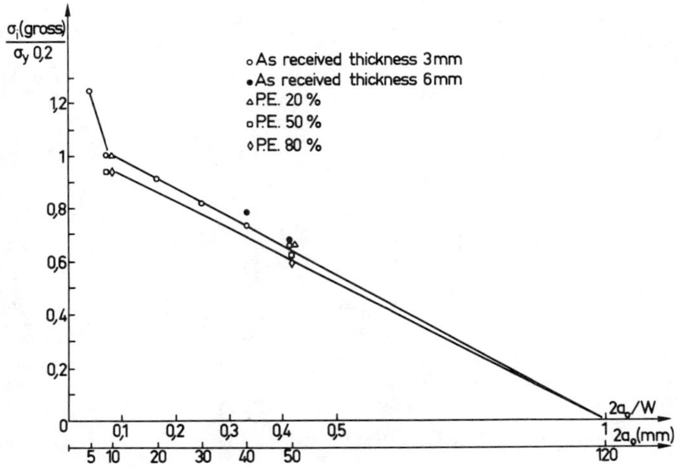

Figure 3 Center Cracked Plates. Relation Between the Gross Initiation Stress and the Crack Length

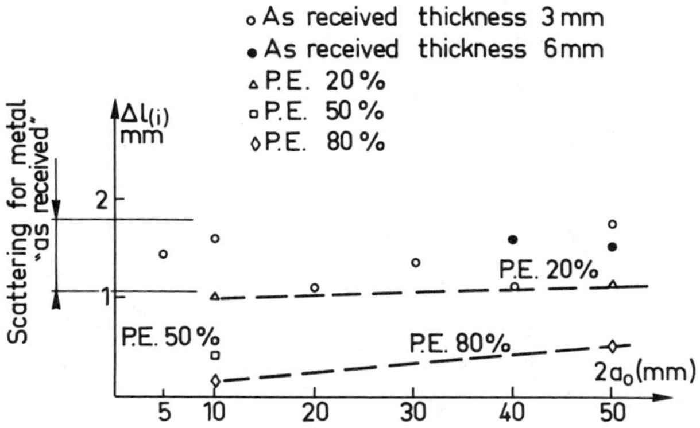

Figure 4 Center Cracked Plates. Relation Between the Central Opening and the Crack Length

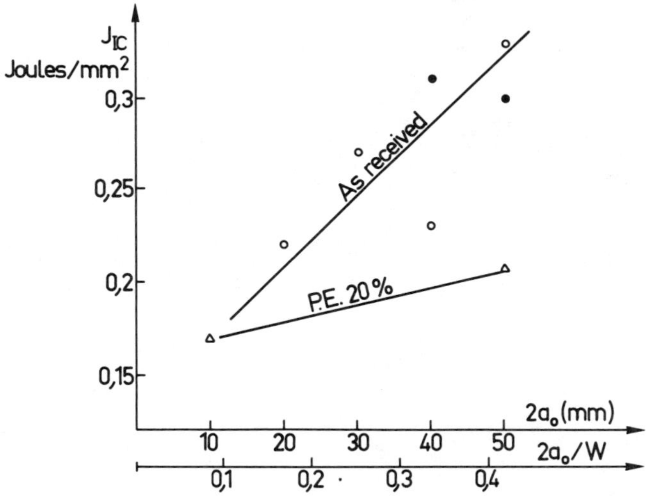

Figure 5 Center Cracked Plates. Relation Between J_{IC} and the Crack Length

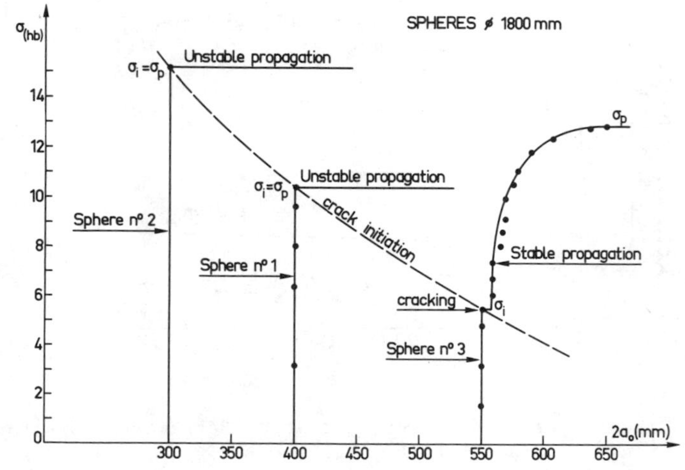

Figure 6 Spheres ∅ 1800. Curves R

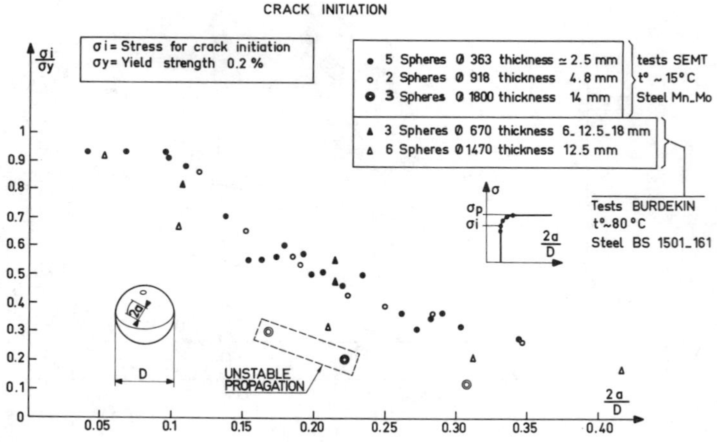

Figure 7 Spheres. Relation Between the Initiation Stress and the Crack Length

Figure 8 Sphere ∅ 1800 (n° 2) after Failure

Fracture 1977, Volume 3, ICF4, Waterloo, Canada, June 19 - 24, 1977

INITIAL STAGES OF CRACK EXTENSION IN TIME-DEPENDENT AND/OR DUCTILE SOLIDS

M. P. Wnuk*

Failure of a volume element located on the prospective path of the crack front is linked to the incremental work dissipated within the process zone just prior to the collapse of this zone. If t denotes the instant at which the control volume element breaks down, then the incremental accumulation of damage occurs within the time interval $t-\delta t<\tau<t$, where the increment δt (= $\Delta/\dot{a}$) corresponds to the time used by the crack front to traverse its own process zone. Size of such hypothetical zone, over which an intensive straining occurs before the crack may advance, is characterized by the length Δ which is assumed to be a structural constant. In metals Δ is determined by metallurgical parameters (Δ is supposed here to be much smaller than the plastic zone extent, and may for example be thought of to correspond to the distance from the tip at which strongly localized deformation sets in because of tunnelling or necking).

The rate of damage accumulation is given by the product of stress restraining separation of new surfaces, S, and the rate $\dot{u}$ at which these surfaces are being separated at a certain fixed control point P. The integral of this product taken over the time interval δt represents the damage accumulated in the material element adjacent to the crack tip while it undergoes the final stages of straining preceding failure. Thereby it is shown that the incremental criterion for crack extension.

$$\int_{t-\delta t}^{t} S\left[x_p,\tau\right] \dot{u}\left[x_p,\tau\right] d\tau = \text{material property} \tag{1}$$

can be reduced to a form containing the entities well-known in linear elastic fracture mechanics, such as COD or J-integral. From equation (1) for the small scale yielding range we obtain the following non-linear differential equations of the first order governing the slow crack growth process:

(a) in terms of the tip opening displacement,

$$\Delta \frac{d\delta}{da} + \frac{\Delta}{2} \log\left(\frac{2\beta\delta}{\Delta}\right) + \delta\psi\ (\delta) = \delta_o \tag{2}$$

(b) in terms of J-integral,

$$\Delta \frac{dJ}{da} + \frac{\Delta}{2} \sigma_{ys} \log\left(\frac{2\beta}{\Delta\sigma_{ys}} J\right) + \delta\psi\ (J) = J_o \tag{3}$$

---

*Mechanical Engineering Department, South Dakota State University, Brookings, South Dakota, U. S. A. 57006

The material constants δ_o and J_o, appearing on the right hand sides of the equations above can be related to the experimentally determined slope of either δ-curve or J-curve as follows

$$\hat{\delta}_o = (d\delta/da)_{ini}\ \Delta\ + \frac{\Delta}{\beta}\{\log(2\beta\delta_{ini}/\Delta)\ -1\}$$
$$\hat{J}_o = (dJ/da)_{ini}\ \Delta\ + \frac{\sigma_{ys}\Delta}{\beta}\ \{\log(2\beta J_{ini}/\sigma_{ys}\Delta)\ -1\} \qquad (4)$$

Symbol σ_{ys} denotes the yield stress encountered at the crack front, while the coefficient β relates the extent of plastic zone R preceding the crack tip and the tip opening δ, namely $R = (\beta/2)\delta$. For the line-plasticity model and within the small scale yielding range $\beta = \pi E/4\sigma_{ys}$. The other material property introduced here is the dimensionless compliance function

$$\psi(t) = D_{creep}(t)/D(0)$$

characterizing the time-dependent behaviour of a solid (D_{creep} denotes the creep compliance as defined in rheology). The increment $\delta\psi$ corresponds to the change of $\psi(t)$ during the time interval δt; therefore

$$\delta\psi = \dot{\psi}(0)\delta t$$

Both equations (2) and (3) have been derived on the basis of line-plasticity model (although the form of the criterion (1) is suitable for a continuum mechanical approach). One should also emphasize that the validity of the results discussed here is restricted to the small scale yielding range. Two limiting cases are investigated, namely

Case I. Inviscid ductile solid in which the material moduli are time-independent, and thus $\delta\psi = 0$. Then the equations (3) and (4) reduce correspondingly

$$\frac{d\delta}{da} = \frac{1}{\beta}\log\left(\delta_{max}/\delta\right), \qquad \delta = \delta(\Delta a) \qquad (5)$$

$$\frac{dJ}{da} = \frac{\sigma_{ys}}{\beta}\ \log\left(J_{max}/J\right), \qquad J = J(\Delta a) \qquad (6)$$

These equations allow theoretical predictions of the resistance curves (either δ or J-curves) for quasi-brittle solids. The R-curves can be obtained through a numerical integration of equations (5) and (6) if the initial values of δ and J (at the onset of failure) are known. When the maximum attainable J does not differ greatly from the initial threshold ("flat" R-curve), but the foregoing equations can be integrated in a closed form, yielding these solutions.

$$\delta(\Delta a) = \delta_{max}\left\{1 - \left(1 - \frac{\delta_{ini}}{\delta_{max}}\right)\exp\left(- \frac{\Delta a}{\beta\delta_{max}}\right)\right\} \qquad (5a)$$

$$J(\Delta a) = J_{max}\left\{1 - \left(1 - \frac{J_{ini}}{J_{max}}\right)\exp\left(- \frac{\sigma_{ys}\,\Delta a}{\beta J_{max}}\right)\right\} \qquad (6a)$$

Symbol Δa is used to denote the amount of slow growth, $\Delta a = a - a_o$. Equations (5a) and (6a) give a fair correlation with the experimental data reviewed recently by Mai, Atkins and Caddell [3]

Case II. For a linear viscoelastic solid (with no plastic flow accounted for) one may show that the dominant terms in the governing equations (2) and (3) are the first and the last on the right hand side. They can be combined to yield equations of motion for a crack traversing a visco-elastic matrix, i.e.

$$(1 + \delta\psi)\delta = \delta_o$$
$$(1 + \delta\psi)J = J_o \qquad (7)$$

or

$$\psi(\delta t)\delta = \delta_o$$
$$\psi(\delta t)J = J_o \qquad (8)$$

These equations imply the following "equivalent" viscoelastic entities

$$J = J_{el}\ \psi(\Delta/\dot{a})$$
$$G = G_{el}\ \psi(\Delta/\dot{a}) \qquad (9)$$
$$K = K_{el}\ \sqrt{\psi(\Delta/\dot{a})}$$

in where the subscript "el" emphasizes the fact that the quantity in question is obtained from the theory of elasticity. The last two forms above are identical to those suggested by the equations of motion for cracks in linear viscoelastic media, as derived by Knauss and Dietmann [1] and Wnuk [4]. From equations (9) it becomes evident that "effective" or "equivalent" energy release rate for a viscoelastic solid is a function of crack growth rate. The viscoelastic R-curves of the type G vs $\dot{a}$ or J vs $\dot{a}$ are, therefore, analogous to the resistance curves suggested for ductile solids by McClintock and Irwin [2]. An experimental verification of these equations is briefly discussed.

The second part of this work is concerned with the study of structure of an integral equation which describes path-dependent failure of an element adjacent to the tip of a crack propagating through an elastic-plastic or a viscoelastic-plastic isotropic solid capable of large deformation. In such a case the integrand of equation (1) has to be studied in a greater detail. Preliminary investigation indicates that the form

$$\delta\phi = S\left[x_p,\tau\right]\ \dot{u}\left[x_p,\tau\right]d\tau$$

is not an exact differential but a Pfaffian and thus the integral appearing in equation (1) is path-dependent. Therefore, failure of an element adjacent to the crack tip depends on the previous states of this element. In this way the dependence of the final state on the loading path (so called "history dependence") is accounted for in the theory developed on the basis of the incremental criterion of failure.

Effects of the rate of loading and time-sensitivity of the material response on fracture propagation are incorporated in the theory through the Crochet constitutive equations involving time-dependent yielding, i.e.

$$\sigma_{ys}(t) = A + B\exp(-C)$$

$$\chi = \left\{\left(\varepsilon^{v}_{ij} - \varepsilon^{e}_{ij}\right)\left(\varepsilon^{e}_{ij} - \varepsilon^{e}_{ij}\right)\right\}^{1/2}$$

$$\varepsilon_{ij} = e_{ij} + (1/3)e\delta_{ij}\ , \ \sigma_{ij} = s_{ij} + (1/3)\sigma_{kk}\delta_{ij}$$

$$e_{ij} = e^{e}_{ij} + e^{v}_{ij}\ , \ e = e^{e} + e^{v}\ , \ e = \varepsilon_{kk}\ , \ s = \sigma_{kk}$$

$$e_{ij}(\underline{x},t) = \int_{-\infty}^{t} J_1(t-\tau)\,\frac{\partial s_{ij}(\underline{x},\tau)}{\partial\tau}\,d\tau$$

$$e(\underline{x},t) = \int_{-\infty}^{t} J_2(t-\tau)\,\frac{\partial s(\underline{x},\tau)}{\partial\tau}\,d\tau$$

Here A, B, C are material constants (for example A = 55.2 MPa, B = 15.5 MPa, C = .0771 for polycarbonate) while χ is a function of the strain state. The last two equations shown above are valid for the "linear" regions in which the effective stress reduced from the multiaxial stress state according to Tresca criterion is below the yield point. Summation is implied by repeated indices; the superscripts v and e denote the viscoelastic and purely (short-time) elastic components of the strain. For strains increasing with time first of the foregoing equations assents that faster loading corresponds to a higher yield stress, while under constant stress it implies that yield occurs at a time which is longer the lower the stress. For initially elastic response under rapid loading $\varepsilon^{v}_{ij} = \varepsilon^{e}_{ij}$ and $\sigma_{ys}(0) = A + B$, while the minimum yield value is given by $\sigma_{ys}(\infty) = A$, provided $\varepsilon^{v}_{ij} - \varepsilon^{e}_{ij}$ is sufficiently large as may be the case for viscoelastic soft polymers.

REFERENCES

1. KNAUSS, W. G. and DIETMANN, H., Int. J. Engng. Sci., 8, 1970, 643.
2. McCLINTOCK, F. A. and IRWIN, G. R., ASTM STP 381, 1964.
3. MAI, Y. W., ATKINS, A. G., CADDELL, R. M., Int. J. Fract., 12, 1976, 391-407.
4. WNUK, M. P., Int. J. Fract. Mech., 7, 1971, 383-407.

Fracture 1977, Volume 3, ICF4, Waterloo, Canada, June 19 - 24, 1977

ON CONSTITUTIVE PROPERTIES AT SINGULAR CRACK BORDERS

H. C. Strifors*

INTRODUCTION

Is the tip of a sharp notch always a crack tip? Or, more precisely, is the singular field of stress and/or strain at the border of a crack-like surface in a body always modelling a crack border, which implies that this border line is able, under specific conditions, to propagate in the body? If not, this singular line simply constitutes a localization of stress and/or a strain concentration and does not deserve attention from the fracture mechanics point of view.

Linearly elastic materials possess features that can be interpreted within a theory of fracture. The first crack problems solved were also formulated for Hookean materials and the stress intensity factors were interpreted as measures of stress at singular crack borders. During the last decades attention has turned considerably to materials capable of plastic deformation. Much effort has been made to describe fracture properties by formal mathematical generalization to infinite strain of constitutive equations which have proven useful for a description of plastic behaviour at small and moderate strains. One main problem discussed in this regard is which measure of stress intensity in plastic materials should replace the stress intensity factors for elastic materials in a fracture criterion. However, the still more important question of whether the resulting theories also fit into a sound fracture theory has not been properly considered.

In this paper, consequences for constitutive properties of materials containing cracks bounded by singular lines will be discussed. For this purpose the general thermomechanical theory of fracture in simple solids developed by Strifors [1] is employed. For convenience the discussion will be based on the fracture theory resting upon use of the linearized strain tensor, ε_{ij}, and the stress tensor, σ_{ij}, which is symmetric by definition. Both tensors are referred to the undeformed configuration according to classical fracture mechanics theory. The qualitative results of the discussion, however, remain valid also within the physically more reasonable theory that duly accounts for large strain.

SURFACE ENERGY

The observation that energy is required to fracture real bodies is accounted for by introduction of the concept of surface energy. The rate of change of surface energy at a propagating singular crack border can be derived by consideration of the energy flow through a control volume in the shape of a narrow tube surrounding the crack border and moving with the same propagation velocity. The net supply of energy through the tube in the limit when the diameter of the tube approaches zero is defined as the rate of change

* The Royal Institute of Technology, S-100 44 Stockholm 70, Sweden

of surface energy. Recently, using a similar technique Freund [2] considered a special case of crack extension in elastic materials.

Without going into mathematical details we formulate the equation for the energy balance at a smooth singular line modelling a propagating crack border. The rate of change of surface energy per unit length of the border line at a particle instantaneously located at a point on that line is given by

$$\dot{\gamma} = \lim_{d \to 0} \oint_{c(d)} \left\{ \left(W + \frac{1}{2} \rho \, u^i{}_{,j} \, u_{i,k} \, w^j \, w^k \right) w^l - w^i \, u^k{}_{,i} \, \sigma_k{}^l - h^l \right\} n_l \, dc \, , \qquad (1)$$

where W is the internal energy per unit volume, ρ is the mass density, u is the displacement field and h is the heat flux. The integration path, c, is a closed curve surrounding the singular line in the normal plane of the line in which the outward unit normal n and the propagation velocity w of the crack border are also situated, and d denotes an upper bound for the distance between points on the integration path and the singular line.

Since heat, as well as mechanical work, may contribute to the surface energy, it is necessary to be able to specify the heat content in the surface energy in a consistent thermomechanical theory. Analogous to the case of internal energy, surface entropy is introduced as the heat specifying parameter for the surface energy. By consideration of the entropy flow through a narrow tube surrounding the crack border it can be shown that the specific production of surface entropy, β, per unit length of the crack border is given by

$$\beta = \dot{\zeta} + \lim_{d \to 0} \oint_{c(d)} \left(\frac{h^i}{\theta} - \rho \, \eta \, w^i \right) n_i \, dc \, , \qquad (2)$$

where ζ is the surface entropy, η the internal entropy, i.e., the heat specifying parameter for the internal entropy, and θ is the temperature.

The equations given should be considered as definitions of $\dot{\gamma}$ and β. Then the question arises as to whether the defining intergrals exist and are unique.

If a circular cylindrical coordinate system (r, ϕ, z) is introduced in such a way that the z-axis is directed along the tangent of the border line, it is readily shown that the integrands must be of the order of r^{-1} to give non-trivial and bounded contributions to the integrals. As regards established fracture mechanics one apparently non-trivial general conclusion follows. The existence of (1) implies that a stronger singularity than that of $r^{-1/2}$ is impossible in the displacement gradient, or, equivalently, in the strain tensor.

From here on we will consider the consequences of the requirement of uniqueness of the given definitions. Uniqueness means that the integrals must yield the same value independent of the shape of the curve surrounding the crack tip. The integrals must in a specified sense be path-independent in the singular region, i.e., in the region where the functions can be approximated by the leading terms of their series expansions. The most far-reaching consequence arising from this requirement combined with the dissipation principle is that the production of internal entropy, (i.e., the rate of change of internal entropy that is not due to radiation or conduction of heat), must be of inferior order to r^{-2}, while the rates of change of internal energy and internal entropy are of the order of r^{-2}. When the quantities under consideration are confined to a small enough region close to the crack border it is apparent that internal entropy production can be disregarded. This means that the material must be *non-dissipative in the singular region.* In other words, the material must respond thermo-elastically in the singular region, or elastically if thermal effects are neglected at the crack border.

Viscoelastic constitutive equations with instantaneous elastic response automatically satisfy the condition of non-dissipation at singular crack borders, since such materials respond elastically at unbounded strain-rates which occur in the singular region during deformation whether the crack is extending or not.

ELASTIC-PLASTIC MATERIALS

For plastic or elastic-plastic materials the condition of non-dissipation in the singular region at crack borders implies restrictions on ordinary constitutive equations.

A first attempt to take the condition of non-dissipation into account may be made as follows. For the sake of simplicity the discussion is confined to a constitutive equation of the Prandtl-Reuss type. Accordingly, the strain tensor is written as the sum of the elastic and plastic strain tensors,

$$\varepsilon_{ij} = \varepsilon^{e}_{ij} + \varepsilon^{p}_{ij} \ . \tag{3}$$

Here, ε^{e}_{ij} is given by Hooke's law, for instance, and ε^{p}_{ij} is given by

$$\dot{\varepsilon}^{p}_{ij} = \begin{cases} \lambda \ \partial f/\partial\sigma^{ij} & \text{during loading} \\ 0 & \text{during neutral loading and unloading} \end{cases} , \tag{4}$$

where the parameter λ depends on the yield function, f, and the current stress-rate.

By introduction of an effective plastic strain, ε^{p}_{eff}, being a functional of the plastic strain history, loading conditions satisfying fracture mechanics requirements can be specified. Besides the usual conditions for loading that the stress tensor shall be situated on the yield surface and λ be positive we impose the additional condition that the effective plastic strain shall be less than some limit value ε^{p}_{lim}.

Other theories for plastic behaviour may be restricted in a similar way. In this regard it is important that the constitutive equations employed allow for elastic strain superposed on the completed plastic strain.

The properties of the elastic-plastic constitutive theories necessary to render a fracture theory meaningful may be given a satisfactory physical basis in terms of crystal lattices and dislocations.

To make a rough sketch of a process of initiation of crack growth that is consistent with the theory discussed let us consider a body with a crystal structure. At low stress atom layers in the crystal lattices will respond by a small displacement which may be interpreted as elastic deformation. When the stress reaches the yield limit, the deformation is accomplished by dislocation movements on a large scale. If such deformation continues, unlimited plastic deformation would result and the material would consequently never fracture by formation of new boundary surfaces.

If, on the other hand, dislocation movements are prevented, only a further (elastic) displacement of lattice atoms remains possible until the separation of atom layers reaches such a magnitude that the material finally fractures. Thus, from the viewpoint of this kind of model it seems reasonable that plastic deformation must reach a limit before fracture can occur. The process of plastic flow must turn into a process of fracture.

CONCLUSIONS

Consideration of the energy supply through a control surface surrounding a singular crack border leads to a natural definition of the rate of change of surface energy at such a crack border. An analogous treatment of entropy makes it possible to describe the quality of the surface energy in accordance with classical thermodynamics. Then, the requirement of the uniqueness of the definitions together with the dissipation principle lead to the consequence that the material response in singular regions must be thermo-elastic, or elastic in isothermal theories.

From this viewpoint elastic-plastic materials are discussed and a simple elastic-plastic theory is proposed as a first attempt at meeting fracture mechanics requirements. Although no reference is made to any specific micro-mechanism of crack extension, it turns out that plastic flow and fracture are two distinct physical phenomena which may be interpreted within traditional dislocation theories.

REFERENCES

1. STRIFORS, H. C., to be published.
2. FREUND, L. B., J. of Elasticity, 2, 1972, 341.

THE CONCEPT OF MATERIAL DIVAGATION AND ITS APPLICATION TO FRACTURE

D. C. Stouffer and A. M. Strauss*

INTRODUCTION

Material Divagation is an expression used to describe processes or material behaviour that are characterized by changes or "wandering" of the material functionals that are used to characterize the mechanical properties of the material in the reference configuration.

The most applicable aspect of the theory of material divagation lies in the interpretation of material divagation as damage. The formalism of the theory provides the machinery necessary for understanding and predicting the (macroscopic) failure of materials.

Measures of time-to-failure and cycles-to-failure are the most popular methods used to predict the failure of materials [1]. The relationship between these measures and the mechanical behaviour of a material has never been adequately established. Thus the use of these measures as a fundamental criterion is not based on rational mechanical principles. Mechanics researchers have taken note of this and have started to work on the problem [2-7].

THE CONCEPT

The basic concept underlying the theory of divagation can be presented by a simple experiment. Consider two test samples that are physically and chemically identical in every aspect. Subject one sample to a cyclic deformation process so that after the process is complete the sample is in the original geometric configuration but with an altered microstructure. At this point subject both samples to identical deformation histories. The observed response of the test samples would, in general, be different. This difference reflects the change in the material properties induced by the cyclic deformation history applied to the first test sample. This observed divagation of the material response properties is identified as the mechanical damage or enhancement due to deformation.

The measure of divagation is developed directly from the constitutive equation that characterizes the response of the material as a functional of the history of the deformation. To this end, let us assume that the first Piola-Kirchhoff Stress Tensor, [8] $\tilde{S}(\tilde{x},t)$, at position $\tilde{x}$ and time t is given by a functional, $\tilde{Q}$, of the history of the deformation gradient, $\tilde{F}^t(s) = \tilde{F}(t-S)$; ie,

*Department of Engineering Analysis, University of Cincinnati, Cincinnati, Ohio 45221.

$$\underset{\sim}{S}(\underset{\sim}{x},t) = \underset{\sim}{Q}[\underset{\sim}{F}^t(S)] \tag{1}$$

for S in $(0,\infty)$.

To measure the divagation of the material properties using $\underset{\sim}{Q}$, it is necessary to separate the direct effect of the deformation from the intrinsic material properties characterized by the functional $\underset{\sim}{Q}$. To accomplish this, let $\underset{\sim}{F}_1(\tau)$ represent some reference deformation gradient history applied at time $\tau=0$ with the properties that $\underset{\sim}{F}_1(\tau) = \underset{\sim}{I}$ for τ in $(-\infty,0)$ and $\underset{\sim}{F}_1(\tau) \neq \underset{\sim}{I}$ for τ in $(0,t)$. The stress $\underset{\sim}{S}_1$ at time t resulting from $\underset{\sim}{F}_1(S)$ is determined from (1) as

$$\underset{\sim}{S}_1(\underset{\sim}{x},t) = \underset{\sim}{Q}[\underset{\sim}{F}_1{}^t(S)] \tag{2}$$

Denote a second deformation gradient $\underset{\sim}{F}_2(\tau)$ with the property that $\underset{\sim}{F}_2(\tau)$ is arbitrary for τ in $(-\infty,0)$ and such that the stress

$$\underset{\sim}{S}_2(t) = \underset{\sim}{Q}[\underset{\sim}{F}_2{}^t(S)] = \underset{\sim}{0} \text{ for } t > 0. \tag{3}$$

Further we require that full recovery occurs by time $\tau=0$, that is $\underset{\sim}{F}_2(0) = \underset{\sim}{F}(t)$ and is constant for all τ in $(0,t)$.

Define a deformation history $\underset{\sim}{F}_c{}^t(S)$ resulting from the composition of $\underset{\sim}{F}_1{}^t(S)$ with the 'preworking' deformation history $\underset{\sim}{F}_2{}^t(S)$ that is adjusted to $\underset{\sim}{0}$ at time t; that is, let

$$\underset{\sim}{F}_c{}^t(S) = \underset{\sim}{F}_1{}^t(S) + \underset{\sim}{F}_{d2}{}^t(S) \tag{4}$$

for all s in $(0,\infty)$. The quantity $\underset{\sim}{F}_{d2}{}^t(S) = \underset{\sim}{F}_2{}^t(S) - \underset{\sim}{F}_2(t)$ is the difference history and represents a measure of the deformation relative to the current configuration; ie, $\underset{\sim}{F}_{d2}(t) = 0$ for $t \geq 0$.

The stress $\underset{\sim}{S}_c(t)$ due to the deformation history $\underset{\sim}{F}_c{}^t(S)$ can be determined from (1) as

$$\underset{\sim}{S}_c(t) = \underset{\sim}{Q}[\underset{\sim}{F}_c{}^t(S)] \tag{5}$$

for t in $(0,t)$. In general, $\underset{\sim}{S}_c$ is different from the stress $\underset{\sim}{S}_1(t)$ due to the prior deformation history $\underset{\sim}{F}_{d2}{}^t(S)$. The difference in the response as observed by $\underset{\sim}{S}_c(t)$ and $\underset{\sim}{S}_1(t)$ for $t > 0$ the MATERIAL DIVAGATION TENSOR $\underset{\sim}{V}(t)$; ie,

$$\underset{\sim}{V}(t) \equiv \underset{\sim}{S}_c(t) - \underset{\sim}{S}_1(t) = \underset{\sim}{Q}[\underset{\sim}{F}_1{}^t(S) + \underset{\sim}{F}_{d2}{}^t(S)] = \underset{\sim}{Q}[\underset{\sim}{F}_1{}^t(S)] \tag{6}$$

The tensor $\underset{\sim}{V}(t)$ is a measure of the relative change in the material properties due to the predeformation $\underset{\sim}{F}_{d2}{}^t(S)$. Observe that the deformation $\underset{\sim}{F}_{d2}{}^t(S)$ does not contribute directly to the divagation $\underset{\sim}{V}(t)$ (since $\underset{\sim}{S}_2(t) = 0$ for $\tau > 0$) but only through history effects which are manifested by changes in the material microstructure. It should be noted that the material divagation $\underset{\sim}{V}$ is a tensor valued functional. This reflects the fact, for example, that plastic tensile and shear deformations would induce different microstructure changes in the material.

REPRESENTATIONS

A representation for $\underset{\sim}{V}(t)$ can be written down immediately by direct application of Taylor's theorem to functionals [11]. However, the Frechet derivative, $\delta\underset{\sim}{Q}$, of the functional $\underset{\sim}{Q}$ with respect to the norm $||\underset{\sim}{F}_{d2}(\ )||_h$ is given by

$$\underset{\sim}{Q}[\underset{\sim}{F}_1{}^t(S) + \underset{\sim}{F}_{d2}{}^t(S)] = \underset{\sim}{Q}[\underset{\sim}{F}_1{}^t(S)] + \delta\,\underset{\sim}{Q}[\underset{\sim}{F}_1{}^t(S)]$$

$$|\underset{\sim}{F}_{d2}{}^t(S)] + o(||\underset{\sim}{F}_{d2}{}^t(\ )||_h) \qquad (7)$$

where $\delta\underset{\sim}{Q}$ is linear in $\underset{\sim}{F}_{d2}{}^t(S)$ and continuous in both hostories [9,10]. An alternative formulation for the divagation tensor, $\underset{\sim}{V}(t)$, can be developed by observing for any time $t > 0$ that

$$\underset{\sim}{V}(t) = \int_0^t \frac{d}{d\tau}\underset{\sim}{V}(\tau)d\tau$$

$$= \int_0^t \{\delta\underset{\sim}{Q}[\underset{\sim}{F}_c{}^\tau(S)\ \dot{\underset{\sim}{F}}_c{}^\tau(S)] = \delta\underset{\sim}{Q}[\underset{\sim}{F}_1{}^\tau(S)\ \dot{\underset{\sim}{F}}_1{}^\tau(S)]\}d\tau \qquad (8)$$

since $\underset{\sim}{V}(0) = \underset{\sim}{0}$ and $O(||\underset{\sim}{F}_{d2}{}^t(\ )||_h)$ vanishes by definition.

RESULTS

The above definition and representation of the divagation tensor $\underset{\sim}{V}$ is used to demonstrate the following:

1) A material is designated as ideal if $\dot{\underset{\sim}{V}}(\tau) = \underset{\sim}{0}$ for all τ in $(0,t)$. The consequences of this statement are sufficient to show that any viscoelastic constitutive equations, with integral kernel functions that depend only on time, cannot predict material divagation.

2) For an isothermal cyclic process the observed change in dissipation due to the 'preworking' deformation history $\underset{\sim}{F}_{d2}{}^t$, is linear in the material divagation tensor $\underset{\sim}{V}(t)$.

3) If the constitutive functional $\underset{\sim}{Q}$ is extended to include an arbitrary number (i) of independent histories $\underset{\sim}{N}_i{}^t(S)$, then the material damage or enhancement due to a particular history $\underset{\sim}{N}_\alpha{}^t(S)$ will vanish if $\dot{\underset{\sim}{N}}_\alpha$ vanishes for all t in $(-\infty,\infty)$. However, the constant value of $\underset{\sim}{N}_\alpha$ does remain coupled to the response.

4) If the response $\underset{\sim}{S}$ of the functional $\underset{\sim}{Q}$ is related to a second parameter $\underset{\sim}{M}$ by $\underset{\sim}{M} = \underset{\sim}{f}(\underset{\sim}{S})$, where $\underset{\sim}{f}$ is a constitutive function. The divagation in $\underset{\sim}{M}$, denoted by $\underset{\sim}{V}_{\underset{\sim}{M}}$, is related to the divagation in $\underset{\sim}{S}$, denoted by $\underset{\sim}{V}_{\underset{\sim}{S}}$, by

$$\underset{\sim}{V}_{\underset{\sim}{M}} = \int_0^t (\underset{\sim}{\nabla}_{\underset{\sim}{S}} \cdot \underset{\sim}{f})\ \dot{\underset{\sim}{V}}(\tau)d\tau$$

where $\underset{\sim}{\nabla}_{\underset{\sim}{S}}$ is a generalized gradient operator.

5) Finally, the divagation tensor can be used as a method of developing the constitutive functional $\underset{\sim}{Q}$, since previous history effects are included through the Frechet derivative of $\underset{\sim}{Q}$.

ACKNOWLEDGMENTS

The research was supported by the National Science Foundation under grant no. ENG75-06619 at the University of Cincinnati.

REFERENCES

1. CONWAY, J. B., STENTZ, R. H. and BERLING, J. T., National Technical Information Service, U.S. Dept. of Commerce, Springfield, Virginia, 1975.
2. BERKOVITS, A., NASA Technical Note, NASA-TN-D6937, September, 1972.
3. BROCKWAY, G. S. and ALLEN, G. D., Mechanics and Materials Research Center, Texas A&M University, report MM2764-74-3, May 1974.
4. BROCKWAY, G. S. and ALLEN, G. D., Mechanics and Materials Research Center, Texas A&M University report MM3168-75-1, April, 1975.
5. FITZGERALD, J. E. and PENNY, D. N., N.S.F. Workshop on Applied Thermoviscoelasticity, Northwestern University, October 1975.
6. QUINLAN, M. H. and FITZGERALD, J. E., University of Utah Report Number UTEC-CE-74-062, June 1974.
7. HUFFERD, W. L. and FITZGERALD, J. E., Proc. 1971, Intern. Conf. Mechanical Behavior of Materials, Vol. III, Japan, 1972, 530.
8. TRUESDELL, C. and NOLL, W., Hanbuch der Physik, Band III/3, Ed. W. Flugge, Springer Verlag, 1965.
9. COLEMAN, B. D., Archive Rational Mechanics and Analysis, 17.1, 1964.
10. VOLTERRA, Vito, Theory of Functionals and of Integral and Integro-Differential Equations, Blackie & Son Limited, London, 1930.
11. GREEN, A. E. and RIVLIN, R. S., Archive Rational Mechanics and Analysis, Vol. I, 1957. 1.

Fracture 1977, Volume 3, ICF4, Waterloo, Canada, June 19 - 24, 1977

SUDDEN TWISTING OF A PENNY-SHAPED CRACK IN A FINITE ELASTIC CYLINDER

E. P. Chen*

INTRODUCTION

In the field of fracture dynamics, much of the previous investigation have been concerned with the determination of the nature of the dynamic stresses near a crack in a body of infinite extent. For axisymmetric geometry, a number of papers [1-5] have discussed the inertia effect on the dynamic stress intensity factors. However, due to the mathematical complexities involved, the interaction between scattered waves by a penny-shaped crack and reflected waves from finite boundaries has not been considered.

In this paper, the torsional impact response of a penny-shaped crack in a finite elastic cylinder is considered. Application of the Laplace and Hankel transforms reduces the problem to the solution of a pair of dual integral equations. These equations are solved by using an integral transform technique and the result is expressed in terms of a Fredholm integral equation of the second kind. The time dependence is recovered by applying the Laplace inversion theorem. Numerical solutions are obtained for the amplitude of the local stress field near the crack; that is, the dynamic stress intensity factor. The influence of the interaction between scattered waves by the crack and reflected waves from the boundaries are discussed. Although only stress-free cylinder surface condition is considered, other boundary conditions can be treated in a similar manner without additional difficulties.

PROBLEM FORMULATION

Let the axis of an elastic cylinder coincide with the z axis of a cylindrical polar coordinate system (r,θ,z). The cylinder is made of a homogeneous and isotropic material and its radius is denoted b. A penny-shaped crack of radius a is lying on the plane z=0 with its center at the origin of the coordinate axes. The geometry of the problem is shown in Figure 1.

The displacements in the r, θ and z directions are denoted, respectively, by u_r, u_θ and u_z. The cylinder is under the action of a rapidly applied torque such that the material particles experience only an angular displacement. Hence, u_r and u_z vanish throughout the body and from symmetry considerations, u_θ is a function of r, z and time t only. The displacement field can thus be written as

$$u_r = u_z = 0 \ , \ u_\theta = u_\theta(r,z,t) \tag{1}$$

*Lehigh University, Bethlehem, Pennsylvania, U.S.A.

Corresponding to equations (1), all stress components except $\tau_{r\theta}$ and $\tau_{\theta z}$ vanish and the shear stresses are given by the following expressions:

$$\tau_{r\theta}(r,z,t) = \mu\left(\frac{\partial u_\theta}{\partial r} - \frac{u_\theta}{r}\right) \tag{2}$$

$$\tau_{\theta z}(r,z,t) = \mu\frac{\partial u_\theta}{\partial z} \tag{3}$$

in which μ is the shear modulus of elasticity. Substituting equations (2) and (3) into the equations of motion of elasticity, it follows that two of them are identically satisfied and the remaining one renders

$$\frac{\partial^2 u_\theta}{\partial r^2} + \frac{1}{r}\frac{\partial u_\theta}{\partial r} - \frac{u_\theta}{r^2} + \frac{\partial^2 u_\theta}{\partial z^2} = \frac{1}{c_2^2}\frac{\partial^2 u_\theta}{\partial t^2} \tag{4}$$

The shear wave velocity c_2 is given by $c_2 = (\mu/\rho)^{1/2}$ with ρ being the mass density of the material.

The cylinder is assumed to be initially at rest. A torque T of magnitude $T = \pi b^4\tau_o/(2a)$, with τ_o having the dimension of stress, is suddenly applied to the elastic body generating torsional waves normally incident on the crack. By the principle of superposition, the equivalent boundary conditions for which the wave passes across the crack plane at t=0 can be written as

$$\tau_{\theta z}(r,o,t) = -\tau_o\left(\frac{r}{a}\right)H(t), \quad 0<r<a \tag{5}$$

$$u_\theta(r,o,t) = 0, \quad r \geq a \tag{6}$$

In addition, the traction-free condition at the cylinder surface requires

$$\tau_{r\theta}(b,z,t) = 0 \tag{7}$$

Equation (4) is to be solved under the constraint of equations (5), (6) and (7).

Applying a Laplace transform pair

$$f^*(p) = \int_o^\infty f(t)e^{-pt}dt, \quad f(t) = \frac{1}{2\pi i}\int_{Br} f^*(p)e^{pt}dp \tag{8}$$

to equation (4) yields

$$\frac{\partial^2 u_\theta^*}{\partial r^2} + \frac{1}{r}\frac{\partial u_\theta^*}{\partial r} - \frac{u_\theta^*}{r^2} + \frac{\partial^2 u_\theta^*}{\partial z^2} = \frac{p^2}{c_2^2}u_\theta^* \tag{9}$$

For a finite cylinder, the solution to (9) may be written as

$$u_\theta^* = \int_o^\infty A(s,p,J_1(rs)e^{-\gamma z}ds + \int_o^\infty B(s,p)I_1(\gamma r)\sin(sz)ds \tag{10}$$

where J_n and I_n are, respectively, the nth order Bessel and modified Bessel function of the first kind. The function γ is given by

$$\gamma = \sqrt{s^2 + \frac{p^2}{c_2^2}} \tag{11}$$

Making use of equations (2), (10) and the Laplace transform of equations (5) and (6) renders a pair of dual integral equations:

$$\int_0^\infty \gamma A(s,p) J_1(rs) ds = \int_0^\infty s B(s,p) I_1(\gamma r) ds + \frac{\tau_o r}{\mu a p}, \quad r<a \tag{12}$$

$$\int_0^\infty A(s,p) J_1(rs) ds = 0, \quad r \geq a \tag{13}$$

The relationship between A(s,p) and B(s,p) is found, by taking Laplace and Fourier sine transforms on equation (7), as

$$\gamma B(s,p) I_2(\gamma b) = \frac{2}{\pi} s \int_0^\infty \frac{\eta A(\eta,p) J_2(b\eta)}{\eta^2+\gamma^2} d\eta \tag{14}$$

A solution to equations (12) and (13) satisfying equation (14) may be obtained by a procedure described in [5] and the result is

$$A(s,p) = \frac{\sqrt{2sa}}{\pi} \frac{2\tau_o a^2}{3\mu p} \int_0^1 \sqrt{\eta}\, \Phi^*(\eta,p) J_{3/2}(sa\eta) d\eta \tag{15}$$

where the function $\Phi^*(\eta,p)$ is governed by a Fredholm integral equation of the second kind:

$$\Phi^*(\xi,p) + \int_0^1 K(\xi,\eta,p)\Phi^*(\eta,p) d\eta = \xi^2 \tag{16}$$

The symmetric kernel $K(\xi,\eta,p)$ is defined by

$$K(\xi,\eta,p) = \sqrt{\xi\eta}\, \{ \int_0^1 (\gamma'-s) J_{3/2}(\xi s) J_{3/2}(\eta s) ds$$

$$- \frac{2}{\pi} \int_0^\infty \frac{s^2}{\gamma'} \frac{K_2(\gamma' \frac{b}{a})}{I_2(\gamma' \frac{b}{a})} I_{3/2}(\xi\gamma') I_{3/2}(\eta\gamma') ds \} \tag{17}$$

in which

$$\gamma' = \sqrt{S^2 + \frac{p^2 a^2}{c_2^2}} \tag{18}$$

and K_2 is the second order modified Bessel function of the second kind.

NUMERICAL RESULTS AND DISCUSSION

Following the same procedures as in [3], the stresses local to the penny-shaped crack in the Laplace transform domain may be obtained as

$$\tau^*_{r\theta}(r_1,\theta_1,p) = -\frac{4}{3\pi}\tau_o\sqrt{a}\,\frac{\Phi^*(1,p)}{p}\,\frac{1}{\sqrt{2r_1}}\sin\frac{\theta_1}{2} + 0(r^o) \tag{19}$$

$$\tau^*_{\theta z}(r_1,\theta_1,p) = \frac{4}{3\pi}\tau_o\sqrt{a}\,\frac{\Phi^*(1,p)}{p}\,\frac{1}{\sqrt{2r_1}}\cos\frac{\theta_1}{2} + 0(r^o) \tag{20}$$

where r_1 and θ_1 are the polar coordinates centered at the crack border in the plane z=0. Applying the Laplace inversion theorem to equations (19) and (20) renders

$$\tau_{r\theta}(r_1,\theta_1,p) = -\frac{K_3(t)}{\sqrt{2r_1}}\sin\frac{\theta_1}{2} + 0(r^o) \tag{21}$$

$$\tau_{\theta z}(r_1,\theta_1,p) = \frac{K_3(t)}{\sqrt{2r_1}}\cos\frac{\theta_1}{2} + 0(r^o) \tag{22}$$

The dynamic stress intensity factor $k_3(t)$ is defined as

$$k_3(t) = \frac{4}{3\pi}\tau_o\sqrt{a}\,\frac{1}{2\pi i}\int_{Br}\frac{\Phi^*(1,p)}{p}e^{pt}dp \tag{23}$$

In order to obtain numerical values of $k_3(t)$, the Fredholm integral equation (16) is first solved on an electronic computer. Once Φ^* is determined, the dynamic stress intensity factor may be obtained by a numerical Laplace inversion scheme described in [3].

Note that by letting $p\to 0$, the solution presented here reduces to that for the corresponding static case of the same geometry. Figure 2 depicts the variation of the normalized static stress intensity factor $\bar{k}_3 = k_3/(4\tau_o\sqrt{a}/3\pi)$ versus the radius ratio a/b. It can be seen that as the ratio a/b is increased, the stress intensity factor also becomes higher. The rate of increase is very small when a/b is less than 0.5 and becomes significantly larger when a/b exceeds 0.5. This suggests that the effect of finite boundaries is serious only when the radius of the cylinder is less than two times of that of the penny-shaped crack.

The normalized dynamic stress intensity factor $\bar{k}_3(t)$ is plotted in Figure 3 against the time variable c_2t/a for various b/a ratios. The dynamic stress intensity factor increases quickly, reaching a peak, and then decreases in magnitude oscillating about the corresponding static value. This type of time dependence has also been observed in [3] for an infinite geometry problem. As b/a decreases, the peak value of $\bar{k}_3$ becomes larger and occurs at a later time. For b/a = 1.1, the interaction between inertia and finite geometry can increase the dynamic stress intensity factor by 65% over its corresponding static value in an infinite medium. Hence, it is obvious that this interaction effect is quite significant and cannot be ignored.

In summary, the interaction effect between scattered waves by a penny-shaped crack and reflected waves from finite boundaries on a cylinder under torsional impact has been determined. The dynamic stress intensity is found to be a function of time and the geometrical parameter b/a. The dynamic stress intensity factor reaches a peak very quickly and then decreases in magnitude oscillating about its corresponding static value. The peak value of the amplitude of the dynamic stresses is higher when the ratio b/a is reduced.

ACKNOWLEDGEMENTS

This work is part of a research program sponsored by the Army Research Office under Grant No. DAAG29-76-G-0091.

REFERENCES

1. SIH, G.C. and LOEBER, J.F., J. Acoust. Soc. America, 44, 1968, 1237.
2. SIH, G.C. and LOEBER, J.F., J. Acoust. Soc. America, 46, 1969, 711.
3. SIH, G.C. and EMBLEY, G.T., J. Appl. Mech., 39, 1972, 395.
4. EMBLEY, G.T. and SIH, G.C., Proceedings of the 12th Midwestern Mechanics Conference, Development in Mechanics, 6, 1972, 473.
5. KASSIR, M.K. and SIH, G.C., Three Dimensional Crack Problems, Volume 2, Mechanics of Fracture Series, edited by G.C. Sih, Noordhoff International Publishing Company, The Netherlands, 1975.

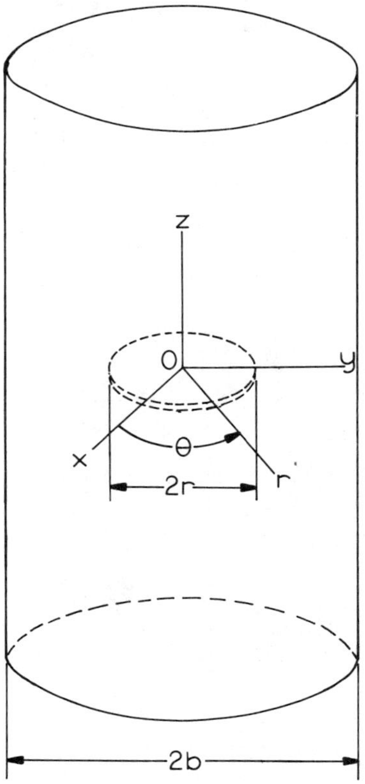

Figure 1   Finite cylinder containing a penny-shaped crack

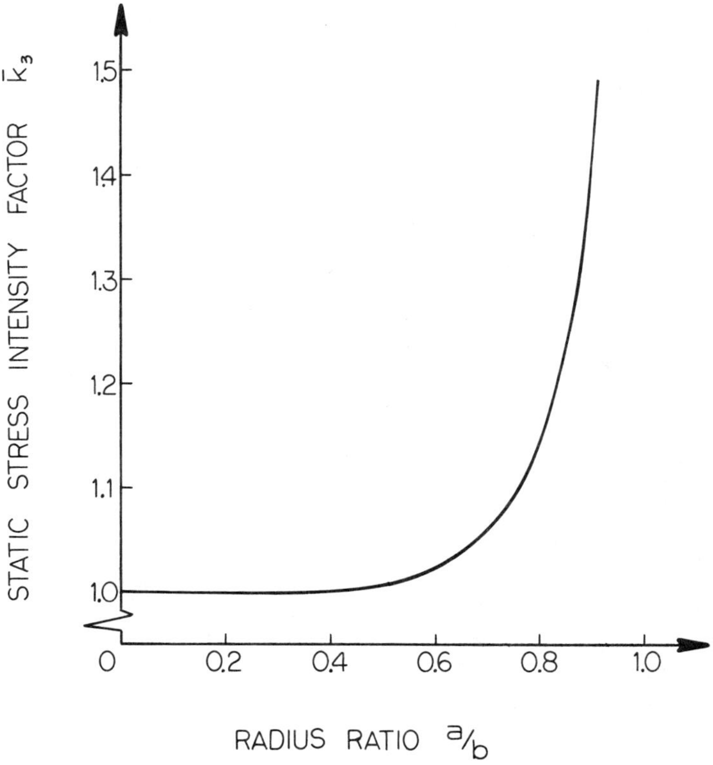

Figure 2 Normalized static stress intensity factor versus a/b

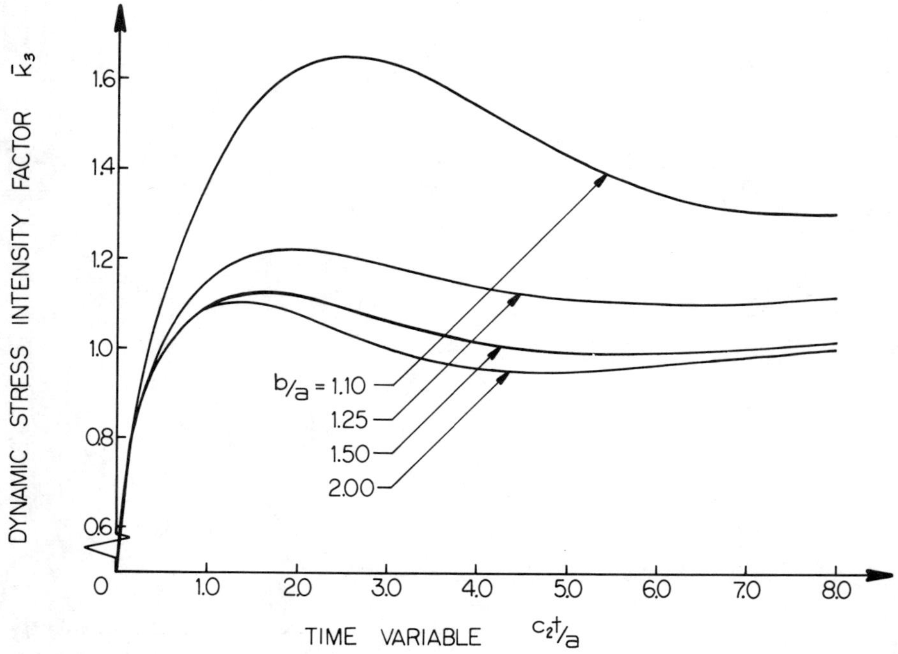

Figure 3 Normalized dynamic stress intensity factor versus time

Fracture 1977, Volume 3, ICF4, Waterloo, Canada, June 19 - 24, 1977

DYNAMIC FINITE ELEMENT AND DYNAMIC PHOTOELASTIC ANALYSES OF AN IMPACTED PRETENSIONED PLATE

A. S. Kobayashi, S. Mall and A. F. Emery*

INTRODUCTION

Two popular test specimens used in studying crack arrest potential of structure steel are the ESSO and Robertson specimens [1,2] in which dynamic crack propagation is initiated through impacting a wedge in the crack of a subcritically loaded single-edged notch tension plate. Crack arrest is achieved by regions of either higher fracture toughness generated by higher local temperature in low carbon steel specimen and/or lower stress intensity factors generated by lower local stress field.

Dynamic photoelastic analysis of ESSO type test specimens modeled by Homalite-100 plates [3] show that the dynamic effects of the propagating crack combined with that of the impacting projectile have considerable effect on the dynamic stress intensity factor and hence on crack propagation. Unfortunately, these results neither provide a unique relation between the crack velocities and dynamic fracture toughness nor a definitive conclusion regarding the basic mechanism of crack arrest. In addition, the results are not in complete agreement with the more recent experimental results obtained on thicker Homalite-100 plates [4,5].

In order to verify, by an independent procedure, some of the controversial results obtained during our past seven-year efforts in fracture dynamics, the authors have used a relatively simple dynamic finite element code to duplicate some of their past work in dynamic photoelasticity [6,7,8]. Encouraged by the reasonable agreements between the numerical and experimental results obtained through this series of studies involving single-edged notch specimens loaded to criticality, the same dynamic finite element code was used to analyze the previous dynamic photoelastic results on the ESSO type test specimens [3].

DYNAMIC PHOTOELASTIC ANALYSIS

The dynamic photoelastic experiments in this paper involve subcritically loaded single-edged notch tension specimens where crack propagation was initiated by an impacted flat-nosed projectile or a 65° wedge. The test specimens consisted of a 9.53mm thick Homalite-100 plate with a 0.254 x 0.254m test section loaded in a fixed gripped condition with uniform grip displacement, and with a single-edged starter crack approximately 9.53mm in length. The dynamic properties of Homalite-100 were obtained following the procedure of Clark and Sanford [9], which yielded an average dynamic modulus of elasticity, Poisson's ratio and stress optic coefficient of 4.65 GPa, 0.345 and 27.15 MPa-mm/fringe, respectively. The averaged static fracture toughness, which was obtained through separate tests using SEN specimens, was 0.64 $MPa \cdot m^{1/2}$.

---

*University of Washington, Department of Mechanical Engineering, Seattle, Washington 98195, U.S.A.

Dynamic stress intensity factors, K_D, were determined by Bradley's two parameter procedure [10] and the dynamic energy release rate, $\mathcal{G}_D$, was computed using Freund's equation [11] from the dynamic stress intensity factors. Further details of these data reduction schemes can be found in references [6] and [8].

DYNAMIC FINITE ELEMENT ANALYSIS

The dynamic finite element code, Hondo [12], used in this investigation is based on an explicit time integration scheme and constant strain quadrilateral elements. The crack tip motion was modeled by discontinuous jumps where crack tip moved from one finite element node to another at discrete time intervals. This discrete propagation of the crack tip generated significant oscillations in the states of stress and displacement surrounding the crack tip. The numerical noise was filtered by computing directly the dynamic energy released by the discrete crack tip advancement from the time averaged normal stress ahead of the advancing crack tip and the corresponding time-averaged crack opening displacement after crack advance. Details of this numerical procedure as well as an accuracy check of the procedure are described in reference [6].

Figure 1 shows the finite element breakdown involving a total of 532 elements and 585 nodes used in this analysis. Impacted wedge-loading was simulated by two simultaneously applied vertical and horizontal forces at the crack mouth without the wedge-shape and the impact forces for the flat nose projectile and 65° wedge were assumed to vary with impact duration. Large plastic deformations at the impact sites were assumed to dissipate about 66 percent and 43 percent of the impact energies for the flat nose and 65° wedge impacts, respectively. Estimates of these energy losses as well as impact durations were made by comparing the calculated dynamic maximum shear stress patterns of a given impulse with the associated dynamic isochromatic patterns as shown in Figure 2.

PRETENSIONED SINGLE-EDGED NOTCH PLATE IMPACTED BY FLAT NOSE PROJECTILE

In the series of dynamic photoelastic experiments reported in reference [3], the crack propagated in some pretensioned single-edged notch plates while it did not run in others. These stop-or-go results potentially provided information for estimating the static fracture toughness under stress-wave loadings but unfortunately the dynamic photoelastic patterns prior to crack propagation were not recorded in these experiments. A combination of dynamic finite element analysis and dynamic photoelasticity results provided a procedure in which the dynamic state prior to triggering of the dynamic polariscope could be estimated by some trial and error. Table 1 shows such maximum dynamic stress intensity factors due to impact for the stationary crack in Test No. W012172 and prior to crack propagation in Test Nos. W020672 and W090711.

It is interesting to note that this combined dynamic photoelastic-dynamic finite element analysis results in Table 1 indicate that the dynamic fracture toughness, K_c, under this combined static and stress wave loadings is close to, within experimental scatters, the static fracture toughness of $K_c = 0.64\ \mathrm{MPa{\cdot}m^{1/2}}$. Perhaps such coincidence may be expected in view of the recent work by G. C. Smith [13] who found that the variations in fracture toughness of 4.76mm thick Homalite-100 plates is approximately equal to the static fracture toughness for the time interval to failure of 20 microseconds. The 30-50 percent increase in stress intensity factor due to impulse loading falls within the rapidly changing dynamic fracture toughness at this time interval to failure.

PRETENSIONED SINGLE-EDGED NOTCH PLATE IMPACTED BY A 65° WEDGE

The dynamic photoelasticity record of Test No. W012472 was analyzed then by the dynamic finite element method using the idealized crack velocity shown in Figure 3. Figure 4 shows a comparison of the dynamic energy release rates due to static preload on the specimen. The rapid fluctuations in the FEM results at an approximate crack length of a/b = 0.15 is due to the momentary drop in crack velocities at this location. Otherwise good agreement between the measured and computed dynamic energy release rate are noted.

DISCUSSION

Our conclusion that fracture toughness, which did not differ with its static counterpart, under the combined static and impulse loadings is in agreement with that discussed in reference [7] involving simulated dynamic tear tests. As mentioned previously, these findings are in agreement with those in reference [13] because of the relatively low strain rate effects in these tests.

The dynamic energy release rate at crack arrest was much lower than those measured in non-impact experiments [7] which again reinforces our postulate that $\mathcal{G}_D$ at crack arrest is not a material property. The average dynamic energy release rate, which is obtained by dividing the sum of the total dynamic energy release rates by the newly created crack surface by crack propagation, for Test W012472, yielded $\mathcal{G}_D]_{ave}/\mathcal{G}_c$ = 2.33 and 2.28 from the dynamic photoelasticy and dynamic FEM analysis, respectively. The large $\mathcal{G}_D]_{ave}$ generated by elastic analysis for a prescribed crack propagation history probably indicates the larger dissipation in dynamic energy due to viscous damping and at the flexible edge grips under high impact loading.

ACKNOWLEDGEMENT

The results reported in this paper were obtained in a research contract funded by the Office of Naval Research under Contract No. N00014-76-C-0060 NR 064-478. The authors wish to acknowledge the continuous support and encouragement of Drs. N. R. Perrone and N. Basdegas of ONR.

REFERENCES

1. BLUHM, J. L., Fracture, ed. H. Liebowitz, Vol. V, Academic Press, 1969, 1-63.
2. KANAZAWA, T., Dynamic Crack Propagation, ed. G. C. Sih, Noordhoff International, 1973, 565-597.
3. KOBAYASHI, A. S. and WADE, B. G., ibid loc cit, 663-677.
4. KOBAYASHI, T. and FOURNEY, W. L., Proc. of 12th Annual Meeting of the Society of Engineering Sciences, University of Texas at Austin, Oct. 20-22, 1975, 131-140.

5. KOBAYASHI, T. and DALLY, J. W., "The Relation Between Crack Velocity and the Stress Intensity Factor in Birefringenet Polymers", presented at ASTM Symposium on Fast Fracture and Crack Arrest, Chicago, Illinois, June 28-30, 1976.
6. KOBAYASHI, A. S., EMERY, A. F. and MALL, S., "Dynamic Finite Element and Dynamic Photoelastic Analyses of Two Fracturing Homalite-100 Plates", to be published in Experimental Mechanics.
7. KOBAYASHI, A. S. and MALL, S., Proc. of the Int. Conf. on Dynamic Fracture Toughness, London, July 5-7, 1976, 259-272.
8. KOBAYASHI, A. S., EMERY, A. F. and MALL, S., Presented at the ASTM Symposium on Fast Fracture and Crack Arrest, Chicago, Illinois, June 28-30, 1976.
9. CLARK, A. B. J. and SANFORD, R. J., Proc. of the Soc. for Experimental Stress Analysis, Vol. 20, No. 2, 1973, 148-151.
10. BRADLEY, W. B. and KOBAYASHI, A. S., Engineering Fracture Mechanics, Vol. 3, 1971, 317-332.
11. FREUND, L. B., J. of Mechanics and Physics of Solids, Vol. 20, 1972, 141-152.
12. KEYS, S. W., Sandia Laboratories Report SLA-74-0039, April 1974.
13. SMITH, G. C., PhD thesis submitted to California Institute of Technology, April 1975.

Table 1 Dynamic Stress Intensity Factor Prior to Crack Propagation in an Impacted Pretensioned Plate

Test No.	Prescribed Displacement u_{top} u_{bottom}	Projectile Velocity	Stress Intensity Factor Due to Pre-loading	Dynamic Stress Intensity Factor Due to Impact at 20 μsec*	Resultant Stress Intensity Factor	Remarks
W012172	0.0572 mm 0.0572 mm	Flatnose, 12.34 gm 24 m/sec	0.30 $MPa\cdot m^{1/2}$	6.41 $MPa\cdot m^{1/2}$	0.70 $MPa\cdot m^{1/2}$	Crack did not run
W020672	0.0254 mm 0.0762 mm	Flatnose, 12.34 gm 46 m/sec	0.40 $MPa\cdot m^{1/2}$	0.40 $MPa\cdot m^{1/2}$	0.80 $MPa\cdot m^{1/2}$	Crack ran
W090771	0.0889 mm 0.0889 mm	Flatnose, 12.34 gm 26 m/sec	0.47 $MPa\cdot m^{1/2}$	0.40 $MPa\cdot m^{1/2}$	0.87 $MPa\cdot m^{1/2}$	Crack ran

$K_c = 0.64\ MPa\cdot m^{1/2}$

*Dynamic S.I.F. for same impact pulse, regardless of differences in muzzle velocity.

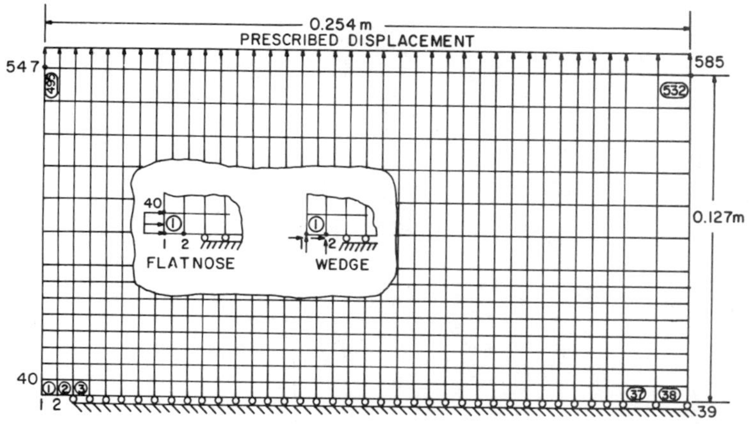

Figure 1 Finite Element Breakdown with Applied Impact Forces in Flatnose and 65° Wedge Projectile

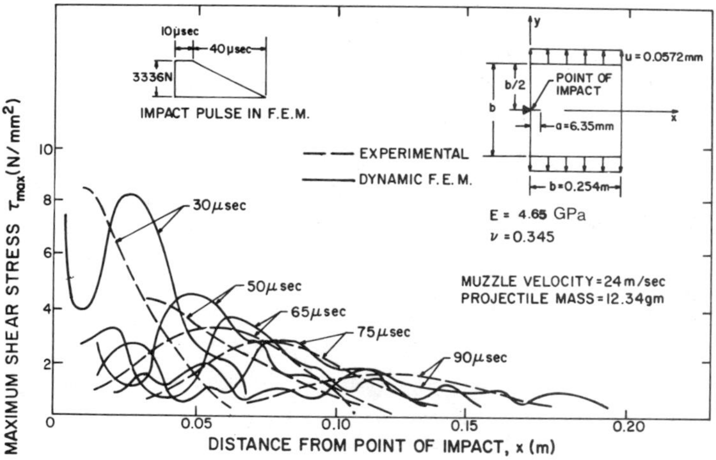

Figure 2 Dynamic Maximum Shear Stresses Along the Axis of Symmetry in Single-Edged Crack Pretensioned Plate Impacted by a Flatnose Projectile, Test No. W012172

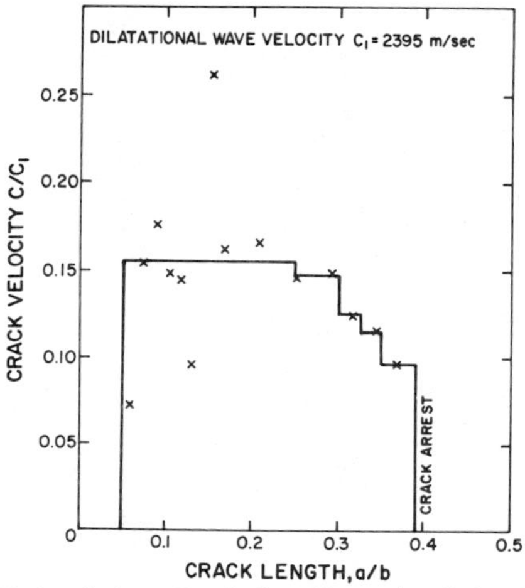

Figure 3 Crack Velocities Used for Numerical Analysis Along with Experimental Data, Test No. W012472

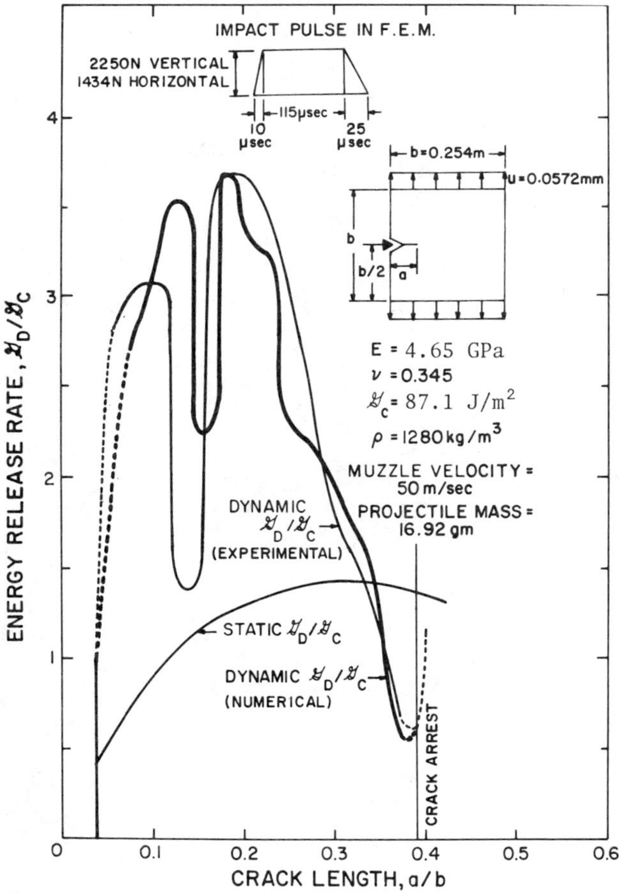

Figure 4 Energy Release Rates in a Single-Edged Crack Pretensioned Plate Impacted by a 65° Wedge, Test No. W012472

Fracture 1977, Volume 3, ICF4, Waterloo, Canada, June 19 - 24, 1977

A FINITE-ELEMENT ANALYSIS OF AN IMPACT TEST

J. A. Aberson, J. M. Anderson and W. W. King*

INTRODUCTION

In 1974 Madison and Irwin [1] published results of a fracture test programme begun at Lehigh in 1966. The purpose of the programme was to determine fracture toughness (K_c) values for structural steels at temperatures and loading rates representative of service conditions. The tests employed precracked three-point-bend specimens measuring 76 mm deep, 30 mm long and 25 mm thick. The supported span was 250 mm, and the fatigue crack length, including a starter notch, was approximately 25 mm. Fracture toughness values were computed using the observed maximum load and the initial crack length adjusted to account for plastic-zone size. Loading times as brief as 0.50 ms were judged by Madison and Irwin to be "...small enough for evaluation of minimum dynamic toughness and long enough to permit static stress anslysis of the specimen." It is with this contention that the present paper takes issue.

For a simply supported beam of flexural stiffness EI, mass M and span S, the fundamental frequency of vibration ω is given (see, for example, [2] p.331) by

$$\omega = \pi^2 \sqrt{\frac{EI}{MS^3}} \tag{1}$$

For steel of dimensions appropriate to the test specimen (1) yields a fundamental period of about 0.37 ms -- a figure much too near the least loading period to warrant neglecting inertia effects. The elementary modal analysis producing (1) neglects shear deformation and rotary inertia, beam overhang and the presence of the crack. But since these are all effects tending to increase the computed fundamental period, a static analysis seems all the more suspect. This suspicion is later confirmed by employing a finite-element model having over 300 degrees of freedom and capable of representing the neglected effects mentioned above.

MADISON-IRWIN PROCEDURE FOR DETERMINING K_c

In the Madison-Irwin exerpiments, the specimen was loaded by an instrumented striking tup mounted in a freely falling weight. The instrumentation provided an oscilloscope trace of the applied load. Two-peak load histories were reported for some of the tests, which Madison and Irwin attributed to obscuring inertia effects. They associated the first peak with inertia effects, while the second peak was judged to be the significant specimen-load record. By placing loading cushions between the specimen and the striking tup, Madison and Irwin obtained a load record with a single peak. This was accepted as evidence that inertia effects had

*Georgia Institute of Technology, Atlanta, Georgia, U.S.A.

been eliminated, and they supposed in their calculations that the peak recorded load was the specimen load at the onset of crack propagation. Figure 1 shows a best-estimate reproduction of a published oscilloscope trace recorded in a -40°C test of a 25 mm thick specimen. The load history depicted in Figure 1 is the type deemed acceptable by Madison and Irwin for static analysis of the problem.

Madison and Irwin used a slightly modified Gross-Srawley [3] formula in conjunction with the peak load from the specimen-load record to obtain a first-estimate value of K_c. Their formula,

$$K_1 = \frac{3PS\sqrt{a}}{2BW^2}\left[1.93 - 3.12\,\frac{a}{w} + 14.7\left\{\frac{a}{w}\right\}^2 - 25.3\left\{\frac{a}{w}\right\}^3 + 25.9\left\{\frac{a}{w}\right\}^4\right], \quad (2)$$

gives K_1 in MPa$\sqrt{m}$ for P in MN with beam span S, thickness B, depth W and crack length a all m. For the peak load (55.6 KN) obtained from Figure 1, (2) yields a first-estimate K_c of 43.7 MPa$\sqrt{m}$.

Such figures were subsequently revised upward by adjusting for plastic-zone size. Briefly this amounted to increasing the fatigue crack length by the plastic-zone radius

$$r_Y = \frac{1}{2\pi}\left\{\frac{K_1}{\sigma_Y}\right\}^2, \quad (3)$$

in which σ_Y is the yield stress. Equations (2) and (3) were then used repeatedly until the iteration scheme produced practically constant values for r_Y and K_c. Since the thrust of the present paper has to do with assessing inertia effects rather than plasticity effects, we shall make no plasticity adjustments to either our results or those of Madison and Irwin.

FINITE-ELEMENT ANALYSIS OF A MADISON-IRWIN TEST

Figure 2 shows a finite-element representation of a Madison-Irwin test specimen. Due to symmetry about the plane of the crack, only the left half of the specimen is modelled. The model consists of 163 nodes, 273 constant-strain triangles and 1 eight-node crack-tip singularity element. The singularity element ABCDE is required to accurately represent the locally severe stress gradients in the neighbourhood of the fatigue-crack tip at D. Consistent with symmetry requirements, nodes along the crack's prolongation DG are restrained against horizontal displacement. A vertical force equal to half the specimen load is applied at G. A vertical restraint at H simulates the specimen support. The fatigue-crack starter notch was not represented, and the two-dimensional idealization of the problem was taken to be the one corresponding to plane stress.

The singularity element used to numerically characterize the near-tip stress field has been successfully employed in many and varied static applications [4] and has performed satisfactorily in a number of problems involving transient stresses near the tip of a stationary crack [5, 6]. These problems generally represent more severe tests of the finite-element analysis than does the present problem. Details of the singularity element's formulation are given in these references. Briefly, it incorporates as generalized coordinates for stiffness and inertial characterization the first thirteen symmetric Williams' [7] eigenfunctions and the three parameters associated with plane rigid-body motion. The Newmark - β method with $\beta = 1/4$ is the time-integration scheme used for dynamic applications.

Figure 3 shows the time dependence of the stress-intensity factor K_I for three different numerical representations of a Madison-Irwin experiment. The solid line indicates the quasi-static response of the model shown in Figure 2; i.e. $K_I(t)$ appropriate to a massless specimen subjected to the load as taken from the oscilloscope trace (Figure 1). The computed value of K_I at peak load (42 MPa$\sqrt{m}$) is in reasonable agreement with the Madison-Irwin estimate (43.7 MPa$\sqrt{m}$) obtained using (2), but based on the results of previous confirmed static applications, the 4% discrepancy is somewhat more than can be attributed to the numerical method. Notwithstanding the small difference, the quasi-static response shown in Figure 3 is used as a basis for assessing inertia effects in the two companion dynamic executions.

The locus of empty circles in Figure 3 is $K_I(t)$ for a model with inertia characteristics corresponding to steel and subjected to the time-dependent load of Figure 1. The integration time step was 10^{-2} ms and K_I was computed at each time step. This is the order of the transit time of longitudinal waves through the depth of the specimen and consequently the transmission and reflection of individual stress waves is not represented. However, the interest here is in the response over a relatively longer time during which the lower modes of vibration dominate. The shape of the response confirms the expected dominance of the first mode and the earlier estimate of its period. The considerable difference between the dynamic and quasi-static responses is exclusively the result of specimen inertia. When specimen inertia is included, the peak stress-intensity factor is elevated by more than 8% above the maximum quasi-static value. More importantly, the peaks occur at different times. So, for the particular geometry and loading rate under consideration, the dynamic result is in clear conflict with an assumption that the crack begins to propagate at the peak load registered by the oscilloscope. Such a conclusion, of course, rests on the tacit assumption that the oscilloscope trace is in fact an accurate time record of the contact force between the specimen and striking tup.

To illustrate the importance of hammer-tup mass and stiffness, the finite-element programme was executed for the model in Figure 2 with a lumped mass of 45.4 kg attached at G. The lumped mass was given an initial velocity corresponding to a free-fall drop of 0.152 m. It is not clear from a reading of the Madison-Irwin paper that these values for mass and drop height are appropriate for the oscilloscope trace in Figure 1, but the paper does imply that these are probably minimum values for the test programme. The solid circles in Figure 3 indicate computed values of K_I for this representation. These stress-intensity factors are unrealistically high as might be anticipated from the use of such a model in a time span in which non-rigid motions of the hammer are likely to be significant.

CONCLUSIONS

The writers do not claim a successful prediction of time-dependent stress-intensity factors for the impact test that has been discussed. Rather the analyses which have been presented call attention to the danger of ignoring specimen inertia or of an oversimplified model of the hammer. It is the writers' opinion that for relatively high-velocity impact, involving a hammer and specimen of similar materials, an analytical model which accounts for the elastodynamics of the specimen and at least that portion of the hammer tup between the specimen and the load transducer is required.

ACKNOWLEDGEMENTS

This research was supported by the Air Force Armament Development and Test Centre at Eglin Air Force Base under Contract F08635-76-C-0136.

REFERENCES

1. MADISON, R. B. and IRWIN, G. R., Proc. ASCE, 100, 1974, 1331.
2. TIMOSHENKO, S., "Vibration Problems in Engineering" (3rd ed.), D. Van Nostrand, N.J., 1955, 331.
3. GROSS, B. and SRAWLEY, J. E., NASA-TN D-3092, 1965.
4. ABERSON, J. A. and ANDERSON, J. M., NASA-TMX-2893, 1973, 531.
5. MORGAN, J. D., III, ANDERSON, J. M. and KING, W. W., AIAA Journal, 12, 1974, 1767.
6. ANDERSON, J. M., ABERSON, J. A. and KING, W. W., "Computational Fracture Mechanics", editors E. F. Rybicki and S. E. Benzley, ASME, N.Y., 1975, 173.

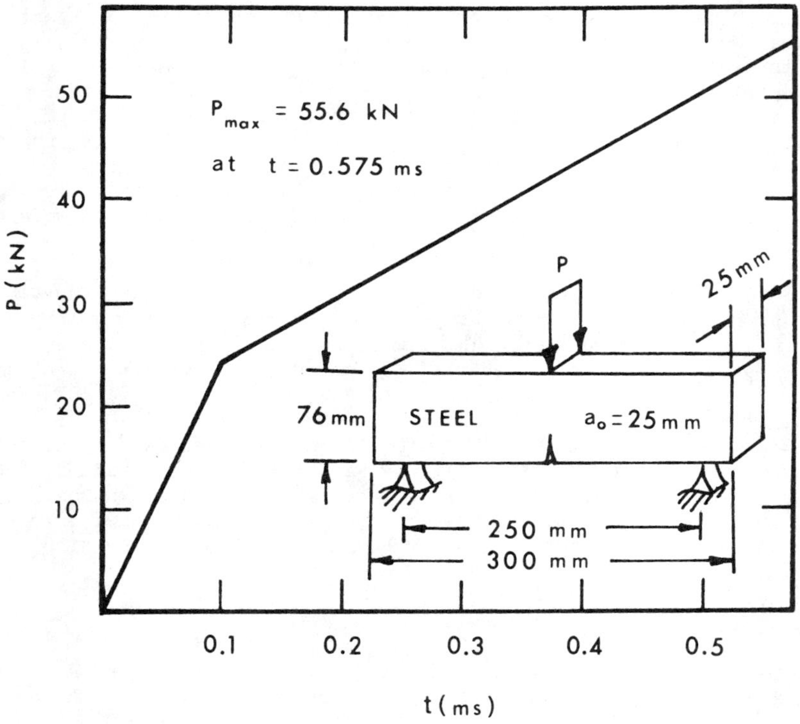

Figure 1

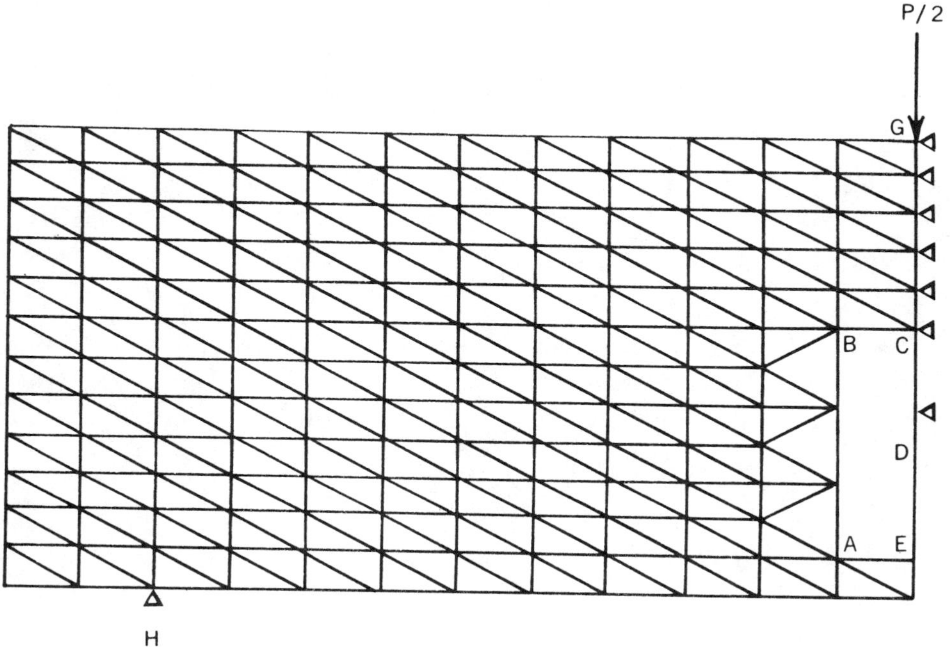

Figure 2

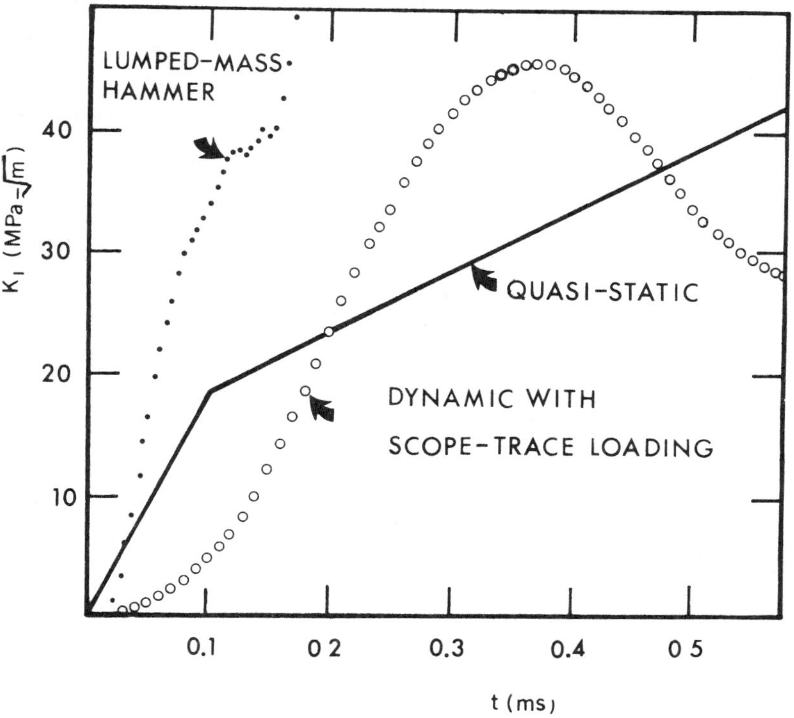

Figure 3

STRESS AND CRACK-DISPLACEMENT INTENSITY FACTORS IN ELASTODYNAMICS

H. D. Bui*

This paper deals with several remarks concerning stress-intensity factors in elastodynamics, for the steady-state and the transient problems of a moving crack with a constant velocity V. It discusses the equivalence between dynamic fracture criterions and introduces the new J-integral for moving crack. The present paper will bring out the same importance of the stress-intensity factor and the crack-displacement intensity factor, in the presentation of fracture criterions.

In the plane strain condition, the components of the displacement fields u,v, with respect to the fixed cartesian axes Oxy, can be expressed in terms of two scalar functions $\phi(x,y,t)$ and $\psi(x,y,t)$ which satisfy respectively the two-dimensional wave equations with velocities $c_L=\{(\lambda+2\mu)/\rho\}^{1/2}$ and $c_L=\{\mu/\rho\}^{1/2}$ where λ,μ are the elastic Lamé's constants and ρ is the mass density. Expressions of the displacement and the stress fields in terms of ϕ and ψ or their well known complex representations can be found in [1]. We introduce the velocity parameters:

$$\beta_1^2 = 1 - \left(\frac{V}{c_L}\right)^2 \qquad \beta_2^2 = 1 - \left(\frac{V}{c_T}\right)^2$$

THE STEADY-STATE PROBLEM

The steady-state problem of a moving crack with constant length has solutions which can be found in [1]. The load is assumed to be symmetric with respect to the crack line Ox. In the moving axes x'=x-Vt, y'=y with polar coordinates, r,θ the displacement and stress fields near the crack-tip are known. We report here only the expressions for v and σ_{yy} with a change on notations from those given in [1]:

$$v = \frac{K_I}{\mu}\left(\frac{r}{2\pi}\right)^{1/2} \mathrm{Im}\Big\{-2\beta_1(1+\beta_2^2)(\cos\theta+i\beta_1\sin\theta)^{1/2} + 4\beta_1(\cos\theta+i\beta_2\sin\theta)^{1/2}\Big\}\Big\{4\beta_1\beta_2-(1+\beta_2^2)\Big\}^{-1} \tag{1}$$

$$\sigma_{yy} = \frac{K_I}{(2\pi r)^{1/2}} \mathrm{Re}\left\{-\frac{(1+\beta_2^2)^2}{(\cos\theta+i\beta_1\sin\theta)^{1/2}} + \frac{4\beta_1\beta_2}{(\cos\theta+i\beta_2\sin\theta)^{1/2}}\right\}\Big\{4\beta_1\beta_2-(1+\beta_2^2)^2\Big\}^{-1} \tag{2}$$

---

*Electricité de France and Ecole Polytechnique, 91120 Palaiseau, France.

Equations (1) and (2) are Yoffé's work [2]. Instead of K_I, introduce the notations K_I^σ for the dynamic stress-intensity factor. It is defined as in the static case by the asymptotic expression of the normal stress $\sigma_{yy}(\theta=0)$:

$$K_I^\sigma = \lim_{r\to o} \sigma_{yy}(r,o)(2\pi r)^{1/2} \tag{3}$$

If we represent the crack-opening displacement $v(r,\pi)$ by similar formula known from the plane strain static case, we must introduce a crack-displacement factor:

$$K_I^V = \lim_{r\to o} \frac{2\mu}{\varkappa+1}\left(\frac{2\pi}{r}\right)^{1/2} v(r,\pi) \tag{4}$$

where $\varkappa=(\lambda+3\mu)/(\lambda+\mu)=3-4\nu$ (ν = Poisson ratio). From (1)-(4) it results:

$$K_I^V / K_I^\sigma = \frac{4}{\varkappa+1} \; \frac{\beta_1(1-\beta_2^2)}{4\beta_1\beta_2-(1+\beta_2^2)^2} \tag{5}$$

This ratio varies monotonically from unit, at V=0, to infinity at the Rayleigh velocity c_R (defined by the vanishing of the denominator of (5)).

A TRANSIENT PROBLEM

Let us consider the particular problem of a small cut which extends symmetrically with the velocity V. The crack tip's coordinates in the fixed axes are x=Vt, y=0 (and x=-Vt, y=0). The crack is subject to opposite normal impulses $\sigma_{yy}=\delta(x)\delta(t)$. This problem is solved by Afanasev and Cherepanov [3] who gave the stress-intensity factor, as defined by equation (3) in the following form:

$$K_I^\sigma(t) = -S(1/V)V^{1/2}(V^{-2}-c_L^{-2})^{-1/2}(c_T^{-2}-c_L^{-2})^{-1}\,\frac{1}{2t^{3/2}\pi^{1/2}} \tag{6}$$

where:

$$S(\tau) = (c_T^{-2}-2\tau^2)^2 + 4\,\tau^2(c_L^{-2}-\tau^2)^{1/2}(c_T^{-2}-\tau^2)^{1/2}$$

Here, the Rayleigh velocity is defined by the root of the equation S(1/V)=0. Afanasev and Cherepanov did not calculate the crack displacement factor. But from their solution, we may obtain the latter quantity as:

$$K_I^V(t) = \frac{2}{\varkappa+1}\,V^{-1/2}c_T^{-2}(c_T^{-2}-c_L^{-2})^{-1}\,\frac{1}{t^{3/2}\pi^{1/2}} \tag{7}$$

From (6) and (7) the ratio K_I^V/K_I^σ is found to be the same as in the steady-state case. This agrees with the general result obtained by Achenbach and Bazant [4] who stated that the near-tip fields are of the same form for steady-state and transient crack-propagations. The result (5) can be found in many works. Here, we only introduce new notations and terminologies in order to make more distinction between the two intensity factors. Remark that for $V=c_R$, the stress-intensity factor (6) vanishes, while the crack-displacement factor (7) does not. So, the latter factor has some importance on the characterization of the crack-tip just when the notion of stress-intensity factor falls.

SOME REMARKS ON FRACTURE CRITERIONS IN SYMMETRIC LOADING

In the static case, the usual criterions can be reduced to the K_I criterion. Whatever is chosen as a criterion, the stress criterion, the COD criterion, the G-theory and the J-integral are equivalent. Let us state that the equivalence between two criterions means that the relationship between their critical values involves only the material's properties, not the velocity dependence. For example, the equation $J_{Ic}=(1-\nu^2)K_{Ic}^2/E$ (E: Young modulus) establishes the equivalence between the J-integral and the K_I criterion, for opening mode in static case. In dynamic crack-propagation, we can see that the K_I^σ criterion has no equivalence with the usual other criterions. As a first example, suppose that the critical value K_{Ic}^σ is a material constant independent of the velocity, then the value K_{Ic}^V computed from K_{Ic}^σ through equation (5) is not, and vice versa. Thus, the criterions K_I^σ and K_I^V are not equivalent in the sense stated above.

Let us consider other parameters.
The G-parameter is defined by the Griffith's energy balance (See Erdogan [5] and Achenbach [6]):

$$G\delta a \equiv -\delta W_{elas} + \delta W_F - \delta W_{kin} = 2D\delta a \tag{8}$$

where $\delta a=V\delta t$, δW_{elas} is the elastic energy variation, δW_F is the work done by given external forces, δW_{kin} the kinetic energy variation, and D the dissipative energy rate in fracture. For brittle material: $D=\gamma_s$ (Specific surface energy). The computation of the left side of equation (8) requires the knowledge of the dynamic fields in the whole body. If G has a representation by mean of some path-independent integral, the computation would be possible with the near-tip fields. If not, the direct computation of G is not easy. Nevertheless, the relationship between G and K_I^σ is expected to involve the velocity dependence.

The G'-parameter is the crack-closure energy (See [5] and [6]):

$$G' = \lim_{\delta a\to o} \frac{1}{2\delta a} \int_{crack} \sigma_{yy}(a)\ v(a+\delta a)\ ds \tag{9}$$

This parameter is obviously easy to compute once the near-tip fields (1) and (2) are known. It may be obtained as:

$$G'(t) = \frac{1-\nu^2}{E} K_I^\sigma(t) K_I^V(t) \tag{10}$$

With different notations, equation (10) can be found in Atkinson and Eshelby [7], in Freund [8] and also in [4]. The interpretation of (9) as the dissipative energy rate is given in [5]. (See in the latter reference the comparison between G' and the strain energy release rate $G_e = -dW_{elas}/da$; in dynamic crack-propagation $G' \neq G_e$). We remark again that the fracture criterions based upon G,G' or any other parameter derived by the energy's consideration are not equivalent to the stress criterion, due to the velocity dependence. For example, if the critical value G'_c is independent of the velocity, the value K^{σ}_{Ic} computed from G'_c through equation (1) is not.†

In what follows, we consider another parameter given by a path-independent integral.

THE J-INTEGRAL FOR MOVING CRACK

Let us consider the fixed axes Ox_i ($x_1=x$, $x_2=y$, $u_1=u$, $u_2=v$). The conservation law given by Fletcher [10] is:

$$\frac{\partial}{\partial t}(\rho \dot{u}_j u_{j,i}) + \frac{\partial}{\partial x_k}\left\{-u_{j,i}\sigma_{kj} + \left(W - \frac{1}{2}\rho \dot{u}_h \dot{u}_h\right)\delta_{ik}\right\} = 0 \qquad (11)$$

where W is the elastic energy density, $\dot{u}_i = \partial u_i/\partial t$. Consider a contour Γ_V joining two points on opposite sides of the crack's surface while going around the tip, and moving with the velocity V, and let A(Γ) be the area within the contour Γ_V. The Rice's J-integral [11] is extended to moving crack in transient loading as follows:

$$J = \int_{\Gamma_V}\left\{Wn_1 - \sigma_{jk}n_k u_{j,1} - \frac{1}{2}\rho \dot{u}_h \dot{u}_h n_1\right\}ds + \frac{d}{dt}\int_{A(\Gamma)} \rho \dot{u}_j u_{j,1}\, d\omega$$

$$- \int_{\partial A} \rho \dot{u}_j u_{j,1} V n_1 \, ds \qquad (12)$$

where n_i is the unit outward normal to the contour and d/dt is the time derivative of integral over moving domain A(Γ). The J-integral is not a line integral, due to the second term. However, it results from (11) that J is independent of the path. The value of the J-integral may be obtained by a contour flattened on the crack line. It is easy to see that only the second singular term in the first integral contributes to J. Thus, it is sufficient to consider the near-tip fields (1) and (2) for the computation. For the steady-state case, we find a very simple formula:

$$J = \frac{1-\nu^2}{E} K^{\sigma}_I K^{V}_I \qquad (13)$$

†In fact, for some material, the experimental value of K_{Ic} depends on the velocity [9]. This raises the question as to the validity of the K_I criterion in such a case. Perhaps, a better choice would be some parameter X such that the theoretical ratio X/K_I multiplied by $K_{Ic}(V)$ has a nearly constant value.

Equation (13) is the generalization of that obtained by Rice for his J-integral in the static opening mode. It has the same form as equation (10). Consequently, there is equivalence between the G' and the J criterions. The result (13) is exactly the flux of energy into the crack tip, as discussed in [6], [7] and [8].

It should be noted that equation (13) can be extended to transient crack-propagation by the use of the near-tip fields obtained by Achenbach and Bazant. Another proof of the above extension to transient loading can be found in [12].

REFERENCES

1. SIH, G. C., "Handbook of Stress Intensity Factors", Lehigh University, Bethlehem, 1973.
2. YOFFE, E. H., Phil. Mag. Vol. 42, 7, 1951, 739.
3. AFANASEV, E. E. and CHEREPANOV, G., PMM 37, 4, 1973, 618.
4. ACHENBACH, J. D. and BAZANT, Z. P., J. Appl. Mech., 42, 1975, 183.
5. ERDOGAN, F., Fracture II, edited by Liebowitz, Acad. Press, 1968, 497.
6. ACHENBACH, J. D., Mechanics Today, ed. Nemat-Nasser, Vol. 1, Pergamon Press, 1974, 1.
7. ATKINSON, C. and ESHELBY, J. D., Int. J. Fracture, 4, 1968, 3.
8. FREUND, L. B., J. of Elasticity, Vol. 2, 1972, 341.
9. KRAFFT, J. M. and SULLIVAN, A. M., ASM Trans. 56, 1962, 847.
10. FLETCHER, D. C., Arch. Rat. Mech. Anal. 60, 4, 1976, 329.
11. RICE, J. R., Fracture Vol. II, ed. Liebowitz, Acad. Press, 1968, 191.
12. BUI, H. D., "Mécanique de la rupture fragile", Séminaire de Mécanique theorique et appliquée, Université Paris-Nord, 1976.

Fracture 1977, Volume 3, ICF4, Waterloo, Canada, June 19 - 24, 1977

ELASTODYNAMIC EFFECTS ON CRACK BRANCHING

J. D. Achenbach*

INTRODUCTION

If a homogeneous, isotropic, linearly elastic solid containing a plane crack is loaded so that the analytically computed singular parts of the near-tip stresses are symmetric relative to the plane of the crack, one might perhaps expect the crack to propagate in its own plane, when the pertinent stress intensity factor reaches a critical value. Experimental evidence often shows, however, the phenomena of skew crack propagation and crack bifurcation, especially for rapidly propagating cracks. Although it has been suggested by several authors that elastodynamic effects play an important role in crack branching, analytical investigations have only recently become available for antiplane strain, see references [1] and [2]. The computation of the elastodynamic fields has presented the principal obstacle.

The general nature of elastodynamic near-tip fields for the case that the tip of a crack propagates rapidly along a rather arbitrary but smooth trajectory in a two-dimensional geometry, was discussed by Achenbach and Bazant [3]. Let a crack be propagating in its own plane with speed v(t), and let a system of moving polar coordinates be affixed to the moving crack tip. For symmetric opening up of the crack (Mode I) we have in the vicinity of the crack tip.

$$\tau_\theta \sim \frac{1}{(2\pi)^{1/2}} \frac{1}{r^{1/2}} k_I(t,v)\ T_\theta^I(\theta,v) \tag{1}$$

In equation (1), $T_\theta^I(0,v) = 1$, and $k_I(t,v)$ is the elastodynamic stress intensity factor. The function $T_\theta^I(\theta,v)$, which is complicated, is shown in Figure 1. It is of note that the maximum value of $T_\theta^I(\theta,v)$ bifurcates out of the plane $\theta = 0$ (the plane of crack propagation) as v(t) increases beyond a certain value.

The curves of Figure 1 could be used to suggest an explanation for crack bifurcation, if it is assumed that a crack tip follows the maximum value of the stress intensity factor. If this would happen, bifurcation should be expected at a crack tip speed somewhat higher than 0.6 c_T. One then would expect the crack branches to curve gradually out of the original plane, since the maximums gradually move out of $\theta = 0$. Experimental results do, however, not substantiate this explanation. They show that the experimentally observed pre-bifurcation speed is lower than 0.6 c_T, and that bifurcation happens with a specific half-angle, in between 10 and 20° Thus, the results of Figure 1 do not offer a direct explanation of crack bifurcation, and further study is necessary. Such further study is the topic of this paper.

---

*Northwestern University, Evanston, Illinois, USA.

APPROACH

In the work presented here we take the view that branching of a running crack is an instability phenomenon, and that a necessary condition for branching can be determined by comparing states prior to branching and after branching has taken place. The comparison requires expressions for the elastodynamic fields near the tips of the branches. An analytical study of skew crack propagation or crack bifurcation thus consists of two parts. In the first part an expression is derived for the elastodynamic stress intensity factor for the pertinent geometry. In the second part a necessary condition for the particular type of crack propagation is established on the basis of the fracture criterion of the balance of rates of energies.

Details can best be explained by the relatively simple case of deformation in antiplane strain. Let us consider a semi-infinite crack propagating at velocity v(t). The near-tip elastodynamic stress is of the form

$$\tau_{\theta z} \sim \frac{1}{(2\pi)^{1/2}} \frac{1}{r^{1/2}} k_{III}(t,v) T_{\theta z}^{III}(\theta,v) \qquad (2)$$

where $T_{\theta z}^{III}(0,v) = 1$. For a semi-infinite crack we have

$$k_{III}(t,v) = (1 - v/c_T)^{1/2} K_{III}(t) \qquad (3)$$

where $K_{III}(t)$ is the stress intensity factor for the corresponding quasi-static problem. Equation (3) is also valid for a crack of finite length, but only for very small times after crack propagation has started. It is noted that $k_{III}(t,v) \to 0$ as $v \to c_T$, i.e., as the speed of crack propagation approaches the velocity of transverse waves.

A propagating crack tip acts as an energy sink. It is quite simple to compute the flux of energy into the crack tip. For Mode III fracture the result is

$$F = \frac{v}{2\mu\left[1 - (v/c_T)^2\right]^{1/2}} \left[k_{III}(t,v)\right]^2 \qquad (4)$$

The energy release rate G and the flux of energy into the crack tip, F, are related by F = Gv. The balance of rates of energies provides the following necessary condition for fracture

$$F = 2\,\Gamma\,v \qquad (5)$$

where Γ is the specific energy of crack extension, i.e., the energy required to produce one unit of fracture surface. Equation (5) is not only a necessary condition for fracture, but it also provides an equation for the computation of v.

At time $t = t_b$ the crack branches. This process is thought of as the arrest of the primary crack, instantaneously followed by the emanation of the branch or branches. If the branches propagate with velocities $v < c_T$, they propagate into fields that have already received signals from the arrest of the primary crack, since the latter propagate with velocity c_T.

The singular part of the stress field radiating from the arresting crack is obtained from equation (2) by setting $v \equiv 0$:

$$\tau_{\theta z} \sim \frac{1}{(2\pi)^{1/2}} \frac{1}{r^{1/2}} \cos\left(\frac{\theta}{2}\right) K_{III}\left(t_b\right) H\left(t-t_b\right) \tag{6}$$

This stress field must be removed from the surfaces of the branches, to render the branches free of shear stresses. Note that here we are interested only in very small times after branching, so that additional terms in equation (6) do not enter.

The computation of the stress-singularities at the tips of the branches is complicated. If the loading conditions are of a special type, the elastodynamic fields for skew crack propagation or crack bifurcation of a semi-infinite crack are, however, self-similar. These fields can then be analyzed in a relatively simple manner. Elastodynamic fields that are not self-similar can subsequently be obtained by approximate superposition considerations, see reference [2].

SOME RESULTS

The elastodynamic field which is generated when a branch emanates assymmetrically from the tip of a stationary semi-infinite crack, when the surfaces of the crack are subjected to shear tractions $\tau_{\theta z} = -\tau_o H(t)$ is first investigated. The shear tractions give rise to plane waves and a cylindrical diffracted wave centred at the original crack tip. The semi-infinite crack propagates at an angle $\varkappa\pi$ and with velocity v, where $v < c_T$, at the instant that the surface tractions are applied. At time $t > 0$ the crack tip is located at point D, see Figure 2. For this problem the particle velocity is self-similar. For a similar problem, details can be found in reference [1]. Relative to the system of moving coordinates shown in Figure 2 we find near the tip

$$\tau_{\theta z} \sim \frac{1}{(2\pi)^{1/2}} \frac{1}{r^{1/2}} k_{III}(t,v,\varkappa)\, T^{III}_{\theta z}(\theta,v) \tag{7}$$

where

$$k_{III}(t,v,\varkappa) = 2\,\pi^{1/2}\left(1 - \frac{v^2}{c_T^2}\right)^{1/4}\left(\frac{t}{v}\right)^{1/2} K(\varkappa) \tag{8}$$

The function $K(\varkappa)$ follows from equation (3.8) of reference [1] by setting $\alpha = 0$ and $W_o = \tau_o c_T/\mu$.

The results obtained above can now be used to analyze the conditions for the emanation of a single branch from a running crack. After branching of the *running* crack, the shear stress near the branch tip is of the general form

$$\tau^*_{\theta z} \sim \frac{1}{(2\pi)^{1/2}} \frac{1}{r^{1/2}} k^*_{III}(\bar{t},v)\, T^{III}_{\theta z}(\theta,v) \tag{9}$$

where $\bar{t} = t - t_b$. We have found that the near tip stress field for instantaneous skew crack propagation upon the application of equal and uniform antiplane shear tractions to the two semi-infinite surfaces of a stationary crack is given by equation (7). If for that case the crack does not branch, nor propagate in its own plane, the near tip stress field is

$$\tau_{\theta z} = \frac{B}{(2\pi)^{1/2}} \frac{t^{1/2}}{r^{1/2}} \cos\left(\frac{\theta}{2}\right) H(t), \text{ where } B = 2\left(\frac{2c_T}{\pi}\right)^{1/2} \tau_o \tag{10}$$

Clearly, the result (7) can be regarded as being the consequence of removing stresses of the form (10) from the crack branches. Thus, we now have a known stress (7) due to the removal of the known distribution of surface tractions (10), and an unknown stress (9) due to the removal of the known distribution (6). Apart from constants the difference between equations (6) and (10) is, however, only in the time dependence; equation (6) contains a step time dependence, while in equation (10) the dependence on time is as $t^{1/2}$. These results then suggest that at least for small times k^*_{III} and k_{III} are related by superposition considerations as

$$k_{III}(t) = \frac{B}{K_{III}(t_b)} \int_o^t k^*_{III}(t-s)\, d(s^{1/2}) \tag{11}$$

This equation can easily be solved for k^*_{III} as

$$k^*_{III} = \left(1 - \frac{v^2}{c_T^2}\right)^{1/4} \left(\frac{1}{2c_T v}\right)^{1/2} \frac{\pi K(\varkappa)}{\tau_o} K_{III}\left(t_b\right) \tag{12}$$

The corresponding flux of energy into the crack tip can be computed by employing equation (4). The noteworthy result is that the rate of energy flux into a propagating crack tip shows a maximum at $\varkappa = 0$ only for values of v/c_T which are smaller than approximately $v/c_T = 0.27$. Apparently the rate of energy flux into a crack tip can be higher for skew crack propagation than for a crack propagating in its own plane.

The tendency towards skew crack propagation can be examined on the basis of the balance of rates of energies. This fracture criterion is stated by equation (5). For essentially brittle fracture Γ is the specific surface energy, which is independent of $\varkappa$. In a plot of F vs. $\varkappa$ and $2\Gamma v$ vs. $\varkappa$ for specific v/c_T, the term $2\Gamma v$ is then represented by a horizontal line. In accordance with the balance of rates of energies, the values of v and $\varkappa$ are determined by a point of intersection of the curves for F and $2\Gamma v$. Since both v and $\varkappa$ are as yet unknown an additional condition is required. Such an additional condition is that only an intersection where $2\Gamma v$ is tangent to F (i.e., F is a maximum with respect to $\varkappa$) defines a case of stable crack propagation relative to variations of $\varkappa$. Thus, in Figure 3, the maximums of F with respect to $\varkappa$ have been replotted versus v/c_T, and values of $\varkappa$ at which the maximums of F are reached have been indicated. In this figure $2\Gamma v$ is a straight line through the origin. The intersection of $2\Gamma v$ and F defines a case of crack propagation and the pertinent values of v and $\varkappa$ follow from the point of intersection in Figure 3. The foregoing discussion defines Γ as the principal quantity controlling skew crack propagation. For large enough Γ, $2\Gamma v$ is tangential to F at $\varkappa = 0$, and thus v/c_T will be relatively small and the crack will propagate in its own plane. For small values of Γ the relevant intersection is at $\varkappa > 0$, i.e. skew crack propagation can be expected.

An analogous analysis for crack bifurcation in antiplane strain was presented in reference [2].

THE IN-PLANE PROBLEM

Computations of deformations in antiplane strain (Mode III), though important for geophysical situations, are of minor practical significance for engineering problems. Solutions of antiplane problems do, however, frequently suggest the proper steps for the attack on in-plane problems. There are, however, some important differences in the basic fracture mechanics of the antiplane and inplane cases, and these should be kept in mind. For example, for inplane deformation the branches of a primary crack are subjected to both Mode-I and Mode-II fracture conditions. Mixed fracture conditions do not occur for crack bifurcation in antiplane strain. Thus it is necessary to analyze the inplane problem separately.

The case of inplane strain is, of course, must more complicated. In the physical plane the region in which the stress field must be analyzed consists of wedge-shaped segments which are connected ahead of the propagating crack tip(s). For inplane deformations there are two wave equations governing the displacement potentials. By taking advantage of self-similarity, these wave equations can be reduced to Laplace's equations in half-planes. The solutions to these equations are, however, coupled along the real axes by conditions which stem from the conditions along the crack surfaces and along the wavefronts. The coupling conditions give rise to singular integral equations for the displacement potentials. A numerical scheme based on series expansions in terms of Chebyshev polynomials has been developed to obtain numerical solutions. Results are forthcoming.

ACKNOWLEDGEMENTS

The work reported here was carried out in the course of research sponsored by the Office of Naval Research under Contract ONR N00014-76-C-0063.

REFERENCES

1. ACHENBACH, J. D. and VARATHARAJULU, V. K., Quart. Appl. Match, XXXII, 1974, 123.
2. ACHENBACH, J. D., Int. J. Solids and Structures, 11, 1975, 1301.
3. ACHENBACH, J. D. and BAZANT, Z. P., J. Appl. Mech., 42, 1975, 183.

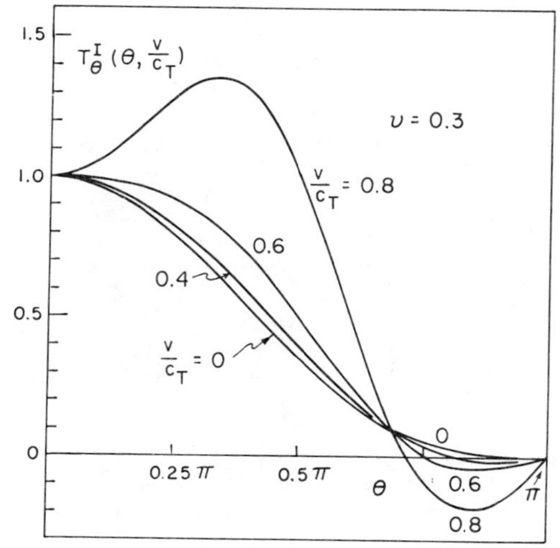

Figure 1 Function T^I_θ (θ, v) versus θ for Various Values of v/c_T.

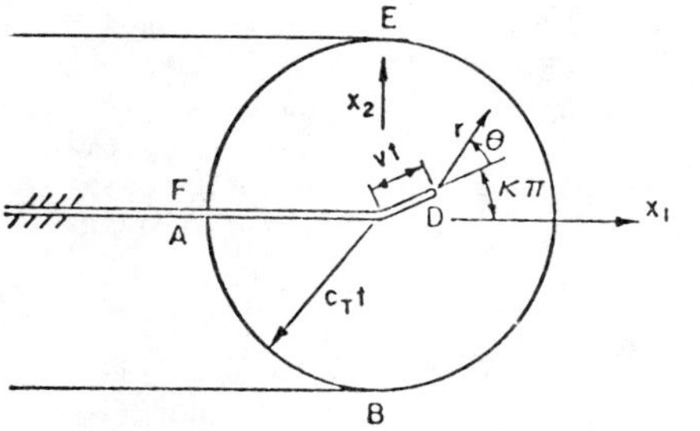

Figure 2 Pattern of Wavefronts and Position of Crack Tip for Crack Branching Under the Influence of a Suddenly Applied Antiplane Shear Traction

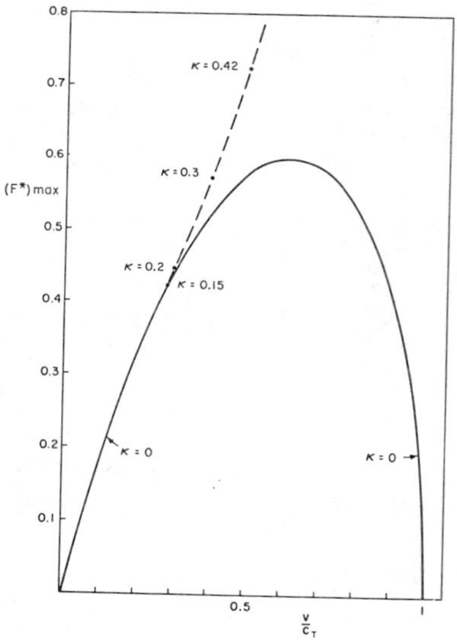

Figure 3 Maximums of F with Respect to κ, Plotted vs. v/c_T; $F^* = 4\mu F/c_T[K_{III}(t_b)]^2$

Fracture 1977, Volume 3, ICF4, Waterloo, Canada, June 19 - 24, 1977

CRACK-TIP STRESS ANALYSIS FROM FIELD VALUES OF THE DISPLACEMENTS USING COMPLEMENTARY ENERGY

J. L. Swedlow[1], M. E. Karabin, Jr.[1] and G. E. Maddux[2]

A variety of techniques is now available for crack-tip stress analysis, and the need for further development along these lines may not at first be evident. For some circumstances, however, none of the present methods is especially applicable so that we have been moved to develop another approach. Concern here derives from having an experimental determination of displacements at points surrounding a crack's tip at the outset; the objective is to find the corresponding stress intensity factor(s). Such measurements are conveniently made at modest distances (a few cm) from the crack's tip; in the tip's immediate vicinity, however, experimental accuracy may be impeded by localized plastic flow, surface roughening, and dimpling. Objections to use of established analytical/numerical methods arise owing to indeterminacy of loads or overall structural complexity. We seek therefore to determine stress intensity factor(s) from *in situ* displacement data at points along a path or surface that surrounds the crack's tip but interior to the structure in which the crack is found.

To this end, *in situ* displacement data are taken to pertain to the difference between two excitation levels, one nominally at rest and the other at load. This pairing is required by whatever experimental method is employed; the analysis proceeds in terms of the net difference and gives either increment(s) in stress intensity factor(s) or total value(s) where the at-rest state is wholly unloaded.

The primary ingredients required are the theorem of minimum complementary energy and a stress function pertinent to a crack. The theorem states that, of all equilibrated stress fields which satisfy prescribed traction boundary conditions (here, the crack's flanks are stress free), the "actual" one is distinguished by a stationary (here, minimum) value of the complementary energy V*, where

$$V^* = \int_D W^*\left(\sigma_{ij}\right) dD - \int_{S_u} \tilde{u}_i t_i \, dS \qquad (1)$$

In (1), integration proceeds over the domain D and that part S_u of its total boundary S where displacements are prescribed; $\tilde{u}_i$ are the prescribed displacements and t_i the corresponding tractions, σ_{ij} is the stress tensor, and W* is the complementary energy density given by

$$W^*\left(\sigma_{ij}\right) = \left[-\nu\left(\sigma_{kk}\right)^2 + (1 + \nu)\sigma_{ij}\sigma_{ji}\right]/2E \qquad (2)$$

---

[1]Carnegie-Mellon University, Pittsburgh, Pennsylvania, USA.
[2]Flight Dynamics Laboratory, Wright-Patterson Air Force Base, Ohio, USA.

for isotropic, Hookean material. (The usual indicial conventions are employed.) Anisotropic, elastic material is treated by suitable alteration of (2) and (3), below. While a fuller account of this theorem may be found in a good text, e.g., [1], our purpose is satisfied by the variational requirement $\delta V^* = 0$.

The appropriate stress function for isotropic planar bodies is due to Williams [2] and is written as

$$\begin{aligned}\chi(r,\theta) = \sum_{m=1}\Big\{&(-1)^{m-1}a_{2m-1}r^{m+1/2}\left[-\cos(m-3/2)\theta + \frac{m-3/2}{m+1/2}\cos(m+1/2)\theta\right]\\ &+ (-1)^{m-1}b_{2m-1}r^{m+1/2}[\sin(m-3/2)\theta - \sin(m+1/2)\theta]\\ &+ (-1)^{m}a_{2m}r^{m+1}[-\cos(m-1)\theta + \cos(m+1)\theta]\\ &+ (-1)^{m}b_{2m}r^{m+1}\left[-\sin(m-1)\theta + \frac{m-1}{m+1}\sin(m+1)\theta\right]\Big\}\end{aligned} \qquad (3)$$

where (r,θ) are coordinates centred at the crack's tip in the usual manner. $\chi(r,\theta)$ satisfies equilibrium by definition and the traction conditions on the crack's flanks by construction. The stresses are

$$\begin{aligned}&\sigma_{11} \to \sigma_r = (1/r)\chi_{,r} + (1/r^2)\chi_{,\theta\theta} && \sigma_{22} \to \sigma_\theta = \chi_{,rr}\\ &\sigma_{12} \to \tau_{r\theta} = -\left[(1/r)\chi_{,\theta}\right]_{,r} && \sigma_{23} \to \tau_{\theta z} = 0\\ &\sigma_{13} \to \tau_{rz} = 0 && \sigma_{33} \to \sigma_z\end{aligned} \qquad (4)$$

and σ_z is either null or given by $\nu(\sigma_r + \sigma_\theta)$ in isotropic plane stress or plane strain, respectively. The stresses of interest are assembled in the form

$$\begin{Bmatrix}\sigma_r\\ \sigma_\theta\\ \tau_{r\theta}\end{Bmatrix} = \{\sigma\} = [S(r,\theta)]\{a\}$$

where [S] is a matrix whose entries depend solely on position as found by inserting (3) into (4), and

$$\{a\}^T = \left\{a_1 b_1 a_2 b_2 a_3 b_3 a_4 b_4 \ldots a_{2m-1} b_{2m-1} a_{2m} b_{2m}\right\}$$

Thus the entries in {a} appear in sets of four so that truncation of (3) at $m = M$ gives $4M$ coefficients to be found.

Using G for the shear modulus and κ for the usual function of Poisson's ratio in planar isotropic elasticity, the compliance matrix [C] is written

$$8G[C] = \begin{bmatrix} \kappa + 1 & \kappa - 3 & 0 \\ \kappa - 3 & \kappa + 1 & 0 \\ 0 & 0 & 8 \end{bmatrix}$$

and (2) is put into the familiar quadratic form

$$W^* = (1/2)\{a\}^T[S]^T[C][S]\{a\}$$

Typically, the displacement data $\tilde{u}_i$ will be resolved in orthogonal directions pertinent to the overall structure, or to the crack's position. Tractions t_i must be resolved in the same manner as $\tilde{u}_i$. Denoting these directions as (ξ,η) we write

$$\begin{Bmatrix} t_\xi \\ t_\eta \end{Bmatrix} = \{t\} = \left[T(r,\theta) \Big|_{S_u} \right]\{a\} \text{ and } \begin{Bmatrix} \tilde{u}_\xi \\ \tilde{u}_\eta \end{Bmatrix} = \{\tilde{u}\}$$

Note that $\{t\}$ is computed from (3) and (4) via Cauchy's formula and, perhaps, Mohr's circle, and that [T] is evaluated along S_u only.

With the foregoing representations (1) becomes

$$V^* = (1/2)\{a\}^T[X]\{a\} - \{Y\}^T\{a\} \tag{5}$$

in which

$$[X] = \int_D [S]^T[C][S]dD, \quad \{Y\}^T = \int_{S_u} \{\tilde{u}\}^T[T]dS \tag{6}$$

Minimization of V* leads to

$$[X]\{a\} = \{Y\} \tag{7}$$

as the algebraic problem statement†. Note that the problem's size is 4M irrespective of the number of data points on S_u. We must, however, carry through the quadratures in (6).

Code for this purpose has been devised. For a number of test problems considered, it was observed that $\{\tilde{u}\}$ does not vary rapidly along S_u but that the entries in [T] can. Hence a limited amount of data in $\{\tilde{u}\}$ is interpolated so that {Y} is determined by a large number of points using Simpson's one-third rule. To determine [X], however, a more elaborate tactic is needed. The interval $-\pi < \theta < \pi$ is divided into a dozen sectors, and Gaussian quadrature (using ten points radially and circumferentially) is employed in each sector. Size of the sectors is not necessarily uniform; we have dealt with rectangular paths S_u and let the corners and mid-points determine actual positions, as in Figure 1.

Accuracy of the quadrature used to find {Y} is essential so that any rigid motion in $\{\tilde{u}\}$ does not affect the result. We have observed that, by adding an arbitrary (and large) rigid motion to a given set of "active"

†This is the complement to the approach outlined in [3] for which traction boundary conditions on S are appropriate.

data $\{\tilde{u}\}$, a small change in $\{Y\}$ could be produced. For this reason, an intuitive scheme has been devised whereby this effect is made negligible. Letting $\xi = r \cos \theta$, $\eta = r \sin \theta$, we construct a new set of displacements,

$$\tilde{u}'_\xi = \tilde{u}_\xi + u_o - \omega\eta$$

$$\tilde{u}'_\eta = \tilde{u}_\eta + v_o + \omega\xi$$

and the quantity

$$\tilde{v}' = -\tilde{u}'_\xi \sin \theta + \tilde{u}'_\eta \cos \theta$$

$$= \tilde{v} - u_o \sin \theta + v_o \cos \theta + \omega r$$

Summing the data along S_u and setting $\Sigma\tilde{u}_\xi = \Sigma\tilde{u}_\eta = \Sigma\tilde{v} = 0$ to approximate the notion of no crack-tip movement gives three equations whose solution is an estimate of u_o, v_o, and ω. The original $\{\tilde{u}\}$ is then recovered. It was found that, where rigid motion is large relative to the active displacements, this scheme virtually negates the rigid motion. For rigid motion of the same magnitude as the active displacements, however, the scheme loses some accuracy. Owing to the care involved in quadrature along S_u, the net effect is negligible in terms of the solution $\{a\}$. Hence, while we cannot rigorously account for rigid motion, we have found means for negating its influence.

Some of our test problems give a sense of performance. We have determined, first, that S_u is best taken as approximately equilateral and, second, that the crack's tip should be near the middle of the domain. Under these conditions, two additional test problems were solved. In the first, S_u was rectangular and oriented such that its edges were parallel and normal to the crack's plane; $\{\tilde{u}\}$ was determined by computing displacements from the series [3], having assumed that

$$a_i = -1.0 \times 10^{1-i} \quad i = 1,8 \tag{8}$$

and that all other coefficients are null. In the second problem, a similar path S_u was rotated clockwise 45 deg and the non-zero coefficients were taken to be

$$a_i = -1.0 \times 10^{1-i} \quad b_i = 1.0 \times 10^{1-i} \quad i = 1,8 \tag{9}$$

(except that† $b_2 = 0$), to obtain a more complicated set of displacements. Using these two data sets at 121 points along S_u, as input, solutions to (7) were obtained as shown in Tables 1 and 2. These coefficients were then used to compute displacements and stresses on S_u; agreement with values derived from (8) and (9) was within 0.013 percent of their respective maximum values.

---

†Close examination of the b_2 term in (1) shows that it does not affect stress. In fact, b_2 is proportional to a rigid rotation, just as a_o and b_o - were they to appear - denote rigid translations. The code automatically sets b_2 to zero.

Other problems used to test the procedure were taken from earlier finite element results (in the elastic range) where stress intensity was known to within a few percent. That is, although highly reliable displacement data could be obtained, there is a modest uncertainty in the value of stress intensity based on the finite element data itself. Nonetheless the present method determined values of stress intensity within the uncertainty band (nominally 5 percent). It should be noted, however, that we found it advisable to interpolate the initial values of $\{\tilde{u}\}$ to provide data at 120 to 160 nearly equispaced points on S_u, and that $M = 7$ was required to achieve this result. The associated CPU time (1108, Exec 2) was about twenty seconds per case, with only modest storage requirements. Thus the procedure appears to be workable and to meet the needs stated at the outset. Moreover, this procedure provides vastly more resolution in the stress and displacement variation near the crack's tip than is usually obtained from finite element analyses.

With this procedure in hand, it is useful to employ speckle photography to measure the displacement data $\{\tilde{u}\}$. This procedure has been developed [4,5] specifically to make *in situ* observations and may briefly be outlined. The diffuse surface of an object and its associated speckle pattern caused by the illuminating laser is imaged onto a sensitized film plate. The result is a recording of a random variation of bright and dark spots whose size depend upon the characteristics of the optical recording system and are usually on the order of a thousand spots per millimeter. If the object is deformed, a new random variation can be recorded which, if superimposed upon the variation corresponding to the undeformed condition will result in a set of "speckle pairs" that can be related to the vector displacement field describing the deformation. If an unexpanded beam from a laser is passed through the developed image in the plate, a circular halo of light with a pattern of parallel fringes similar to Young's Fringes is observed. The distance between these fringes can be related to the displacement which occurred on the test specimen. By moving the laser beam around the image, a displacement field can be calculated.

At this point, the coupling of the two techniques is in progress, and additional results should be reported soon. It is clear from preliminary work, however, that the effort is highly interactive: refinement of the computational procedure has been stimulated by the character of actual displacement data, and the orientation and shape of the path followed by the interrogating laser have been adjusted to meet computational requirements. On this basis an overall procedure for extracting stress intensity value(s) from *in situ* observations is established.

ACKNOWLEDGEMENT

This work is proceeding with the support of the United States Air Force Flight Dynamics Laboratory and Carnegie-Mellon University, for which we are grateful.

REFERENCES

1. FUNG, Y. C., "Foundations of Solid Mechanics", Prentice-Hall, Englewood Cliffs, 1965, Chapter 10.
2. WILLIAMS, M. L., Journal of Applied Mechanics, 24, 1957, 109-114.

3. EWING, P. D., SWEDLOW, J. L. and WILLIAMS, J. G., International Journal of Fracture, 12, 1976, 85-93.
4. ADAMS, F. D. and MADDUX, G. E., "On Speckle Diffraction Interferometry for Measuring Whole Field Displacements and Strains", AFFDL TR-73-123.
5. ADAMS, F. D. and MADDUX, G. E., "Dual Plate Speckle Photography", AFFDL TR-75-57-FBR.

Table 1 Known Coefficients and Computed Values

i	a_i (known)	a_i (computed)	b_i (known)	b_i (computed)
1	-1.0	-0.99999422	0.0	0.00000002
2	-0.1	-0.10004858	0.0	0.0
3	-0.01	-0.00995341	0.0	0.00000010
4	-0.001	-0.00100015	0.0	0.00000005
5	-0.0001	-0.00009944	0.0	0.00000014
6	-0.00001	-0.00001922	0.0	-0.00000019
7	-0.000001	0.00000893	0.0	-0.00000005
8	-0.0000001	0.00000250	0.0	0.00000001

Table 2 Known Coefficients and Computed Values (Crack at 45 Deg)

i	a_i (known)	a_i (computed)	b_i (known)	b_i (computed)
1	-1.0	-0.99998467	1.0	0.99999324
2	-0.1	-0.10008076	0.0	0.0
3	-0.01	-0.00987082	0.01	0.00999148
4	-0.001	-0.00098246	0.001	0.00100222
5	-0.0001	-0.00004756	0.0001	-0.00008448
6	-0.00001	0.00003787	0.00001	-0.00001321
7	-0.000001	-0.00003930	0.000001	-0.00002506
8	-0.0000001	-0.00000196	0.0000001	0.00002333

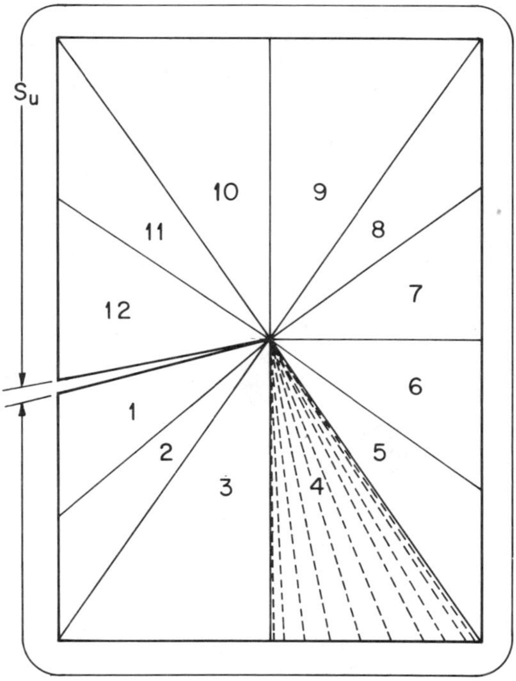

Figure 1 Geometry, Showing Data Path S_u and Interior Quadrature Regions

Fracture 1977, Volume 3, ICF4, Waterloo, Canada, June 19 - 24, 1977

ANALYTICAL DETERMINATION OF STRESS INTENSITY FACTORS OF ECCENTRIC CRACKS BY VARIATIONAL METHOD

H. Kitagawa and H. Ishikawa*

INTRODUCTION

A direct application of variational methods to the analysis of a crack has not been found, except for analysis by finite element methods. A series of researches [1 - 4] has shown that a variational method can serve as a useful tool in the following cases:

(1) Analyses of a cracked finite plate with good accuracy and rapid convergence [1, 2].

(2) Analyses of a cracked plate with complicated or unknown boundary conconditions, being combined with experimental stress analysis techniques [3, 4].

Previously the authors proposed an analytical method for determination of the stress intensity factors of a crack in a finite plate by a variational method [1, 2]. In these papers an analytical solution for a crack in an infinite plate was applied to the analysis of a crack in a finite plate. In the previous papers [1, 2], a rectangular plate with an edge crack was analysed by means of a stress function for a semi-infinite crack [5, 6] and the principle of minimum potential energy as a variational principle; accurate numerical results were obtained. However, there is a possibility that some reduction of accuracy might occur when the method is applied to such cases as internally cracked plates or mixed boundary value problems; some improvements or developments of the method may therefore be required.

In this paper, the formulation of the variational principle is improved and extended into a general form more convenient for the analysis of various crack problems. To examine the accuracy of results and the applicability of the present method, two types of rectangular plates with an internal and eccentric crack, that is, a plate pulled by uniaxial uniform tension and a plate pulled by rigid clamped ends (i.e. uniform displacement at the ends), are analysed. In the former case, the numerical results are directly compared with the results obtained by a collocation technique. The latter is analysed for the purpose of clarifying the following phenomenon with regard to the behaviour of an eccentric crack, which has become of general interest and has remained unsolved. The phenomenon is that a slightly eccentric crack in a plate grows in such a fashion that this eccentricity is decreased. In this paper, it is found that the phenomenon is well explained by the concept of stress intensity factors.

By two numerical examples described above, the availability of the new method proposed in this paper is thought to be assured.

Moreover, the new method can effectively be applied to the edge crack problems analysed in the previous paper [1].

---

*Institute of Industrial Science, University of Tokyo, Tokyo, Japan.

FORMULATION OF VARIATIONAL PRINCIPLE

Consider a mixed boundary value problem with prescribed boundary traction $\overline{T}_i$ over the boundary Γ_σ, and prescribed displacement $\overline{u}_i$ over the boundary Γ_u. For effective application of the variational principle to the analysis of mixed boundary value problems, the Hellinger-Reissner formulation [7, 8] is chosen. The functional Π_R of the Hellinger-Reissner formulation is given by

$$\Pi_R = \int_S \left[\frac{1}{2}\sigma_{ij}(u_{i,j}+u_{j,i})-B(\sigma_{ij})-\overline{F}_i u_i\right]dS-\int_{\Gamma_\sigma}\overline{T}_i u_i d\Gamma$$

$$-\int_{\Gamma_u} T_i(u_i-\overline{u}_i)d\Gamma \qquad (1)$$

where σ_{ij} is stress; u_i is displacement; $B(\sigma_{ij})$ is the complementary energy function expressed in terms of the stress; $\overline{F}_i$ is prescribed body force; S is the area of the cracked plate; and Γ is the outer boundary of S, composed of Γ_σ and Γ_u; T_i is the traction force. Equation (2) is the definition of the traction force.

$$T_i = \sigma_{ij}n_j \qquad (2)$$

where n_j is the direction cosine of the unit normal drawn outwards on Γ. The Euler equations for equation (1) are

$$\frac{1}{2}(u_{i,j}+u_{j,i}) = S_{ijkl}\sigma_{kl} \qquad (3)$$

$$\sigma_{ij,j}+\overline{F}_i = 0 \qquad (4)$$

where S_{ijkl} are the compliance coefficients. In equation (1), σ_{ij}, T_i and u_i can be assumed to be independent of each other. However, in the case of crack problems, they have to be chosen so that the Euler equations (equations (3) and (4)) are exactly satisfied, by means of the analytical solution which satisfies the stress free condition along the crack surfaces.

In this paper, the body forces $\overline{F}_i$ in equation (1) are assumed equal to zero. Applying Gauss' theorem to equation (1) and substituting equation (4) into it,

$$\Pi_R = \frac{1}{2}\int_{\Gamma_\sigma} T_i u_i d\Gamma - \frac{1}{2}\int_{\Gamma_u} T_i u_i d\Gamma - \int_{\Gamma_\sigma} u_i \overline{T}_i d\Gamma + \int_{\Gamma_u} T_i \overline{u}_i d\Gamma \qquad (5)$$

or, in matrix form,

$$\Pi_R = \frac{1}{2}\int_{\Gamma_\sigma} \underline{T}^T\underline{u}d\Gamma - \frac{1}{2}\int_{\Gamma_u} \underline{T}^T\underline{u}d\Gamma - \int_{\Gamma_\sigma} \underline{u}^T\underline{\overline{T}}d\Gamma + \int_{\Gamma_u} \underline{T}^T\underline{\overline{u}}d\Gamma \qquad (6)$$

where the superscript T is the transpose of a matrix or a vector. It is noted that this modified form of the Hellinger-Reissner formulation can be conveniently employed for various classes of crack problems, since one can avoid the area integral in S that includes the region of the high gradient of stress near a crack tip, thus avoiding the problems of accuracy associated with numerical integration in such regions. Instead, it is sufficient to carry out the line integrals on Γ using equation (5) or (6), which is expected to be comparatively simple and accurate.

PROCEDURES OF ANALYSIS

To illustrate the method, the results of analytical calculation for two types of internally and eccentrically cracked rectangular plates will be shown. One is pulled under uniform tension (shown in Figures 1 and 2) and the other is pulled under uniform displacement (at its clamped ends) (shown in Figure 3).

Since the former problem has been solved by Terada and Isida [9] with the collocation method, the accuracy of the results obtained by the variational method can be examined by comparison with their results.

Analytical Solution for an Internally Cracked Plate

In the theory of two-dimensional isotropic elasticity, the stresses (σ_x, σ_y, τ_{xy}) and the displacements (u_x, u_y) are generally expressed in terms of two analytical functions $\phi(z)$ and $\Omega(z)$ of the variable $z = x + iy$. They are

$$\sigma_x+\sigma_y = 2[\phi(z)+\overline{\phi(z)}] \tag{7}$$

$$-\sigma_x+\sigma_y+2i\tau_{xy} = 2[(\bar{z}-z)\phi'(z)-\phi(z)+\bar{\Omega}(z)] \tag{8}$$

$$2\mu(u_x+iu_y) = \eta\int\phi(z)dz-\int\Omega(\bar{z})d\bar{z}+(\bar{z}-z)\overline{\phi(z)} \tag{9}$$

where i is the imaginary number; $\eta = (3-\nu)/(1+\nu)$ for plane stress and $\eta = 3-4\nu$ for plane strain; $\mu = E/2(1+\nu)$; E and ν are Young's modulus and Poisson's ratio, respectively; $\overline{(\;)}$ or $(\;)'$ denotes conjugation or differentiation, respectively, with respect to z.

When $\phi(z)$ and $\Omega(z)$ satisfy the conditions of traction free crack surfaces, they are expressed [10, 11] as

$$\phi(z) = \frac{\sum_{n=0}^{N} A_n z^n}{\sqrt{z^2-a^2}} + \sum_{n=0}^{N} B_n z^n \qquad \Omega(z) = \frac{\sum_{n=0}^{N} A_n z^n}{\sqrt{z^2-a^2}} - \sum_{n=0}^{N} B_n z^n \tag{10}$$

where A_n and B_n are unknown complex coefficients; N is a finite integer; 2a is crack length; and the coordinates are shown in Figures 1, 2 and 3.

Moreover, $\phi(z)$ and $\Omega(z)$ have to satisfy the conditions of single-valuedness of the displacements, that is, the following equation has to be satisfied for an arbitrary closed curve L around the crack.

$$\eta\int_L\phi(z)dz-\int_L\Omega(\bar{z})d\bar{z} = 0 \tag{11}$$

Application of Variational Principle

Rectangular plates with an eccentric crack, as shown in Figures 1, 2 and 3, are analysed. On account of the symmetry of the problem with respect to the x-axis, the unknown complex coefficients (A_n and B_n) in equation (10) have the property that A_n and B_n (n = 0,1,2,...,N) are real. From the condition of single-valuedness of the displacements, equation (11), the following relation between unknown coefficients is obtained.

$$A_o + \sum_{n=1}^{N} A_{2N}\, M(2n) = 0 \qquad (12)$$

where $M(2n) = (2n-1)a^2M(2n-2)/(2n)$

$M(0) = 1$

Then, from equations (2), (7), (8), (9), (10), and (12), the traction forces $\underline{T}$ and displacements $\underline{u}$ are given in a matrix form by

$$\underline{T} = \begin{Bmatrix} T_x \\ T_y \end{Bmatrix} = \underline{R_\alpha}\, \underline{\alpha} \qquad (13)$$

$$\underline{u} = \begin{Bmatrix} u_x \\ u_y \end{Bmatrix} = \underline{U_\alpha}\, \underline{\alpha} \qquad (14)$$

where

$$\underline{\alpha}^T = [A_1 \ldots A_N B_1 \ldots B_N] \qquad (15)$$

$$\underline{R_\alpha} = \begin{bmatrix} g_x^1 \ldots g_x^N g_x^{N+1} \ldots g_x^{2N} \\ g_y^1 \ldots g_y^N g_y^{N+1} \ldots g_y^{2N} \end{bmatrix} \qquad (16)$$

$$\underline{U_\alpha} = \begin{bmatrix} h_x^1 \ldots h_x^N h_x^{N+1} \ldots h_x^{2N} \\ h_y^1 \ldots h_y^N h_y^{N+1} \ldots h_y^{2N} \end{bmatrix} \qquad (17)$$

and all of g_x^m, g_y^m, h_x^m and h_y^m (m = 1,2,...,2N) are the functions of z.

[Example 1] "A rectangular plate with an eccentric crack under uniform tension"

In this case, the terms including Γ_u vanish in equation (5) or (6). Furthermore, only the case for which the body force $\overline{F}$ is equal to zero is addressed. Then, substituting equations (13) and (14) into equation (5) or (6),

$$\Pi_R = \frac{1}{2}\underline{\alpha}^T\, \underline{H}\, \underline{\alpha} - \underline{\alpha}^T\, \underline{G} \qquad (19)$$

where

$$\left.\begin{aligned} \underline{H} &= \frac{1}{2}\int_\Gamma (\underline{R_\alpha^T}\, \underline{U_\alpha} + \underline{U_\alpha^T}\, \underline{R_\alpha})\, d\Gamma \\ \underline{G} &= \int_{\Gamma_\sigma} \underline{U_\alpha^T}\, \overline{\underline{T}}\, d\Gamma \end{aligned}\right\} \qquad (20)$$

From the stationary condition of equation (19) with respect to $\underline{\alpha}$, we have

$$\underline{\alpha} = \underline{H}^{-1}\,\underline{G} \tag{21}$$

Then, from equation (21), all of coefficients in equation (15) are obtained.

When the coordinates are given as shown in Figures 1, 2 and 3, stress intensity factors are defined by

$$K_j = K_{Ij} - iK_{IIj} = 2\sqrt{2\Pi}\,\lim_{z\to z_j}\left[\sqrt{z-z_j}\;\phi(z)\right] \quad (z_j = \pm a) \tag{22}$$

Therefore, stress intensity factors are given by equations (21) and (22). Figures 1 and 2 show the numerical results of dimensionless stress intensity factors, F_{IA} and F_{IB}, at the tips A and B of the crack, respectively. These present results by the variational method coincide with the results [9] by a collocation technique up to three or four figures, as shown in Table 1.

Next, to check the accuracy, the variations of F_{IA} and F_{IB} with increase of the number of terms 2N in equation (10) are examined. Table 1 shows a typical example for the cracked plate with e/w = 0.3 and λ = 0.3 (see Figure 1). The numerical convergence is quite excellent and the errors are less than one per cent when 2N is more than 6.

[Example 2] "A rectangular plate with an eccentric crack under uniform displacement (clamped ends)"

Also, in this case, the body forces $\overline{F}$ are assumed equal to zero. Substituting equations (13) and (14) into equation (5) or (6), we have

$$\Pi_R = \frac{1}{2}\underline{\alpha}^T\,\underline{H}_\sigma\,\underline{\alpha} - \frac{1}{2}\underline{\alpha}^T\,\underline{H}_u\,\underline{\alpha} - \underline{\alpha}^T\,\underline{G}_\sigma + \underline{\alpha}^T\,\underline{G}_u \tag{23}$$

where

$$\left.\begin{aligned}
\underline{H}_\sigma &= \frac{1}{2}\int_{\Gamma_\sigma}(\underline{R}_\alpha^T\,\underline{U}_\alpha + \underline{U}_\alpha^T\,\underline{R}_\alpha)d\Gamma \\
\underline{H}_u &= \frac{1}{2}\int_{\Gamma_u}(\underline{R}_\alpha^T\,\underline{U}_\alpha + \underline{U}_\alpha^T\,\underline{R}_\alpha)d\Gamma \\
\underline{G}_\sigma &= \int_{\Gamma_\sigma}\underline{U}_\alpha^T\,\overline{\underline{T}}\,d\Gamma \\
\underline{G}_u &= \int_{\Gamma_u}\underline{R}_\alpha^T\,\overline{\underline{U}}\,d\Gamma
\end{aligned}\right\} \tag{24}$$

From the stationary condition of equation (23) with respect to $\underline{\alpha}$,

$$\underline{\alpha} = (\underline{H}_\sigma - \underline{H}_u)^{-1}\,(\underline{G}_\sigma - \underline{G}_u) \tag{25}$$

And then, stress intensity factors are obtained from equations (22) and (25).

Numerical results are shown in Figure 3. It is found that under the appropriate condition of crack length and eccentricity of the crack, the stress intensity factor K_{IB} at crack tip B (with smaller distance from the centre of this plate) can be slightly greater than the stress intensity factor K_{IA} at crack tip A.

The uniform displacement condition given to an asymmetrically cracked plate causes a negative in-plane bending moment which acts so that the crack tip A closes. An unsolved phenomenon, that a slightly eccentric crack in a plate grows such that the eccentricity decreases, can be explained by a difference of the stress intensity factors. A similar argument can probably be applied to a double edge cracked plate with clamped ends or a pin-loaded eccentric plate.

CONCLUDING REMARKS

Based on a variational principle, a new analytical method for determination of the stress intensity factors of a crack in a finite plate is proposed. By means of a modified Hellinger-Reissner formulation as presented above, mixed boundary crack problems can be solved. The numerical results indicate that by the present method an accurate evaluation of the stress intensity factors can easily be done with rapid convergency.

A phenomenon, that a slightly eccentric crack in a plate pulled by cyclic loads at the clamped ends grows so that the eccentricity decreases, is well explained by the stress intensity factors obtained by the present method.

ACKNOWLEDGEMENTS

The authors would like to express their hearty acknowledgements to the kind and valuable suggestions or discussions given by Professor T. H. H. Pian of MIT, USA, and Assistant Professor K. Watanabe of the University of Tokyo.

REFERENCES

1. KITAGAWA, H. and ISHIKAWA, H., Proceedings of the 19th Japan Congress on Materials Research, Kyoto, 1976, 251.
2. KITAGAWA, H. and ISHIKAWA, H., Preprint of Japanese Society for Strength and Fracture on Materials, May 1976.
3. KITAGAWA, H. and ISHIKAWA, H., Preprint of Japan Society of Mechanical Engineers, No. 750-11, October 1976, 191.
4. KITAGAWA, H. and ISHIKAWA, H., Preprint of Japan Society of Mechanical Engineers, No. 760-1, April 1976, 175.
5. WILLIAMS, M. L., J. of Appl. Mech., 24, September 1957, 109.
6. SIH, G. C. and Liebowitz, H., Fracture II, H. Liebowitz, Ed., Academic Press, New York, 1968.
7. REISSNER, E., J. Mathe. Phys., 29, 2, July 1950, 90.
8. WASHIZU, K., "Variational Methods in Elasticity and Plasticity", Pergamon Press, Oxford, 1968.
9. TERADA, H. and ISIDA, M., Preprint of Japan Society of Mechanial Engineers, No. 720-10, August 1972, 29.
10. VOOREN, J. V. D., Trans. ASME, Series D, 89, 1, March 1967, 236.
11. WILSON, W. K., Trans. ASME, Series D, 93, 4, 1971, 685.

Table 1 Convergence and Comparison of Dimensionless Stress Intensity Factors of an Eccentric Crack in a Square Plate under Uniform Tension

$e/w=0.3,\ \lambda=0.3$

2N	F_{IA}	F_{IB}
6	1.056	1.054
8	1.059	1.055
12	1.060	1.057
16	1.063	1.060
20	1.064	1.061
24	1.065	1.061
26	1.065	1.062
28	1.065	1.062
30	1.065	1.062
32	1.065	1.062
collocation method*	1.066	——

*reference [9]

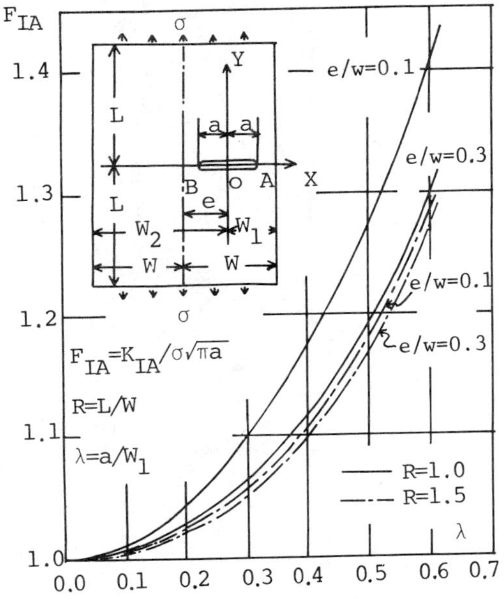

Figure 1 Dimensionless Stress Intensity Factor (F_{IA}) of the Tip (A) of an Eccentric Crack in a Plate Under Uniform Tension

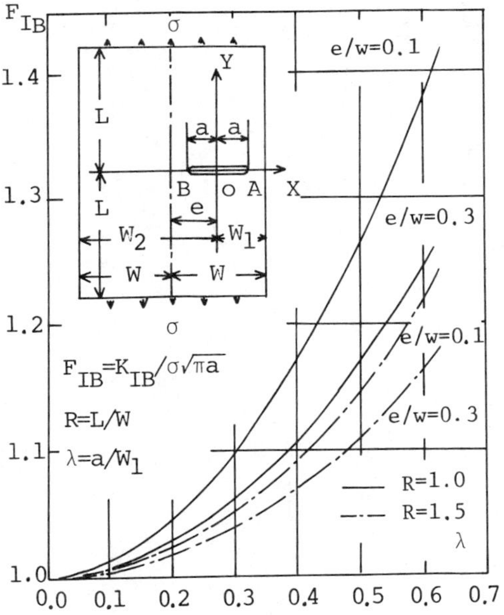

Figure 2 Dimensionless Stress Intensity Factor (F_{IB}) of the Tip (B) of an Eccentric Crack in a Plate Under Uniform Tension

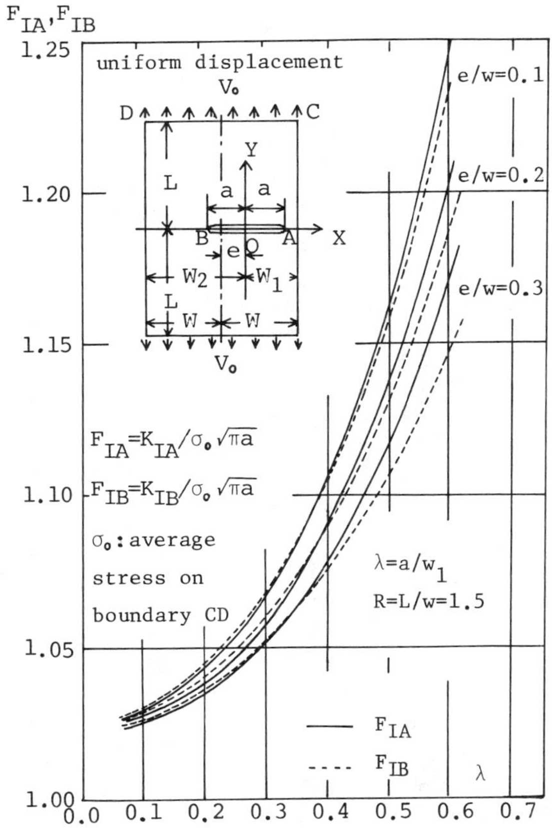

Figure 3 Dimensionless Stress Intensity Factors (F_{IA} and F_{IB}) of an Eccentric Crack in a Plate Under Uniform Displacement

Fracture 1977, Volume 3, ICF4, Waterloo, Canada, June 19 - 24, 1977

A FINITE ELEMENT ANALYSIS FOR DETERMINING DUGDALE MODEL SOLUTIONS OF CRACKED BODIES

K. J. Lau* and C. L. Chow**

INTRODUCTION

In recent papers [1,2,3], the authors have introduced a conic-section simulation method of finding the stress intensity factor; the displacements on the crack surface calculated by the finite element analysis are used to match each segment on the crack surface with a segment on an equivalent ellipse corresponding to an opening crack in an infinite sheet under uniform tension [4]. That segment on the crack surface is then taken to provide an estimate for the stress intensity factor as quantified by the infinite sheet configuration. Mathematically, this may be expressed as

$$K = p' \sqrt{a_e'} \qquad (1)$$

where p' = uniform pressure in infinite sheet containing the simulating crack

a_e' = half-crack length of the simulating crack

The matching can be effected either (i) by direct fitting of the displacements onto an elliptic curve with the crack tip as an apex [1,2], or (ii) by first fitting these displacements onto another conic-section (parabola or hyperbola) and then determining an equivalent ellipse which provides the same value for K through a relation between crack surface displacements and the strain energy release rate as described by Key [3,5]. The method has been proved to be of acceptable accuracy even when a relatively coarse finite element mesh and the simple constant-strain triangular elements are used in the finite element analysis.

Also based on the same infinite sheet crack configuration is the Dugdale strip yield model [6] characterized by the crack opening displacement [7]. In this model (Figure 1) an ideal elastic-plastic material is considered and the crack is assumed to deform elastically under the action of externally applied uniform tension p_o with a tensile stress σ_y acting over a hypothetical extension of length s at each end of the crack. σ_y can be identified with the yield stress of the material. By superposition of the two stress fields and consideration of the finiteness of the stresses at the location of the hypothetical crack tip, it can be shown that the load/yield-stress ratio is given by [6]

$$\frac{p_o}{\sigma_y} = \frac{4}{\pi} \sin^{-1} \left(\frac{s}{2\,a_e}\right)^{1/2} = \frac{2}{\pi} \cos^{-1} \frac{a}{a_e} \qquad (2)$$

---

*Hong Kong Polytechnic, Hung Hom, Hong Kong.
**University of Hong Kong, Pokfulam Road, Hong Kong.

and the COD is given by

$$\delta = \frac{8}{\pi}\varepsilon_y \; a \log \frac{a_e}{a} = \frac{8}{\pi}\frac{\sigma_y}{E} \; a \log \left(\sec \frac{\pi p_o}{2\sigma_y} \right) \tag{3}$$

where ε_y is the elastic yield strain. By the relation between ε_y and the overall strain on a given gauge length, critical COD-values can be applied to real structures for which the overall strains are easily measurable.

The present paper deals with the determination of the COD in cracks of arbitrary shape using the ellipse parameters in the K-determination through the conic-section approach. The advantage of the method is that both models have been built on the same Griffith crack geometry so that correspondence between p' and p_o, a_e' with a_e in equations (1) and (2) makes the determination of COD just a convenient extension of the K-finding process.

NUMERICAL DETERMINATION OF DUGDALE MODEL SOLUTION

In the original Dugdale model a crack in an infinite sheet was considered and the load/yield-stress ratio was obtained through the finiteness of the stresses at the hypothetical crack-tip region, which is equivalent to a cancellation effect between the elastic stress intensity factors due to the two stress fields p_o and σ_y. In a cracked body of arbitrary shape, the same principle can be applied. Thus if the stress intensity factor K_{Ip} due to the uniform pressure p_o is expressed as

$$K_{Ip} = p_o \; a_e^{1/2} \; k_{Ip} \tag{4}$$

and the stress intensity factor K_{Iy} due to a crack opening σ_y at the crack tip region is expressed as

$$K_{Iy} = \sigma_y \; a_e^{1/2} \; k_{Iy} \tag{5}$$

where k_{Ip} and k_{Iy} are geometrical correction factors such that

$$k_{Ip} = k_{Ip}\left(a_e \right)$$

and $$k_{Iy} = k_{Iy}\left(a, \; a_e \right),$$

then according to the Dugdale's theory

$$K_{Ip} - K_{Iy} = 0,$$

or, from (4) and (5),

$$\frac{p_o}{\sigma_y} = \frac{k_{Iy}}{k_{Ip}} \tag{6}$$

The crack opening displacement can be found by superposition of the displacements obtained from the two finite element analyses, involving p_o and σ_y as

$$\delta_d^* = \delta_p^* \frac{k_{Iy}}{k_{Ip}} - \delta_y^* \tag{7}$$

where δ_p^* and δ_y^* are the crack surface separations due to unit magnitude of p_o and σ_y respectively, while δ_d^*, the superposed value, is the overall COD corresponding to unit magnitude of σ_y.

The above method of analysis was carried out by Hayes and Williams [8] using Bueckner's formulation [9] in the treatment of finite element analysis results, which is essentially an energy approach requiring at least two sets of finite element calculations for each loading system so that the Dugdale model solution for one particular combination of a and a_e values necessitates four sets of finite element calculations. On the other hand, using the conic-section simulation approach, only two finite element analyses are necessary for any particular combination of geometry and loading system. Furthermore since both the Dugdale model and the conic-section analysis methods are built on the Griffith crack geometry, it is possible that closed-form solutions can be applied to reduce the amount of the finite element calculations. Specifically if the solution for the externally applied load for a particular value of a_e is obtained, then the values of COD for different values of s can be calculated through the modifications of equations (2) and (3) so that no finite element calculation involving the σ_y geometry is necessary. This is described in the following section.

DUGDALE MODEL SOLUTIONS FROM SIMULATION ELLIPSE PARAMETERS

Three ways of implementing the ellipse parameters approach for the finding of COD are considered:

Method A: In the conic-section analysis applied to the displacements of two nearby points on the crack surface, the segment between these two points is matched with a segment on the Griffith crack opened by uniform pressure and having the same stress intensity factor K (Figure 2). The actual value of the surface displacement at the middle of the segment is preserved in the resulting simulation ellipse. Within this ellipse model, the superposition of a uniform tensile stress over a length s on the crack surface adjacent to the crack tip to produce zero value of resultant K can be quantified according to (2) as

$$\frac{s}{a_e'} = 2 \sin^2 \frac{\pi p_o'}{4\sigma_y'} \tag{8}$$

while the corresponding COD produced is given by (3) as

$$\delta = \frac{8}{\pi} \frac{\sigma_y'}{E} a' \log \frac{a_e'}{a'} \tag{9}$$

where various terms in the above equations are defined in Figure 2. Returning to the actual crack under consideration, the necessary value of the stress σ_y in the region s to produce the stress intensity factor will be different from σ_y', but the value of the COD should remain the same, subject to the validity of two assumptions; firstly, the displacements of the points P_1 and P_2 (Figure 2) must be truly representative of the crack tip stress intensity factor; secondly, the crack shape produced by uniform stress σ_y in region s in the real crack system should be functionally similar to that produced by a uniform stress σ_y' in the region s of the model system.

Method B: Each simulation ellipse corresponding to a segment on the actual crack surface is considered in turn with respect to a given value of s. The effective change in crack surface separation at r = s produced by σ_y' over the length s is calculated as

$$\delta_y = \frac{4p_o'a_e'}{E}\left(1 - \frac{a'^2}{a_e'^2}\right)^{1/2} - \frac{8}{\pi}\frac{\sigma_y'}{E}a' \log\frac{a_e'}{a'} \tag{10}$$

where the first term on the right hand side is the separation due to p_o' alone and the second term is the COD as given by equation (3). An average is taken of the δ_y-values from all the simulating ellipses and the crack opening displacement is then evaluated as the difference between the crack surface separation produced by the actual load applied as calculated by the finite element analysis, (interpolated if necessary) and the average δ_y-value. This method is in principle the same as method A except that the effective stress intensity factor is now taken as the average value represented by the whole crack profile. Furthermore this method is more convenient for treating values of s that do not match with the nodal positions of the finite element mesh used. There is however one minor restriction in that the a_e'-values of the simulation ellipses at points close to the hypothetical crack tip may sometimes fall short of s so that in some extreme cases, sufficient δ_y-values may not be available to give a good average value.

Method C: Both methods A and B are subject to the assumption that equation (10) governs the change in crack surface separation due to σ_y on the actual crack surface. To check whether this equation based on the original Dugdale model adequately represents the crack profile produced, a more practical assumption is considered - that the ratio of the displacements due to the two sets of loadings in the actual system remains the same as that in the model system. Since the displacements are to be produced by the loading systems giving the stress intensity factors of equal magnitude as reflected in an opening displacement, the effects of finite width, etc., on these representative displacements should be equal. Hence in this third method, a ratio of displacements is obtained as

$$\frac{\delta_y}{\delta_p} = 1 - \frac{\dfrac{8\sigma_y'}{\pi E}a' \ln\dfrac{a_e'}{a'}}{\dfrac{4p_o'a_e'}{E}\left(1 - \dfrac{a'^2}{a_e'^2}\right)^{1/2}} \tag{11}$$

whence the COD can be found as the product of $(1 - \delta_y/\delta_p)$ and the actual displacement δ_p obtained by the finite element analysis.

In the above methods the COD-values are found with no reference made to the load/yield-stress ratio in the actual system. An approximation process is now used to transfer the value of σ_y' in the model system to that of σ_y in the actual system by using the correction factors of some standard configurations. Thus σ_y is estimated as

$$\frac{\sigma_y'}{\sigma_y} = k_1\ (s,\ W - a) \tag{12}$$

where k_1 corresponds to the mode I correction factor function $k_I(a_e,\ W)$ of the fully loaded crack in a plate of finite width 2W.

DISCUSSIONS AND CONCLUSIONS

Due to the space restriction in this presentation, only the results obtained for a centrally cracked finite-width plate geometry are presented. Results for other configurations can be found in [10] and the COD-values by the methods A and B are compared in Figure 3 in dimensionless form as

$$\delta^* = \frac{E}{p_o W}\,\delta \tag{13}$$

It can be seen that remarkable agreement in the computed results has been achieved except for the combinations of the high a_e/W and s/a_e values. The reason for the disparities is probably that finite-width effect in these cases renders the displacements of the points remote from the crack tip inadequate in the representation of the crack-tip stress intensity factors. Method B is expected to produce more reliable results in these cases provided that sufficient δ_y-values are made available for the averaging process. This provision is found to have been largely satisfied except for only a few extreme cases. The results from method C are found to be practically co-incident with those from method B for most cases so that, for clarity of the plotting, they are not shown in the same graph, but are given in Table 1 together with those of method B and the results from reference [8]. Also shown in the table are the estimated values of the load/yield-stress ratio. Since satisfactory agreement can be observed between the results of the present method and that of Hayes and Williams [8], it may be inferred that conceptually similar processes have been executed in relating the stress ratio to the plastic zone size through equating the magnitudes of the K-values from the two stress systems. Comparison of the COD values reveals that appreciable disagreement between the methods B and C exists only at the high s/a_e values and $a/s = 0.1$ or 0.2, i.e. when the plastic zone length is much larger than that of the actual crack length. This establishes the validity of equation (10) as mentioned earlier. Comparison with the COD-values given by Hayes and Williams show rather disappointing discrepancies. One major problem in both approaches is that the COD is seldom a large fraction of the displacement due to the external applied stress system so that small percentage error in the displacement calculations may become magnified in the COD-values. From the nature of the finite element mesh used by Hayes and Williams, which consists of coarse grids with constant element width along the line of the crack length, underestimates of the displacements are expected. This probably explains the discrepancies between the COD-values for the central crack case. Nevertheless agreement in the results may be considered as satisfactory when the plastic zone size is not too large as compared with the actual crack size while the accuracy of each method depends heavily on the accuracy of the displacements obtained. The present methods via the conic-section analysis follow closely the original idea of representing a crack-tip strain field by an opening displacement. Agreement among the three methods of implementation lends credibility to the principle used and the practically constant values of σ_y' and δ_y from different simulation ellipses for different sections of the crack surface in practically all cases under consideration, as exemplified by Table 2, further confirm the validity of the method.

REFERENCES

1. CHOW, C. L. and LAU, K. J., J. Strain Analysis 11, 1976, 1.
2. CHOW, C. L. and LAU, K. J., Int. J. Fracture 12, 1976, 1.
3. CHOW, C. L. and LAU, K. J., Int. J. Fracture 12, 1976, 5.
4. SNEDDON, I. N. and ELLIOT, H. A., Quart. Appl. Math. 4, 1946.
5. KEY, P. L., Int. J. Frac. Mech. 5, 1969.
6. DUGDALE, D. S., J. Mech. Phys. Sol. 8, 1960.
7. WELLS, A. A., Proc. Cranfield Crack Propagation Symposium, Vol. 1, 1961.
8. HAYES, D. J. and WILLIAMS, J. G., Int. J. Frac. Mech. 8, 1972.
9. BUECKNER, H. F., Trans. A.S.M.E. 80, 1958.
10. LAU, K. J., Ph.D. Thesis, University of Hong Kong, 1975.

Table 1 COD for Centrally Cracked Plate in Uniform Tension, H/W = 3.5

a/W	s/a	δ^* Method B	δ^* Method C	δ^* Hayes & Williams 1972	p_o/σ_y Present Method	p_o/σ_y Hayes & Williams 1972
0.1	0.5000	0.2666	0.2317	0.2471	0.6344	0.6406
	0.6667	0.3712	0.3996	0.3449	0.7457	0.7662
	0.7500	0.4455	0.5390	0.4109	0.8140	0.8208
	0.8000	0.5669	0.6321	0.4623	0.8304	0.8559
	0.8333	0.6116	0.7702	0.5054	0.8523	0.8690
	0.8572	0.6977	0.9017	0.5421	0.8686	0.8808
	0.8750	0.5853	0.7795	0.5768	0.8898	0.8887
	0.8889	0.9805	1.2950	0.6120	0.9388	0.8944
0.2	0.3333	0.3876	0.3825	0.3680	0.4879	0.4996
	0.5000	0.5760	0.5997	0.5247	0.6124	0.6322
	0.6400	0.7290	0.7639	0.6366	0.6728	0.6981
	0.6667	0.8750	0.9637	0.7269	0.7112	0.7367
	0.7143	1.0620	1.1840	0.8051	0.7408	0.7610
	0.7500	1.1676	1.3316	0.8775	0.7699	0.7772
	0.7778	1.0604	1.1653	0.9540	0.8290	0.7889
0.3	0.2500	0.5016	0.5091	0.4725	0.4024	0.4125
	0.4000	0.7621	0.7686	0.6821	0.5295	0.5359
	0.5000	0.9837	1.0238	0.8364	0.5751	0.6008
	0.5714	1.2638	1.3169	0.9663	0.6074	0.6400
	0.6250	1.5420	1.6323	1.0845	0.6450	0.6654
	0.6667	1.6700	1.6936	1.2093	0.7128	0.6835
0.4	0.2000	0.6090	0.6163	0.5786	0.3376	0.3477
	0.3333	0.9670	0.9738	0.8432	0.4384	0.4574
	0.4386	1.3100	1.3368	1.0452	0.4923	0.5166
	0.5000	1.6840	1.7558	1.2237	0.5305	0.5529
	0.5556	2.1171	2.1211	1.4079	0.5924	0.5776
0.4	0.1667	0.7471	0.7550	0.6980	0.2852	0.2935
	0.2857	1.2239	1.2307	1.0290	0.3690	0.3879
	0.3750	1.7363	1.7646	1.2941	0.4164	0.4392
	0.4444	2.4278	2.4276	1.5564	0.4632	0.4717
0.6	0.1428	0.9281	0.9314	0.8492	0.2383	0.2444
	0.2500	1.6165	1.6163	1.2717	0.3066	0.3226
	0.3333	2.5103	2.4953	1.6484	0.3606	0.3655
0.7	0.1250	1.2234	1.2279	1.0644	0.1917	0.1970
	0.2222	2.3148	2.2529	1.6485	0.2519	0.2585
0.8	0.1111	1.8185	1.8127	1.4364	0.1466	0.1483

Table 2 σ_y' and δ_y* Values Obtained from Method B

r/a_e	a_e'/a_e	σ_y'/p_o	δ_y*
0.1667	2.4593	1.6960	1.581
0.2250	1.3611	1.6933	1.656
0.3000	1.1268	1.6967	1.704
0.3917	1.0401	1.7018	1.734
0.4958	1.0066	1.7061	1.751
0.6083	0.9841	1.7111	1.764
0.7500	0.9720	1.7152	1.773
0.9167	0.9761	1.7130	1.769

$s/a = 1.4$, $a_e/W = 0.6$, $\delta_p^* = 2.700$

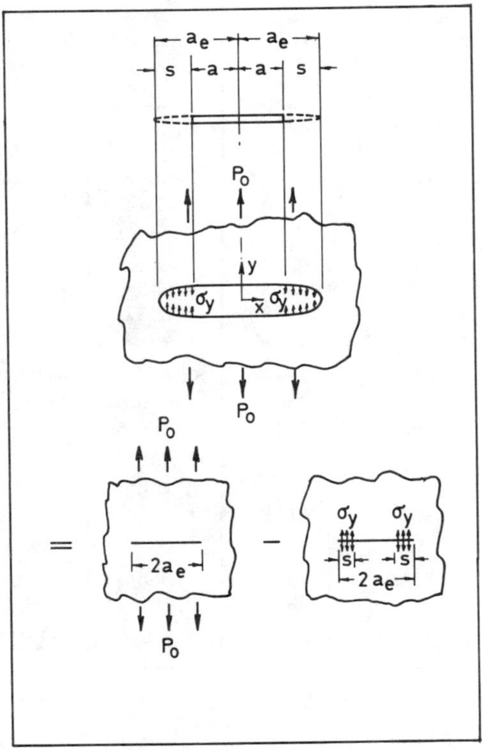

Figure 1 Geometry of Slit and Equivalent System for Dugdale Model

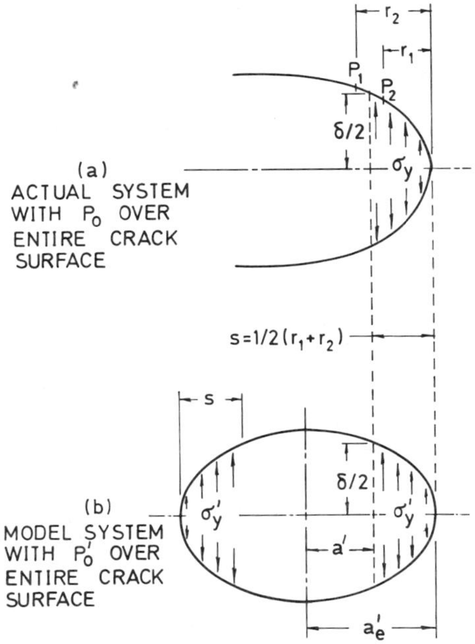

Figure 2 Actual Crack Shape and Simulation Model System

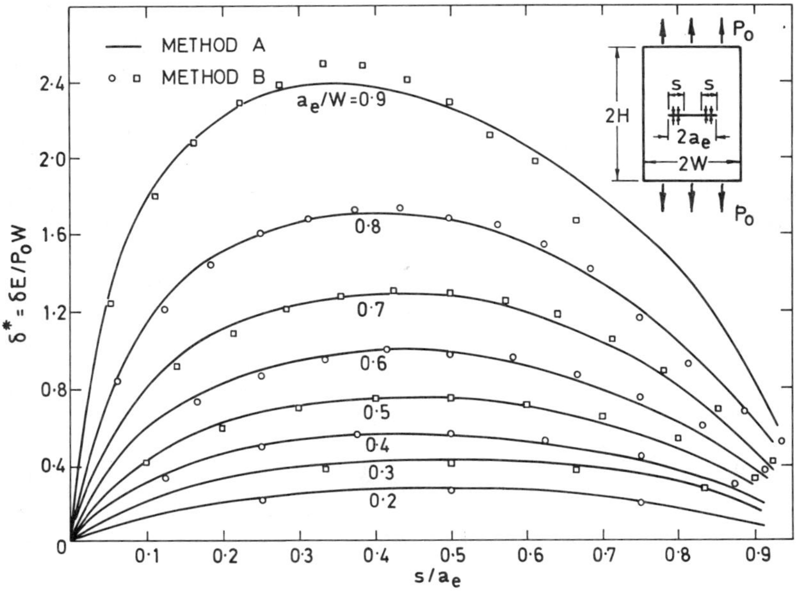

Figure 3 COD for Central Crack Under Uniform Tension, H/W = 3.5

Fracture 1977, Volume 3, ICF4, Waterloo, Canada, June 19 - 24, 1977

ANALYSIS FOR THE PROBLEM OF MISFITTING INCLUCION AND CIRCULAR ARC CRACKS IN A SHEET

Ram Narayan*, R. K. Pandey** and B. N. Ganguli***

INTRODUCTION

Tamate [1] combined together the inclusion and straight crack problem and obtained the solution of the problem in series form. An effort of combining together the misfitting inclusion of different materials and circular arc crack problems forms the subject matter of the present paper. Analytical solution of the problem is given in series form and complex variable approach of Muskhelishvili [2] is used throughout.

BASIC EQUATIONS AND STATEMENT OF PROBLEM

Let the region S be the entire plane, cut along the arc L_s(S=1, 2) of a circle of radius R with the centre at the origin of $z=x+iy=r\exp(i\theta)$. The arc L_s is assumed to lie symmetrically on the x-axis and subtend an angle 2α at the centre. By using the complex potentials $\phi(z)$, $\psi(z)$ which are defined in S, we define a new function W(z) in the following manner:

$$W(z) = \overline{\phi}(R^2/z)-(R^2/z)\overline{\phi}'(R^2/z)-(R^2/z^2)\overline{\psi}(R^2/z) \qquad (1)$$

Whence $\psi(z)$ can be expressed in terms of $\phi(z)$ and W(z) as

$$\psi(z) = (R^2/z^2)\phi(z)-(R^2/z^2)\overline{W}(R^2/z)-(R^2/z)\phi'(z) \qquad (2)$$

where the bar denotes the complex conjugate.

In the absence of body forces, the stress components in polar coordinates σ_r, σ_θ, $\tau_{r\theta}$ and the displacement components U_r, U_θ for the elastic sheet occupying the regions S are expressed in terms of $\phi(z)$ and $\psi(z)$ as

$$\sigma_r+\sigma_\theta = 2[\phi(z)+\overline{\phi(z)}] \qquad (3)$$

$$\sigma_r+i\tau_{r\theta} = \phi(z)+\overline{\phi(z)}-\overline{z\phi'(z)}-(\overline{z}/z)\overline{\psi(z)} \qquad (4)$$

$$2\mu\frac{\partial}{\partial\theta}\{(U_r+iU_\theta)\exp(i\theta)\} = iz[\kappa\phi(z)-\phi(z)$$

$$+\overline{z}\,\overline{\phi'(z)}+(\overline{z}/z)\,\overline{\psi(z)}] \qquad (5)$$

* Department of Mathematics, Bararas Hindu University, India.
** Department of Applied Mechanics, I.I.T., Delhi, India.
***Department of Mathematics, Harish Chandra Degree College, Varanasi, India.

where μ is the shear modulus and $\kappa=3-4\nu$ for plane deformation, $\kappa=(3-\nu)/(1+\nu)$ for generalised plane stress, ν being Poisson's ratio. The subscript following a comma stands for partial differentiation. Equations (4) and (5) one may write as follows:

$$\sigma_r+i\tau_{r\theta} = \phi(z)+W(R^2/\bar{z})+\bar{z}(\bar{z}/R^2-1/z)\overline{\psi(z)} \quad (6)$$

$$2\mu \frac{\partial}{\partial\theta} \{(U_r+iU_\theta)\exp(i\theta)\} = iz[\kappa\phi(z)-W(R^2/\bar{z})-\bar{z}(\bar{z}/R^2-1/z)\overline{\psi(z)}] \quad (7)$$

Let an infinite elastic plate, isotropic and homogeneous, occupy the aforementioned regions S and be cut out by a circular hole of radius c with its centre at any point M(M may be complex). An inclusion of different elastic material of radius $(c+\eta)$, (η is of the order of displacements admissible in elasticity theory) is supposed to be inserted and bonded to the hole. Further it will be assumed that the stresses vanish at infinity and the edges of the crack are free from external tractions. The crack and the hole do not overlap.

The boundary condition of the problem can be expressed as follows:
(i) At infinity,

$$\phi(z) = 0(z^{-2}),\ \psi(z) = 0(z^{-2}) \quad (8)$$

whence at the origin,

$$W(z) = 0(1) \quad (9)$$

(ii) on the rims of the crack L_s, $\sigma_r^{\pm}+i\tau_{r\theta}^{\pm} = o$;

$$\phi^+(\zeta)+W^-(\zeta) = 0, \phi^-(\zeta)+W^+(\zeta) = o \quad (10)$$

where ζ is the coordinate of the point on the cut L_S and superscripts + and - refer to the boundary values of the functions as z approaches from the inside and the outside of the arc L_S respectively.
(iii) On the common circle r=c, when origin is considered as M,

$$(\sigma_r+i\tau_{r\theta}) = (\sigma_r+i\tau_{r\theta})_i \quad (11)$$

$$(U_r+iU_\theta)-(U_r+iU_\theta)_i = \eta \quad (12)$$

where the subscript i refers to the inclusion.

COMPLEX POTENTIALS FOR THE PLATE AND FOR THE INCLUSION

Since the equation (10) are dual homogeneous Hilbert problems for two functions $\phi(z)$ and $W(z)$, which are analytic in the entire plane cut along L_S we can readily construct the complex potentials $\phi(z)$ and $W(z)$ for the infinite plate which satisfy the conditions in (8 - 10) by the use of Muskhelishvili's technique. Taking into account the fact that $\phi(z)$ and $W(z)$ could have poles of various orders at z=M, we can write them as follows:

$$\left.\begin{matrix}\phi(z)\\ W(z)\end{matrix}\right\} = \pm 1/2 \sum_{j=2}^{\infty} F_{-j} \left\{\frac{1}{z-M}\right\}^{j} \pm 1/2 \sum_{j=o}^{\infty} F_j (z-M)^j$$

$$+ 1/2\, X(z)\left[\sum_{j=1}^{\infty} H_{-j} \left\{\frac{1}{z-M}\right\}^{j} + \sum_{j=o}^{\infty} H_j (z-M)^j\right] \tag{13}$$

where

$$X(z) = [z-R \exp(-i\alpha)]^{-1/2}[z-R \exp(i\alpha)]^{-1/2}[z+R \exp(i\alpha)]^{-1/2}$$
$$[z+R \exp(-i\alpha)]^{-1/2} \tag{14}$$

means the branch, analytic in the entire plane cut along L_s such that $X(z) \to (1/z^2)$ for $|z| \to \infty$. The coefficients $F_{\pm j}$ and $H_{\pm j}$ are to be determined.

The origin of the coordinate system is now shifted to M. The functions $\{\phi(z), \psi(z)\}$ transform to new functions $\{\phi_1(z_1), \psi_1(z_1)\}$. We drop the suffix 1 for the convenience but remember these are the potentials obtained after shifting the origin to M. By the conditions (8) and (9), we get

$$F_{-j} = \sum_{s=o}^{\infty} L_s H_{-(j+s)} (j \geq 2),$$

$$F_j = - \sum_{s=1}^{\infty} L_{-s} H_{j+s} (j \geq o), \tag{15}$$

$$\sum_{s=o}^{\infty} L_s H_{-(s+1)} = o, \qquad \sum_{s=1}^{\infty} L_{-s} H_{s-1} = o$$

where the constants $L_{\pm s}$ are known quantities determined from the relation,

$$[z+m-R \exp(-i\alpha)]^{-1/2}[z+M-R \exp(i\alpha)]^{-1/2}[z+M+R \exp(i\alpha)]^{-1/2}$$

$$[z+M+R \exp(-i\alpha)]^{-1/2} = \sum_{j=o}^{\infty} L_j z^j,$$

$$L_{-(2j+1)} = L_{2j+1} = o, \; L_{-(2j+2)} = - L_{2j}, \; (j \geq o) \tag{16}$$

The function $\phi(z)$ in (13) and the corresponding function $\psi(z)$, obtained from (2) can be expressed in the following Laurent series in the region $c < |z| < R$:

$$\phi(z) = 1/2 \sum_{j=-\infty}^{\infty} C_j z^j, \; \psi(z) = 1/2 \sum_{j=-\infty}^{\infty} D_j z^j \tag{17}$$

where,

$$\left.\begin{aligned} C_{-1} &= o, \\ C_{-j} &= F_{-j} + \sum_{s=o}^{\infty} L_s H_{-(j+s)} (j \geq 2), \\ C_j &= F_j + \sum_{s=o}^{\infty} L_s H_{j-s} (j \geq o), \end{aligned}\right\} \quad (18)$$

$$\left.\begin{aligned} D_{-1} &= o \\ D_{-j} &= (j-1)C_{-(j-2)} - 2\sum_{s=1}^{\infty} L_{-s} \overline{H}_{j+s-2} (j \geq 2), \\ D_j &= \overline{F}_{-(j+2)} - (j+1)C_{j+2} - \sum_{s=1}^{\infty} L_{-s} \overline{H}_{-(j-s+2)} (j \geq o) \end{aligned}\right\} \quad (19)$$

Thus the form of the function $\phi(z)$ and $\psi(z)$ are determined for the infinite elastic sheet which satisfy the condition at infinity as well as along the rims of the crack L_s.

The potentials $\phi_i(z)$ and $\psi_i(z)$ for inclusion are analytic functions for the region $|z| \leq c$, hence can be written as

$$\phi_i(z) = \sum_{j=o}^{\infty} Y_j z^j, \; \psi_i(z) = \sum_{j=o}^{\infty} S_j z^j, \quad (20)$$

where the coefficients Y_j and S_j have to be determined by the conditions (11 - 12).

By the conditions (11) and (12), one may obtain

$$\left.\begin{aligned} C_o+\overline{C}_o-\overline{D}_{-2}c^{-2} &= Y_o+\overline{Y}_o, \\ \left[(\kappa C_o-\overline{C}_o)-\overline{D}_{-2}c^{-2}\right](\mu_i/\mu) &= \kappa Y_o-\overline{Y}_o+4\eta\mu_i/c \end{aligned}\right\} \quad (21)$$

$$\left.\begin{aligned} C_j c^j+(j+1)\overline{C}_{-j}c^{-j}-\overline{D}_{-j-2}c^{-j-2} &= Y_j, \quad (j \geq 1) \\ \left[\kappa C_j c^j-(j+1)-\overline{C}_{-j}c^{-j}+\overline{D}_{-j-2}c^{-j-2}\right](\mu_i/\mu) &= \kappa_i Y_j \end{aligned}\right\} \quad (22)$$

$$\left.\begin{aligned} C_{-j}c^{-j}-(j-1)\overline{C}_j c^j-\overline{D}_{j-2}c^{j-2} &= (1-j)\overline{Y}_j-\overline{S}_{j-2}, \quad (j \geq 2) \\ \left[\kappa C_{-j}c^{-j}+(j-1)\overline{C}_j c^j+\overline{D}_{j-2}c^{j-2}\right](\mu_i/\mu) &= (j-1)\overline{Y}_j+\overline{S}_{j-2} \end{aligned}\right\} \quad (23)$$

When (21 - 23) are solved, the following expressions are obtained

$$Y_j = \frac{\kappa\mu_i/\mu+1}{\kappa_i+1} c^j C_j - \frac{\mu_i/\mu-1}{\kappa_i+1}\left[(j+1)c^{-j}\overline{C}_{-j} - c^{-j-2}\overline{D}_{-j-2}\right]$$

$$- \frac{4\mu_i}{(\kappa_i+1)c}\,\delta_{o,j} \quad (j \geq o), \tag{24a}$$

$$S_j = (j+1)c^{-j-4}D_{-j-4} + c^j D_j - (j+2)^2 c^{-j-2}\overline{C}_{-j-2} (j \geq o), \tag{24b}$$

$$\left.\begin{aligned} C_{-j} &= -\frac{\mu_i/\mu-1}{\kappa\mu_i/\mu+1}\left[(j-1)c^{2j}\,\overline{C}_j + c^{2j-2}\overline{D}_{j-2}\right](j \geq 2), \\ D_{-2} &= \frac{8\mu_i c}{2\mu_i/\mu+\kappa_i-1} - \frac{\kappa\mu_i/\mu-\kappa_i-\mu_i/\mu+1}{2\mu_i/\mu+\kappa_i-1}\, c^2(C_o+\overline{C}_o), \\ D_{-j-2} &= (j+1)c^2 C_{-j} - \frac{\kappa\mu_i/\mu-\kappa_i}{\mu_i/\mu+\kappa_i}\, c^{2j+2}\overline{C}_j (j \geq 1), \end{aligned}\right\} \tag{25}$$

where $\delta_{o,j}$ is Kronecker delta.

Equations (18), (19) and (25) give the following sets of infinite linear equations of H_j and H_{-j}:

$$\sum_{s=o}^{\infty} L_s\, H_{-(s+1)} = o, \tag{26a}$$

$$\sum_{s=o}^{\infty} L_s\, H_{-(j+s)} = 1/2\,\frac{\mu_i/\mu-1}{\kappa\mu_i/\mu+1}\left[(j-1)(1-c^2)\left\{\sum_{s=o}^{\infty} L_s\, \overline{H}_{-(j-s)}\right.\right.$$

$$\left.\left. - \sum_{s=1}^{\infty} L_{-s}\, \overline{H}_{j+s}\right\} + \sum_{s=1}^{\infty} L_{-s}\, H_{-(j-s)} - \sum_{s=o}^{\infty} L_s\, H_{-(j+s)}\right] c^{2j-2},$$

$$(j \geq 2), \tag{26b}$$

$$\sum_{s=1}^{\infty} L_{-s}\, \overline{H}_{s-1} = o \tag{27a}$$

$$\sum_{s=1}^{\infty} L_{-s}\, \overline{H}_s = -\frac{4\,c\,\eta\,\mu_i}{2\mu_i/\mu+\kappa_i-1} + 1/2\left\{\sum_{s=o}^{\infty} L_s\, H_{-s} - \sum_{s=1}^{\infty} L_{-s}\, H_s\right\}$$

$$+ 1/2\,\frac{\kappa\mu_i/\mu-\kappa_i-\mu_i-\mu_i/\mu+1}{2\mu_i/\mu+\kappa_i-1}\left\{\sum_{s=o}^{\infty} L_s\, H_{-s} - \sum_{s=1}^{\infty} L_{-s}\, H_s\right.$$

$$\left. + \sum_{s=o}^{\infty} L_s\, \overline{H}_{-s} - \sum_{s=1}^{\infty} L_{-s}\, \overline{H}_s\right\}, \tag{27b}$$

$$\sum_{s=1}^{\infty} L_{-s}\,\overline{H}_{j+s} = (j+1)(1-c^2)\sum_{s=o}^{\infty} L_s\,H_{-(j+s)}$$

$$+1/2\,\frac{\kappa\mu_i/\mu-\kappa_i}{\mu_i/\mu+\kappa_i}\left\{\sum_{s=o}^{\infty} L_s\,\overline{H}_{j-s} - \sum_{s=1}^{\infty} L_{-s}\,\overline{H}_{j+s}\right\}c^{2j+2}$$

$$(j \geq 1), \qquad (27c)$$

The constants H_i and H_{-j} are determined from the equations (26) and (27) by assigning values to μ_i/μ, κ, K_i, α, c, R. The values of $F_{\pm j}$ are determined from (15) using the values of $H_{\pm j}$ determined earlier. Thus the potentials $\{\phi(z), \psi(z)\}$ are completely known. The coefficients Y_j and S_j are determined by (24) with the help of the values of $H_{\pm j}$ and $F_{\pm j}$ determined previously. Hence $\{\phi_i(z), \psi_i(z)\}$ for inclusion is also known. The stress field for matrix and inclusion can now be determined with the help of (3 - 5).

STRESS INTENSITY FACTORS

By using the definition given by Sih, Paris and Erdogan [3], the stress intensity factors at the crack-tip for the case when the inclusion and crack are concentric can be expressed as $K = (K_1 - iK_2)$.

$$K_1 = \sqrt{\frac{R}{2\sin 2\alpha}}\left[\left\{H_o^{(i)} + \sum_{j=1}^{\infty}\left\{H_j^{(r)} - H_{-j}^{(r)}\right\}\sin j\alpha + \sum_{j=1}^{\infty}\left\{H_j^{(i)} + H_{-j}^{(i)}\right\}\cos j\alpha\right\} x \cos\alpha - \left\{H_o^{(r)} + \sum_{j=1}^{\infty}\left\{H_j^{(r)} + H_{-j}^{(r)}\right\}\cos j\alpha - \sum_{j=1}^{\infty}\left\{H_j^{(i)} - H_{-j}^{(i)}\right\}\sin j\alpha\right\}\sin j\alpha\right],$$

$$K_2 = \sqrt{\frac{R}{2\sin 2\alpha}}\left[\left\{H_o^{(i)} + \sum_{j=1}^{\infty}\left\{H_j^{(r)} - H_{-j}^{(r)}\right\}\sin j\alpha + \sum_{j=1}^{\infty}\left\{H_j^{(i)} + H_{-j}^{(i)}\right\}\cos j\alpha\right\}\sin\alpha - \left\{H_o^{(r)} + \sum_{j=1}^{\infty}\left\{H_j^{(r)} + H_{-j}^{(r)}\right\}\cos j\alpha - \sum_{j=1}^{\infty}\left\{H_j^{(i)} - H_{-j}^{(i)}\right\}\sin j\alpha\right\}\cos\alpha\right].$$

Numerical results for the stress intensity factors have been calculated under plane deformation for the following values: $\alpha = 10°$, $\mu_i = \mu = 5.60\text{x}10^{11}$ gm/cm^2, $\nu_i = \nu = .339$, $c = (1+\varepsilon)$, R=1.5(.25)2.5. The results are given in table form.

It is clear from Table 1 that stress intensity factor numerically decreases as radius of the cracks increases.

REFERENCES

1. TAMATE, O., Int. J. Frac. Mech., 4, 1968, 257.
2. MUSKHELISHVILI, N. I., "Some Basic Problems of the Mathematical Theory of Elasticity", P. Noordhoff, Groningen, 1953.
3. SIH, G. C., PARIS, P. C. and ERDOGAN, F., J. Appl. Mech., 29, 1962, 651.

Table 1

R = Radius of the Crack	$K_1 = \dfrac{\text{Symmetric Stress Intensity Factor}}{10^{11} \times \eta}$	$K_2 = \dfrac{\text{Skew Symmetric Stress Intensity Factor}}{10^{11} \times \eta}$
1.5	-1.00150	-0.04382
1.75	-0.79525	-0.03474
2.00	-0.65105	-0.02843
2.25	-0.54567	-0.02383

Fracture 1977, Volume 3, ICF4, Waterloo, Canada, June 19 - 24, 1977

GREEN'S FUNCTION FOR THRU-CRACK EMANATING FROM FASTENER HOLES

T. M. Hsu* and J. L. Rudd**

INTRODUCTION

The application of fracture mechanics to fatigue crack growth and residual strength analyses has resulted in much progress during the last decade. Yet the presence of cracks in engineering structures still poses many serious research problems which remain to be solved. One such problem is a crack emanating from an inelastic field near a fastener hole, such as produced by an interference-fit fastener or a cold-worked hole.

Fatigue cracks usually originate in the regions of high stress concentration, which exist notwithstanding careful detail-design procedures. Hardly any assembled structure is free of geometric discontinuities, such as fastener holes and access holes. Since a hole is a source of stress concentration, and since there may be many holes involved in any one structure, it can be anticipated that fatigue cracks will start at some of these holes during its service life. A review of U.S. Air Force aircraft structural failures [1] revealed that cracks emanating from fastener holes represent the most common origin of these failures.

The stress-intensity factor, which generally depends upon crack length, remote loading and structural geometry, has been employed to characterize the severity of the crack-tip stress field. To date, there has been much useful work done on the problem of determining reliable stress-intensity factors for cracks emanating from fastener holes. Almost all of these analytical determinations are based upon modifications of a solution obtained by Bowie [2] for cracks emanating from a circular hole in an infinite elastic sheet. For cracks emanating from an inelastic field near a fastener hole, the stress intensity factors could be estimated by using the weight function approach as discussed by Bueckner [3 - 6] or the reciprocal theorem proposed by Rice [7]. Both techniques require a knowledge of the unflawed stress distribution in the region of the hole. Paris et al [8] has combined these techniques with the finite-element method to develop a weight function for the single edge cracked strip.

The closed form expressions for the weight function for edge cracks [4, 9], centre cracks [10] and collinear cracks [6] in a wide panel are available. But, the closed form weight function for cracks emanating from a fastener hole is not available. Development of such a function will be very difficult, if not impossible. Therefore, the weight function for a straight crack has sometimes been used to estimate the stress-intensity factor for radial cracks emanating from a circular hole, [11 - 13]. For a large crack, where the influence of a fastener hole on the stress intensity factor is small, such an approximation gives good results. However, for the case of a small crack, say $a/r \leqq 1.0$, such an approximation could be

*Lockheed-Georgia Company, Marietta, Georgia, U.S.A.
**Air Force Flight Dynamics Laboratory, Wright-Patterson AFB, Ohio, U.S.A.

significantly in error. Grandt [14] used the reciprocal theorem due to Rice [7] to develop the following equation which estimates the Mode I stress intensity factor, K_I, for cracks emanating from any type of circular fastener hole:

$$K_I = \int_\Gamma p.h \, d\Gamma = \frac{H}{K_B} \int_o^a p(x) \frac{\partial \eta}{\partial a} dx \quad (1)$$

where p is the stress vector on the boundary; h is the weight function; η is the y-component of the crack surface displacements; K_B is the Bowie solution for the stress intensity factor; and H is an appropriate elastic modulus: it is $E/1 - \nu^2$ for plane strain and E for generalized plane stress. Since the closed form expression for η as a function of the crack length a is not available, it was determined by fitting an equation to the discrete displacements computed using the finite-element method for a series of crack lengths ranging from a/r = 0.4 to a/r = 2.8. The stresses and strains computed using the conventional finite-element method may be fairly accurate. But the differentiation of an approximate expression obtained by curve fitting finite-element results may not be warranted.

Two high order singularity elements have been developed at Lockheed-Georgia. One takes only the symmetric terms in the Williams' series and, hence, is applicable only to symmetric problems ($K_{II} = 0$); the other makes use of both symmetric and antisymmetric terms and is applicable to unsymmetric or mixed mode (K_I and K_{II}) problems. The efficiency and accuracy of these elements has been demonstrated in reference [15]. In order to obtain more accurate solutions for cracks emanating from a hole, the high order singularity element for symmetric problems was used to compute the Mode I stress intensity factor for a double-radial crack emanating from an open hole and subjected to concentrated loads on and perpendicular to the crack surface. The computed stress intensity factor was used to develop the Green's function (equivalent to the nondimensional stress-intensity factor) for a double-radial crack emanating from a circular hole. In the case of mixed mode conditions, the corresponding Green's function analogy to the symmetric case can be developed from the stress intensity factor computed using the unsymmetric crack element for the same cracked hole subjected to a pair of concentrated forces (equal and opposite in direction) on and parallel to the crack surface. However, in this paper, only the symmetric problem will be considered. Once the Green's functions are available, the Mode I stress-intensity factors for cracks emanating from any type of fastener hole can be calculated from a knowledge of the unflawed stress distribution in the region of the hole.

DEVELOPMENT OF THE GREEN'S FUNCTION

Figure 1 shows the scheme of the linear superposition method. The stress intensity factor of problem 1a is equivalent to the sum of that of problems 1b and 1c. Since problem 1b is crack free, the stress-intensity factor of problem 1a is equivalent to that of problem 1c. By idealizing the stress in problem 1c as N discrete loads, $P_1, \ldots, P_N$, then the stress-intensity factor, for a given crack length a, can be computed from the following equation:

$$K(a) = \sum_{i=1}^{N} K_i = \sum_{i=1}^{N} k_i\left(x_i, a\right) P_i\left(x_i\right) \quad (2)$$

where $k_i(x_i,a)$ is the normalized stress-intensity factor due to the ith load, P_i, applied at location x_i. For arbitrary distributed stress, σ, instead of discrete forces, P_i, equation (20) becomes

$$K(a) = \int_0^a k(x,a) \cdot \sigma(x)\, dx \tag{3}$$

By defining $G = k(a/\pi)^{1/2}$ and $\xi = x/a$ and substituting them into equation (3), one obtains

$$K(a) = \sigma_0 \sqrt{\pi a} \int_0^1 \bar{\sigma}(\xi)\, G(a,\xi) \cdot d\xi \tag{4}$$

where σ_0 is the uniform far-field stress and $\bar{\sigma} = \sigma/\sigma_0$ is the normalized unflawed stress distribution on the prospective crack surface.

For a straight crack subjected to two pairs of concentrated forces on the crack surface as shown in Figure 2, the corresponding Green's function, G, is

$$G\left(a, \frac{b}{a}\right) = \frac{K}{P}\left(\frac{a}{\pi}\right)^{1/2} = \left[\left(\frac{a+b}{a-b}\right)^{1/2} + \left(\frac{a-b}{a+b}\right)^{1/2}\right]/\pi \tag{5}$$

The Green's function G, for a double-radial crack emanating from a circular hole and subjected to two pairs of concentrated forces on the fracture surface, as shown in Figure 3a, can be obtained from the computed stress-intensity factor using finite-element analysis with inclusion of the singularity element for various crack lengths a/r and b/a ratios as follows:

$$G\left(\frac{a}{r}, \frac{b}{a}\right) = \frac{K}{P}\sqrt{\frac{a}{\pi}} \tag{6}$$

Due to the limitation of finite element methodology, when the concentrated forces were applied close to the crack tip, say b/a > 0.9, the corresponding Green's function was obtained using the central crack solution by idealizing the hole as a portion of a straight crack as shown in Figure 3b. The Green's function corresponding to this case is

$$G\left(\frac{a}{r}, \frac{b}{a}\right) = \left[\frac{4\left(1 + \frac{r}{a}\right)}{\left(1 - \frac{b}{a}\right)\left(1 + \frac{b}{a} + \frac{2r}{a}\right)}\right]^{1/2}/\pi \tag{6a}$$

The computed Green's functions were then tabulated as a function of a/r and b/a. For any a/r ratio different from those tabulated values, an interpolation or extrapolation technique was used to obtain the corresponding Green's functions.

With a knowledge of the Green's functions, G, and the stress, σ, on the prospective crack surface with the crack absent, one can compute from equation (4) the corresponding stress-intensity factor for any radial crack from a hole.

When crack face overlapping occurs or the applied force P_i is in compression, the computed K_i in equation (2) will become negative. Physically, it means the crack surfaces are closed and react against each other. Occurrences of such cases are illustrated in examples 3 and 4, where the computed negative K_i were set equal to zero.

For a case where there is only one crack emanating from a hole, instead of redeveloping the associated Green's function, it was found that the following equation will give a good estimation of the stress-intensity factor:

$$K_I \Big|_{\text{one crack}} = \frac{B_1}{B_2} K_I \Big|_{\text{two cracks}} \tag{7}$$

where B_1 and B_2 are Bowie's factors for single and double cracks, respectively.

EXAMPLE PROBLEMS

1. Open Holes

To check the validity and accuracy of the present solution, Bowie's [2] solution for a double radial crack emanating from an open hole in an infinite plate was employed. By approximating the unflawed stress distribution as

$$\sigma = \sigma_o\left[1 + \frac{1}{2}\left(\frac{r}{r + x}\right)^2 + \frac{3}{2}\left(\frac{r}{r + x}\right)^4\right] \tag{8}$$

the computed non-dimensional stress-intensity factors using equation (5) for a crack emanating from an open hole subjected to uniaxial and biaxial uniform far-field loading are presented in Figure 4. The corresponding Bowie's solutions are also included in the figure. As can be seen, the current results are within 2 percent of Bowie.

2. Neat-Fit Hole with Fastener Load Transfer

The normalized unflawed tangential stress distributions along the plane perpendicular to the load-line in the hole of a 7075-T6 aluminum plate fitted with a Ti-6Aℓ-4V titanium fastener are given in Figure 5 for various percentages of fastener load transfer. The non-dimensional stress-intensity factors computed using those unflawed stresses and equation (4) for a double crack emanating from the neat-fit hole are presented in Figure 6. From this figure, one can see that the stress-intensity factor for a neat-fit hole without fastener load transfer is lower than that of an open hole (shown as dotted line). However, when the amount of fastener load transfer increases the stress-intensity factor increases rapidly, especially for short crack lengths. The computed non-dimensional stress-intensity factor at the edge of the hole is approximately equal to 1.12 times the normalized unflawed stress at that location. This agrees very well with the edge crack solution. When the crack length is larger than 2½ times the hole radius, the effect of fastener load transfer becomes negligible.

3. Interference-Fit Fastener Holes

The Green's function approach is also used to compute the stress-intensity factor for a double radial crack emanating from an interference-fit fastener hole. Figure 7 shows the unflawed stress distributions for an aluminum plate with a steel interference-fit fastener caused by 0.010 centimeters interference and subsequent edge loading and unloading [16]. The computed stress-intensity factors are shown in Figure 8. From this figure, one sees that for $a/r < 0.5$, when the far-field loading (172 MN/m^2) is removed, the computed K is less than zero. Physically, it means that the fracture surfaces are completely closed and compress each other. The effective stress-intensity factor range equals the difference between curves 2 and 3. For a similar plate with an open hole subjected to 172 MN/m^2 far-field loading, the corresponding K (also ΔK) is plotted in the same figure as dotted lines for comparison purposes. For small crack lengths, the computed ΔK is much smaller for an interference-fit fastener hole than an open hole. This explains why the crack emanating from an interference-fit fastener hole grows much slower than the corresponding crack in an open hole when the crack length is small. When the crack length is longer than 3 times the radius of fastener hole, the growth rates are about the same, since the effective stress-intensity factor ranges are about the same. This indicates that the influence of the interference-fit fastener is negligible when $a/r > 3$.

4. Cold-Worked Holes

Figure 9 shows the unflawed stress distributions in the region of a 4.4% cold-worked hole in a 7075-T6 plate caused by 110 MN/m^2 edge loading and subsequent unloading [17]. After the edge loading is removed, a residual compressive tangential stress remains at the edge of the hole ($a/r < 1$). The computed stress intensity factors using the current approach is presented in Figure 10 as dotted lines. Curve A is the computed K_{max} corresponding to 110 MN/m^2 edge loading while Curve B is the stress-intensity factor range $K_{max} - K_{min}$. K_{min} was computed using the unflawed stress corresponding to 5.5 MN/m^2 edge loading. For the same level of cold-working (4.4%) and edge loadings ($\sigma_{max} = 110$ MN/m^2, $\sigma_{min} = 5.5$ MN/m^2), the stress-intensity factors obtained from crack growth tests reported in references [18] and [19] and the one predicted using the linear superposition method [18] are also included in the figure. As can be seen, the current analysis gives an excellent correlation with the experimental data.

CONCLUSIONS

The Green's function for a double-radial crack emanating from a fastener hole was developed from the computed stress-intensity factor for such a crack loaded with concentrated forces by using the finite element method with the inclusion of a high-order singularity element. The stress-intensity factors for cracks emanating from any type of fastener hole were able to be computed from a knowledge of the unflawed stress distributions in the region of the hole and the developed Green's function. Stress-intensity factors computed using this approach agree very well with known solutions for open holes and neat-fit holes and correlate excellently with data generated using cold-worked hole specimens. The approach can also be used to estimate the stress-intensity factors for cracks emanating from interference-fit fastener holes.

ACKNOWLEDGEMENTS

This work was performed under contract to the United States Air Force Flight Dynamics Laboratory under Contract F33615-75-C-3099.

REFERENCES

1. GRAN, R. J., ORAZIO, F. D., PARIS, P. C., IRWIN, G. R. and HERTZBERG, R., Technical Report AFFDL-TR-70-149, March 1971.
2. BOWIE, O. L., J. Math. and Phys., 35, 1956, 60.
3. BUECKNER, H. E., Zeitschrift für Angewandte Mathematik und Mechanik, 50, 1970, 529.
4. BUECKNER, H. F., Zeitschrift für Angewandte Mathematik und Mechanik, 51, 1971, 97.
5. BUECKNER, H. F., Chapter 5, Methods of Analysis and Solutions of Crack Problems edited by G. C. Sih, Noordhoff International Publishing, 1973.
6. BUECKNER, H. F., International J. of Fracture, 11, 1975, 71.
7. RICE, J. R., International J. of Solids and Structures, 8, 1972, 751.
8. PARIS, P. C., McMEEKING, R. M. and TADA, H., ASTM STP 601, 1976, 471.
9. EMERY, A. F., WALKER, G. E., Jr, and WILLIAMS, J. A., J. Basic Engng., 91, 1969, 618.
10. PARIS, P. C. and SIH, G. C., ASTM STP381, 1965, 30.
11. CREWS, J. H. and WHITE, N. H., NASA TND-6899, Sept. 1972.
12. SCHMIDT, R. A., ASME, ASM 13th Annual Symposium on Fracture and Flaws, Albuquerque, N.M., 1973.
13. SHAH, R. C., ASTM STP 590, 1976, 429.
14. GRANDT, A. F., International J. of Fracture, 11, 1975, 283.
15. ABERSON, J. A. and ANDERSON, J. M., Third NASTRAN User's Colloquium NASA TMX-2893, 1973, 531.
16. BROMBOLICH, L. J., AIAA paper no. 73-252, 1973.
17. ADLER, W. F. and DUPREE, D. M., Technical Report AFML-TR-74-44, July 1974.
18. GRANDT, A. F., Jr. and HINNERICHS, T. D., Proceedings of the Army Symposium on Solid Mechanics, Watertown, Ma., September 1974.
19. CATHEY, W. H., Technical Report AFML-TR-74-283, May 1975.

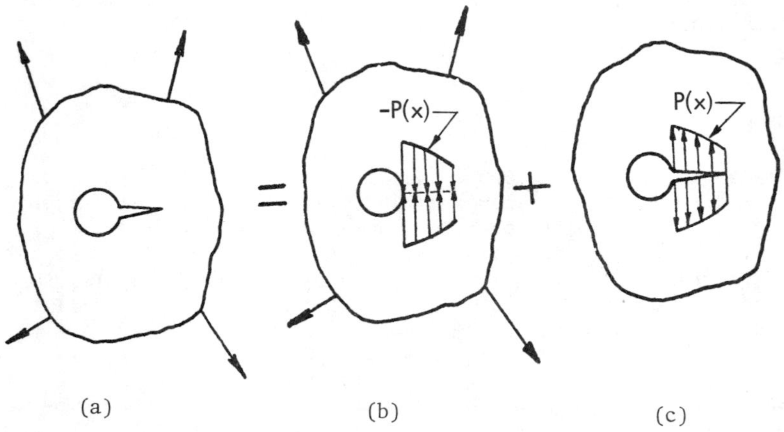

Figure 1 Schematic of Linear Superposition Method

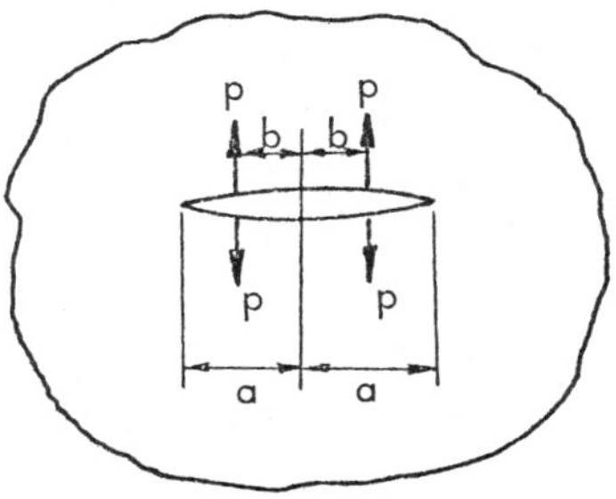

Figure 2

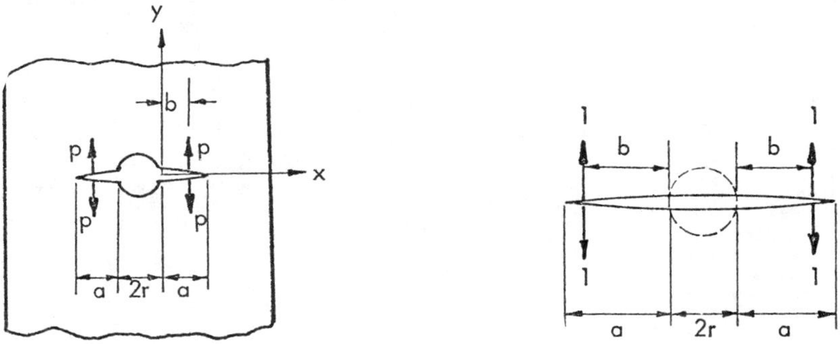

(a) (b)

Figure 3

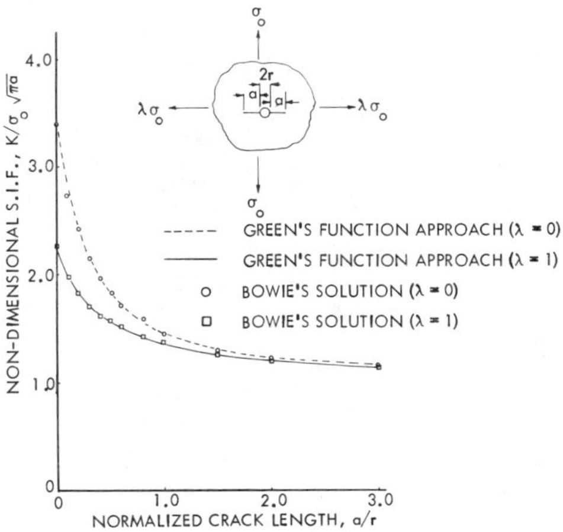

Figure 4 Normalized Stress-Intensity Factors for a Double-Crack Emanating from an Open Hole

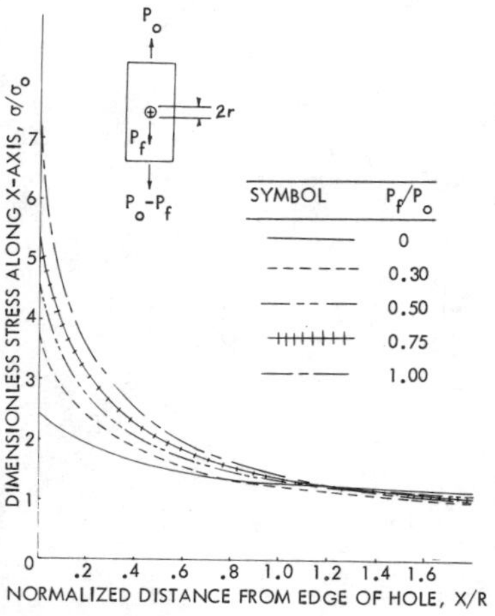

Figure 5 Normalized Unflawed Stress Distributions Along x-Axis in the Plate with Various Amounts of Fastener Load Transfer

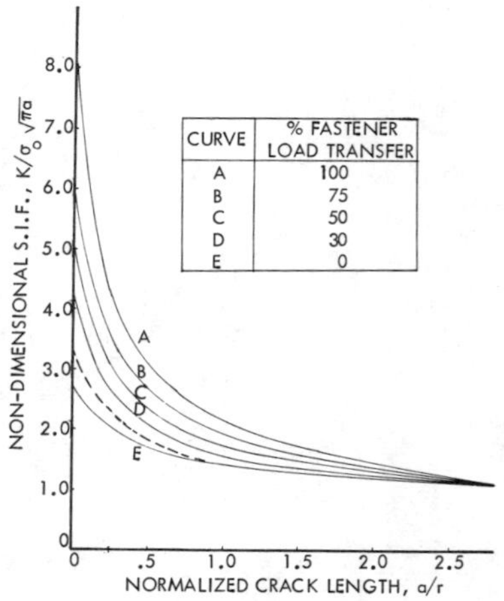

Figure 6 Normalized Stress Intensity Factors for a Double-Crack Emanating from Neat-Fit Hole with Fraction of Fastener Load Transfer

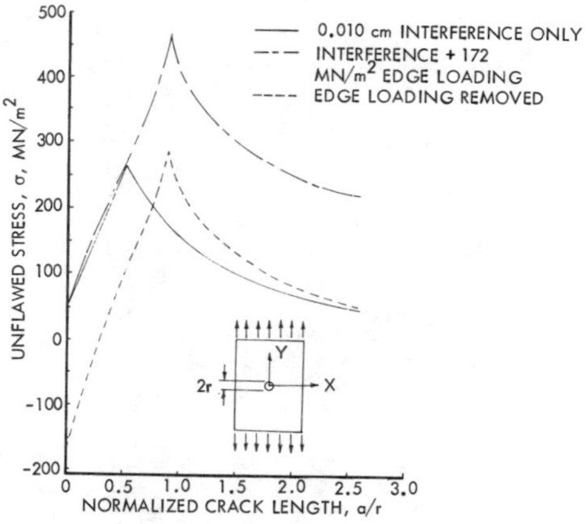

Figure 7 Stress σ_Y at Y = 0 in an Aluminum Plate with a Steel Interface Fit Fastener Caused by Interference, Edge Loading and Unloading

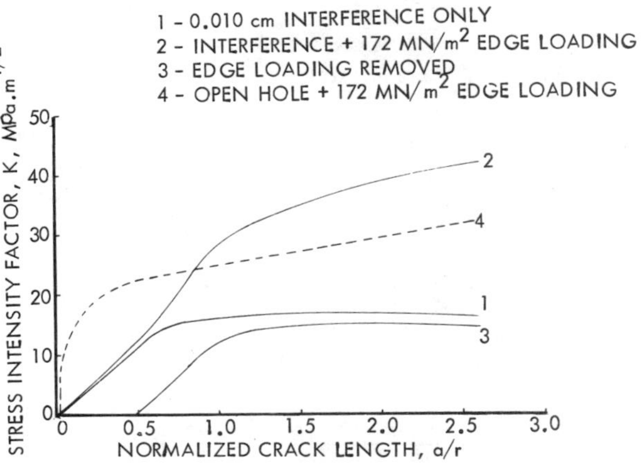

Figure 8 Stress-Intensity Factors for a Double-Crack Emanating from an Interference-Fit Fastener Hole

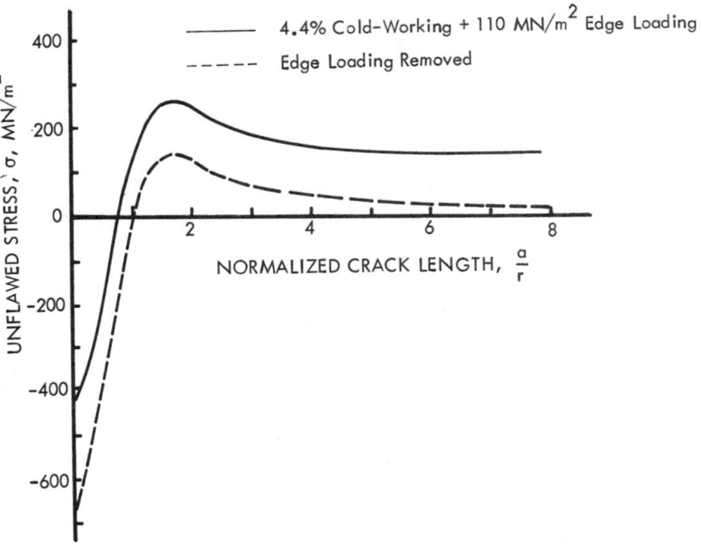

Figure 9 Stress at the Region of 4.4% Cold-Worked Hole in 7075-T6 Aluminum Plate Caused by 110 MN/m^2 Edge Loading and Subsequent Unloading

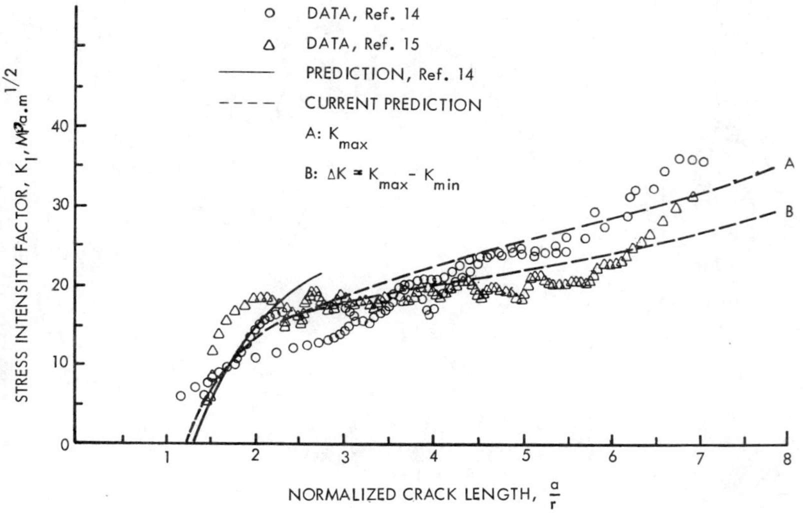

Figure 10 Stress-Intensity Factors for a Single-Crack Emanating from Cold-Worked Hole

Fracture 1977, Volume 3, ICF4, Waterloo, Canada, June 19 - 24, 1977

ELASTIC PLASTIC FRACTURE ANALYSIS OF A CRACKED THICK WALLED CYLINDER UNDER DISPLACEMENT CONTROLLED LOADING

A. Muscati*

INTRODUCTION

Many engineering components are subjected to high thermal and residual stresses that are strain or displacement controlled. Under extreme operating conditions, it is sometimes necessary to show that a large level of plastic deformation is tolerable. For cases where plasticity is limited to a small zone at the crack tip, linear elastic fracture mechanics (L.E.F.M.) can be used to assess the fracture risk in such components; the method is applicable to load, displacement and strain controlled loading. However, as the plastic zone spreads through the structure, the validity of L.E.F.M. becomes questionable and the behaviour of load and displacement controlled problems can vary considerably. There are two main methods of analysis for yielding fracture mechanics problems, the J contour integral [1] and the crack opening displacement [2]. The applicability and limitations of these methods have been discussed in the literature [3] and are beyond the scope of this paper.

Whilst, in principle, elastic-plastic stress analysis is possible using advanced finite element computer programmes, such computations require a large amount of computer time and it is often impracticable to obtain an accurate solution for real engineering problems. In such cases, a pseudo-elastic approximation with some correction for the effect of plasticity may be used as an empirical alternative. For load controlled problems, pseudo-elastic fracture analysis, with no plasticity correction, predicts higher fracture loads compared with an elastic-plastic model such as J [4 and 5]. This suggests that pseudo-elastic analysis is unconservative and could lead to unsafe predictions of failure load [6].

In the present computations, elastic-plastic finite element programmes were used to determine the behaviour of a cracked thick walled cylinder subjected to axial loading resulting in a large level of plasticity with both load and displacement controlled boundary conditions. By using the J contour integral as a fracture criterion, a comparison is made between the pseudo-elastic and elastic-plastic computations for both cases. Obviously, the application of an elastic analysis to problems involving net section yielding cannot be justified on theoretical grounds, but as one is often restricted to an elastic analysis, such a comparison is useful to quantify the error. For strain controlled problems, there are computational difficulties in evaluating the J contour integral as its path independence becomes more questionable [7].

*Central Electricity Generating Board, South West Region, England, U.K.

THE COMPUTATION MODEL

The model was a thick walled cylinder with an axisymmetric crack extending from the bore at mid-axial position. The dimensions were as follows:

> inner diameter 254mm, outer diameter 508mm, axial length = 1.270m and the radial crack depth was 25.4mm.

An elastic-plastic finite element programm (BERSAFE) was used in the present computations [8]. The plasticity part of the programme [9] is based on the incremental theory using the initial stress method for numerical iterations. As the crack is in a plane of symmetry, only half the structure was modelled in the mesh. The crack was simulated by restraining the axial movement of all nodes at the crack plane, ahead of the crack, leaving those at the crack surface free from constraint. The mesh consisted of one hundred (10 x 10) quadrilateral axisymmetric elements with midside nodes as shown in Figure 1.

Both load and displacement controlled boundary conditions were considered. In the load controlled case, a uniformly distributed axial load was applied at the end, whilst for the displacement controlled case both axial and radial displacements were applied at the same position. The radial displacements were small compared with the axial ones (about 10%), they were applied to reproduce the same displacement field created by uniform axial loading.

The J contour integral was calculated by the same computer programme for both elastic and elastic-plastic computations, the path used to determine the integral is shown in Figure 1.

The material data used for these calculations were as follows:

	Yield stress	= 138 MPa
	Youngs modulus	= 207 GPa
and	Poisson's ratio (elastic)	= 0.3

A small amount of work hardening was introduced for the plasticity computations corresponding to a power hardening coefficient of about 0.04.

RESULTS AND DISCUSSION

The results are presented using the non dimensional ratio of J_1/J_{1e}, where J_1 is the contour integral for the opening mode based on the elastic-plastic computations whilst J_{1e} is the corresponding integral based on pseudo-elastic analysis for the same applied load and displacement.

For the load controlled case the ratio of J_1/J_{1e} is plotted against P/P_e in Figure 2, whilst it is plotted against δ/δ_e in Figure 3 for the displacement controlled case, where P_e and δ_e are the load and displacement which cause yielding at any point in the structure for the first time. These parameters (P_e and δ_e) have no fundamental value and are dependent on the mesh size; they were used as a convenient method of presentation. In theory, there is always a small plastic zone at the crack tip and hence there is no first yield event.

For the load controlled case, it was not practicable to obtain a solution beyond P/P_e = 1.71, which corresponds to net section yielding, due to the slow speed of convergence. For the displacement controlled case, there was no convergence problem and a solution was obtained well beyond net section yielding. The plastic zone contours for different displacements are shown in Figure 1 and it is interesting to note that once plasticity spreads through the section, a further increase in the displacement would not have a large effect on the size of the plastic region but would increase the level of plastic strains. As the assumed level of work-hardening is small, the corresponding increase in load beyond net section yielding is also small. Table 1 gives the relation between load and displacement increments for the load controlled case. It can be seen that until the plasticity level approaches net section yielding, the non dimensional displacement and load increments are approximately the same but the ratio of displacement to load increments rises rapidly at net section yielding.

As plasticity spreads across the section, the load controlled case will approach the unstable plastic collapse condition. A small increase in load could lead to large displacements and a rapid increase in J_1 with load, but the pseudo-elastic calculations predict a small change in J_{1e} and the ratio of J_1/J_{1e} would increase rapidly as shown in Figure 2. In the displacement controlled case, the situation is different as the unstable plastic collapse condition is unattainable. The present results show that, beyond net section yielding, the increase in J_{1e} with displacement can be faster than the increase in J and the ratio of J_1/J_{1e} decreases after reaching a maximum value as shown in Figure 3.

For any particular load and displacement, the value of J_1, as determined from elastic-plastic calculations, would be the same whether it is displacement or load controlled, but the value of the elastic J_{1e} could vary considerably. In an elastic analysis the load is proportional to the displacement and for the case of widespread plasticity, the analysis will underestimate the displacement in the load controlled case, whilst it will overestimate the load in the displacement controlled case, which explains the effect shown in Figures 2 and 3. As strain controlled problems are similar, in terms of plastic collapse, to displacement controlled loading, they are expected to show a similar behaviour to that shown in Figure 3.

CONCLUSIONS

Elastic-plastic finite element computations were carried out on a cracked thick walled cylinder under axial loading. When using the J contour integral as a fracture criterion, the error involved in pseudo-elastic analysis has been quantified under different levels of plasticity and for both load and displacement controlled cases. In the load controlled case the elastic analysis underestimated J and the error increased with the increase in plasticity level, but for the displacement controlled case the apparent error was reduced after reaching a maximum value. It is suggested that these results could be used in a qualitative manner to estimate the errors in using elastic fracture analysis for cases of widespread plasticity.

ACKNOWLEDGEMENTS

The permission of Central Electricity Generating Board (C.E.G.B.) for publication is acknowledged. The computer programme used in the analysis was developed with C.E.G.B. at Berkeley Nuclear Laboratories.

REFERENCES

1. RICE, J. R., "Fracture, An Advanced Treatise", ed. H. LIebowitz, Academic Press, 2, 1968, 191.
2. WELLS, A. A., Cranfield Crack Propagation Symposium, 1, 1960, 210.
3. TURNER, C. E., Journal of Strain Analysis, 10, 1975, 207.
4. SUMPTER, J. D. G. and TURNER, C. E., Second Inter. Conf. on Pressure Vessel Technology, San Antonio, 1973, 1095.
5. MUSCATI, A. and TURNER, C. E., Conference on Fracture Mechanics in Engineering Practice, Sheffield, September 1976.
6. DOWLING, A. R. and TOWNLEY, C. H. A., Int. Journal of Pressure Vessel and Piping, 3, 1975, 77.
7. BLACKBURN, W. S., HELLEN, T. K. and JACKSON, A. D., to be published in the Int. J. of Fracture Mechanics.
8. HELLEN, T. K. and PROTHEROE, S. J., Computer Aided Design, 6, 1974, 15.
9. HELLEN, T. K. and HARPER, P. G., C.E.G.B. Report No. RD/B/N3765, Berkeley Nuclear Laboratories, 1976.

Table 1 Load and Displacement Increments for the Load Controlled Case

Inc. No.	Load Inc. $\Delta P/P_e$	Displacement Inc. $\Delta\delta/\delta_e$	Inc. No.	Load Inc. $\Delta P/P_e$	Displacement Inc. $\Delta\delta/\delta_e$
0	1	1	12	.027	.029
1	.051	.051	13	.028	.029
2	.051	.051	14	.027	.030
3	.051	.051	15	.028	.032
4	.051	.051	16*	.032	.050
5	.051	.051	17	.004	.010
6	.051	.051	18	.005	.010
7	.051	.052	19	.004	.012
8	.051	.052	20	.005	.013
9	.051	.052	21	.004	.014
10	.051	.052	22	.005	.015

*net section yielding.

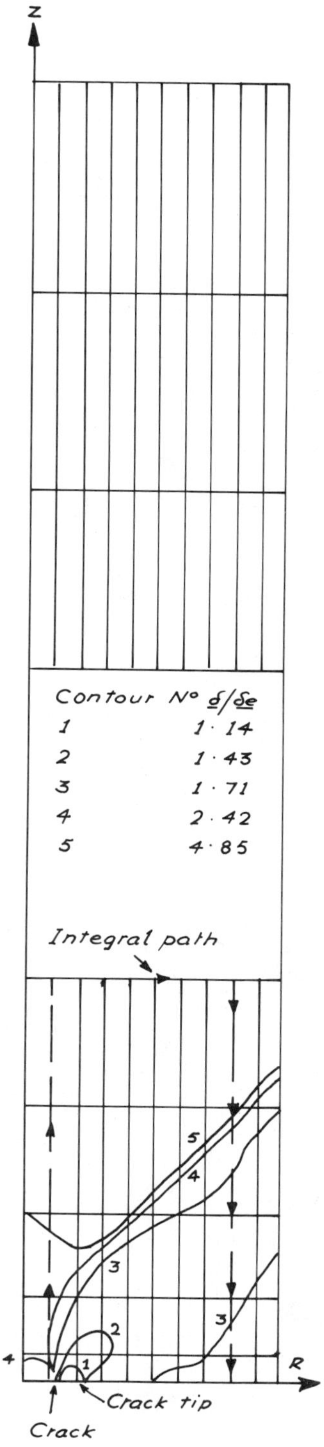

Figure 1 Plastic Zone Contours

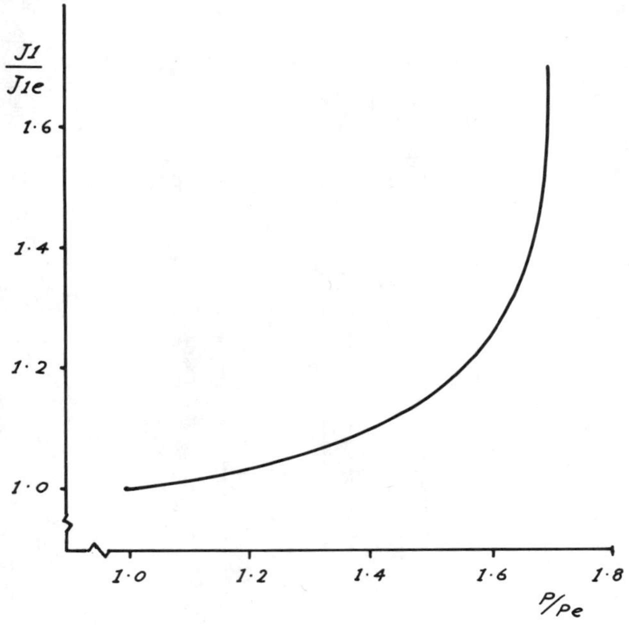

Figure 2 J Results for the Load Controlled Case

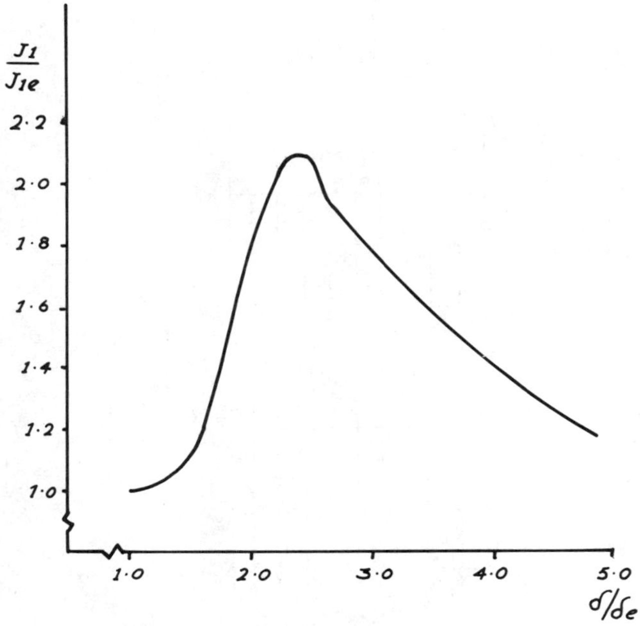

Figure 3 J Results for the Displacement Controlled Case

Fracture 1977, Volume 3, ICF4, Waterloo, Canada, June 19 - 24, 1977

THE EVALUATION OF THE RESISTANCE AGAINST DUCTILE CRACK EXTENSION

H. C. van Elst, H. Wildschut, M. A. Lont and F. H. Toneman*

INTRODUCTION

Single edge notched specimens were torn to complete fracture in a ductile way by excentrical pin loading under constant displacement speed; 3-points bend notched specimens were fractured in the ductile region by impact loading. The load P/deflection f-diagram was recorded and the cracklength a simultaneously filmed.

The resistance R against crack extension was evaluated, using an energy balance analysis, which appears applicable in the elastic, elastic-plastic and fully plastic situation. The importance of a distinction between energy dissipated for crack extension and energy dissipated in an accompanying but alternative way as determined by the test and specimen conditions for indicating the actual crack resistance of the material is argued.

The influences of initial crack tip radius, finiteness of ligament length (w-a) and height h of the specimens are briefly discussed to account for the observed anomalous R-curve behaviour.

This investigation was performed with special reference to the ductile shear fracture as can unstably occur in steel gas pipes if work is performed by the gas on the fracturing vessel wall and aimed at evaluation of the resistance against large scale crack extension as far as determined by the material properties. For the relevant case of sufficiently low crack velocity the displacement speed controlled tests allow to determine R after "unstability" in a quasi static way.

If work A is done on a precracked specimen, as in a "constant displacement speed" controlled experiment, instability of the crack implies that the rate of work can decrease for its further extension. For a sufficiently slow crack velocity $\dot{a}$ the load P can quasi-statically adjust itself to the compliance change and thus decreases. As a consequence the generation of kinetic energy V is then negligible. The energy absorbed during fracture (the fracture energy) contains (recoverable) elastic energy U and dissipated energy W. The latter can for a part W_a be connected to the effective surface energy, mostly consumed to propagate the plastic zone size at the crack tip. However a dissipation of energy W_d can also occur by processes not to be linked directly with the actual crack extension, but occurring where the stress is raised above yield by other influences than the crack tip field (below general yield this is revealed by non-coherent plastic regions possibly involving compressed ones). For this situation the l e f m approach including plastic zone corrections will not be amenable. The total resistance against fracture:

---

*Metal Research Institute TNO, Apeldoorn, Netherlands.

$$R = \frac{dW}{da} = \frac{dW_a}{da} + \frac{dW_d}{da} = R_a + R_d \tag{1}$$

then contains besides the crack resistance R_a or the effective surface energy (this being the material property of primary interest), also a term R_d.

Obviously, R can only be a constant, if $R_d = 0$ or constant, conditioning validness of $J_{Ic} = R_a$ at first physical crack growth, when J_{Ic} is evaluated from an integration of the load/deflection diagram.

Assumedly R_a will be dependent on $\dot{a}$ and possibly on a, while as elaborated below there will be an influence of the finite specimen dimensions on R_a. This is contrary to current (l e f m) material concepts, which claim that a suitable geometrical correction factor in the corresponding mechanical fracture mechanics concepts, allows for a critical fracture mechanics material value not dependent on dimensions or geometry. However, the anticipation that a R_a-value only dependent on $\dot{a}$ can be indicated, characterizing the material for suitably chosen large dimensions with absence of free surface influences, can be considered an instigation to this investigation.

THEORETICAL

In the following an evaluation of fracture resistance during large scale extension - thus not restricted to the quasi static crack growth after first physical crack propagation up to instability, but also for the state thereafter - is suggested from an energy balance analysis.

For the deflection of a loaded specimen one can write:

$$f = f_e + f_p \;;\; f = f(a,P) \tag{2}$$

f_e is the reversible, f_p is the irreversible part of f.

$$CP = C_e P + C_p P \;;\; C_e = C_e(a) \;;\; C_p = C_p(a,P) \tag{3}$$

$$\Delta f = \Delta f_e + \Delta f_p = C_e \Delta P + C_p \Delta P + P\frac{dC_e}{da}\Delta a + P\left(\frac{\partial C_p}{\partial a}\right)_P \Delta a + P\left(\frac{\partial C_p}{\partial P}\right)_a \Delta P \tag{4}$$

C = compliance; index e refers to elastic, index p to plastic.

The energy balance will read:

$$Pdf = \bar{d}A = \bar{d}Q + dW + dU + dV = TdS + Rda + dU + dV \tag{5}$$

T = absolute temperature, S = entropy.

For the case of interest here, dV = 0 as elaborated above. It will moreover be understood that TdS is incorporated in R, or rather with the right proportion in R_a and R_d.

$$Pdf = \left\{PC_e + PC_p + P^2\left(\frac{\partial C_p}{\partial P}\right)_a\right\}\frac{dP}{da}\,da + \left\{P^2\,\frac{dC_e}{da} + P^2\left(\frac{\partial C_p}{\partial a}\right)_P\right\}da \tag{6}$$

and also for unit thickness:

$$Pdf = dW + dU = Rda + d\left(\frac{C_e P^2}{2}\right) = Rda + \left\{C_e P\,\frac{dP}{da} + \frac{P^2}{2}\,\frac{dC_e}{da}\right\}da \tag{7}$$

$$R = \frac{P^2}{2}\,\frac{dC_e}{da} + \left\{PC_p + P^2\left(\frac{\partial C_p}{\partial P}\right)_a\right\}\frac{dP}{da} + P^2\left(\frac{\partial C_p}{\partial a}\right)_P \tag{8}$$

Note I:

1° $$\frac{dU}{da} = C_e P\,\frac{dP}{da} + \frac{P^2}{2}\,\frac{dC_e}{da} \tag{9}$$

2° If P = constant, $$R = \frac{P^2}{2}\,\frac{dC_e}{da} + P\left(\frac{\partial C_p}{\partial a}\right)_P \tag{10}$$

3° If l e f m applies, i.e.

$$C_p = 0\ ;\ C = C_e \text{ and } R_e = \frac{P^2}{2}\,\frac{dC}{da} \tag{11}$$

4° $$R = R_e + R_p \tag{12}$$

5° $$\bar{d}A = Pdf = dU + R_e da + R_p da \tag{13}$$

R has only to be interpreted as R_a, when $W_d = 0$ of $dW_d/da = 0$. Recordings of a load/displacement curve for a specimen of interest e.g. an excentrical pin loaded single edge notched (e p l s e n) specimen, while unloading and reloading again for certain cracklengths obtained by increasing crack extension, allow to estimate $C_e = C_e(a)$. This can be compared with a numerically evaluated $C_e = C_e(a)$ using finite element method technique. If the agreement is satisfactorily verified the latter can substitute the former (cf. however Results). R can now be correspondingly determined.

Note II:

If no kinetic energy is generated, i.e. the load adjusts itself to the compliance at each moment:

$$\dot{A} = P\dot{f} = R + \frac{dU}{da}\,\dot{a} \tag{14}$$

For dU/da << R this implies:

$$\dot{a} = \frac{\dot{A}}{R_a + R_d} \tag{15}$$

EXPERIMENTAL

Single edge machine notched specimens were torn to complete fracture by excentrical pin tensile loading, sometimes wedge loading, using a 250 tons MTS-closed loop system equipment in a displacement speed controlled way. The crack proceeded in a ductile way at velocities, which were so small (10 cm/sec) that at each moment the load could adjust itself to the increasing compliance in a quasi static way.

Three line pipe steels, with properties as indicated in Table 1, were tested in 3-fold, using e p l s e n-specimens at displacement speeds of 9 and 25 mm/sec; one material was moreover tested in 3-fold using wedge loaded single edge notched (w l s e n) specimens at the same displacement speeds. The load-deflection diagram was recorded; the extending crack was filmed.

As the w l s e n-specimen showed a severe liability to buckling, all specimens were adapted with stiffener beams, bolted at both sides of the specimen and meeting as a hinge just outside the specimen opposite the advancing crack tip. The applied specimen dimensions are shown in Figure 1. Initial machined notches of mostly 60, also 100 and 140 mm were introduced. For each material one extra specimen was used for finding the elastic compliance as a function of cracklength. This proceeded by unloading after some crack extension, then loading again, and repeating this after some additional crack extensions, several times, till the ligament was completely torn.

A small hysteresis was observed by unloading and reloading, slightly increasing with increasing cracklength.

This did not interfere with a sufficiently accurate estimate of the elastic compliance C_e as a function of a. The dependence of the elastic compliance on cracklength both with and without stiffeners to the specimen as observed was compared with a numerical evaluation using the finite element method, applying a substructuring technique. The relevant elements division is illustrated in Figure 2.

This allows in principle to interpret the cracklength as observed at the surface in a for the compliance effective one, in this way taking into account the protruding of the crack tip in the core of the plate specimen.

Single edge machine notched specimen (with a height of 100 mm and initial notch length of 15 mm; cf. Figure 1c) were dynamically tested in 3-points bending by a falling weight of 280 kg mass from a height of 2.60 m. The striker was straingauge instrumented, allowing to record load in time. The crack propagation was filmed (using a Dynafax camera).

RESULTS

For the e p l s e n specimens R/a-curves as inferred from the load/deflection diagram and corresponding $\dot{a}$/a and R/$\dot{a}$ curves are shown in Figures 3a, 3b, 3c respectively. The low $\dot{a}$ and its small variations in these experiments obviously did not influence R. After a preambulary phase the R-curves appear rather independent of a_0, the initial notch length. A recording for determining the elastic compliance dependence on a is shown in Figure 4a. The deduced C_e(a)-curve (cf. Figure 4b) did not quite agree with the numerically evaluated one using the finite element method for the stiffened specimens. The dissipated energy W to accomplish a certain crack extension was determined from the load/deflection diagram as well, duly taking into account the elastic energy correction. Cf. Figure 5. The W/a curve was polynomium fitted as:

$$W = W_0 + W_1(a-a_0) + W_2(a-a_0)^2 + W_3(a-a_0)^3 + W_4(a-a_0)^4 \quad (16)$$

with a_0 = initial notch length.

The R-value at $a = a_R$ for which $d^3W/da^3 = 0$, was taken as R-reference for the material (cf. Discussion). For these and other data of the investigated line pipe steels cf. Table 1. Results of R/a, $\dot{a}$/a and R/$\dot{a}$-curves for the striker instrumented drop weight tear tests are given in Figure 6.

DISCUSSION

The start of physical crack extension as observed at the surface will lag behind to that actual according in the core of the material thickness. If the protruding of the crack tip still increases, the real crack extension will be underestimated by a surface crack observation, implying an overestimate of R. Recordings of crack extension by an electrical potential method can overcome this difficulty, but due to influences of the plastic development on the electrical resistance the physical crack extension effect tends to get blurred. If the initial notch is machined as in most of these experiments one might expect that the finite tip radius will cause first physical crack extension to occur for a R_a substantially larger than J_{Ic}. The development of a possible sharper crack tip after some crack extension will have a corresponding lowering effect on R. However, experiments on fatigue cracked specimens showed only a minor decreasing effect for the initial R-values as compared to those occurring for the machine notched specimens.

A dissipation of energy W_d not connected with the crack extension will be more pronounced in the first phase of the experiment as well, when dealing with a high load. It is due to bending of the separated parts, deformation of the bolt holes etc.

When general yield occurs, possibly already at the onset of real crack growth, less plastic work will be dissipated pro cm crack extension the more the ligament decreases, thus decreasing R with increasing a. The deformation has to comply with the displacements as enforced by the stiffeners, implying at each moment a displacement, where these are localized, linearly approaching zero with the distance to the hinge. The resulting constraint on plastic development will progressively decrease R the more the crack tip has penetrated through the ligament.

The above mentioned effects all contribute to the anomalous R-curve behaviour showing instead of the initial increase, merging into a rather constant level, a decrease with an inflexion point, i.e. with a $dR/da < 0$ showing a maximum, for which thus $d^3W/da^3 = 0$ (and $d^4W/da^4 < 0$).

If the disturbing influence of the initial phase at real crack extension implying an overestimate of R, overlaps the region of influence of the terminal phase for which R tends to get underestimated, an anticipated constant R-level will be obscured, but still introduces this inflexion point in the R/a-curve. This was used to approximate the undisturbed R-level. The impact test results obtained until now suggested that a rather constant R-level might become realized in the advanced phase of crack extension, cf. Figure 6a. The small differences between the line pipe materials could be revealed by separation of the R/a-curves for the e p l s e n-specimens in a significant way. Cf. Figure 7.

The influence of dimensions on estimated R-values is yet not quite clear. Besides the finite ligament width also the height of the specimen might exercise constraining influences on the plastic flow development connected with the crack extension.

Instead of aiming at a plastic flow development as free as possible by increase of dimensions, one could also try to achieve a high constraint. Therefore sidegrooving along the anticipated crack path was introduced. These experiments are still in progress. While in general can be referred to [1], as recently appeared related work can be mentioned [2], [3] and [4].

CONCLUSION

Contrary to J_{Ic}-evaluations, requiring the detection of the onset of first physical crack growth and rejecting (large scale) crack extension, the R/a-determination can proceed in a quasi static way for large scale ductile crack extension with velocities up to 1-100 m/sec. The only restriction appears a possible dissipation of energy not to be identified with that required for crack extension. (This will unfavourably interfere with all other fracture mechanics determinations). Estimates of G_{Ic} or J_{Ic} from small specimens, using the R/a-determination appear quite feasible, (extrapolation from small crack extensions to zero offering) the J_{Ic}-values, showing only slight discrepancies with R. As the irreversibility of the deflection is incorporated in the analysis for determining R there is no need to discuss "pseudo potential energy".

ACKNOWLEDGEMENT

This work was carried out in the phase IV part of the European Pipeline Research Group (EPRG)-programme on ductile fracture in gas pressurized pipelines - and sponsored by the European Community for Steel and Carbon under contract nr. 6210.46/6/601 and the EPRG-members: Nederlandse Gasunie British Gas Corporation and British Steel Corporation. We are also indebted for fruitful discussions on the results by these sponsors.

REFERENCES

1. ASTM STP 526, "Fracture Toughness Evaluation by R-curve Methods".
2. GARWOOD, S. J., ROBINSON, J. N. and TURNER, C. E., Int. Journal of Fracture 11, 1975, 528.
3. MAI, Y. W., ATKINS, A. J. and CADDELL, R. M., Int. Journal of Fracture 12, 1976, 391.
4. FEARNEHOUGH, G. D., DUCKSON, D. T. and JONES, D. G., Int. Conf. on Dynamic Fracture Toughness (Paper 29), London, July 5 - 7, 1976.

Table 1

line pipe steels cast	yield strength [MN/m^2]	ultimate tensile strength [MN/m^2]	area reduction in tensile test [%]	2/3 Charpy V at 0°C [MN/m]	R [MN/m]	a_R [mm]
4338	427	524	69.0	1.09	2.5	144
7898	460	564	72.0	1.71	3.5	130
8380	420	502	76.5	2.12	4.5	125

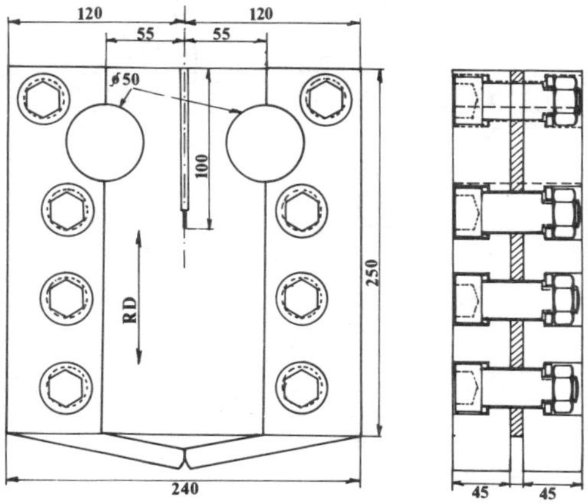

Figure 1a EpIsen - Specimen, Provided with Stiffeners Hinge

continued

continued

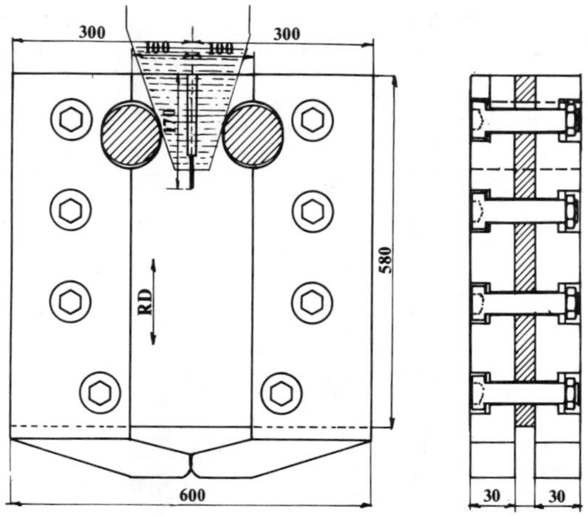

Figure 1b Wlsen - Specimen, Provided with Stiffeners Hinge

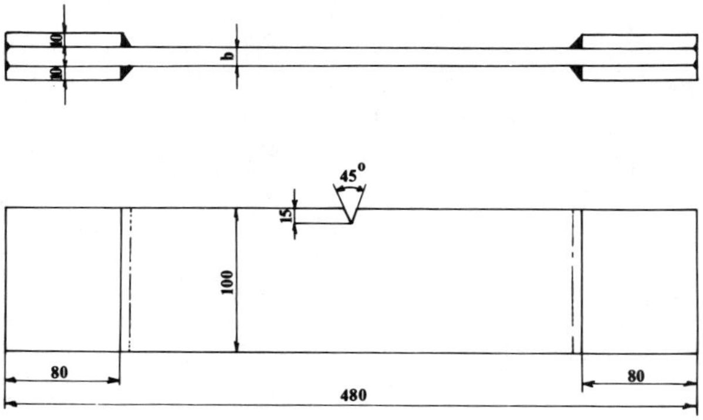

Figure 1c Three-Points Bend Sen-Specimen for Impact Testing

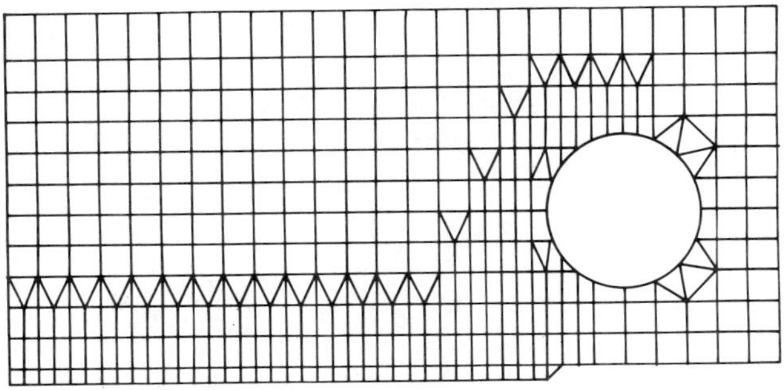

Figure 2 Elements Distribution for Numerical Evaluation of Dependence of Elastic Compliance on Cracklength, Using the Finite Element Method with a Substructuring Technique

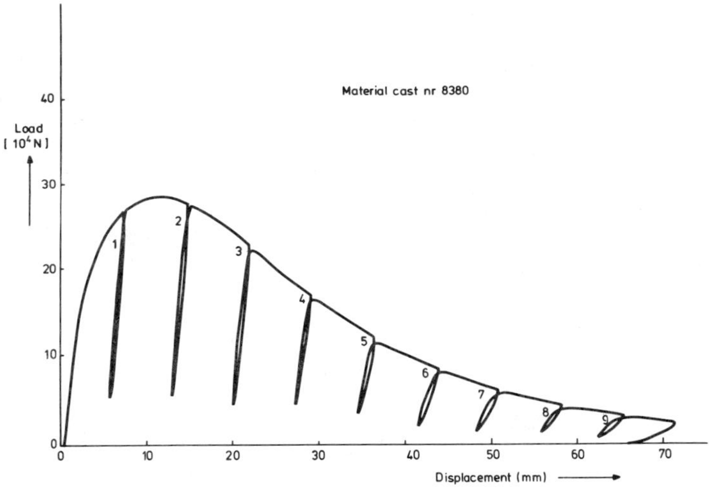

Figure 3a Recordings of Load/Deflection Diagram for Eplsen-Specimen at Repeated Unloading and Loading After Subsequent Crack Extension for Evaluation of Dependence of Elastic Compliance on Cracklength

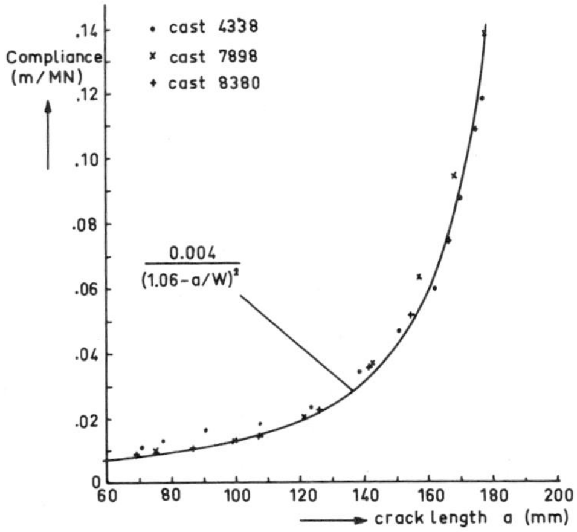

Figure 3b Elastic Compliance Dependence on Cracklength According to Recordings Shown in Figure 3a

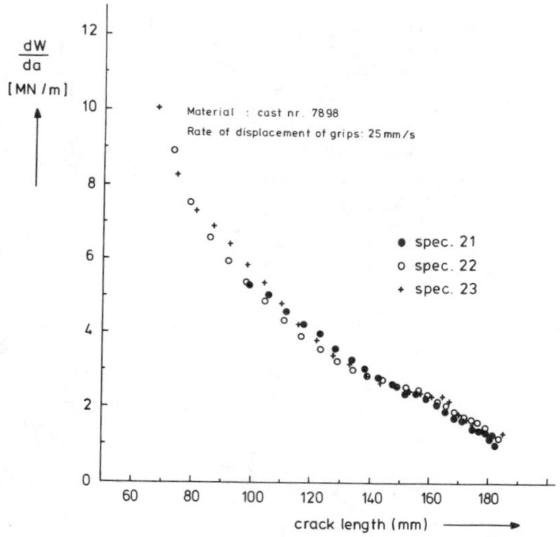

Figure 4a Fracture Resistance Dependence on Cracklength in Tested Eplsen-Specimens at Displacement Speed of 25 mm/sec

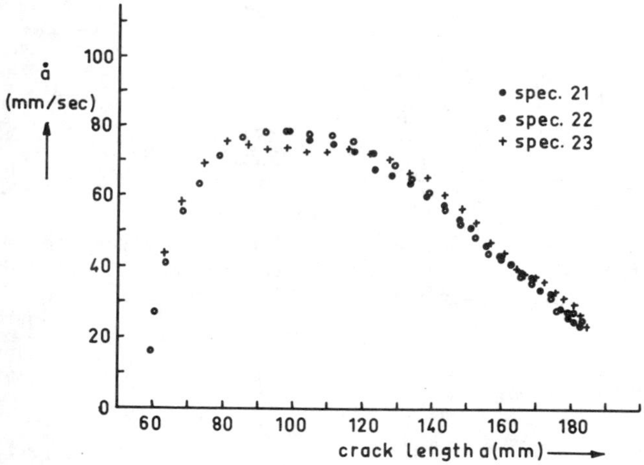

Figure 4b Fracture Velocity Dependence on Cracklength in Tested Eplsen-Specimens at Displacement Speed of 25 mm/sec

continued

continued

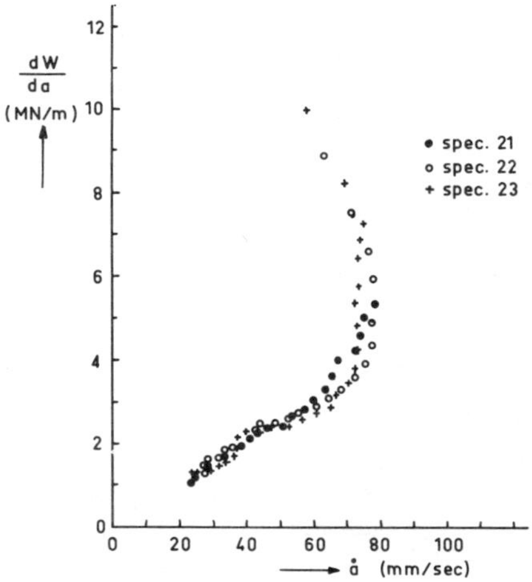

Figure 4c Fracture Resistance Dependence on Crack Velocity in Tested Eplsen-Specimens at Displacement Speed of 25 mm/sec

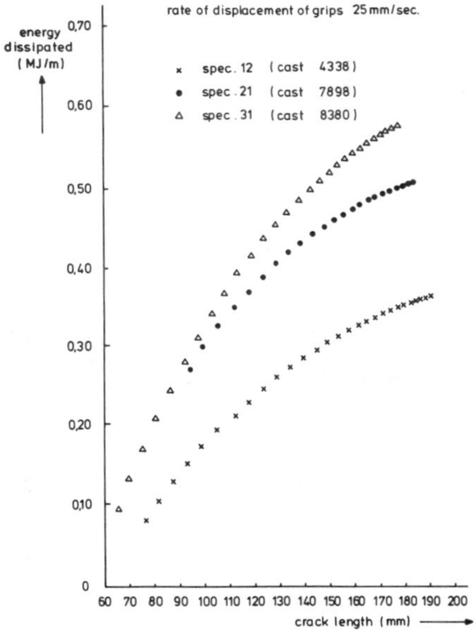

Figure 5 Energy Absorbed as a Function of Cracklength for Extending Crack in Eplsen-Specimens at Displacement Speed of 25 mm/sec

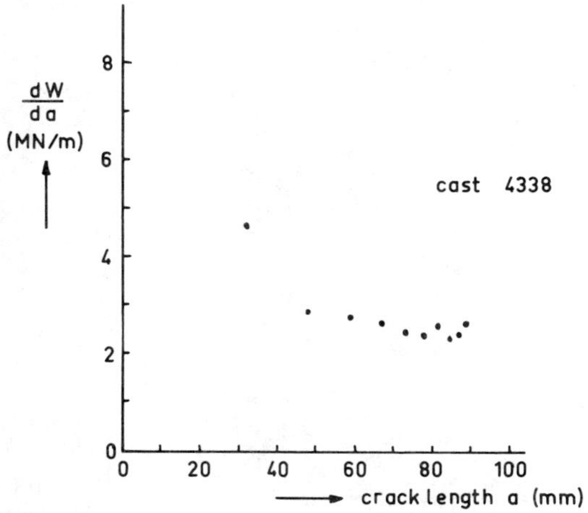

Figure 6a Fracture Resistance Dependence on Cracklength in Impact Tested Three-Points Bend Sen-Specimen

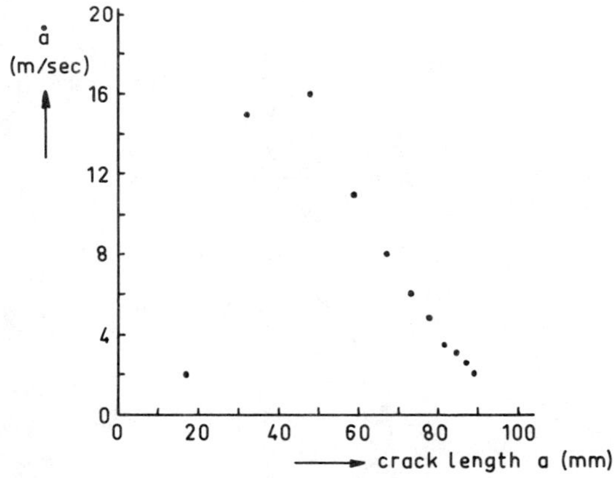

Figure 6b Fracture Velocity Dependence on Cracklength in Impact Tested Three-Points Bend Sen-Specimen

continued

continued

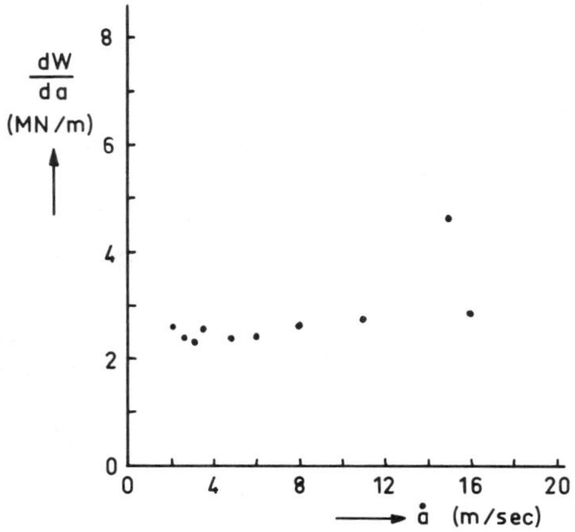

Figure 6c Fracture Resistance Dependence on Crackspeed in Impact Tested Three-Points Bend Sen-Specimen

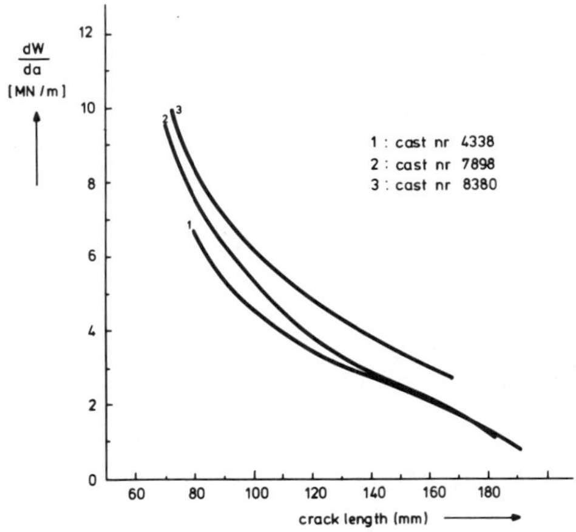

Figure 7 Average R/a Curves for Three Line Pipe Steels, According to Testing of Eplsen Specimens

A DESCRIPTION OF THE DEPENDENCE OF FRACTURE RESISTANCE ON CRACKLENGTH

H. C. van Elst[1]

INTRODUCTION

By evaluation of the absorbed plastic energy at the cracktip up to the moment of instability an expression for the fracture resistance in critical cracklength, critical gross stress and average reciprocal modulus over the plastic zone proves possible. Taking into account some conditions relevant for instability (like e.g. the Orowan conditions) and using the relevant expression for the energy release rate the fracture resistance curve appears from a small number of instability tests.

It is assumed according to Kraaft [1] that the absorbed energy pro cm crack extension at stable crack growth does not depend on initial crack-length, but only on the increase of cracklength a. By observing the stable crack extension at any time one can find dU/da = elastic energy rate and for an infinite plate of unit thickness, uniaxially loaded transversal to a crack a up to a gross stress σ_0 for the elastic case with E = Young's modulus:

$$R = \frac{dW}{da} = \frac{dU}{da} = \frac{2\pi\sigma_0{}^2 a}{E} \tag{1.1a}$$

For a finite plate $R = \tfrac{1}{2}\sigma_0{}^2 w^2 (dC/da)$ with w = width and C = compliance, and for the infinite plate $C = 1/EwB + 2\pi a^2/Ew^2 B$ with B = thickness, yielding (1.1a).

When the increasing $2\pi\sigma_0{}^2 a/E$ for increasing a and σ_0 can be no longer matched by an increase of R = dW/da of the material, instability occurs and (cf. Figure 1, Orowan [2]):

$$R_c = \left\{\frac{dW}{da}\right\}_c = \frac{2\pi\sigma_{0c}{}^2 a_c}{E}, \text{ while } \left\{\frac{d^2W}{da^2}\right\}_c = \frac{2\pi\sigma^2_{0c}}{E} \tag{1.1b}$$

The R-curve can thus be obtained by a series of tests revealing instability at some cracklengths. If a reliable R vs. a description proves possible, the number of required instability tests might be greatly reduced.

CRITICAL ENERGY EVALUATION NEAR CRACKTIP AT INSTABILITY

For the absorbed energy W (cf. Figure 2) is written:

$$\frac{W}{2} = \int_{-\pi}^{\pi} d\Theta \int_0^{R(\Theta)} r(\Theta)dr(\Theta) \int_{0{\cdot}2}^{\varepsilon^*_c\{r(\Theta)\}} Y[\varepsilon^*\{r(\Theta)\}]\, d\varepsilon^*\{r(\Theta)\} \tag{2.1}$$

---

[1]Metal Research Institute TNO, Apeldoorn, Netherlands.

with:

- $r(\Theta)$ = radius from notchtip to considered point in plastic zone
- Θ = angle between notch direction and radius
- $R(\Theta)$ = elastic-plastic boundary at instability
- $\varepsilon\{r(\Theta)\}$ = effective deformation in $r(\Theta)$
- c = index referring to instability
- $Y[\varepsilon^*\{r(\Theta)\}]$ = uniaxial yield stress for effective deformation

Take:

$Y[\varepsilon^*\{r(\Theta)\}] = \sigma_0\, f(s)$, with

$\sigma_0 = \zeta[\varepsilon^*\{r(\Theta)\}]\sigma_0$ = gross stress, when in $r(\theta)$ the effective deformation is $\varepsilon^*\{r(\theta)\}$

$0 < \zeta < 1$

$\zeta \rightarrow 1$ for $\varepsilon^* \rightarrow \varepsilon_c^*$ and $\sigma_0 \rightarrow \sigma_{0c}$

$$f(s) = \sum_{k,j} \sqrt{1/2\left[s_k\,[\varepsilon^*\{r(\Theta)\}] - s_j\,[\varepsilon^*\{r(\Theta)\}]\right]^2}$$

$k, j = 1, 2, 3\ ;\ k \neq j$

$s_k[\varepsilon^*\{r(\Theta)\}]\ \zeta\ [\varepsilon^*\{r(\Theta)\}]\ \sigma_{0c} = \sigma_k\ [\varepsilon^*\{r(\Theta)\}]$

k = 1, 2, 3

σ_k = principal stress

If:

$E^* = E^*[\varepsilon^*\{r(\Theta)\}]$ = local modulus of the Y vs. ε^* curve

$$\frac{W}{2} = \sigma_{0c}^{\ 2} \int_{-\pi}^{\pi} d\Theta \int_0^{R(\Theta)} r(\Theta)dr(\Theta) \int_0^{\varepsilon_c^*\{r(\Theta)\}} \zeta f(s)\, d\left\{\frac{\zeta f(s)}{E^*}\right\} \tag{2.2a}$$

$$\frac{W}{2} = \sigma_{0c}^{\ 2} \int_{-\pi}^{\pi} d\Theta \int_0^{R(\Theta)} r(\Theta)dr(\Theta) \int_0^{\varepsilon_c^*\{r(\Theta)\}} \left[\zeta^2\left\{\frac{f}{E^*} - \frac{f^2}{E^{*2}}\frac{dE^*}{df}\right\}\frac{df}{d\varepsilon^*} + \frac{\zeta f^2}{E^*}\frac{d\zeta}{d\varepsilon^*}\right] d\varepsilon^* \tag{2.2b}$$

Taking:

$$1/2\,\beta\left[\varepsilon_c^*\ \{r(\Theta)\}\right] \equiv \int_0^{\varepsilon_c^*\{r(\Theta)\}} \left[\zeta^2\left\{\frac{f}{E^*} - \frac{f^2}{E^{*2}}\frac{dE^*}{df}\right\}\frac{df}{d\varepsilon^*} + \frac{\zeta f^2}{E}\frac{d\zeta}{d\varepsilon}\right] d\varepsilon^* \tag{2.3}$$

$$\frac{W}{2} = \sigma_{0c}^{\ 2} \int_{-\pi}^{\pi} d\Theta \int_0^{R(\Theta)} r(\Theta)dr(\Theta)\ 1/2\,\beta\left[\varepsilon_c^*\{r(\Theta)\}\right] \tag{2.4a}$$

$$\frac{W}{2} = 1/2\bar{\beta}_c \, \sigma_{oc}^2 \int_{-\pi}^{\pi} \frac{R(\Theta)^2}{2} d\Theta = 1/2\bar{\beta} \, \sigma_{oc}^2 \int_{-\pi}^{\pi} \frac{\rho(\Theta)^2}{2} d\Theta \tag{2.4b}$$

THE EVALUATION OF R FROM W_c IN THE ELASTIC CASE

In the case that the plastic zone is sufficiently small an easy calculation of the plastic zone boundary can proceed from the goniometric Sneddon approximations of Westergaard's description of the stresses near the cracktip in the elastic case.

Then one has, using reduced stresses $s = \sigma/\sigma_o$:

$$s_x = \sqrt{\frac{a}{2r}} \cos\frac{\Theta}{2}\left\{1 - \sin\frac{\Theta}{2}\sin\frac{3\Theta}{2}\right\}$$

$$s_y = \sqrt{\frac{a}{2r}} \cos\frac{\Theta}{2}\left\{1 + \sin\frac{\Theta}{2}\sin\frac{3\Theta}{2}\right\}$$

$$t_{xy} = \sqrt{\frac{a}{2r}} \sin\frac{\Theta}{2}\cos\frac{\Theta}{2}\cos\frac{3\Theta}{2}$$

with as principal reduced stresses:

$$s_{1,2} = \sqrt{\frac{a}{2r}} \cos\frac{\Theta}{2}\left\{1 \pm \sin\frac{\Theta}{2}\right\};$$

$$s_3 = 0 \text{ for plane stress} \tag{3.1}$$

$$= 2\nu\sqrt{\frac{a}{2r}} \cos\frac{\Theta}{2} \text{ for plane strain}$$

Combining this with the Von Mises-Hencky yield criterion:

$$2Y_0^2 = (\sigma_1 - \sigma_2)^2 + (\sigma_1 - \sigma_3)^2 + (\sigma_2 - \sigma_3)^2 \tag{3.2}$$

one arrives at:

$$R(\Theta) = \frac{K^2}{2rY^2} \cos^2\frac{\Theta}{2}\left\{1 + 3\sin^2\frac{\Theta}{2} - p\right\} \tag{3.3}$$

$p = 0$ for plane stress; $p = 4\nu(1 - \nu)$ for plane strain with K referring to the mathematical cracklength a_e.

$$a_e = a + \alpha\frac{K^2}{2\pi Y^2} \tag{3.4}$$

$\alpha = 1$ for plane stress; $\alpha = 1/3$ for plane strain.

At plane stress:

$$\rho(\Theta) \equiv \frac{R(\Theta)}{a} = \frac{\sigma_{oc}^2}{2Y_0^2}\left\{1 + \frac{\sigma_{oc}^2}{2Y_0^2}\right\}\cos^2\frac{\Theta}{2}\left\{1 + 3\sin^2\frac{\Theta}{2}\right\} \tag{3.5}$$

and

$$W = \bar{\beta}_c \, \sigma_{oc}^2 \, a^2 \frac{\sigma_{oc}^4}{4Y_0^4}\left\{1 + \frac{\sigma_{oc}^2}{2Y_0^2}\right\}^2 \int_{-\pi}^{\pi} \cos^4\frac{\Theta}{2}\left\{1 + \sin^2\frac{\Theta}{2}\right\}^2 d\Theta \tag{3.6}$$

$$= \bar{\beta}_c \, \sigma_{oc}^2 \, a^2 \left\{\frac{f_0^4}{4} + \frac{f_0^6}{6} + \frac{f_0^8}{16}\right\}\frac{123\pi}{64} = CY_0^2 a^2 \bar{\beta}_c F_0 \tag{3.7}$$

with: $f_0 = \frac{\sigma_{0c}}{Y_0}$; $C = \frac{123\pi}{64} = 6.038$; $F_0 = \frac{f_0{}^6}{4} + \frac{f_0{}^8}{6} + \frac{f_0{}^{10}}{16}$

$cF_0a^2/f_0{}^2$ obviously represents the surface (volume) of the plastic zone (cf. Table 1).

$\overline{\beta}_c$ = "plastic zone averaged" reciprocal modulus.

$$\overline{\beta}_c Y^2 f_0{}^2 = \overline{\beta}_c \sigma_{0c}{}^2 \tag{3.8}$$

represents the average energy density.

At instability

$$R_a = \left\{\frac{dW}{da}\right\}_c = \frac{dU}{da} \text{ and } \left\{\frac{d^2U}{da^2}\right\}_c = \left\{\frac{d^2W}{da^2}\right\}_c, \text{ (Orowan [2])} \tag{3.9}$$

It is moreover assumed that for the critical stress σ_{0c} at instability:

$$\frac{d\sigma_{0c}}{da} = 0, \text{ implying } \frac{df_0}{da} = 0 \text{ and } \frac{dF_0}{da} = 0 \tag{3.10}$$

Then:

$$R_a = \left\{\frac{dW}{da}\right\}_c = CaF_0Y_0{}^2 \left\{2\overline{\beta}_c + a\frac{d\overline{\beta}}{da}\right\} \tag{3.11}$$

For an infinite plate this equals:

$$\frac{2\pi}{E} Y_0{}^2 f_0{}^2 a \tag{3.12}$$

while

$$2CF_0Y_0{}^2 \left\{\overline{\beta}_c + 2a\frac{d\overline{\beta}_c}{da} + a^2\frac{d^2\overline{\beta}_c}{da^2}\right\} = \frac{2\pi}{E} Y_0{}^2 f_0{}^2 \tag{3.13}$$

Then:

$$2\overline{\beta}_c + a\frac{d\overline{\beta}_c}{da} = 2\left\{\overline{\beta}_c + 2a\frac{d\overline{\beta}_c}{da} + a^2\frac{d^2\overline{\beta}_c}{da^2}\right\} \tag{3.14}$$

or

$$\frac{d^2\overline{\beta}_c}{da^2} \Big/ \frac{d\overline{\beta}_c}{da} = -\frac{3}{2a} \rightarrow \frac{d\overline{\beta}_c}{da} = Aa^{-3/2}; \quad \frac{d^2\overline{\beta}_c}{da^2} = -\frac{3}{2}Aa^{-5/2} \tag{3.15}$$

$$\overline{\beta}_c = -2Aa^{-1/2} + \beta_0 \tag{3.16}$$

(A and β_0 constants)

It might be noted that for:

$$\frac{a_0}{\Delta a_0} = \frac{R_a}{\left\{\frac{dR}{da}\right\}\Delta a_0} - 1 \qquad (3.17)$$

$$\left.\begin{aligned} R_a &= cF_0a(2\beta_0 - 3Aa^{-1/2}) \\ \frac{dR_a}{da} &= 2cF_0(\beta_0 - 3/2\ Aa^{-1/2}) \end{aligned}\right\} \rightarrow R_a \Big/ \frac{dR_a}{da} = a \qquad (3.18)$$

For a cracked infinite plate in the elastic case:

$$f_0^2 = \frac{\sigma_0^2}{Y_0^2} = \frac{ER}{2\pi a Y_0^2} \left\{ \text{as } \frac{dU}{da} = \frac{2\pi\sigma_0^2 a}{E} \right\} \qquad (3.19)$$

$$F_0 = \frac{(f_0^2)^3}{4} + \frac{(f_0^2)^4}{4} + \frac{(f_0^2)^5}{16} = 1/4\left\{\frac{ER}{2\pi a Y_0^2}\right\}^3 + 1/4\left\{\frac{ER}{2\pi a Y_0^2}\right\}^4 + \frac{1}{16}\left\{\frac{ER}{2\pi a Y_0^2}\right\}^5 \qquad (3.20)$$

Thus:

$$R = CaF_0Y_0^2\left\{2\bar{\beta}_c + a\frac{d\bar{\beta}_c}{da}\right\} = C\ F_0Y_0^2a\ (2\beta_0 - 3Aa^{-1/2}) \qquad (3.21)$$

$$R = 1/4\,CY_0^2a(2\beta_0 - 3Aa^{-1/2})\left[\left\{\frac{ER}{2\pi a Y_0^2}\right\}^3 + \left\{\frac{ER}{2\pi a Y_0^2}\right\}^4 + 1/4\left\{\frac{ER}{2\pi a Y_0^2}\right\}^5\right] \qquad (3.22)$$

or:

$$\frac{EC}{8\pi} = (2\beta_0 - 3Aa^{-1/2})\left[\left\{\frac{ER}{2\pi a Y_0^2}\right\}^2 + \left\{\frac{ER}{2\pi a Y_0^2}\right\}^3 + 1/4\left\{\frac{ER}{2\pi a Y_0^2}\right\}^4\right] = 1 \qquad (3.23)$$

Incorporating the geometrical correction factor for the finite plate, one has:

$$\frac{dU}{da} = \frac{2\phi^2\sigma^2 a}{E} + \frac{2\sigma^2 a}{E}\phi\frac{d\phi}{da} \text{ with } \phi = \frac{K}{\sigma\sqrt{a}} \qquad (3.24)$$

$$f_0^2 = \frac{\sigma_0^2}{Y_0^2} = \frac{ER}{\left\{2\phi^2 a + 2a^2\phi\frac{d\phi}{da}\right\}Y_0^2} = \frac{ER}{w\left\{2\phi^2\lambda + 2\lambda^2\phi\frac{d\phi}{d\lambda}\right\}} \equiv \mathscr{R} \qquad (3.25)$$

$$f_0 = \frac{(f_0^{\,2})^3}{4} + \frac{(f_0^{\,2})^4}{4} + \frac{(f_0^{\,2})^5}{16} = 1/4\,\mathscr{R}\left\{1+\mathscr{R}+\frac{\mathscr{R}^2}{4}\right\} \qquad (3.26)$$

$$R = 1/4\; CY_0^{\,2}a(2\beta_0 - 3Aa^{-1/2})\,\mathscr{R}^3\left\{1+\mathscr{R}+\frac{\mathscr{R}^2}{4}\right\} \qquad (3.27)$$

With that a description of R in a, and in adaptable material parameters β_0, A, Y, E (and width w) appears possible.

For further references cf. [3]. Related ideas were developed in [4], [5], [6].

REFERENCES

1. KRAFFT, J. M. SULLIVAN, A. M. and BOYLE, R. W., Proceedings of the Crack Propagation Symposium, Cranfield 1961, I, 1962, 8.
2. OROWAN, E., Welding Journal, 34, 3, 1955, 157S.
3. ASTM-STP 527, "Fracture Toughness Evaluation by R-Curve Methods".
4. RAJU, K. N., International Journal of Fracture Mechanics 5, 1969, 101.
5. WNUK, M. P., International Journal of Fracture Mechanics 7, 1971, 383.
6. RAJU, K. N., International Journal of Fracture Mechanics 8, 1972, 1.

Table 1

$f_0 = \frac{\sigma_0}{Y_0}$	$\frac{F_0}{f_0} = \frac{f_0^{\,4}}{4} + \frac{f_0^{\,6}}{6} + \frac{f_0^{\,8}}{16}$	$CF_0/f_0^{\,2} = \frac{123\pi}{64}\frac{F_0}{f_0}$
0	0	0
0.1	0.0000252^5	0.000152^5
0.2	0.000416	0.00251
0.3	0.00221	0.0133^5
0.4	0.00746^5	0.0450^5
0.5	0.0198	0.1194
0.6	0.0451	0.2724
0.7	0.0930	0.5618
0.8	0.178	1.0777
0.9	0.324	1.955
1.0	0.5625	3.396

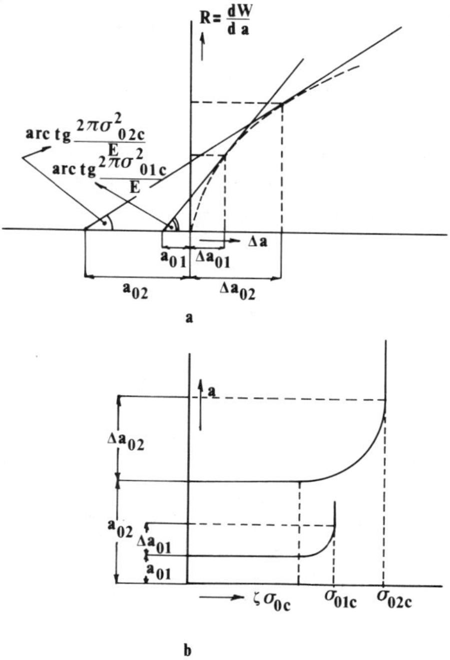

Figure 1 a) Assumed increase of $R = \frac{dW}{da}$ with Δa matched by $\frac{dU}{da}$ up to instability.

b) Assumed gross stress and cracklength behaviour before and after instability

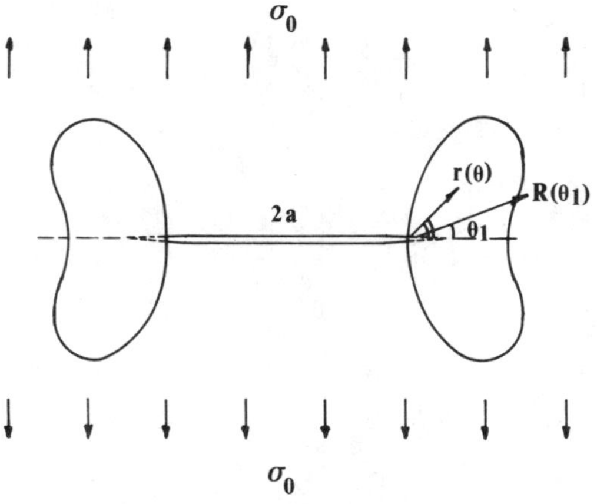

Figure 2   Plastic zone at crack tip in loaded plate, absorbing energy

Fracture 1977, Volume 3, ICF4, Waterloo, Canada, June 19 - 24, 1977

THE STRESS INTENSITY FACTORS FOR X-FORMED ARRAYS OF CRACKS

O. Aksogan*

INTRODUCTION

In the present study, the interaction of arbitrary arrays of cracks located along two intersecting infinite straight lines is considered. The method of analysis is similar to the one used by the author for V-formed arrays of cracks [1]. The new feature of a branched crack at the centre is treated by a special procedure during the application of the numerical method.

The analytical method used in this work consists of the joint use of the Mellin transform and the Green's function technique. The system of singular integral equations, thus obtained, is solved by a special application of an effective numerical method [2].

FORMULATION OF THE PROBLEM

In polar coordinates, in the absence of body forces, the stresses and the displacements in plane elasticity can be given as follows:

$$\begin{aligned}
\sigma_r &= \frac{1}{r^2}\frac{\partial^2\Phi}{\partial\theta^2} + \frac{1}{r}\frac{\partial\Phi}{\partial r}, \\
\sigma_\theta &= \frac{\partial^2\Phi}{\partial r^2}, \qquad \tau_{r\theta} = -\frac{\partial}{\partial r}\left(\frac{1}{r}\frac{\partial\Phi}{\partial\theta}\right), \\
2\mu u_r &= -\frac{\partial\Phi}{\partial r} + (1-\lambda)r\frac{\partial\psi}{\partial\theta}, \\
2\mu u_\theta &= -\frac{1}{r}\frac{\partial\Phi}{\partial\theta} + (1-\lambda)r^2\frac{\partial\psi}{\partial r},
\end{aligned} \tag{1}$$

where the Airy stress function, Φ, and the displacement function, ψ, satisfy the equations

$$\nabla^4\Phi = 0\ , \qquad \nabla^2\psi = 0\ , \qquad \frac{\partial}{\partial r}\left(r\frac{\partial\psi}{\partial\theta}\right) = \nabla^2\Phi\ . \tag{2}$$

In (1) and (2), $\lambda = \nu/(1+\nu)$ for plane stress and $\lambda = \nu$ for plain strain, μ is the shear modulus and ∇^2 is the harmonic operator.

---

*Middle East Technical University, Ankara, Turkey.

The Mellin transform of a function f(r), defined and suitably regular in $(0 < r < \infty)$, and its inverse are defined by

$$\tilde{f}(s) = \int_0^\infty f(r) r^{s-1}\, dr \ , \qquad f(r) = \frac{1}{2\pi i} \int_{c-i\infty}^{c+i\infty} \tilde{f}(s) r^{-s}\, ds \ , \tag{3}$$

where c is such that $r^{c-1}f(r)$ is absolutely integrable in $(0,\infty)$. The transforms of the derivatives can be found by using the relation

$$\int_0^\infty r^j \frac{d^j f(r)}{dr^j}\, r^{s-1}\, dr = (-1)^j \frac{\Gamma(s+j)}{\Gamma(s)} \tilde{f}(s) \ , \tag{4}$$

provided

$$r^{s+\ell-1} \frac{d^{\ell-1} f(r)}{d\, r^{\ell-1}} \to 0 \text{ as } r \to (0,\infty) \ , \qquad \ell=1,\ldots,j \ . \tag{5}$$

In plane elasticity problems, in which polar coordinates are used, the solution of (2) for each infinite wedge gives:*

$$\tilde{\Phi}(s,\theta)=Z_1\exp(is\theta)+\bar{Z}_1\exp(-is\theta)+Z_2\exp[i(s+2)\theta]+\bar{Z}_2\exp[-i(s+2)\theta],$$

$$(\widetilde{r^2R})=2i(s+1)\{Z_1 s \exp(is\theta)+Z_2(s+1)\exp[i(s+2)\theta]-\bar{Z}_2\exp[-i(s+2)\theta]\},$$

$$(\widetilde{r^2V})= - \frac{s+1}{\mu}\{Z_1 s \exp(is\theta)+Z_2(s+1)\exp[i(s+2)\theta]+\kappa\bar{Z}_2\exp[-i(s+2)\theta]\} \tag{6}$$

where

$$R = \tau_{r\theta} + i\sigma_\theta \ , \qquad u = u_r + iu_\theta \ ,$$
$$V = \frac{\partial u_r}{\partial r} + i\frac{\partial u_\theta}{\partial r} \ , \qquad \kappa = 3 - 4\lambda \ , \tag{7}$$

and Z_1 and Z_2, with their complex conjugates $\bar{Z}_1$ and $\bar{Z}_2$, are independent of θ.

In the present work, the isotropic homogeneous infinite plane is separated into four infinite wedges along the four lines of cracks (see Figure 1). Let the union of all the straight line segments representing the cracks along one radial line be called L and the remainder L', the former being finite and the latter infinite. The singular part of the solution may be formulated with the following boundary conditions:**

---

* The complex notation used here is only for convenience.

**The crack surface tractions, considered here, are the reversed self-equilibrating stresses along the crack lines for the medium without the cracks under the actual loading.

$$
\begin{aligned}
R_1(r,0) &= R_4(r,2\pi) && \text{on} \quad L_1+L_1', \\
V_1(r,0) &= V_4(r,2\pi) && \text{on} \quad L_1', \\
R_1(r,0) &= w_1(r) && \text{on} \quad L_1, \\
R_{j-1}(r,\theta_j) &= R_j(r,\theta_j) && \text{on} \quad L_j+L_j', \\
V_{j-1}(r,\theta_j) &= V_j(r,\theta_j) && \text{on} \quad L_j', \\
R_j(r,\theta_j) &= w_j(r) && \text{on} \quad L_j, \quad j=2,3,4
\end{aligned}
\tag{8}
$$

where the subscripts show the region or the boundary to which a certain quantity pertains. Needless to say, $w_j(r)$ are the complex tractions on the surfaces of the cracks.

Making use of the Mellin transform and the Green's function technique following the method used in [1], the stress expressions for the wedge-shaped domains are found in the following form:

$$
R_\ell(r,\theta) = \sum_{j=1}^{\ell} \int_{L_j} H_{1j}(r,\theta,\rho_j)d\rho_j + \sum_{j=\ell+1}^{4} \int_{L_j} H_{2j}(r,\theta,\rho_j)d\rho_j, \ell=1\ldots,4 \tag{9}
$$

where

$$
\begin{aligned}
H_{\alpha j}(r,\theta,\rho_j) = \frac{1}{\pi}\int_{c-i\infty}^{c+i\infty} \frac{\mu ds}{\rho_j(\kappa+1)(\beta-1)}\left(\frac{\rho_j}{r}\right)^{s+2} \Big\{ & \beta^{\alpha-1}\exp\left[is(\theta-\theta_j)\right]\Big[-sg_j(\rho_j) \\
& +i(s+2)f_j(\rho_j)\Big] + (s+1)\beta^{\alpha-1}\exp\left[i(s+2)(\theta-\theta_j)\right]\left[g_j(\rho_j)-if_j(\rho_j)\right] \\
& +\beta^{2-\alpha}\exp\left[-i(s+2)(\theta-\theta_j)\right]\left[g_j(\rho_j)+if_j(\rho_j)\right]\Big\}, \alpha=1,2, \; j=1,\ldots,4
\end{aligned}
\tag{10}
$$

in which $\beta = \exp(2is\pi)$ and*

$$
\begin{aligned}
g_1(r)+if_1(r) &= V_1(r,+0) - V_4(r,2\pi-0), && r \text{ on } L_1+L_1', \\
g_j(r)+if_j(r) &= V_j(r,\theta_j+0) - V_{j-1}(r,\theta_j-0), && r \text{ on } L_j+L_j', \; j=2,3,4.
\end{aligned}
\tag{11}
$$

Applying the stress expressions (9) to the third and sixth of equations (8), the integral equations of the problem are found as

---

*The unknown functions, f and g, are the densities of the dislocations of opening and edge-sliding modes, respectively.

$$\sum_{j=1}^{\ell} \int_{L_j} H_{1j}(r,\theta_\ell,\rho_j)d\rho_j + \sum_{j=\ell+1}^{4} \int_{L_j} H_{2j}(r,\theta_\ell,\rho_j)d\rho_j = w_\ell(r),$$

$$r \text{ on } L_\ell,\quad \ell=1,\ldots,4. \tag{12}$$

For continuity of the displacements, from the second and fifth of equations (8):

$$\int_{c_{j\ell}}^{d_{j\ell}} \left[g_j(\rho_j)+if_j(\rho_j)\right]d\rho_j = 0\ ,\quad j=1,\ldots,4\ ,\quad \ell=1,\ldots,n_j\ , \tag{13}$$

where n_j are the numbers of the cracks on the corresponding radial lines (see Figure 1 for the integration limits).

During the solution of (12), the kernels H_{1j}, H_{2j} are evaluated making use of the residue theory by a special procedure (see reference [1]). After lengthy but straightforward computations, (12) takes the following form, in terms of real variables:

$$\frac{1}{\pi}\sum_{j=1}^{4}\int_{L_j}\left\{\left[\frac{\delta_{\alpha j}}{\rho_j - r} + \frac{1-\delta_{\alpha j}}{2\rho_j} H_1^*\left(\frac{\rho_j}{r}\ ,\ \theta_{\alpha j}\right)\right]g_j(\rho_j)\right.$$

$$\left. + \frac{1}{2\rho_j} H_2^*\left(\frac{\rho_j}{r}\ ,\ \theta_{\alpha j}\right) f_j(\rho_j)\right\}d\rho_j = \frac{\kappa+1}{2\mu} q_\alpha(r) \tag{14}$$

$$\frac{1}{\pi}\sum_{j=1}^{4}\int_{L_j}\left\{\frac{1}{2\rho_j} H_3^*\left(\frac{\rho_j}{r}\ ,\ \theta_{\alpha j}\right) g_j(\rho_j)\right.$$

$$\left. + \left[\frac{\delta_{\alpha j}}{\rho_j - r} + \frac{1-\delta_{\alpha j}}{2\rho_j} H_4^*\left(\frac{\rho_j}{r}\ ,\ \theta_{\alpha j}\right)\right]f_j(\rho_j)\right\}d\rho_j = \frac{\kappa+1}{2\mu} p_\alpha(r)$$

$$r \text{ on } L_\alpha,\quad \alpha=1,\ldots,4,$$

where

$$w_\alpha(r) = q_\alpha(r) + i\, p_\alpha(r)\ ,$$

$$\theta_{\alpha j} = \theta_j - \theta_\alpha = 2M_{\alpha j}\ \pi/N_{\alpha j}\ ,\qquad \alpha,j=1,\ldots,4$$

and $H_\ell^*(x,2M\pi/N)$, $\ell=1,\ldots,4$, are defined in the Appendix.

NUMERICAL RESULTS

A special application [1] of an effective numerical method [2] is used for the solution of (14) subject to the continuity conditions (13). The nodal points for collocation are chosen at the zeros of the Chebyshev polynomials, in the ranges pertaining to each and every crack in the medium (see reference [1] for details). It must be noted that, when two or more cracks meet at the origin, the continuity conditions (13) do not apply to them separately. In that case, if there is a central symmetry in the crack setting and a central symmetry or antisymmetry in the loading (crack surface tractions), the consequent central symmetry or antisymmetry in the unknown functions renders the application of the numerical technique possible. What needs to be done, in that case, is to choose the collocation points as if each crack, terminating at the origin, is extended to the other side of the origin by a reflection.

Although the numerical solution yields the values for any elasto-mechanic quantity, we will be concerned with the stress intensity factor only. Besides the closed form solutions for two and three collinear cracks, comparisons of the numerical results of the present work were also made with the graphical presentations of Isida [3] for other arrays of isolated cracks. The results matched perfectly. Surprisingly enough, even the cases of parallel cracks have been treated (the results matching with those of [3]), just by taking the angle between the cracks small enough. (It must be noted that, the origin being at infinity, the case of exactly parallel cracks cannot be treated by the present method).

The special case of four symmetrically situated radial cracks under constant internal pressure was also treated and the results were in good agreement with those of Tweed and Rooke [4].

The results of the present work for cross-shaped cracks with two pairs of unequal arms, loaded with unequal constant normal tractions were compared with those given by Sneddon and Das [5]. The results obtained for the case of nonuniform internal pressure for a cross-shaped crack with equal arms were compared with the results of Stallybrass [6]. There was a mismatch of about 0.1 per cent for both cases when 20 points of collocation were taken along each branch.

Some other cases, which cannot be found in the literature, have been considered. Choosing a suitable parameter for each case, the two types of stress intensity factors at all crack tips were presented in graphical form (Figures 2 - 4). For these computations sixteen collocation points were taken along each isolated crack and each branch of the X-shaped crack. The results for the limiting cases of single and two or three collinear cracks were observed to coincide with those in the literature [3]. Because of limited space, loadings which cause partial closures could not be included here, although a number of such cases were treated, making use of the procedure in references [7, 8]. For the same reason an X-formed array of cracks with an X-shaped crack at the centre, having a central symmetry or antisymmetry could not be exposed here.

ACKNOWLEDGEMENT

The author wishes to acknowledge the valuable comments of the reviewers towards the improvement of this paper. This research has been supported by the Scientific and Technical Research Council of Turkey under grant MAG-428.

APPENDIX

$$H_1^*(x,2M\pi/N) = (x^2-1)C_1 + x\cos(2M\pi/N) ,$$

$$H_2^*(x,2M\pi/N) = (x^2-1)S_1 - x^2S_2 + x\sin(2M\pi/N) ,$$

$$H_3^*(x,2M\pi/N) = (x^2-1)S_1 - S_2 - x\sin(2M\pi/N) ,$$

$$H_4^*(x,2M\pi/N) = (1-x^2)C_1 + (x^2+1)C_2 - x\cos(2M\pi/N) + 2$$

where

$$\begin{Bmatrix} C_j \\ S_j \end{Bmatrix} = \sum_{\ell=1}^{N} \left[\left(\frac{N}{2(x^N-1)} + \frac{\ell}{2} \right)^{2-j} \frac{2x^{N-\ell}}{x^N-1} \begin{Bmatrix} \cos \\ \sin \end{Bmatrix} (2\ell M\pi/N) \right], \quad j=1,2 .$$

REFERENCES

1. AKSOGAN, O., J. Engng. for Ind., Trans. ASME, 98, 1976, 1086.
2. ERDOGAN, F., GUPTA, G. D. and COOK, T. S., "Methods of Analysis and Solutions of Crack Problems", ed. G. C. Sih, Noordhoff, Leyden, 1973, 368.
3. ISIDA, M., Bulletin JSME, 13, 1970, 635.
4. TWEED, J. and ROOKE, D. P., Int. J. Engng. Sci. 12, 1974, 423.
5. SNEDDON, I. N. and DAS, S. C., "Trends in Elasticity and Thermo-elasticity", Witold Nowacki Anniversary Volume, Wolters-Noordhoff, Groningen, 1971, 233.
6. STALLYBRASS, M. P., Quart. J. Mech. Appl. Math., 23, 1970, 35.
7. AKSOGAN, O., Int. J. Frac., 11, 1975, 659.
8. AKSOGAN, O., Int. J. Frac., 12, 1976, 223.

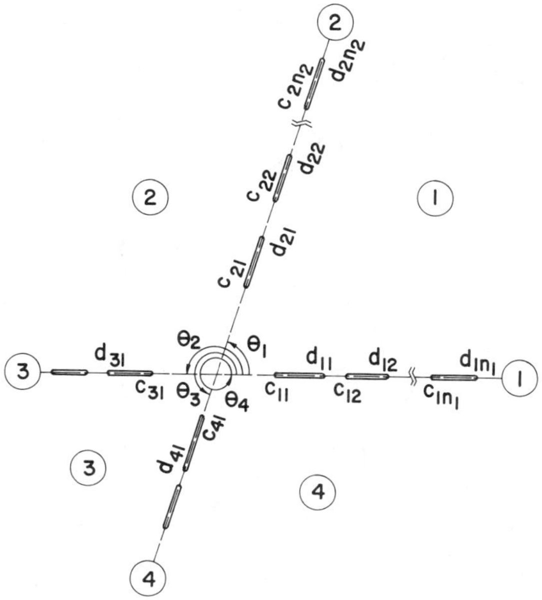

Figure 1 The Geometry and Notation

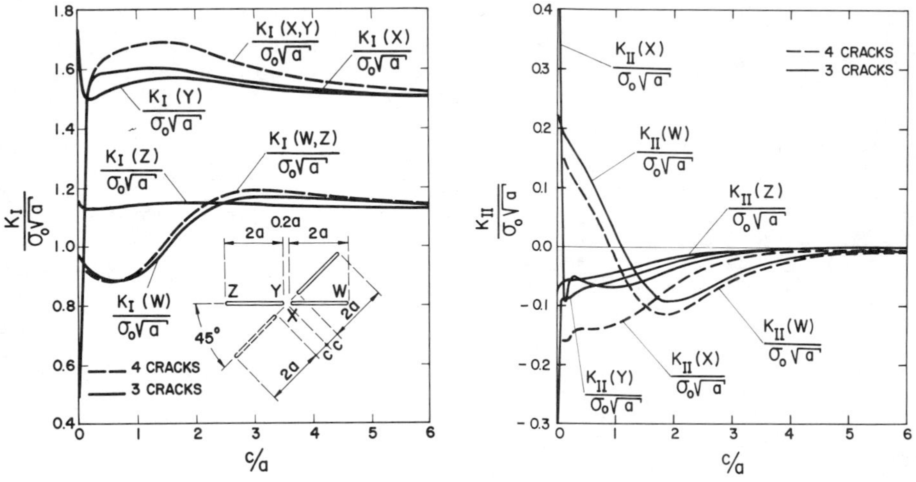

(a) (b)

Figure 2 (a) Opening Mode Stress Intensity Factors for 3 and 4 Radial Cracks Under Uniform All-Round Tension, σ_0

(b) Edge-Sliding Mode Stress Intensity Factors for 3 and 4 Radial Cracks Under Uniform All-Round Tension, σ_0

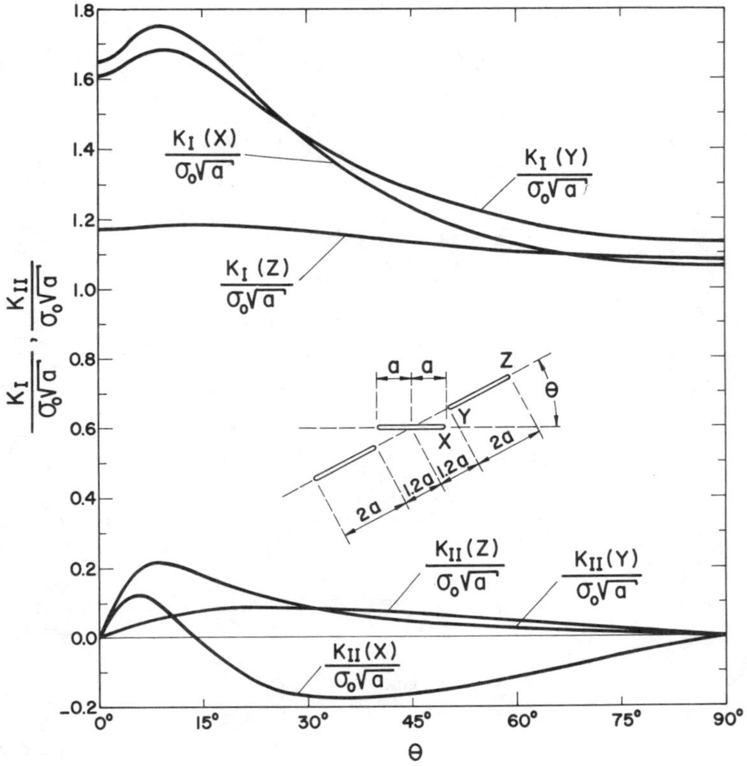

Figure 3 Stress Intensity Factors for an Array of Cracks with Central Symmetry Under Uniform All-Round Tension, σ_0

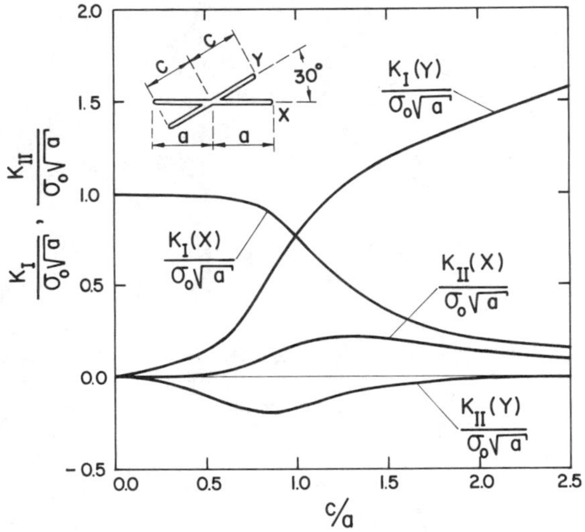

Figure 4 Stress Intensity Factors for an X-Shaped Crack Under Uniform All-Round Tension, σ_0

Fracture 1977, Volume 3, ICF4, Waterloo, Canada, June 19 - 24, 1977

THE MODIFIED WESTERGAARD EQUATIONS*

R. de Wit**

INTRODUCTION

Sih [1] and Eftis and Liebowitz [2] have pointed out that the Westergaard method, which applies to a certain class of plane problems in linear elasticity and is most frequently used in fracture mechanics [3], suffers from a restriction. The restriction is essentially the following: In symmetric problems (mode I), the Westergaard function allows only a hydrostatic tension for the remote state of stress. This means, for example, that the simple case of uniaxial tension is excluded from this formulation (with or without cracks).

The above authors corrected this shortcoming by appending constant terms to Westergaard's stress equations. They did this by appealing to the Goursat-Kolosov and MacGregor complex formulations of the problem. The present note shows how those additions to the Westergaard functions can be made in a more straightforward way without reference to the more sophisticated representations. It is done by simply adding the real part of a term in z^2 to the Airy stress function of Westergaard.

WESTERGAARD'S FUNCTION

The classical problem of plane isotropic elasticity is set up in terms of the Airy stress function Φ, which satisfies the biharmonic equation [3, 4]

$$\nabla^2(\nabla^2\Phi) = 0 \tag{1}$$

and from which the stresses can be derived as follows

$$\sigma_{xx} = \frac{\partial^2\Phi}{\partial y^2}, \quad \sigma_{yy} = \frac{\partial^2\Phi}{\partial x^2}, \quad \sigma_{xy} = -\frac{\partial^2\Phi}{\partial x \partial y} . \tag{2}$$

For mode I Westergaard then gave the Airy stress function in terms of the analytic function $Z_I(z)$ of the complex variable $z = x+iy$ as follows:

$$\Phi = \mathrm{Re}\, \bar{\bar{Z}}_I + y\, \mathrm{Im}\, \bar{Z}_I , \tag{3}$$

where $d\bar{\bar{Z}}/dz = \bar{Z}$ and $d\bar{Z}/dz = Z$. For mode II he gave it in terms of another analytic function $Z_{II}(z)$ as follows:

---

* Contribution of U. S. National Bureau of Standards, not subject to copyright.

**National Bureau of Standards, Metallurgy Division, Washington, D. C. 20234, U. S. A.

$$\Phi = -\ y\ \mathrm{Re}\ \overline{Z}_{II}\ . \tag{4}$$

Most of the classical results in fracture mechanics have been derived from these equations by suitable choices of Westergaard's analytic functions Z.

THE MODIFICATION

Muskhelishvili [4] has shown that it is possible and also convenient to write the Airy stress function as the real part of a complex but not necessarily analytic function, as follows:

$$\Phi = \mathrm{Re}\left\{z^*\phi(z) + \chi(z)\right\}\ , \tag{5}$$

where the Goursat functions $\phi(z)$ and $\chi(z)$ are analytic. Now, if the function $\chi(z)$ includes a term in the analytic function z^2, then we see from (2) that this would add at most constant terms to the stresses. Hence, this provides a very simple way of making the correction proposed by Sih and by Eftis and Liebowitz.

If we write the Goursat functions in terms of the Westergaard functions as follows:

$$\phi = \frac{1}{2}\left(\overline{Z}_I - i\ \overline{Z}_{II}\right)\ , \tag{6a}$$

$$\chi = \overline{\overline{Z}}_I - \frac{1}{2}\left(\overline{Z}_I - i\ \overline{Z}_{II}\right) z - \frac{1}{2}\ (A + iB)z^2\ , \tag{6b}$$

then the resulting Airy stress function is by (5)

$$\Phi = \mathrm{Re}\overline{\overline{Z}}_I + y\mathrm{Im}\overline{Z}_I - y\mathrm{Re}\overline{Z}_{II} - \frac{1}{2}\ A(x^2-y^2)+Bxy\ . \tag{7}$$

The first two terms in this expression are the same as (3), the third is (4), and the fifth and sixth are the correction terms. The stresses follow from (2)

$$\sigma_{xx} = \mathrm{Re}Z_I - y\mathrm{Im}Z_I' + 2\mathrm{Im}Z_{II} + y\mathrm{Re}Z_{II}' + A\ , \tag{8a}$$

$$\sigma_{yy} = \mathrm{Re}Z_I + y\mathrm{Im}Z_I' - y\mathrm{Re}Z_{II}' - A\ , \tag{8b}$$

$$\sigma_{xy} = -y\mathrm{Re}Z_I' + \mathrm{Re}Z_{II} - y\mathrm{Im}Z_{II}' - B\ . \tag{8c}$$

The displacements in plane-strain can be shown to be

$$2\mu u = (1-2\nu)\ \mathrm{Re}\bar{Z}_I - y\,\mathrm{Im}Z_I + Ax + 2(1-\nu)\ \mathrm{Im}\bar{Z}_{II} + y\,\mathrm{Re}Z_{II} - By\ , \quad (9a)$$

$$2\mu v = 2(1-\nu)\ \mathrm{Im}\bar{Z}_I - y\,\mathrm{Re}Z_I - Ay - (1-2\nu)\ \mathrm{Re}\bar{Z}_{II} - y\,\mathrm{Im}Z_{II} - Bx\ , \quad (9b)$$

where μ is the shear modulus and ν Poisson's ratio. Equations (8) and (9) are the basic results of Sih and Eftis and Liebowitz.

The rotation of the medium is given by

$$\omega = \frac{1}{2}\left(\frac{\partial v}{\partial x} - \frac{\partial u}{\partial y}\right) . \quad (10)$$

By (9) the rotation is then easily expressed in terms of the Westergaard functions as follows

$$\mu\omega = (1-\nu)\ (\mathrm{Im}Z_I - \mathrm{Re}Z_{II}) . \quad (11)$$

APPLICATIONS

(a) Constant Stress Without Cracks

We obtain case of a constant stress field in the medium by taking constant (complex) values for the Westergaard functions

$$Z_I = C + iD\ , \quad (12a)$$

$$Z_{II} = E + iH\ . \quad (12b)$$

The values of the constants C, D, E and H are determined by the boundary conditions. Then the stresses are from (8):

$$\sigma_{xx} = C + 2H + A\ , \quad (13a)$$

$$\sigma_{yy} = C - A\ , \quad (13b)$$

$$\sigma_{xy} = E - B\ . \quad (13c)$$

We see that there are now enough constants available to obtain any arbitrary set of stresses.

However, for mode I, E = H = 0, without the correction terms, A = B = 0, only hydrostatic tension is possible, $\sigma_{xx} = \sigma_{yy} = C$, $\sigma_{xy} = 0$. This is the nature of the restriction on Westergaard's original equations, as we mentioned in the Introduction. The correction term A makes it possible in mode I to have σ_{xx} differ from σ_{yy}. The simple case of uniaxial tension is then given by A = - C.

We note, however, that the correction terms are not strictly necessary for a completely arbitrary choice of stresses, if modes I and II are combined, for the term H in (13a), which comes from mode II in (12b), also allows us to choose σ_{xx} different from σ_{yy} in an arbitrary way.

The constant D in (12a) does not play a role in the stresses (13). It is related to the rotation of the medium, together with the constant E, as can easily be deduced by substituting (12) into (11):

$$\mu\omega = (1-\nu)\ (D-E)\ . \tag{14}$$

(b) Small Crack

This is the classical case of an infinite medium with a central crack of length 2a along the x-axis. The solution to this problem is given by the Westergaard functions

$$Z_I = \sigma(1-a^2/z^2)^{-1/2} + C + iD\ , \tag{15a}$$

$$Z_{II} = \tau(1-a^2/z^2)^{-1/2} + E + iH\ , \tag{15b}$$

where the values of the constants σ, τ, C, D, E and H are determined by the boundary conditions.

One set of boundary conditions is that the crack surface is stress free

$$\sigma_{yy} = \sigma_{xy} = 0 \qquad \text{for} \qquad y = 0,\ |x| < a\ . \tag{16}$$

From (15) and (8) this leads to the relations

$$C = A, \qquad E = B\ . \tag{17}$$

Another set of boundary conditions is given by the asymptotic behaviour of the stresses at remote distances from the crack. From (17), (15) and (8) we find at $|z| = \infty$

$$\sigma_{xx} = \sigma + 2(A+H)\ , \tag{18a}$$

$$\sigma_{yy} = \sigma\ , \tag{18b}$$

$$\sigma_{xy} = \tau\ . \tag{18c}$$

We see therefore that σ and τ in (15) represent the remote normal tensile and shear stress applied to the cracked medium. From (18a), we see that the remote tensile stress parallel to the crack can be arbitrarily adjusted to any desired value, either by fixing C (= A) in mode I or H in mode II. This does not affect the first terms in equations (15), which contain the essence of the crack field.

The constants E (= B) and D play no role in the stress field of the cracked medium. From (11) it can easily be shown that they are related to the rotation of the medium. It fact, for $|z| = \infty$ we find

$$\mu\omega = (1-\nu)\ [D-E-\tau]\ . \tag{19}$$

Our result that B does not appear in (18) contradicts a conclusion by Sih[1] who stated that B cannot vanish for a non-trivial solution. The reason for this is that Sih made the unnecessary assumption that $E = -\tau$ in his equation (14), which corresponds to our equation (15b).

CONCLUSION

We have shown that a correction, proposed by Sih and by Eftis and Liebowitz, to remove a restriction on Westergaard's equations, can be made in a very simple way by adding an elementary term to the Airy stress function.

The result is illustrated by two very simple examples. However, applications to more complex problems, such as those discussed by Eftis and Liebowitz, can of course be made in a similar straightforward manner.

REFERENCES

1. SIH, G. C., Int. J. Fract. Mech., 2, 1966, 628.
2. EFTIS, J. and LIEBOWITZ, H., Int. J. Fract. Mech., 8, 1972, 383.
3. TADA, H., PARIS, P. C. and IRWIN, G. R., The Stress Analysis of Cracks Handbook, Del Research Corporation, 1973.
4. MUSKHELISHVILI, N. I., "Some Basic Problems of the Mathematical Theory of Elasticity", P. Noordhoff Ltd., Groningen-Holland, 1953.

Fracture 1977, Volume 3, ICF4, Waterloo, Canada, June 19 - 24, 1977

AN EQUIVALENT INCLUSION METHOD FOR A THREE-DIMENSIONAL LENS-SHAPED CRACK IN ANISOTROPIC MEDIA

T. Mura and P. C. Cheng*

INTRODUCTION

Micro cracks in materials sometimes take a three-dimensional lens-shape. In this paper crack opening displacements, crack extension forces, stress concentration factors for $a_3 \neq 0$ and stress intensity factors for $a_3 = 0$ are developed through the use of the equivalent inclusion method for an isolated three-dimensional lens-shaped crack under simple tension and pure shear, where a_3 is the smallest principal axis of the ellipsoid.

CRACK OPENING DISPLACEMENT

A three-dimensional lens-shaped crack is given by Figure 1 or by

$$\Omega:\ x_1^2/a_1^2 + x_2^2/a_2^2 + x_3^2/a_3^2 \leq 1 \tag{1}$$

where a_3 is smaller in comparison to a_1, a_2. The elastic constants of domain Ω are zero. An applied stress (σ_{ij}^0 at infinity) becomes $\sigma_{ij}^0 + \sigma_{ij}$ in the neighbourhood of Ω. The stress disturbance σ_{ij} is equivalent to the stress caused by eigenstrains ε_{ij}^* (phase transformation strains) defined in Ω, assuming the elastic constants of Ω to be the same as those of the matrix (denoted by $C_{ijk\ell}^0$). ε_{ij}^* are determined from

$$\sigma_{ij}^0 + \sigma_{ij} = 0,\ \sigma_{ij} = C_{ijk\ell}^0\left(u_{k,\ell} - \varepsilon_{k\ell}^*\right) \text{ in } \Omega, \tag{2}$$

where u_i is the displacement field due to ε_{ij}^*. We found by Green's function technique [1] that when ε_{ij}^* are constant and $a_3 \ll a_1$, a_2,

$$u_{i,k} \approx (1/4\pi)C_{j\ell mn}^0\varepsilon_{mn}^*\left[4\pi G_{ijk\ell}(0,0,1) - a_3\Pi_{ijk\ell}\right] \tag{3}$$

where

$$G_{ijk\ell}(\xi_1,\xi_2,\xi_3) = N_{ij}(\underset{\sim}{\xi})\xi_k\xi_\ell/D(\underset{\sim}{\xi})$$

$$\Pi_{ijk\ell} = \int_{S^2} \frac{a_1a_2}{(a_1^2\cos^2\theta + a_2^2\sin^2\theta)^{3/2}}\,\frac{\bar{\xi}_3}{(1-\bar{\xi}_3^2)^{1/2}}\,\frac{\partial}{\partial\bar{\xi}_3}G_{ijk\ell}(\underset{\sim}{\bar{\xi}})dS(\underset{\sim}{\bar{\xi}}) \tag{4}$$

$$\bar{\xi}_1 = (1-\bar{\xi}_3^2)^{1/2}\cos\theta,\ \bar{\xi}_2 = (1-\bar{\xi}_3^2)^{1/2}\sin\theta,\ dS(\underset{\sim}{\xi}) = d\theta d\bar{\xi}_3.$$

*Materials Research Centre, Northwestern University, Ill., U.S.A.

$N_{ij}(\xi)$, $D(\bar{\xi})$ are the cofactor and the determinant of matrix $(C^0_{ipjq}\bar{\xi}_p\bar{\xi}_q)$ respectively and S^2 is the unit sphere $\bar{\xi}_i\bar{\xi}_i = 1$. Equation (3) is linear with respect to a_3.

When $\sigma^0_{ij} = \sigma^0_{33}$ (simple tension) and the crystalline directions are parallel to the principal axes direction of Ω, equation (2) gives a non-zero component of ε^*_{ij} as

$$a_3\varepsilon^*_{33} = 4\pi\sigma^0_{33}/C^0_{33mn}C^0_{pq33}\Pi_{mpnq} \tag{5}$$

When $\sigma^0_{ij} = \sigma^0_{31}$ (pure shear) and the crystalline directions are parallel to the principal axes directions of Ω, a non-zero component of ε^*_{ij} is obtained as

$$a_3\varepsilon^*_{31} = 2\pi\sigma^0_{31}/C^0_{31mn}C^0_{pq31}\Pi_{mpnq}. \tag{6}$$

From the dislocation theory [2], the crack opening displacements u_3 for (5) and u_1 for (6) (displacements on Ω) are given by $\varepsilon^*_{ij}h$, where h is the half thickness of Ω in the x_3 direction:

$$h = a_3(1 - x_1^2/a_1^2 - x_2^2/a_2^2)^{1/2}. \tag{7}$$

Since the right hand sides in (5) and (6) do not contain a_3, $\varepsilon^*_{ij} \to \infty$ for $a_3 \to 0$.

CRACK EXTENSION FORCES

The interaction energy between σ^0_{ij} and the crack is given by

$$\Delta U = -\frac{2}{3}\pi a_1a_2a_3\varepsilon^*_{ij}\sigma^0_{ij} \;. \tag{8}$$

When the crack is expanding in the x_2 direction, keeping a_1, a_3 constant, the crack extension force G is given by $G = -\partial(\Delta U)/\partial a_2$. For the simple tension it becomes

$$G = 8\pi^2 a_1(\sigma^0_{33})^2(f/g)/C^0_{33mn}C^0_{pq33}\Pi_{mpnq} \tag{9}$$

where

$$f = C^0_{33ik}C^0_{j\ell 33}\int_{S^2} \frac{\bar{\xi}_3}{(1-\bar{\xi}_3^2)^{1/2}}\left\{\frac{\partial}{\partial\bar{\xi}_3}G_{ijk\ell}(\bar{\xi})\right\}\frac{a_2^2\sin^2\theta\; ds(\bar{\xi})}{(a_1^2\cos^2\theta+a_2^2\sin^2\theta)^{5/2}}$$

$$\tag{10}$$

$$g = C^0_{33ik}C^0_{j\ell 33}\int_{S^2} \frac{\bar{\xi}_3}{(1-\bar{\xi}_2^2)^{1/2}}\left\{\frac{\partial}{\partial\bar{\xi}_3}G_{ijk\ell}(\bar{\xi})\right\}\frac{ds(\bar{\xi})}{(a_1^2\cos^2\theta+a_2^2\sin^2\theta)^{3/2}}$$

and $f/g \to 1$ for $a_1 \to \infty$.

A similar calculation can be done for the pure shear case.

The above result can be applied to a flat ellipsoidal crack ($a_3 \to 0$) since (9) does not contain a_3. For the slit-like crack ($a_1 \to \infty$), G agrees with that given by Barnett and Asaro [3].

STRESS CONCENTRATION FACTORS

The stress concentration factor is defined by

$$\kappa = \left(\sigma^0_{ij} + \sigma_{ij}(\text{out})\right)/\sigma^0_{ij} \tag{11}$$

where $\sigma^0_{ij} + \sigma_{ij}(\text{out})$ is defined immediately outside the crack. $\sigma_{ij}(\text{out})$ can be determined by σ_{ij} in Ω (denoted by $\sigma_{ij}(\text{in})$). The stress jump on Ω, $[\sigma_{ij}] \equiv \sigma_{ij}(\text{out}) - \sigma_{ij}(\text{in})$, can be written as (see [4] - [7]),

$$\left[\sigma_{ij}\right] = C^0_{ijk\ell}\left\{-C^0_{pqmn}\varepsilon^*_{mn}G_{kpq\ell}(\underset{\sim}{n}) + \varepsilon^*_{k\ell}\right\} \tag{12}$$

where n is the outward normal vector of the boundary of Ω. Since $\sigma^0_{ij} + \tilde{\sigma}_{ij}(\text{in}) = 0$, (11) is written as

$$\kappa = \left[\sigma_{ij}\right]/\sigma^0_{ij} \tag{13}$$

where ε^*_{ij} are given by (5) and (6) for the two types of applied stress and are inversely proportional to a_3. κ at $\beta = \pi/2$ (see Figure 1) is shown in Figures 2 and 3 with respect to a_1/a_2 for various crystals. a_3 is expressed in terms of ρ_n from geometry,

$$a_3 = (\rho_n a_1 a_2)^{1/2}(a_2^2\cos^2\beta + a_1^2\sin^2\beta)^{-1/4} \tag{14}$$

where ρ_n is the root radius of the curve which is an intersection of Ω and the plane containing n and the line parallel to the x_3 axis (see Figure 1). The values of $\kappa/(a_2/\tilde{\rho_n})^{1/2}$ converge to constant values which agree with the values expected from Lekhnitiskii [8] for simple tension in the plane strain case.

STRESS INTENSITY FACTORS

The stress intensity factor is defined when $a_3 \to 0$ by

$$K = \lim_{\rho\to 0} \sqrt{2\rho}\left(\sigma^0_{ij} + \sigma_{ij}\right) \tag{15}$$

where ρ is the distance from the boundary of the crack.

When a_1 is fixed and a_2 is changed by δa_2 (small variation), the work done by the tension on the new crack surface δA is

$$2(K/\sqrt{2\rho})\varepsilon^*_{33}h\delta A = 2K\varepsilon^*_{33}a_3(\bar{x}_1^2/a_1^4 + \bar{x}_2^2/a_2^4)^{1/4}\pi a_1\delta a_2 \tag{16}$$

where $\bar{x}_1^2/a_1^2 + \bar{x}_2^2/a_2^2 = 1$ and $(1 - x_1^2/a_1^2 - x_2^2/a_2^2)^{1/2}$ is approximated as $\sqrt{2\rho}(\bar{x}_1^2/a_1^4 + \bar{x}_2^2/a_2^4)^{1/4}$. Equation (16) must be equal to $G\delta a_2$ which can be written from (5) as

$$G\delta a_2 = 2\pi a_1 \sigma_{33}^0 a_3 \varepsilon_{33}^* (f/g)\delta a_2. \tag{17}$$

Comparing (16) and (17), we have

$$K = \sigma_{33}^0 (\bar{x}_1^2/a_1^4 + \bar{x}_2^2/a_2^4)^{-1/4} (f/g) \tag{18}$$

For the slit-like crack ($a_1 \to \infty$), $f/g = 1$ and therefore K is independent of the elastic constants. The value of K agrees with Barnett and Asaro [3] for the slit-like crack. Similar discussion can be done for the pure shear case.

CONCLUSION

We have shown that the equivalent inclusion method can provide the crack opening displacements, crack extension forces, the stress concentration factor when $a_3 \neq 0$, and the stress intensity factor when $a_3 = 0$. It is also found that some difficulties are involved in deriving the stress intensity factors from the stress concentration factors through a limiting process. The two concepts of stress intensity and concentration factors appear to belong to separate categories.

In this paper we considered uniform applied stress fields at infinity. However, a similar calculation can be done for a linearly changing applied stress at infinity. Similarly the present method can be applied to any orientation of crystals and any anisotropic material. Comparison with other results about the stress intensity factors obtained by Kassir and Sih [9] and Willis [10] will be possible only numerically and will be reported in this conference.

ACKNOWLEDGEMENT

The authors acknowledge Professor A. S. Kobayashi for his helpful discussion.

APPENDIX

For isotropic materials

$$C^0_{33mn} C^0_{pq33} \Pi_{mpnq} = \frac{1}{a_2} \frac{4\pi\mu}{1-\nu} E(k)$$

$$C^0_{31mn} C^0_{pq31} \Pi_{mpnq} = \frac{1}{a_2} 4\pi\mu \left\{ \frac{1}{1-\nu} \frac{k'^2}{k^2} (F(k)-E(k)) + \frac{k'^2}{k^2}\left(\frac{E(k)}{k'^2} - F(k)\right)\right\}$$

where μ, ν are the shear modulus and Poisson's ratio, respectively, and $k^2 = 1 - a_2^2/a_1^2$, $a_1 > a_2$, $k'^2 = 1 - k^2$. F(k), E(k) are the complete elliptic integrals of the first and second kinds respectively.

REFERENCES

1. KINOSHITA, N. and MURA, T., Phys. Stat. Sol., (a)5, 1971, 759.
2. MURA, T., The Continuum Theory of Dislocations, "Advances in Materials Research", Interscience Pub. 3, 1968.
3. BARNETT, D. M. and ASARO, R. J., J. Mech. Phys. Solids, 20, 1972, 353.
4. ESHELBY, J. D., Proc. Roy. Soc. A241, 1957, 376.
5. GOODIER, J. N., Phil. Mag., 23, 1937, 1017.
6. HILL, R., Progr. Solid Mech., 2, 1969, 245.
7. WALPOLE, L. J., Proc. Roy. Soc., A300, 1967, 270.
8. LEKHNITSKII, S. G., "Anisotropic Plates", Gordon and Breach Science Pub., New York, 1967, 166.
9. KASSIR, M. K. and SIH, G. C., Engineering Fracture Mechanics, 1, 1968, 327.
10. WILLIS, J. R., Int. J. Engr. Sci., 6, 1968, 253.

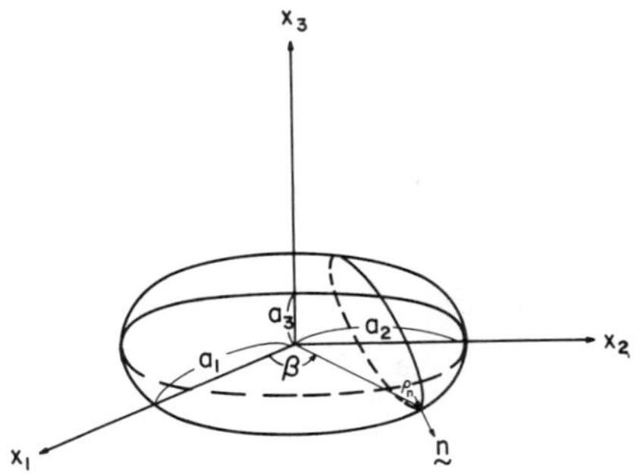

Figure 1 Configuration of a Flat Ellipsoid or a Crack

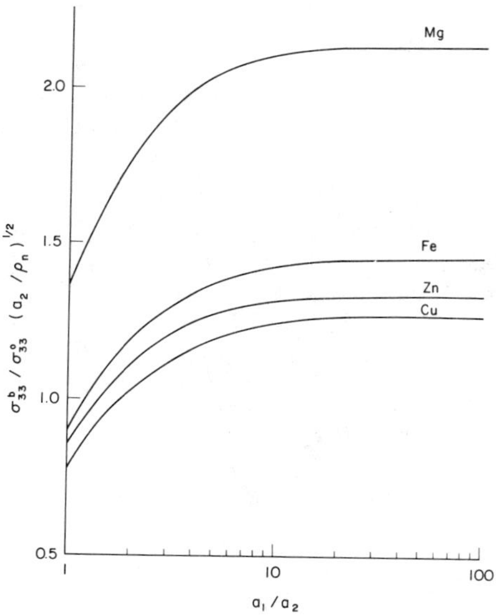

Figure 2 Tensile Stress on the Boundary of an Elliptical Crack at $\beta = 90°$ for Various Ratio of a_1/a_2

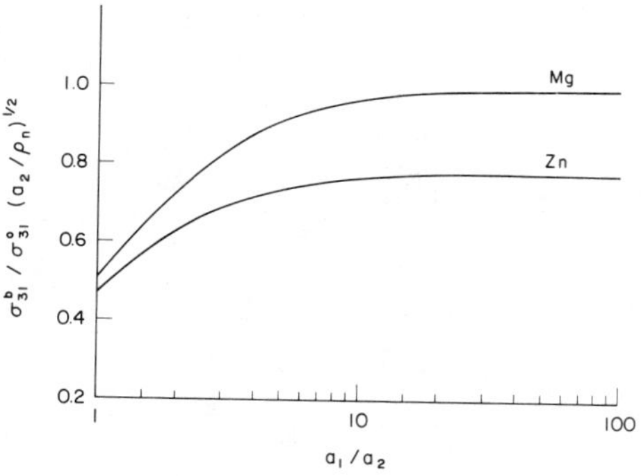

Figure 3 Shear Stress on the Boundary of an Elliptical Crack at $\beta = 90°$ for Various Ratio of a_1/a_2

Fracture 1977, Volume 3, ICF4, Waterloo, Canada, June 19 - 24, 1977

STRESS INTENSITY FACTORS AT THE TIPS OF KINKED AND FORKED CRACKS

B. A. Bilby, G. E. Cardew and I. C. Howard*

INTRODUCTION

Amongst the fundamental problems that form the theoretical basis for studies of the paths of crack propagation and of the stability of cracks are the problems of the elastic fields around a long or semi-infinite crack with either a single kink or a fork at its tip (to discuss the onset of deviation or branching, the case where the main crack is long is of most interest). We present here briefly some of our computations of the relevant stress intensity factors for these two problems. A number of workers have considered problems of this kind; for some references, see [1 - 10]. In making available our own results, which sometimes differ from those previously published by other workers, we hope to contribute to an ultimate consensus of agreement, for these problems have a considerable history of published error, some of which has been acknowledged. In the text we give some indications of why we think this has come about.

THE KINKED CRACK

We image that a semi-infinite crack has, at its tip, a kink of unit length making an angle α with the main crack (Figure 1). The loading is specified by the stress intensity factors K_1 and K_2 of the main crack without the kink. The analysis of [14] enables the stress intensity factors k_1 and k_2 at the tip of the kink to be computed in terms of K_1 and K_2 by quadratures. We find [1]:

$$k_1 = K_{11}(\alpha)\ K_1 + K_{12}(\alpha)\ K_2 \ , \tag{1}$$

$$k_2 = K_{21}(\alpha)\ K_1 + K_{22}(\alpha)\ K_2 \ , \tag{2}$$

where the functions $K_{ij}(\alpha)$ are displayed in Figure 1. (A table of $K_{ij}(\alpha)$ allowing interpolation to within 1% accuracy is available upon request to the authors). We have checked the accuracy of the numerical procedures whereby $K_{ij}(\alpha)$ are computed, and we believe that the results upon which Figure 1 is based are an accurate solution to the problem. Where comparisons can be made, we agree with the results of Chatterjee [5], but disagree slightly with [2]. The appropriate curves in our Figure 1 and the Figure 4 of [2] look very much the same, but our results differ from those of [2] by as much as 20%. We are unable to explain this difference on the basis of an error in our calculation, and suggest that the method of conformal transformation used in [2] may be less reliable than is usually supposed. This supposition is partly born out by the difficulties we experienced in

* Department of the Theory of Materials, University of Sheffield, Sheffield, United Kingdom

attempting to make a related method work for the problem of the forked crack.

THE FORKED CRACK

We have solved this problem by two methods.

(a) Firstly, we used a method of conformal transformation which maps a finite crack with a forked tip into the unit circle. The appropriate stress functions are found by inverting an infinite system of algebraic equations, and we found that the simplest formulation of the problem was that of [9]. Although the method produced results which agree qualitatively with those of [2] we had to do very large amounts of computation to achieve them, and we suspect that the rate of convergence of the method was slow enough for simple numerical rates of convergence to be misleading.

(b) Secondly, we represented the cracks by continuous distributions of dislocations and we solved the resulting singular integral equations by a method similar to that reported in [11]. All our attempts to represent a semi-infinite crack by a continuous distribution of dislocations were unsatisfactory, and so we performed the computations for a finite, but long, main crack. Figure 2 shows the normalized stress intensity factors k_1/K_1, k_2/K_1 at the tip of the upper fork when the main crack is 40 times as long as the kink. K_1 is the stress intensity factor at the tip of the main crack without the kink. Our method is numerically unstable for small α, and we have no reliable results, as yet, for $0^0 < \alpha < 5^0$.

The most important feature of Figure 2 is the zero of k_2 at about 18^0, slightly larger than the value read from Figure 14 of [2]. The existence of this zero is used in theories of the crack forking which occurs both with fast moving cracks [12], and in stress corrosion [2]. This zero appears to arise because of the presence of the fork, and the computations of Kalthoff [12] for a fork without a main crack clearly show its presence. A recent paper [10] reports the results of computations on a forked crack where the main crack is four times the length of a fork; k_2 has no zero. We have repeated out computations for this particular geoemtry and clearly discern a zero somewhere between 10^0 and 20^0. Finally, we note that our Figure 2 disagrees with the results of [4]. We believe that this is because the formulation of the problem given in [4] is incorrect. We find that proper application of the boundary conditions leads to a coupled pair of Wiener-Hopf equations, rather than the separated equations of the authors. The point is a subtle one and has not been noticed by a recent reviewer [13].

ACKNOWLEDGEMENT

We are indebted to Professor J. D. Eshelby for many valuable discussions.

REFERENCES

1. BILBY, B. A. and CARDEW, G. E., Int. J. Fract., 11, 1975, 708.
2. KITAGAWA, H., YUUKI, R. and OHIRA, T., Eng. Fract. Mech., 7, 1975, 515.
3. KITAGAWA, H. and YUUKI, R., "Fracture 1977", editor, D. M. R. Taplin, Volume 3, University of Waterloo Press, Canada, 1977.
4. CHEREPANOV, G. P. and KULIEV, V. D., Int. J. Fract., 11, 1975, 29.

5. CHATTERJEE, S. N., Int. J. Solids and Structures, 11, 1975, 521.
6. MONTULLI, L. T., Ph. D. Thesis, Univ. of California, 1975.
7. HUSSAIN, M. A., PU, S. L. and UNDERWOOD, J., ASTM STP 560, 1973, 2.
7a. VITEK, V., "Plane Strain Stress Intensity Factors for Branched Cracks", CEGB Report No. RD/L/N210/76.
8. PALANISWAMY, K. and KNAUSS, G., Int. J. Fract., 8, 1972, 114.
9. PALANISWAMY, K. and KNAUSS, G., Graduate Aeronautical Laboratories California Inst. of Technology, Report No. 74-8; to appear in Mechanics Today, editor, S. Nemat-Nasser, Pergamon Press.
10. THEOCARIS, P. S. and IOAKIMIDIS, N., J. Appl. Math. and Phys., 27, 1976, 801.
11. CARDEW, G. E. and HOWARD, I. C., Int. J. Eng. Sci., 14, 1976, 403.
12. KALTHOFF, J. F., Proc. Third Int. Congress on Fracture, Munich, Volume X, Paper 325, 1973.
13. FRANCIS, P. H., Appl. Mech. Rev., 29, (6), 1976, Review 5065, 841.
14. KHRAPKOV, A. A., Prikl. Mat. Mek., 35, 1971, 677, Int. J. Fract. Mech., 7, 1971, 373.

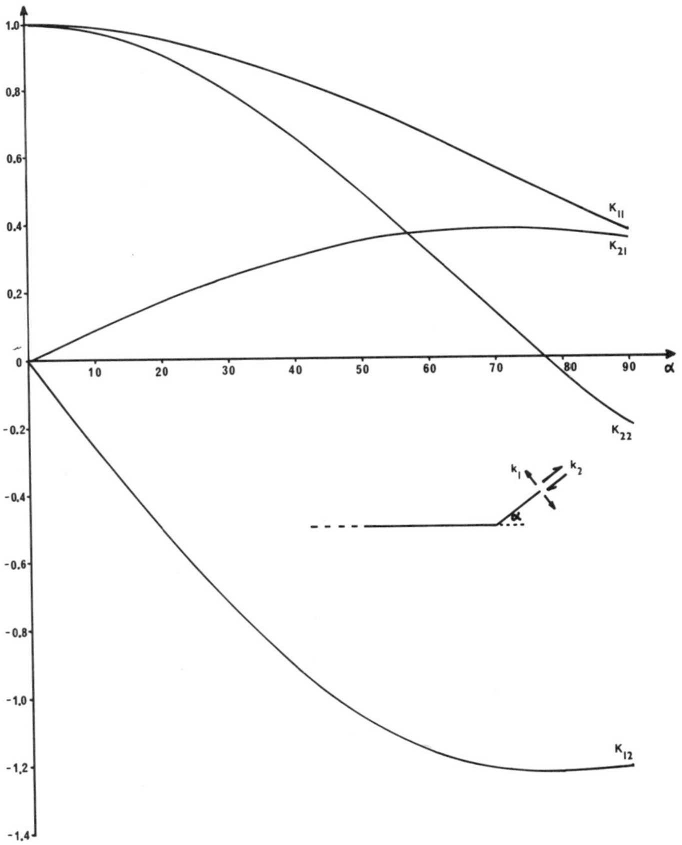

Figure 1

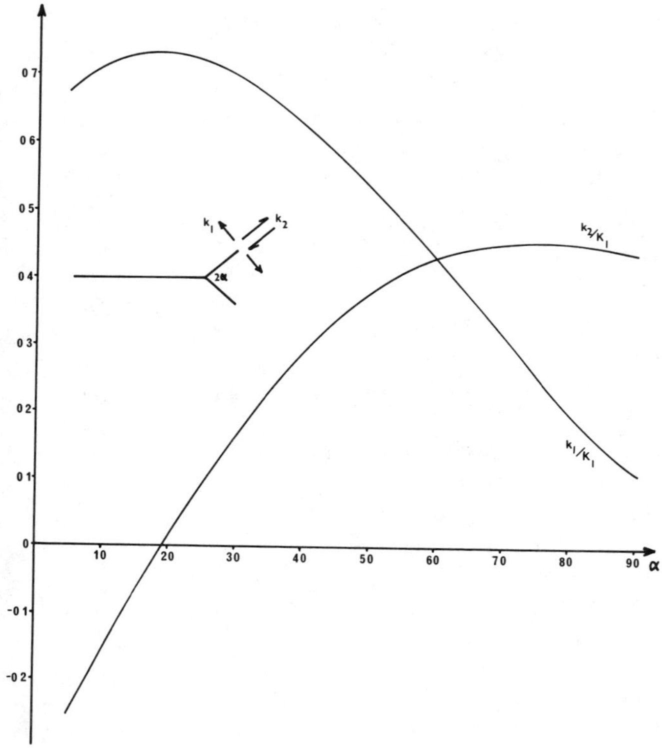
0 7
0 6
0 5
0 4
0 3
0 2
0 1
0
-0 1
-0 2
10
20
30
40
50
60
70
80
90
α
k_1
k_2
2α
k_2/K_1
k_1/K_1

Figure 2

Fracture 1977, Volume 3, ICF4, Waterloo, Canada, June 19 - 24, 1977

ANALYSIS OF BRANCHED CRACKS UNDER BIAXIAL STRESSES

H. Kitagawa and R. Yuuki*

INTRODUCTION

Branched cracks are often observed in brittle fracture and also in stress corrosion cracking. The reasons for these crack branching phenomena in brittle fracture have been explained on the basis of dynamic effects [1]. However, there are many characteristics common to both crack branching behaviour in brittle fracture and in stress corrosion cracking. In order to discuss these phenomena, it seems important to analyse the static stress intensity factors for the branched crack model. However, few solutions for such a crack have been obtained. In another report, a general method analysing some kinds of the branched cracks was formulated using a conformal mapping function. The numerical values of stress intensity factors for a branched crack under uniaxial stress were reported in references [2 - 5].

In this paper, taking into consideration that most engineering structures are often subjected to biaxial loading, the numerical solutions for the stress intensity factors of some kinds of branched cracks under biaxial stress are presented. Moreover, using the results obtained, crack extension behaviour under biaxial stresses is discussed.

ANALYSIS

Conformal mapping functions have been applied to the analyses of crack problems by Muskhelishvili [6], Bowie [7] and others. Muskhelishvili and Savin [8] analysed various shaped holes with the method of polynomial mapping approximation. Bowie applied this method to the analysis of a crack emanating from a circular hole. We applied this method, with various improvements, to the analysis of a branched crack.

In the analysis of a branched crack, due to the complicated crack geometry, the polynomial approximation of the mapping function does not converge easily. Moreoever, the first and second mode stress intensity factors coexist and the crack has dual crack tips at which the stress intensity factors are different. This increases the difficulties of the analysis.

The authors have overcome these difficulties and were able to develop a general method of analysis of branched cracks. In other papers [2 - 5], we analysed a branched crack with one or two branches on one side of the main crack in a uniaxial stress field, utilizing the mapping functions which were used by Andersson [9]. In this paper, we analyse a doubly symmetric branched crack, utilizing the mapping function introduced by the Schwartz-Christoffel transformation with several devices in calculation.

*Institute of Industrial Science, University of Tokyo, Tokyo, Japan.

As a result of the high symmetry of the crack geometry, we obtained more accurate solutions than the solutions of the forked crack obtained in another paper. The outline of our method is described below.

Mapping Function

A doubly symmetric branched crack as shown in Figure 1 is considered. A conformal mapping function which maps the crack into a unit circle is given by a Schwartz-Christoffel transformation. As a result of the symmetry, the conformal mapping function is given by equation (1).

$$Z = \omega(\zeta) = A\int^{\zeta} H(\zeta)(1-2\cos 2\beta\,\zeta^{-2}+\zeta^{-4})d\zeta \tag{1}$$

$$H(\zeta) = \left\{1-\frac{1}{\zeta^2}\right\}^{\frac{2\theta}{\pi}-1}\left\{1-\frac{e^{2i\alpha}}{\zeta^2}\right\}^{-\frac{\theta}{\pi}}\left\{1-\frac{e^{2i(\pi-\alpha)}}{\zeta^2}\right\}^{-\frac{\theta}{\pi}} \tag{2}$$

Where the parameters α, β correspond to the branching points and the crack tips respectively as shown in Figure 1 and A is a real constant. The mapping function $\omega(\zeta)$ must be expanded in a series, because of the determination of the parameters α, β and the stress function as stated in the next section. The function $H(\zeta)$ in equation (1) can be expanded in a binomial series as shown in equation (3).

$$H(\zeta) = 1 + \sum_{n=1}^{\infty} h_n \zeta^{-2n} \tag{3}$$

Substituting equation (3) into equation (1) and integrating equation (1), we get the following equation:

$$\omega(\zeta) = A\left[\zeta + \sum_{n=1}^{\infty} B_n \zeta^{1-2n}\right] \tag{4}$$

For practical reasons the number of terms in equation (4) must be kept finite. If the terms of the polynomial mapping function are truncated at a finite number, the tip of the crack is rounded off. In the analysis of crack problems, the truncation of terms needs some devices to preserve the crack tip geometry without disturbing the overall crack configuration. Referring to Bowie's truncation plan [8], the polynomial mapping function $\omega(\zeta)$ of equation (4) is truncated at a finite term to satisfy the following conditions at the crack tips.

$$\omega'(e^{i\beta}) = 0, \qquad \omega''(e^{i\beta}) = Q \tag{5}$$

Where Q is the exact second derivative of the mapping function of equation (1) at the crack tip and is given by equations (6) and (7).

$$Q = \omega''(e^{i\beta}) = -4A\, H(e^{i\beta})(e^{-2i\beta} - 2\cos 2\beta)e^{-3i\beta} \tag{6}$$

$$H(e^{i\beta}) = \frac{1}{2}\left|\sin\beta\right|^{\frac{2\theta}{\pi}-1}\left|\sin(\alpha+\beta)\,\sin(\alpha-\beta)\right|^{-\frac{\theta}{\pi}}\left\{e^{i\left(\theta+\beta-\frac{\pi}{2}\right)}\right\} \tag{7}$$

To satisfy the above conditions, we add the two correction coefficients to the function $\omega(\zeta)$ which is expanded in series and truncated at the Nth term. Thus we obtain the following mapping function.

$$\omega(\zeta) = A\left[\zeta + \sum_{n=1}^{N+2} B_n^* \zeta^{1-2n}\right] \tag{8}$$

Next, the parameters α and β must be determined. The parameters follow the relations.

$$0 \leq \beta < \alpha \leq \pi/2 \tag{9}$$

For a given value of β:

$$Z = \omega(\zeta) = 0, \qquad \text{on } e^{i\alpha} < \zeta < e^{i(\pi-\alpha)} \tag{10}$$

The value of α for a given value of β is numerically determined to satisfy the above relation. Thus we obtain the mapping function and its polynomial approximation for the doubly symmetric branched crack.

Stress Function and Stress Intensity Factors

The Muskhelishvili complex stress functions $\phi(\zeta)$, $\psi(\zeta)$ are used in our analysis. Taking the polynomial mapping function $\omega(\zeta)$ obtained by equation (8) into consideration, the stress functions can be defined as follows, when the crack is subjected to uniform uniaxial tension σ in the direction ϕ_0 as shown in Figure 2.

$$\phi(\zeta) = A\sigma\left[\frac{\zeta}{4} + \sum_{n=1}^{N+2} d_n \zeta^{1-2n}\right] \tag{11}$$

$$\psi(\zeta) = A\sigma\left[-\frac{e^{2i\phi_0}}{2}\zeta + \sum_{n=1}^{N+2} f_n \zeta^{1-2n}\right] \tag{12}$$

Where d_n and f_n are the coefficients of the stress functions and they have to be determined so that the stress-free condition is satisfied on the crack edge. This condition is given by equation (13).

$$\overline{\phi}(1/\zeta) + \overline{\omega}(1/\zeta)\phi'(\zeta)/\omega'(\zeta) + \psi(\zeta) = 0, \quad \text{on } |\zeta| = 1 \tag{13}$$

Substituting equations (11), (12) and (8) into equation (13), and equating the coefficients of all positive powers of ζ to zero, a set of linear simultaneous equations with respect to the unknown coefficient d_n can be given. Solving these equations, the stress function $\phi(\zeta)$ to satisfy the boundary conditions is obtained. By means of the function $\phi(\zeta)$ obtained above and equations (6) and (7), the stress intensity factors of the branched crack can be calculated by the following equation.

$$K = K_1 - iK_2 = \frac{2\pi^{1/2}\,\phi'(e^{i\beta})}{[e^{i\theta}\omega''(e^{i\beta})]^{1/2}} \tag{14}$$

Where θ is the branching angle as shown in Figure 1. The value of the stress intensity factors for the branched crack under given biaxial stresses can be obtained by the summation of these kinds of solutions.

NUMERICAL RESULTS

Doubly Symmetric Branched Crack

By means of the method mentioned above, the values of the stress intensity factors of the branched crack were calculated. The accuracy is within 1%. They are normalized by the stress intensity factors of the straight crack length 2c as shown in Figure 3. Figure 3 shows the non-dimensional stress intensity factors (F_1, F_2) of the branched crack, subjected to uniform tension along the y axis and Figure 4 along the x axis, respectively. The solutions for the branched crack under arbitrary biaxial stresses can be calculated by simply superimposing the solutions for the cracks under uniaxial stress. Figure 5 shows the stress intensity factors of the branched crack for $b/a = 0.1$ under various biaxial stresses. λ is the ratio of the tensile stress in the x direction to that in the y direction.

It is well known that in brittle fracture or in stress corrosion cracking, crack branching has often been observed. In such a fracture process, the stable crack branching angle can be considered. Figure 5 shows that for various values of λ, the F_2 (or K_2) value changes sign and becomes zero at a chracteristic branching angle. A branched crack with such a branching angle will grow along the extension line of the branch. If we take a criterion such that each branch can grow only in the direction F_2 (or K_2) = 0 [3], the stable branching angles 2θ for the various value of λ are determined by Figure 5. They are about 25°, 33°, 41°, 52° for λ = -1, 0, 0.5, 1, respectively. Thus the branching angle increases with increases of the value of λ. For the case of uniaxial stress ($\lambda = 0$), it is well known that the average macro branching angle observed in experiments is almost 30° - 40°. In this discussion, the results for $b/a = 0.1$ are used. If the b/a value increases, the branching angle decreases slightly. This corresponds to the phenomenon that the branched crack changes direction with the growth of the branches. We can obtain stable branching angles under various biaxial stress states. If a branched crack is observed in a structure, it is possible to infer the stress state to which the branched crack was subjected using these results.

Bent Crack

We also analysed a bent crack, as shown in Figure 6, as a limiting kind of branched crack. In this case, the mapping function and the method of analysis are slightly different from those mentioned above. They have been reported in reference [2]. This bent-crack crack-model is very useful when the extension direction of a crack under a mixed mode stress state is discussed [11]. Several authors have tried to analyse the same crack model as used here [12 - 13]. Only numerical results are presented here in order to discuss crack extension behaviour under biaxial stresses. Figure 6 shows the stress intensity factors at the tip of the bent crack with a branch ($b/a = 0.1$) subjected to various biaxial stress states. It is a very difficult problem to consider the stability of crack extension. However, if the model of the bent crack can be used and the stress intensity factors of this crack are known, it seems that the stability of crack extension can be supposed to some extent. We can deduce that for the case of λ = 0, -1, the crack is very stable because the nondimensional stress intensity factors at the bent crack tip F_{1B}, F_{2B} vary rapidly at the vicinity of the point $\lambda = 0$. This means that a small deviation in the direction of crack extension from the original direction induces a rapid return to the original direction. On the other hand, for the case of $\lambda = 2$, the crack seems to be unstable and cannot grow in a straight line. Figure 7 shows the directions of crack extension from the tip of a bent crack which are calculated by the theory of maximum circumferential stress proposed by Sih, et al. [10]. For the case of $\lambda = 0, -1$, if the crack is bent for some reason, it will be bent again towards the direction of the extension line of the initial crack. On the other hand, for the case of $\lambda = 2$, the crack grows in the direction of the extension line of the bent branch. Thus the crack has different stability for each biaxial stress state. We suppose that this affects the crack path and crack growth law found in biaxial fatigue tests.

SUMMARY

We constructed a general method for analysis of a branched crack in a given biaxial stress state by means of a conformal mapping function and its series expansion. The calculated values of stress intensity factor of a doubly symmetric branched crack and a bent crack under various biaxial stresses are presented. On the basis of these results, the crack extension behaviour under biaxial stresses is discussed from our crack morphological view-point. Some of the interesting points are as follows:

1) In a fracture process such that a symmetric branched crack can grow, there is a stable branching angle particular to a given biaxial stress state, which increases with increase of the lateral biaxial tensile stress.
2) Using the solutions for the bent crack, we discussed the stability of the crack under various biaxial stresses. The crack has a particular stability, which may affect a crack extension behaviour.

REFERENCES

1. YOFFE, E. H., Phil. Mag., 42, 1951, 739.
2. KITAGAWA, H. and YUUKI, R., Trans. JSME, 41-346, 1975, 1641.
3. KITAGAWA, H., YUUKI, R. and OHIRA, T., Eng. Fract. Mech., 7, 1975, 515.
4. KITAGAWA, H. and YUUKI, R., Preprint of JSME, No. 750-1, 1975, 223.
5. KITAGAWA, H. and YUUKI, R., Preprint of JSME, No. 750-11, 1975, 183.
6. MUSKHELISHVILI, N. I., "Some Basic Problems of the Mathematical Theory of Elasticity", Noordhoff, Leiden, 1953.
7. BOWIE, O. L., "Method of Analysis of Crack Problems", Edited by G. C. Sih, Noordhoff, Leiden, 1973, 1.
8. SAVIN, G. N., "Stress Concentration Around Holes", Pergamon Press, Oxford, 1961.
9. ANDERSSON, H., J. Mech. Phys. Solids, 17, 1969, 405.
10. ERDOGAN, F. and SIH, G. C., Trans. ASTM, Ser. D, J. Basic Engr., 85, 1963, 525.
11. BILBY, B. A. and CARDEW, G. E., Int. J. Fract., 11, 1975, 708.
12. PALANISWAMMY, K. and KNAUSS, W. G., Int. J. Fract., 8, 1972, 114.
13. CHATTERJEE, S. N., Int. J. Solids and Structures, 11, 1975, 521.

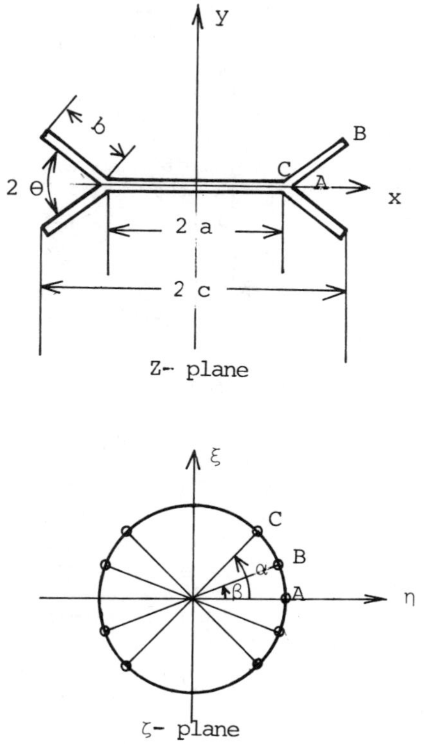

Figure 1 Crack geometry in Z-plane and mapped in mapped ζ-plane

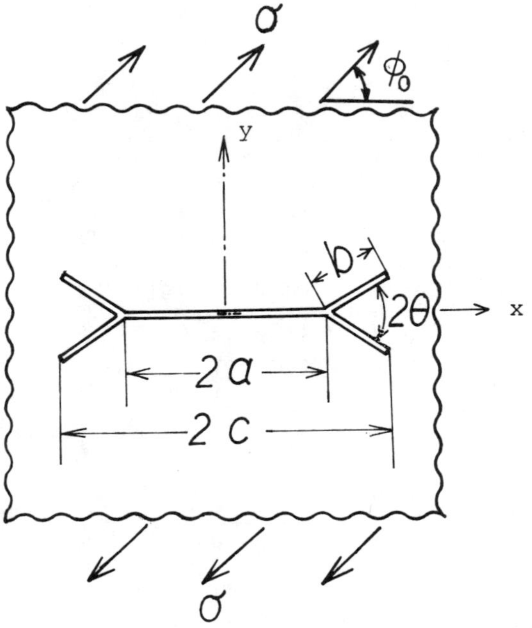

Figure 2 The doubly symmetric branched crack subjected to uniform uniaxial tension

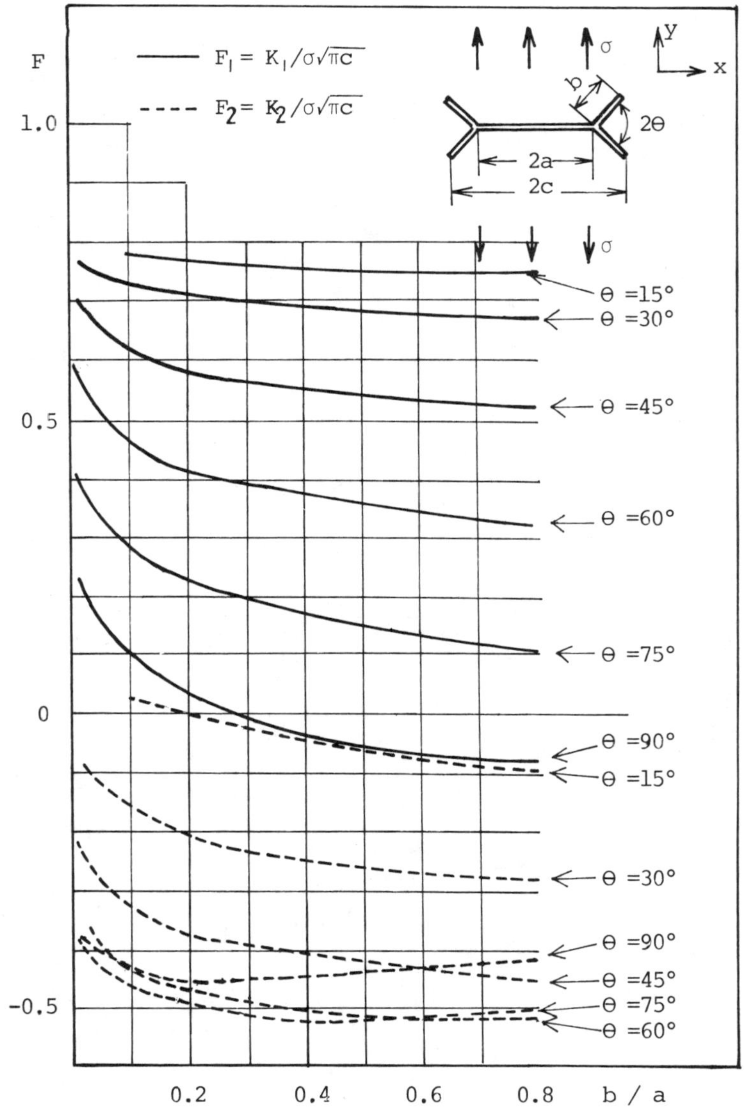

Figure 3 The stress intensity factors of the doubly symmetric branched crack subjected to uniaxial tension in the y-axial direction

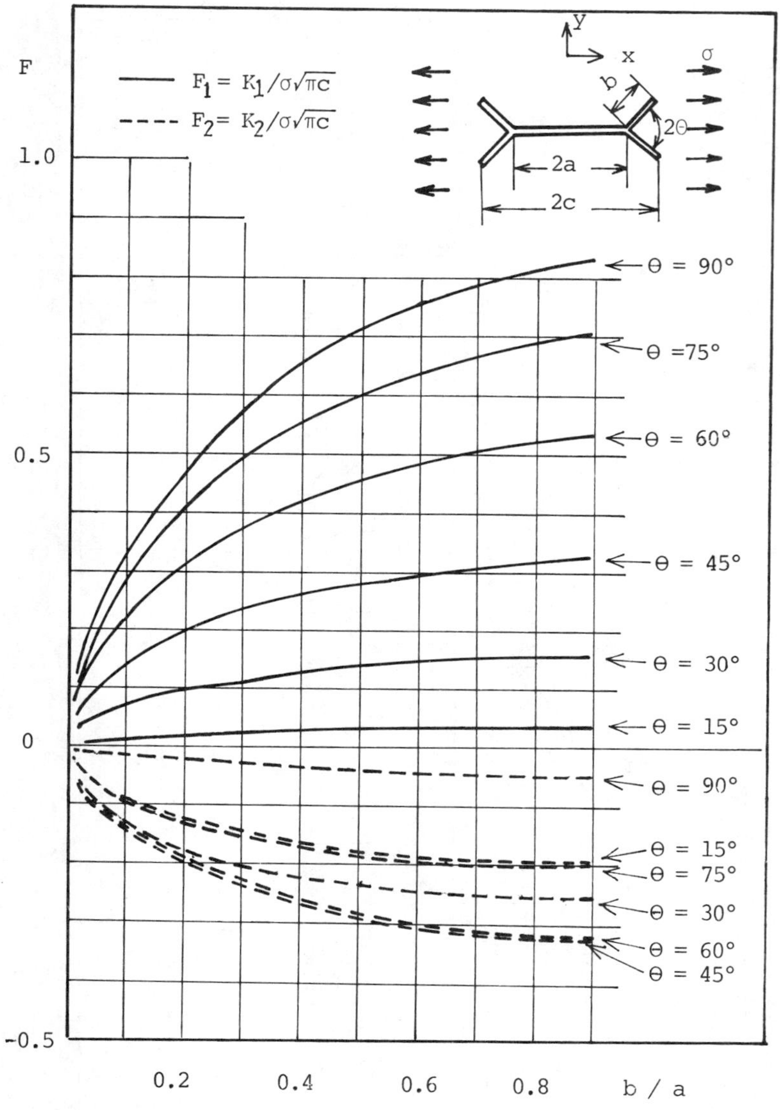

Figure 4 The stress intensity factors of the doubly symmetric branched crack subjected to uniaxial tension in the x-axial direction

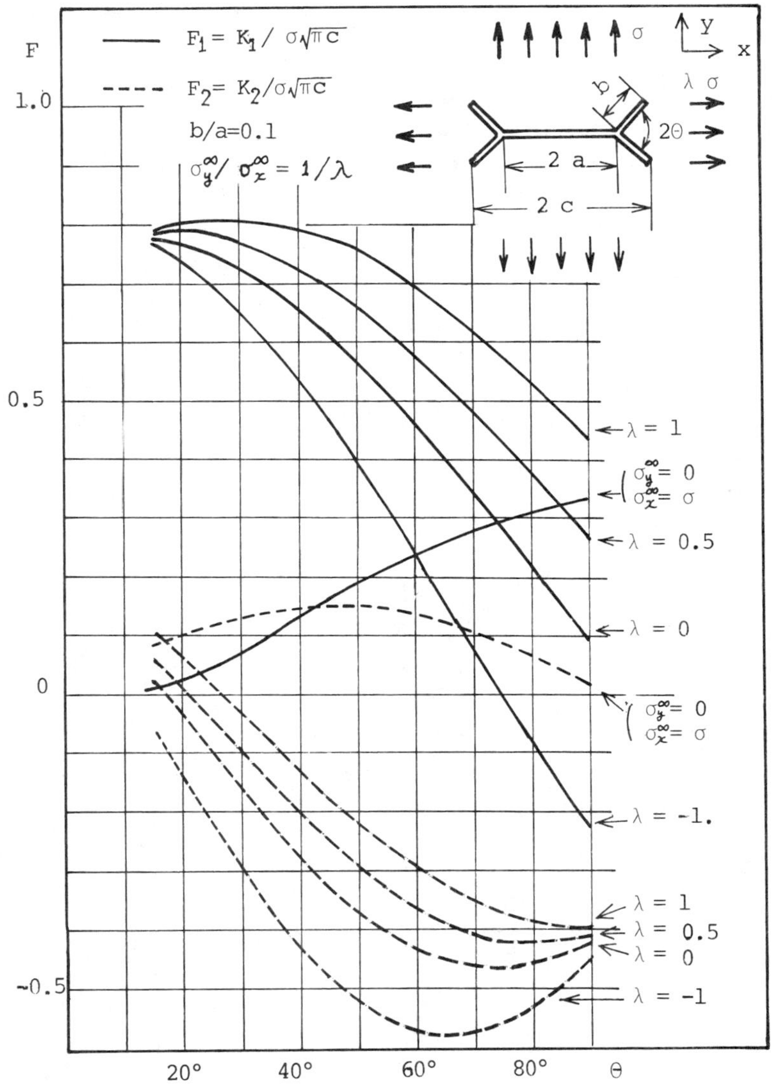

Figure 5 The stress intensity factors of the doubly symmetric branched crack for b/a=0.1 subjected to biaxial stresses

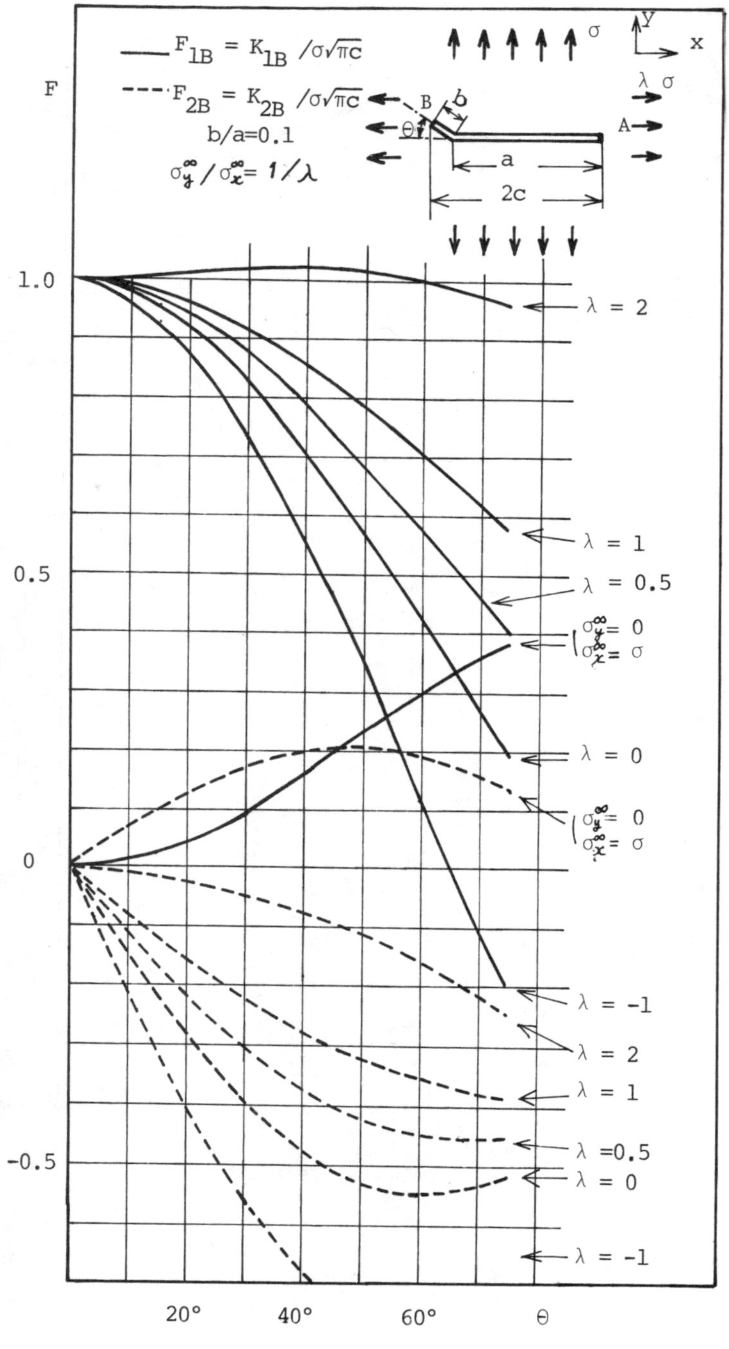

Figure 6 The stress intensity factors of a bent crack for b/a=0.1 subjected to a biaxial stress state

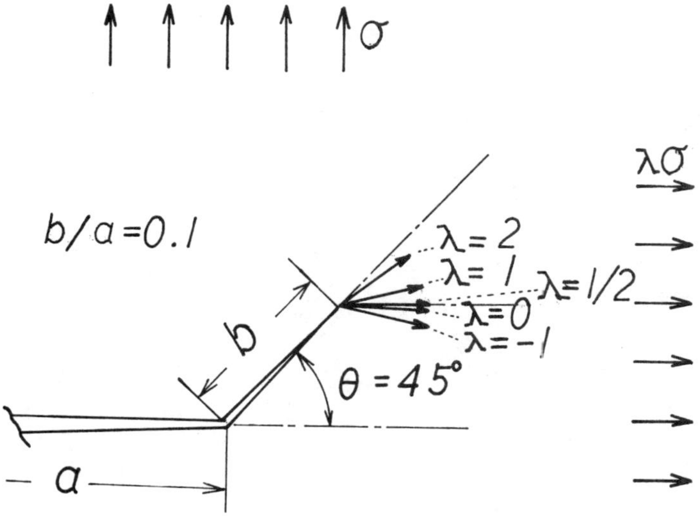

Figure 7 The direction of crack extension from the bent crack tip

Fracture 1977, Volume 3, ICF4, Waterloo, Canada, June 19 - 24, 1977

A SIMPLE ANALYTICAL MODEL FOR THE THICKNESS DEPENDENT TRANSITION FROM PLANE STRESS TO PLANE STRAIN IN BODIES WITH A CRACK

G. Prantl*

INTRODUCTION

From experimental work an influence of the dimensions of specimens containing cracks on their fracture behaviour is apparent. While there is a large amount of data on size effects, and there are empirically derived criteria for the dimensions of laboratory test specimens used to obtain results applicable to full scale structures, the quantitative understanding of the problem and of the effect of the various parameters is not yet perfect. A possible consequence of increasing the dimensions from small specimens to large structures is a transition from a plane stress to a plane strain state of deformation in the critical region of the material and a connected transition in fracture mode.

The purpose of the present paper is to construct a simple analytical model, using continuum mechanics principles, which is able to describe the influence of specimen thickness on the stress state in the neighbourhood of a crack, assuming linear elastic material. For the more important practical case of elastic plastic material behaviour, where yield zones develop, some limited conclusions can be drawn on the basis of an analogy between the elastic and the elastic plastic distribution of the in-plane stresses in front of the crack.

Although it is possible to calculate the stresses and strains in a body of finite dimensions using three dimensional finite element techniques, the present work is justified, because it allows the effect of the various parameters of the problem (e.g., thickness, external load, Poisson's constant, crack length) to be estimated in a direct closed form.

DEVELOPMENT OF AN ELASTIC MODEL

The work described here is restricted to the following class of problems: A disc, made of linear elastic material, contains a crack penetrating through the thickness. The dimensions of the disc in its main plane are large compared to the length of the crack, 2c. The thickness 2t can take any value between zero and infinity. The body is loaded by uniformly distributed tensile stresses, σ_∞, perpendicular to the plane of the crack. Figure 1 explains the coordinates used, as well as the notation of the stresses.

In the vicinity of the crack tip, there is a concentration of the stresses σ_r and σ_ϕ, which is known from the solution of the respective plane problem (1), if t/c is either approaching zero or infinity. In the first case $\sigma_z \equiv 0$, in the second case $\sigma_z = (\sigma_r+\sigma_\phi)$. In real bodies σ_z is within

* Swiss Federal Institute for Reactor Research, CH-5303 Würenlingen, Switzerland

these limits.

It is assumed for the sake of argument, that the disc is divided into two zones, as shown in Figure 2. The inner zone I contains the material under relatively high stresses, while the ringshaped outer zone II includes the rest of the body within the region of the stress concentration. In each of the two zones the sum of the stresses $\sigma_r+\sigma_\phi$ is averaged.

If the zones were free to deform laterally, each would suffer a contraction, caused by the average stresses in it. This individual deformation of each zone results in a misfit at their common boundary. Before they are reassembled, the difference in lateral contraction must be cancelled. This can be done by the application of shear stresses $\tau_{I,II}$ with the proper magnitude and distribution to the common boundary of the zones. Still considering the separated zones, it is noticed that these shear stresses, which are necessary to maintain the continuity of the whole body, induce tensile stresses σ_z in zone I and compressive stresses in zone II. It is assumed in the model, that the stresses σ_z are uniformly distributed over the cross section of the respective zone.

The calculated stresses σ_{Iz} and σ_{IIz} are mean values over the cross section of the respective zone, and therefore depend on the zone size chosen. For the mathematical formulation of this idea, the following conditions are used:

- Equilibrium of forces in direction z on zone I alone.
- Equilibrium of forces in direction z on zone I and zone II together.
- A compatibility relation at the common boundary of the two zones, formulated with a series expansion with respect to r of the displacement in direction z.
- Hooke's law.
- Boundary conditions for σ_z at the median plane and at the surfaces of the disc.

In order to simplify the treatment of the problem, the cross sections of the zones are assumed to be circular. The radius of inner zone I is 'a', the radius of outer zone II is 'b', (see Figure 2).

Using the conditions listed leads to a differential equation for the average stress σ_{Iz} in zone I:

$$\sigma_{Iz}'' - \lambda^2\sigma_{Iz} = - \lambda^2 B(z) . \tag{1}$$

The double prime denotes the second derivative of σ_{Iz} with respect to z. The parameter λ contains the dimensions of the zones and Poisson's constant ν:

$$\lambda = \sqrt{\frac{3}{2(1+\nu)}}\frac{1}{\sqrt{a(b-a)}} \tag{2}$$

B(z) is a function, which consists of the difference of the stresses σ_r and σ_ϕ between the zones, and a term that depends on the radial deformation at the common boundary of the zones. While the first part can be determined from the distribution of the stresses σ_r and σ_ϕ, the second part remains unknown.

The boundary conditions are:

$$\sigma_{Iz}'(0) = 0 \; , \quad \sigma_{Iz}(t) = 0 \; , \quad \sigma_{Iz}'(t) = 0 \; . \tag{3}$$

The prime denotes the first derivative of σ_{Iz} with respect to z. The solution of equation (1), using the first two boundary conditions (3), is:

$$\sigma_{Iz} = \lambda \left\{ \left[-\int_z^t B(z)\sinh \lambda z \, dz + \text{tangh}\, \lambda t \int_0^t B(z)\cosh \lambda z \, dz \right] \cosh \lambda z - \right. \\ \left. - \sinh \lambda z \int_0^z B(z)\cosh \lambda z \, dz \right\} . \tag{4}$$

Equation (4) does not satisfy the third boundary condition. The reason is that equation (1), which results from the simplified one dimensional analysis, is of the second order.

The complete solution, accounting for the radial stresses at the borders of the zones too, would be a differential equation of the fourth order.

In order to introduce the retroaction of σ_z on the stresses σ_r and σ_ϕ, a specific function is proposed for B(z), which makes it possible, to satisfy this boundary condition too:

$$B(z) = \sigma_{Iz_{max}} \left(1 - \frac{2 \cosh \lambda z}{\cosh \lambda t + \dfrac{\lambda t}{\sinh \lambda t}} \right) . \tag{5}$$

Here, $\sigma_{Iz_{max}}$ is the greatest possible value of σ_{Iz} for the problem under consideration, namely:

$$\sigma_{Iz_{max}} = \nu(\Delta\sigma_r + \Delta\sigma_\phi) \cdot \left(1 - \frac{a^2}{b^2} \right) . \tag{6}$$

The differences of the stresses between the two zones, $\Delta\sigma_r$ and $\Delta\sigma_\phi$, can be determined from the respective stress distribution. The form of equation (5) is indicated by still unpublished photoelastic measurements, made by the author. Further, it is based on the assumption, that the z-distribution of the in-plane stresses and deformations is also governed by the parameter λ.

Inserting equation (5) into equation (4) results in the final equation for σ_{Iz}, the average stress in zone I:

$$\frac{\sigma_{Iz}}{\sigma_{Iz_{max}}} = 1 - \frac{(\lambda t \cosh \lambda t + \sinh \lambda t)\cosh \lambda z - \lambda z \sinh \lambda t \sinh \lambda z}{\sinh \lambda t \cosh \lambda t + \lambda t} . \tag{7}$$

From equation (7) the average stress in zone II, σ_{IIz}, and the shear stress $\tau_{I,II}$ at the common boundary can easily be deduced, if necessary.

If the thickness of the disc is large compared to the dimension b, it is also large compared to the half crack length c, because b and c are of the same order of magnitude. In this special case equation (7) reduces to:

$$\frac{\sigma_{Iz}}{\sigma_{Iz_{max}}} \cong 1 - (1+\lambda\xi)\, e^{-\lambda\xi} \,. \qquad (8)$$

In this equation the distance ξ from the surface is used instead of z, where ξ = t-z.

APPLICATION OF THE ELASTIC MODEL

Before the stresses can be calculated, the input data must be derived from the distribution of the in-plane stresses. The parameter λ, equation (2) and the stress $\sigma_{Iz_{max}}$, equation (6) are determined using the two dimensional solution of the crack problem (1). The results are presented in Figure 3 as a function of the relative zone size a/b.

The mean stresses within the two zones and the parameter λ are strongly influenced by a/b. The dimension 'b', according to the model, defines that part of the body which is able to constrain the material in zone I. It is specific to the particular problem treated and it can be estimated from the distribution of the in-plane stresses. It is reasonable to assume 'b' proportional to the fading distance of the perturbation of the stress field. In this work, b = 0.5 c is used. This is an arbitrary choice, suited to the application of Sneddon's elastic solution. The dimension 'a', which fixes the size of zone I, is varied in the numerical analysis. If 'a' approaches zero, then σ_{Iz} becomes infinitely large and the ratio $\sigma_{Iz}/\nu(\sigma_r+\sigma_\phi)_I$ approaches unity. With this data, equation (8) is evaluated. It yields the distribution of σ_{Iz} over the thickness coordinate ξ, according to Figure 4.

Figure 5 illustrates the influence of the choice of 'a', with 'b' kept constant, as determined by the distribution of σ_r and σ_ϕ. The condition of equilibrium of the forces in direction z, formulated on zone I and zone II together, requires that the integral of σ_z over any cross section of the two zones must vanish. Therefore tension in zone I causes compressive stresses in zone II. Finally, Figure 6, derived from equation (7), gives an example of the rise of the out-of-plane stress σ_{Iz} in the mid plane with increasing relative thickness t/c.

TREATMENT OF PROBLEMS WITH PLASTIC ZONES

When real materials are considered, plastic zones are found ahead of the crack tip, even at very small loads. According to the crack model of Dugdale [2] for instance, perfect plastic behaviour of the material confines the stress within the yield zone to the uniaxial yield stress 's'. As this crack model gives no information on the magnitude of the stresses σ_r and σ_ϕ within the plastic zone except at the y-axis ahead of the crack tip, evaluation of the input data, needed to calculate σ_{Iz} analogous to the elastic case, is not possible.

But, setting the radius 'a' of zone I equal to the length of the yield zone according to the Dugdale model enables some conclusions to be drawn concerning the effect of the external load on σ_{Iz}. To our present knowledge, the length of the plastic zone varies through the thickness of the body. The same should reasonably be assumed for the dimension 'a', but the model can deal only with a constant 'a'. As long as the radius 'a' is smaller than or equal to the radius of the plastic zone, it has a very small influence on the stress difference $(\Delta\sigma_r+\Delta\sigma_\phi)$, because the stress is constant within the plastic zone. Therefore the effect of the external load on a given body, which determines the plastic zone size, can be investigated separately from the choice of the radius 'a'.

Figure 7 compares λ and $(\Delta\sigma_r+\Delta\sigma_\phi)$ with the respective values of the elastic case. The term $\lambda t/t/c\sqrt{3/2(1+\nu)}$ is the ratio between the effective thickness and the geometrical thickness of the body. The smaller this term, the smaller is σ_{Iz}. The term $(s-1,2\sigma_\infty)/\sigma_\infty$ is the analogue of the elastic stress difference $(\Delta\sigma_r+\Delta\sigma_\phi)\sigma_\infty$. The smaller its value, the smaller is the maximum stress, $\sigma_{Iz_{max}}$, that can be achieved with a sufficiently large thickness. Using this data, Figure 8 finally demonstrates the fall off of σ_{Iz} with rising external load, when the dimensions of the body remain constant.

It should be pointed out again, that this kind of treatment of elastic-plastic problems cannot be expected to give absolute values of the stress σ_{Iz}. It only indicates the effect of external load on the stress σ_z in a specimen of given dimensions.

ACKNOWLEDGEMENTS

The permission of the Swiss Federal Institute for Reactor Research (EIR) and the Swiss Federal Institute of Technology Zürich (ETHZ) to publish this work, which is part of a dissertation submitted to ETHZ, is acknowledged.

REFERENCES

1. SNEDDON, I. N., Proc. Roy. Soc., A 187, 1946, 229.
2. DUGDALE, D. S., J. Mech. Phys. Solids, 8, 1960, 100.

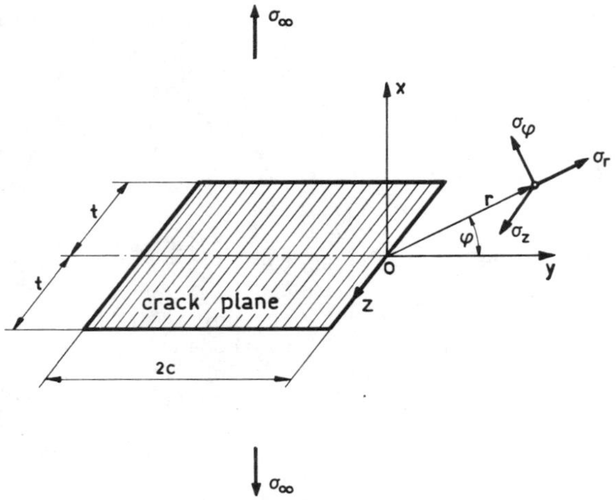

Figure 1 Dimensions, Coordinates and Notation of Stresses

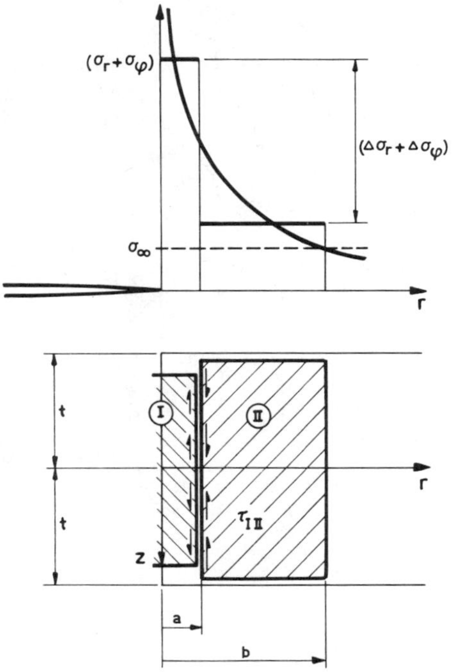

Figure 2 Illustration of the Proposed Model. Half Cross Section of the Separated Zones

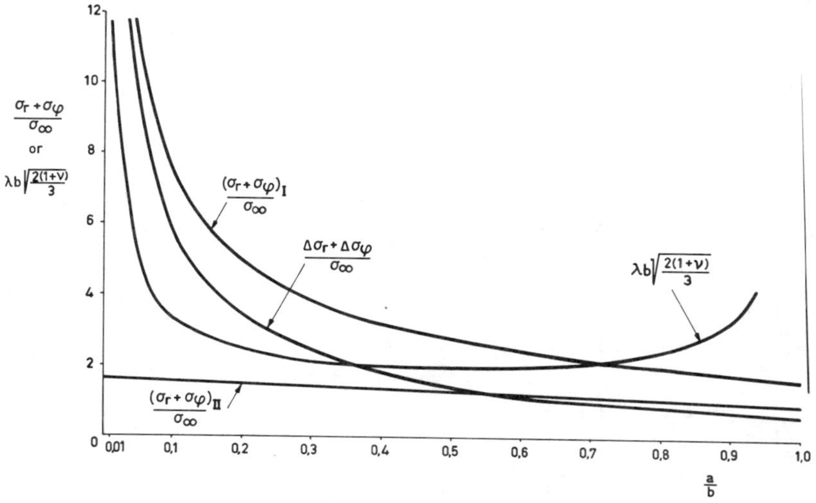

Figure 3 Average In-Plane Stresses Within the Zones and Parameter λ for a Crack in a Large Disc

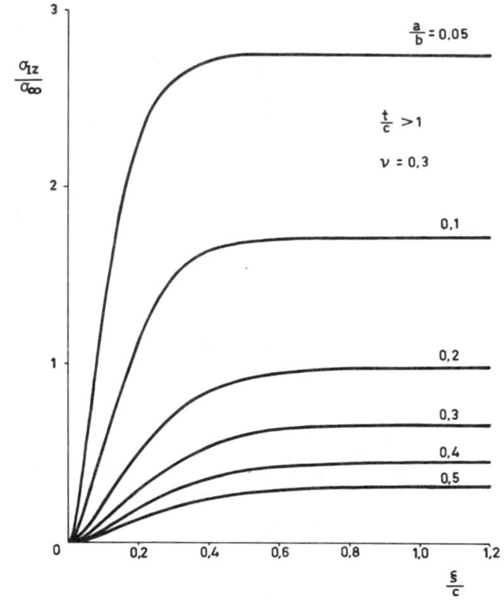

Figure 4 Distribution of σ_{Iz} through the Thickness for Various Values of the Relative Zone Size a/b

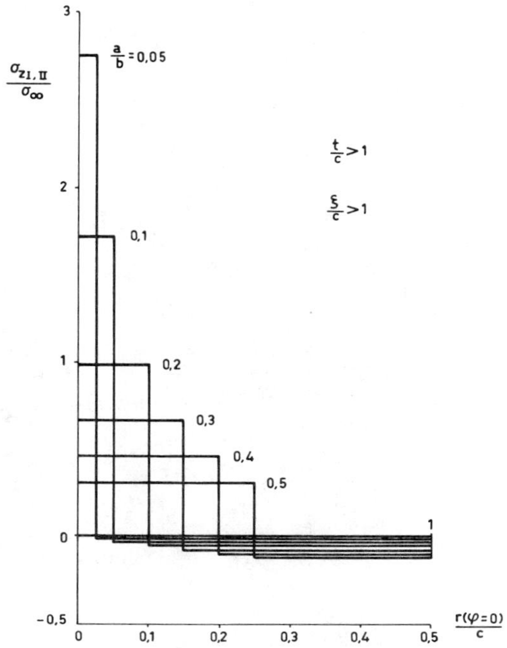

Figure 5 Average σ_z in Zone I and Zone II for Various Values of a/b

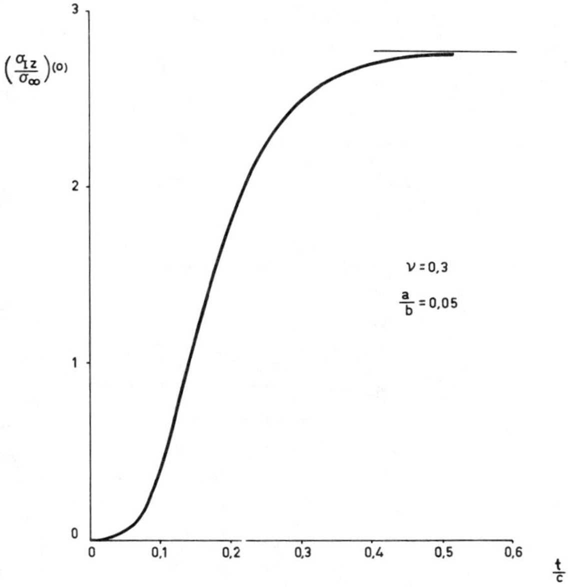

Figure 6 σ_{Iz} in the Mid-Plane as a Function of Relative Thickness

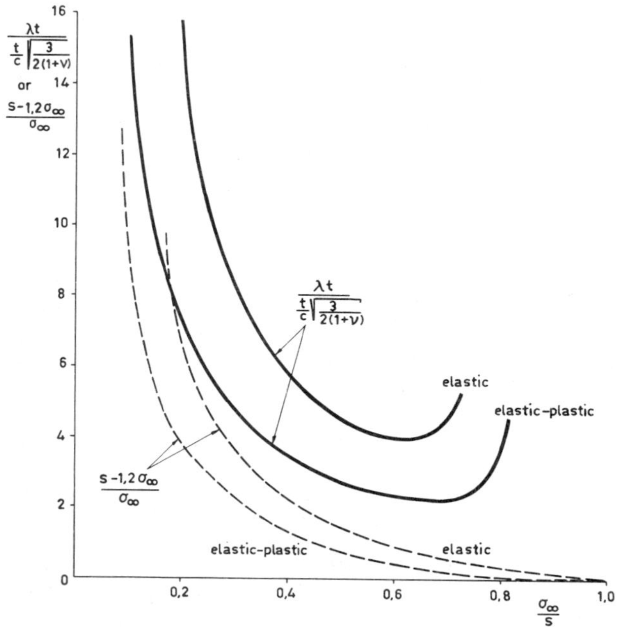

Figure 7 $(\Delta\sigma_r+\Delta\sigma_\phi)$ and λ as a Function of Load for Elastic and Elastic-Perfectly Plastic Material

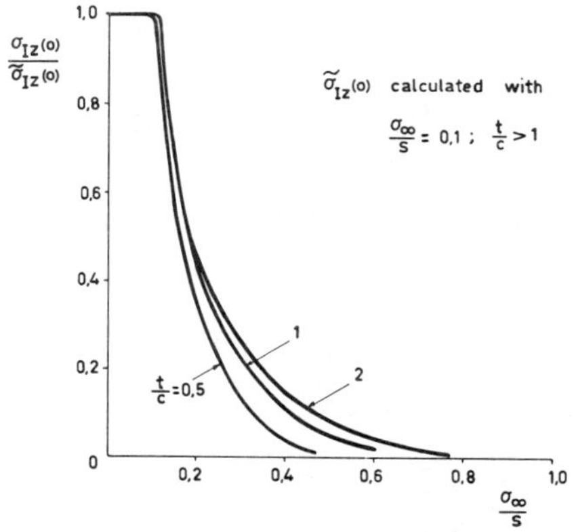

Figure 8 σ_{Iz} at the Median Plane as a Function of Load for Various Relative Thicknesses

Fracture 1977, Volume 3, ICF4, Waterloo, Canada, June 19 - 24, 1977

ENERGY CONSIDERATIONS IN DYNAMIC CRACK PROPAGATION AND ARREST

T. Kanazawa* and S. Machida*

INTRODUCTION

Since the ship building industry has experienced a number of casualties due to brittle fracture, most naval architects have been forced to be interested in the brittle behavior of low and medium strength structural steels. While linear-elastic fracture mechanics was originated and applied to ultra high strength steels and other special alloys in space engineering in the United States, Japanese research groups were trying to use fracture mechanics to interpret the results of brittle fracture propagation arrest tests using wide plate specimens.

Within the limit of a relatively short arrested crack, a static approximation using either a linear fracture mechanics concept or an arrest toughness concept has yielded useful results for both the theoretical interpretation and the design application of currently used brittle fracture propagation arrest tests [1]. Later experimental investigations using very wide specimens (1,300 and 2,500 mm wide plates) have revealed that the above simple interpretation is inconsistent with results for long arrested cracks [2-4].

In order to seek a more reasonable theory of fast fracture and crack arrest and to study how neglect of dynamic aspects affects the interpretation of results of unstable, fast crack propagation arrest tests and the philosophy of crack design, the authors have started with a dynamic fracture mechanics analysis of crack propagation and arrest with the use of a finite-difference method. The results of the analysis were compared with previous data on structural steels and experiments using PMMA specimens, and are discussed in terms of dynamic fracture mechanics analysis.

BRITTLE CRACK PROPAGATION ARREST TEST USING VERY WIDE PLATE SPECIMENS

A simple fracture mechanics concept has been successfully applied to the analysis of the experimental results of several kinds of brittle crack arresters using medium size specimens (500 mm wide) [1]. This analysis is based on the arrest toughness concept and the assumption that the dynamic component of the stress intensity factor or strain energy release rate can be neglected, in so far as the crack length is relatively small, and a crack just before arrest or running at a speed near to the lower critical velocity of a brittle crack (200 $\sim$ 400 m/sec [5]) is considered.

In the case of large welded steel constructions such as ships, the occurrence of a very large brittle crack is to be presumed. When a brittle crack extends to a very large-scale crack, however, it is questionable whether the above simple approach is valid or not. In relation to this

*University of Tokyo, Tokyo, Japan.

problem, extensive experimental research projects have been conducted in Japan, using very wide (max. 2,400 mm) plate specimens (gradient temperature type and hybrid type crack arrest tests) [2-4].

Figure 1 shows an example of test results for a gradient temperature type ESSO test plotted as log K_c vs. 1/T, where K_c is the nominal arrest toughness or static K value at the arrest point and T is absolute temperature. The solid circles denoted as DG are the data obtained from the standard double tension test (500 mm wide specimen, temperature gradient type crack arrest test), and they are considered to represent the material toughness against propagation which can be handled on the basis of a linear fracture mechanics criterion. Effect of load drop is taken into consideration in the sense of statics [6]. Figure 2 shows examples of crack speed measured in various types of arrest tests. It is to be noted that crack speed in the wide plate specimen is much higher than DG specimen for most of the propagation period.

The considerable discrepancy between K_c values obtained from wide specimens and those from standard specimens (DG specimens), as seen in Figure 1 suggest that estimation of K values (parameter characterizing crack driving force) on the basis of a simple static approximation is invalid, possibly due to disregard of dynamic features involved in extensive propagation of brittle crack such as is observed in the wide specimens.

DYNAMIC ANALYSIS USING FINITE DIFFERENCE METHOD

For the analysis of a dynamic crack in a plate, the following two-dimensiona equations of motion in an elastic medium were used

$$\frac{\partial^2 u}{\partial t^2} = C_1^2 \frac{\partial^2 u}{\partial x^2} + (C_1^2 - C_2^2) \frac{\partial^2 v}{\partial x \partial y} + C_2^2 \frac{\partial^2 u}{\partial y^2} \tag{1}$$

$$\frac{\partial^2 v}{\partial t^2} = C_2^2 \frac{\partial^2 v}{\partial x^2} + (C_1^2 - C_2^2) \frac{\partial^2 u}{\partial x \partial y} + C_1^2 \frac{\partial^2 v}{\partial y^2} \tag{2}$$

where C_1 and C_2 are the dilatational and shear wave speed, respectively. u and v are the displacements in the x and y directions, respectively. Taking the x-axis as the crack line, dynamic analysis of a crack in a plate was made by solving equation (2) with relevant boundary and initial conditions using the finite-difference method.

The time increment Δt was chosen so as to meet the stability condition of the numerical solution [7]:

$$\frac{C_1 \Delta t}{h} = 0.5 \tag{3}$$

where h is mesh spacing.

As a preliminary to examining the validity of the present method of analysi several illustrative problems were solved. The width and length of the cracked plates analyzed were 50 $\sim$ 100h. One of the illustrative problems solved was a crack extending in a plate (B=70h, H=70h) at constant speed from an initial crack length of 2a = 2 × 5.5h. The stress distribution along the crack line is shown in Figure 3 compared with the analytical solution obtained by Broberg [8]. On other aspects such as deformation and energy flow, the numerical results compared well with the analytical solution except for the region close to the crack tip, which would be inevitable in using the finite difference method. Thus the present finite-difference method was shown to be useful for analysis of dynamic crack propagation, especially for higher crack speeds which necessitate that the dynamic effect be properly included.

NUMERICAL EXPERIMENT

Several numerical experiments were carried out on a plate (B=70h, H=70h) with an initial crack $2a = 2 \times 5.5.h$ long, using a simple crack growth criterion which specifies the critical breaking stress σ_{Yc} at a fixed distance from the crack tip (point A in Figure 4). Examples of the crack speed obtained are shown in Figure 4. Figure 4(a) shows a result for a crack running through the field of uniform material resistance or constant critical stress for crack growth: σ_{Yc} (A)=3.0T. The crack is accelerated to a terminal speed and this agrees with many experimental observations so far reported. Figure 4(b) shows an example for a crack traversing a field of linearly increasing material resistance such as is the case with a temperature gradient type crack arrest test, the crack will be decelerated and finally be arrested in this case, and this compares qualitatively with experimental observations such as are shown in Figure 2. Figure 4(c) is a result for the case of decreasing applied stress with time, which simulates a crack subjected to a load drop effect and/or a crack running into a decreasing stress field such as occurs in a crack arrest test using a bending or wedge opening specimen with fixed grips.

DISCUSSION ON EXPERIMENTAL RESULTS

Experiment on PMMA

Using 10 mm thick PMMA, a series of crack arrest tests were conducted. The specimens used were single-edge-notched 300 mm wide and 280 mm long plates. The mean velocity of the crack was measured by crack detector gages mounted along the crack propagation line and the load drop was dynamically recorded. The measured crack speeds ranged from 120 to 540 m/sec, that is, relative speeds with respect to a dilatational wave (2.15×10^3 m/sec for the PMMA used) were 0.06 to 0.25. The experimental results were analyzed by the numerical method using the experimentally obtained crack speed and load as the time dependent boundary conditions. Figure 5 shows an example of the variation of the energy components with crack length i.e., work done by external load, W, kinetic energy, K, strain energy, U and dissipated energy, D.

The dissipated energy D, which is interpreted as material fracture energy, is obtained from law of conservation of energy given by

$$D = W - U - K \tag{4}$$

where W, U and K are computed from experimentally obtained a vs. t and (load) vs. t relations. Dynamic fracture toughness K_d, which is defined by

$$K_d = E\left(\frac{dD}{da}\right)^{1/2} \tag{5}$$

was computed taking several points on D vs. a and $\dot{a}$ vs. a curves, and $\dot{a}$ values are plotted against K_d, as shown in Figure 6, compared with the data obtained by Green et al [9]. With the limited experimental results it is hard to determine whether the material property associated with dynamic crack propagation is completely defined by an $\dot{a}$ vs. K_d relation. The increasing trend in toughness with crack speed seems to support the experimental observation that the roughness of fractured surfaces increases with crack speed.

Experiment on ship steel

A similar analysis was made on the previously mentioned experimental result using a 2,400 mm wide steel plate specimen. An example of the energies calculated using experimental crack speed data is shown in Figure 7. This was a temperature gradient type ESSO crack arrest test with an arrested crack about 1,450 mm. The experimental observations showed that the fixed grip condition was approximately realized in this case. Figure 8 shows the variation of the K_d value calculated from the D curve with crack length compared with the static stress intensity factor K. The increasing trend in K_d with crack length is probably due to the same factors. It is to be noted that the K_d value is smaller than the nominal K value at the arrest point. The K_d value at the arrest point is nearly equal to the K_c value obtained from the small size arrest test (DG) on a static basis (Figure 1). However, the numerical evaluation of dD/da may involve considerable error, and further accumulation of data is needed in order to determine whether the present analysis is full enough to expalin such a discrepancy as is shown in Figure 4.

CONCLUSIONS

It is shown that dynamic analysis, using the finite-difference method, applied to the equation of motion in an elastic medium will provide a useful tool for investigation of the dynamic aspects of fast fracture and crack arrest. The preliminary study reported in this paper has shown that considerable refinement is needed in current theories of crack propagation and arrest. One of the main concerns is whether or not crack arrest is defined only by an energy condition.

ACKNOWLEDGEMENTS

The authors gratefully acknowledge partial support of this research by the Ministry of Education of the Japanese Government, and the assistance in experiments and analysis given by Messrs. T. Teramoto and Y. Niimura of the University of Tokyo.

REFERENCES

1. YOSHIKI, M., KANAZAWA, T. and MACHIDA, S., Selected Papers from J. Soc. Nav. Archit., Japan 2, 1969, 63.
2. KIHARA, H., KANAZAWA, T., IKEDA, K., OKABE, T. and YAJIMA, H., J. Soc. Nav. Archit. Japan, 122, 1967, 191; J. Soc. Nav. Archit. Japan 124, 1968, 331.
3. Technical Report of Japan Ship Research Association, Investigation on Brittle Fracture Control in Large Ships, Rep. No. 108, 1970.
4. MACHIDA, S. and AOKI, M., J. Soc. Nav. Archit. Japan, 131, 1972, 367.
5. AKITA, Y. and IKEDA, K., J. Soc. Nav. Archit. Japan, 112, 1963, 153.

6. KANAZAWA, T. MACHIDA, S., YAJIMA, H. and AOKI, M., J. Soc. Nav. Archit. Japan, 130, 1971, 343.
7. SHMUELY, M. and PERETZ, D., Int. J. Solids and Structure, 12, 1976, 67.
8. BROBERG, K. B., Ark. Fys., 35-38, 1960.
9. GREEN, A. K. and PRATT, P. L., Engr. Fract. Mech., 6, 1974, 71.

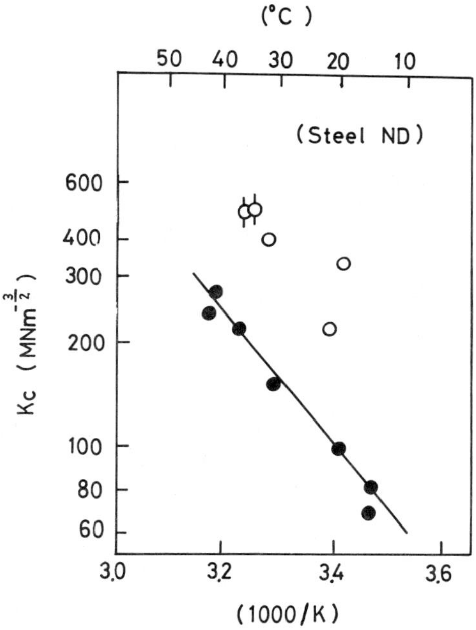

Figure 1 Relation between K_c value and temperature at crack arrest points (normalized D-class ship steel).

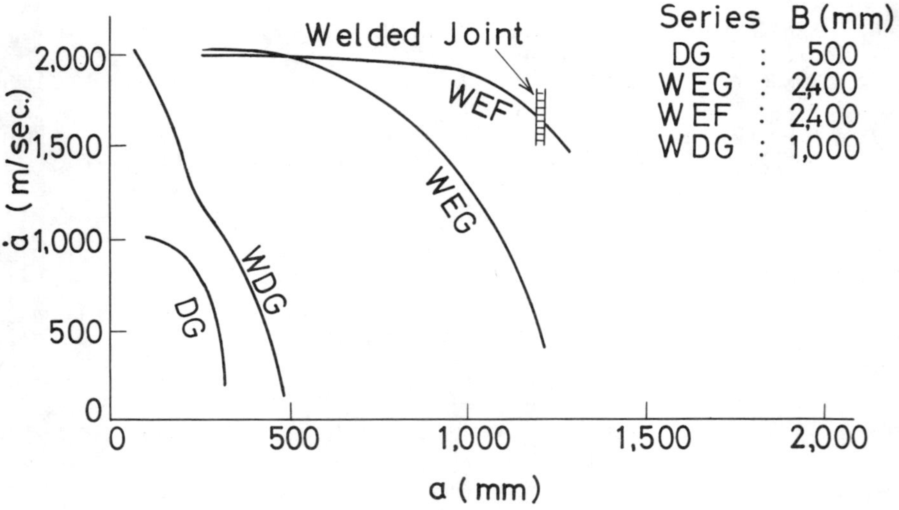

Figure 2 Comparison of crack speeds

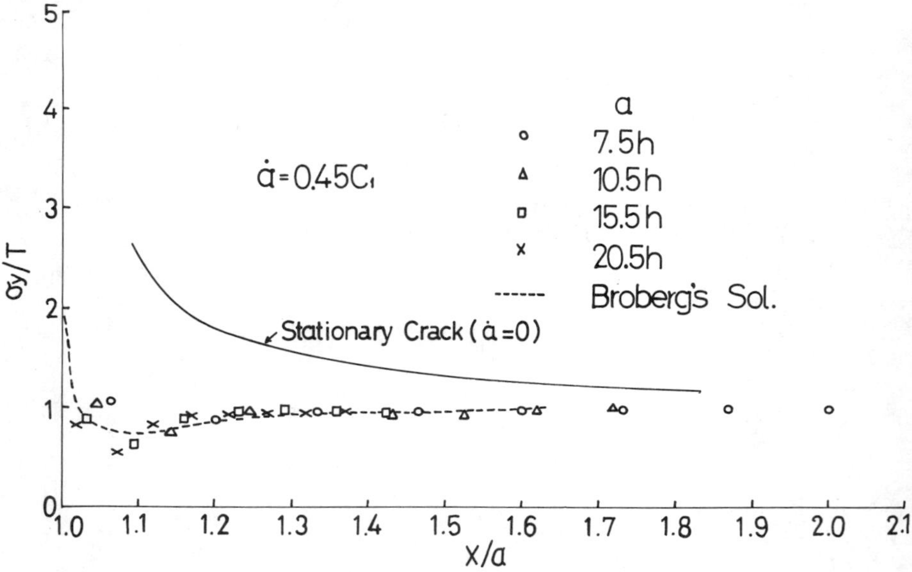

Figure 3 Stress distribution along the crack line of a crack extending at constant speed ($\dot{a}$ = 0 .45C_1)

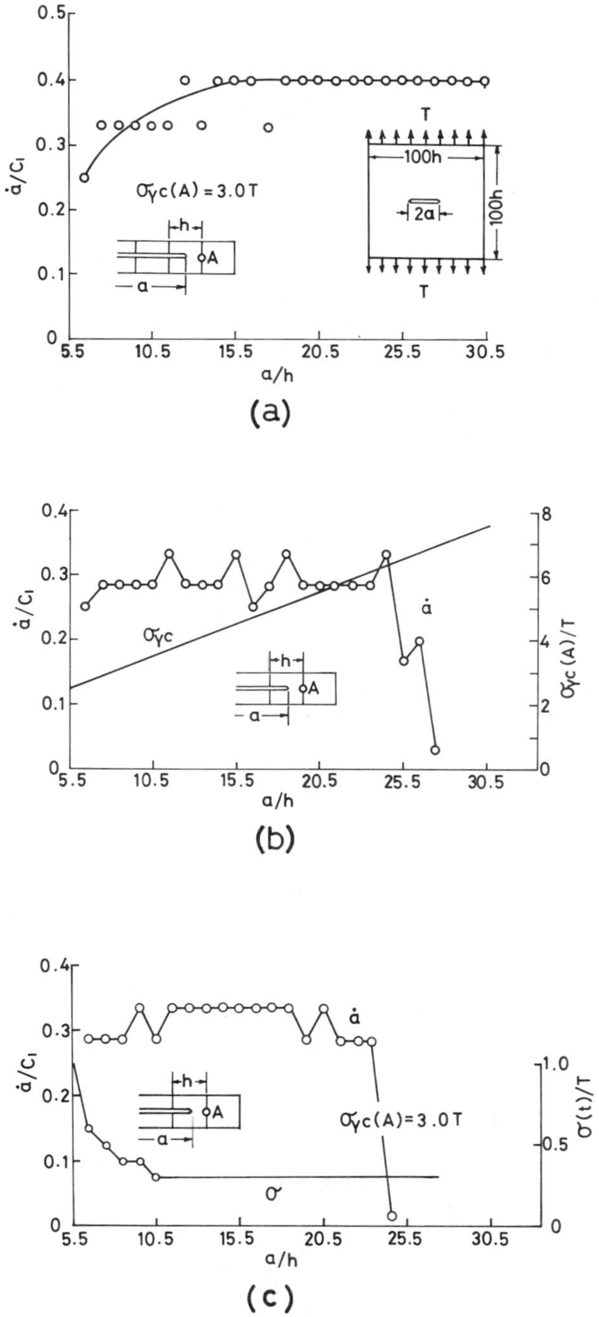

Figure 4 (a) Crack length variation of crack speed for a crack traversing a uniform stress and uniform material resistance field.
(b) Crack length variation of crack speed for a crack running into a linearly increasing material resistance field.
(c) Crack length variation of crack speed subjected to a load drop effect.

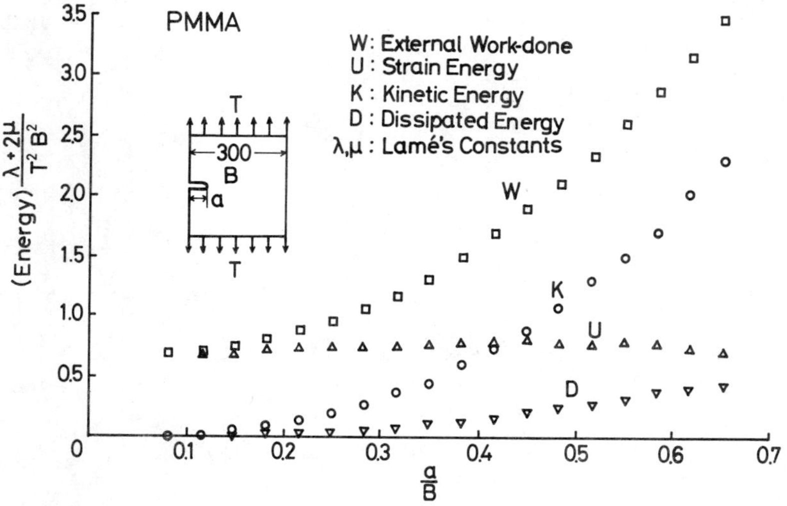

Figure 5 Energy changes with extension of a crack (PMMA)

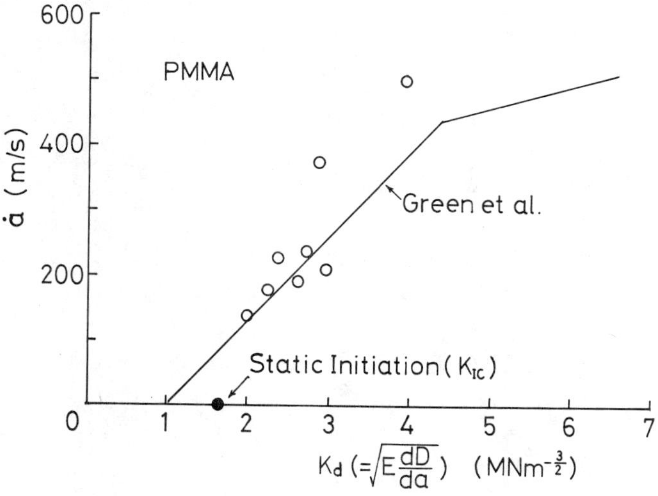

Figure 6 ȧ vs. K_d relation for PMMA

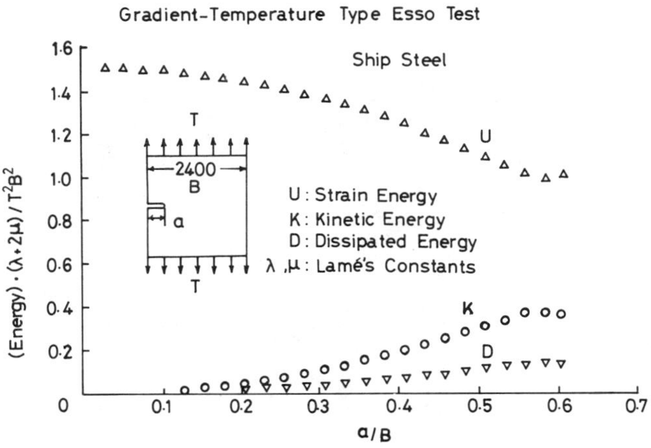

Figure 7 Energy changes with extension of a crack (ship steel)

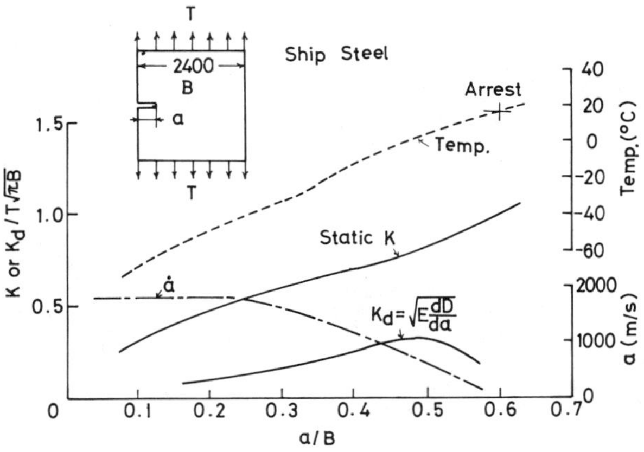

Figure 8 Arrest toughness compared with static K value

Fracture 1977, Volume 3, ICF4, Waterloo, Canada, June 19 - 24, 1977

THE FRACTURE MECHANISM OF MATERIALS HAVING A HETEROGENEOUS STRUCTURE

V. P. Tamuzh, P. V. Tikhomirov and S. P. Yushanov*

INTRODUCTION

Mathematical modelling of the behaviour of a material subjected to loading is commonly based on two main assumptions - on homogeneity of the material on the one hand and on continuity of the medium on the other.

Basically there are hardly any objections to the bringing of these hypotheses into the study of deformation characteristics of a material, although we find a number of articles [1, 2] operating with non-homogeneity of the material in modelling these characteristics. However, after a careful examination of processes of scattered fracture, that is the fracture processes developing more or less uniformly all over the bulk of the material it is believed that the hypothesis of homogeneity is hardly admissible in this case. There is no better proof for such a statement than the fact that almost all processes associated with the accumulation of damage are qualitatively like those expressed by creep curves (with the upward bend of the curve close to fracture not necessarily taking place). To illustrate this one might point out the change of the Young's modulus in cyclic loading [3], the accumulation of submicrocracks and free radicals in polymer materials [4], and the like. An explanation of such processes is easy to find in the simple fact that at the start of testing there is rather rapid fracture of weak bonds followed by slowing down of the process until a more or less constant rate of accumulation of damage is reached.

Concerning the other assumption, in our subsequent discussion we speak of the structure as consisting of elements, i.e., of polycrystalline grains, and by elementary fracture process we mean the fracture of the grain face.

The introduction of inhomogeneity of the material in the mathematical model directly involves a statistical approach to the process of fracture. But since the classical statistical theories of brittle fracture include no time variable in the fracture process they also exclude the description of the long-term and fatigue strength of the material. This is the reason we chose a statistical kinetic model for study of the fracture mechanism. An analogous approach is used in references [5 - 9] which examine the fracture of a homogeneous material consisting of discrete elements, and also in references [10, 11] which discuss the fracture of a homogeneous medium.

* Institute of Polymer Mechanics, Latvian SSR Academy of Sciences, 23 Aizkraukles St., 226006 Riga 6, U. S. S. R.

DISCUSSION OF MODEL

As stated above, the material is composed of grains, and fracture occurs only between their faces. The stresses are calculated as for isotropic elastic bodies and are averaged at the grain faces. It is assumed that the fracture of the grain face is a random process which is defined by the mean time of expectation τ

$$\tau = \tau_o \exp\left[\frac{u_o - \gamma^{\sigma}}{KT}\right] \tag{1}$$

where

τ_o = constant that equals to 10^{-13} - 10^{-12} sec,
u_o = activation energy for the fracture process,
γ = overstress factor,
σ = mean normal stress acting on the grain face.

Inhomogeneity of the structure is represented by the distribution function ψ of the factor γ [12].

For simplicity of calculation we further deal only with the fracture of oriented structures, the external loading force σ_o being applied parallel to the axis of orientation.

We introduce a definition [5]. If the fracture affects a number of adjacent faces, j, the defect obtained we call a j-defect and its area is equal to jF, where F stands for the average area of the grain face. By analogy, the non-fractured grain face assumes the name of 0-defect. In order to calculate the stress concentration round the defect we regard its form as a spheroid with radius

$$R = \sqrt{\frac{jF}{\pi}}$$

and with height H, see Figure 1. It should be noted that we neglect interactions between defects and their coalescence. Now, having assumed that fracture is caused by the mean stress acting on the grain face we have to find the mean stress acting within a ring, the width of which is the average diameter of the grain face, i.e.,

$$2\sqrt{\frac{F}{\pi}} .$$

On the basis of precise solutions from the theory of elasticity dealing with cavities of radial spheroid form [13], we infer the mean stress values expressed as elementary functions. Table I lists the values of $\phi_j = \sigma : \sigma_o$ for two values of r. It should be observed that for 1-defects with r = 0.0012 R we obtain H = 0.05 R, and r = 0.86 R corresponds to H = 0.99 R. Column I gives the values of ϕ_j for the faces directly surrounding the defect and column II shows the values for the faces of the next row. Poisson's ratio is assumed to be 0.3. From Table I we may conclude that:

a) the value of ϕ_j is scarcely dependent on the value of r, and as we are not quite certain about the nature of the latter it is assumed in further discussion that r equals 0, that is, we regard defects as penny-shaped cracks;

b) the mean stresses acting upon the grain faces of the second row are considerably smaller than those acting upon the grain faces of the first row, and consequently, it is reasonable that the stress concentration upon the grains of the second row be neglected.

Now let us consider the kinetics of the accumulation of defects. Through a linear transformation of the random variable γ from equation (1) and the distribution function ψ of the factor γ we come to the distribution of $\lg_e\tau$ for a definite σ :

$$f(\ell n\tau) = \frac{KT}{\sigma}\,\psi\left[\frac{(\ell n\tau_o - \ell n\tau + \frac{u_o}{KT})KT}{\sigma}\right]. \tag{2}$$

The fracture probability of the grain face with definite τ at constant σ is, by analogy with radioactive decay, as follows

$$W(t) = 1 - \exp\left(-\frac{t}{\tau}\right). \tag{3}$$

However, in the case where the value of $f(\lg\tau)$ has a distribution, by generalizing equation (3) we have,

$$W(t) = \int_{-\infty}^{\infty}\left\{1-\exp-\left[\frac{t}{\exp(\ell n\tau)}\right]f(\ell n\tau)d\ell n\tau\right. . \tag{4}$$

Further we determine the probability density of the transition of a j-defect into a (j+1) - defect. If j-defects are surrounded with n 0-defects, the probability of transition from a j-defect into a (j+1) - defect is $[1-W(t)]^n$, and correspondingly the probability of transition from a j-defect into a (j+1) - defect is equal to

$$1 - [1-W(t)]^n . \tag{5}$$

The probability density of a transition from a j-defect into a (j+1) - defect may be derived from equations (4) and (5),

$$P_j^{j+1}(t) = \frac{d\{1-[1-W(t)]^n\}}{dt} = n[1-W(t)]^{n-1}\int_{-\infty}^{\infty}\exp[-x-t\exp(-x)]f(x)dx . \tag{6}$$

We introduce a definition - the value of $W_j(t)$ is the probability that an 0 - defect nucleates a defect having size $\geq$ j. Then the value of $W_1(t)$ is computed from equation (4) whereas the value of $f(x)$ is calculated for $\sigma = \sigma_o$. The expression W_j ($j \geq 2$) may be obtained from equation

$$W_j(t) = \int_o^t W_{j-1}(x)P_{j-1}^{j}(t-x)dx ,$$

where the value of $f(x)$ of equation (6) is calculated for $\sigma = \sigma_o\phi_{j-1}$.

The probability of emergence of at least one defect having size $\geq$ j in the specimen with N-faces may be determined from the equation analogous to equation (5) $W_j^n = 1-[-1-W_j(t)]^n$. When the number of grain faces (N) is considerable then

$$W_j^n = 1 - \exp(-W_j N) \; . \tag{7}$$

RESULTS OF CALCULATION AND DISCUSSION

As reported in [14] highly-oriented capron is characterized by $u_o = 26.7.10^{-20}$J, $\gamma = 14.8.10^{-20}$mm^3. In calculations T is assumed to be 293K. We chose the Weibull distribution for overstresses $\psi(\gamma)$

$$\psi(\gamma) = \frac{h}{s}\left(\frac{\gamma-\mu}{s}\right)^{\eta-1} \exp\left[-\left(\frac{\gamma-\mu}{s}\right)^{\eta}\right] \qquad \gamma \geq \mu$$

$$\psi(\gamma) = 0 \qquad \gamma < \mu$$

with the following parameters $\mu = 6.5.10^{-20}$, $s = 10^{-29}$, $\eta = 0.07$. See Figure 2.

Results of calculations are given in Figures 3 - 6. Figure 3 shows the probability of the transition of a (j-1) - defect into a j - defect. It is apparent that with growth of j the curve tends to the right and that considerable growth of j results in an almost immediate enlargement of the defect. This means that the limit curve, as shown in Figure 4, shows the probability that the grain face will nucleate a defect which causes the ultimate fracture of the specimen. We define this probability as W_f, and similarly W_f^N stands for the probability of emergence of a critical defect in a specimen having N-faces.

The mean size of the structural element of oriented capron is 10 - 25 mm [15]. By assuming the grain face to be 20 mm, a specimen with a volume of 10^3mm^3 will possess $1.25.10^{17}$ faces exposed perpendicular to the applied force. The fracture probability of the whole specimen is assumed to be $W_f^N = 0.5$. Then from equation (7) we determine the probability that any grain face will nucleate a critical defect, i.e., $\lg W_f = -17.3$. Figure 3 allows us to derive the logarithm of the fracture time as given in Figure 6. The value of U_o resulting from Figure 6, $U_o = 25.5.10^{-20}$ J, closely approximates to that obtained experimentally, $U_o = 26.7.10^{-20}$ J. As noted in [16] the value of U_o agrees well with the value of activation energy for the thermodestructive process and might be easily determined from independent physical tests. This means that only constants of distribution, $\psi(\gamma)$, remain undefined. Figure 5 depicts the curves that show accumulation of defects at different stresses. As it stands, the concentration of 1 - defects is approximately 10^6 higher than the concentration of 4 - defects and hence the difficulty arises of revealing large defects.

As seen from graphs of Figure 5 the probability of emergence of critical defects in the wide region of probability might be well characterized by a straight line $\ell n\, W_f = m\ell n t + B$ that corresponds to $W_f = \exp(b)t^m$. By replacing W_f in equation (7) we come to the Weibull distribution

$$W_f^N = 1 - \exp\ [-\exp(b)t^m N]$$

for the fracture time. The above formula and Figure 4 give the size effect.

In conclusion it should be mentioned that the model discussed embraces the whole fracture process and presents a natural transition from the process of scattered fracture to the process of propagation of the macro-crack.

REFERENCES

1. LOMAKIN, V. A., "The Statistical Problems in Mechanics of Deformable Solids", (in Russian), Nauka, Moscow, 1970, 139.
2. BOGACHEV, I. N., VAINSHTEIN, A. A. and VOLKOV, S. D., "Introduction into Statistical Science of Metals", (in Russian), Metallurgica, Moscow, 1972, 216.
3. OLDIREV, P. P. and TAMUZH, V. P., Mekhanika Polimerov, 1967, No. 5, 864.
4. TOMASHEVSKY, E. E., et al, Int. J. Fract., 11, 1975, 803.
5. GOTLIB, Y. J., et al, Physikha Tvjordogo Tela, 15, 1973, 801.
6. GOTLIB, Y. J. and SVETLOV, Y. E., Physikha Tvjordogo Tela, 15, 1973, 2732.
7. PETROV, V. A. and ORLOV, A. N., Int. J. Fract., 11, 1975, 881.
8. PETROV, V. A. and ORLOV, A. N., Int. J. Fract., 12, 1976, 231.
9. PETROV, V. A., Physikha Tvjordogo Tela, 18, 1976, 1290.
10. WLADIMIROV, W. I., PETROV, V. A. and ORLOV, A. N., Phys. Stat. Sol., 42, 1970, 197.
11. WLADIMIROV, W. I., PETROV, V. A. and ORLOV, A. N., Phys. Stat. Sol., (B), 47, 1971, 293.
12. TAMUZH, V. P. and TIKHOMIROV, P. V., Mekhanika Poliverov, 1973, No. 2, 227.
13. LURIE, A. I., "Theory of Elasticity", (in Russian), Nauka, Moscow, 1970, 940.
14. ZHURKOV, S. N. and ABASOV, S. A., Vysokomolek. soedin., 3, 1961, 450.
15. GEZALOV, M. A., KUKSENKO, V. S. and SLUCKER, A. I., Physikha Tvjordogo Tela, 12, 1970, 100.
16. REGAL, V. R., SLUCKER, A. I. and TOMASHEVSKY, E. E., "Kinetic Nature of the Strength of Solids", (in Russian), Nauka, Moscow, 1974, 560.

Table I

j	r = $0.0012\sqrt{F/\pi}$		r = $0.86\sqrt{F/\pi}$	
	I	II	I	II
1	1.21	1.01	1.21	1.01
4	1.51	1.03	1.48	1.04
9	1.80	1.07	1.74	1.09
16	2.07	1.12	1.98	1.15
25	2.31	1.17	2.20	1.21
36	2.55	1.22	2.41	1.28
49	2.77	1.27	2.60	1.34
64	2.97	1.32	2.78	1.41
81	3.17	1.38	2.95	1.47
100	3.36	1.43	3.11	1.53

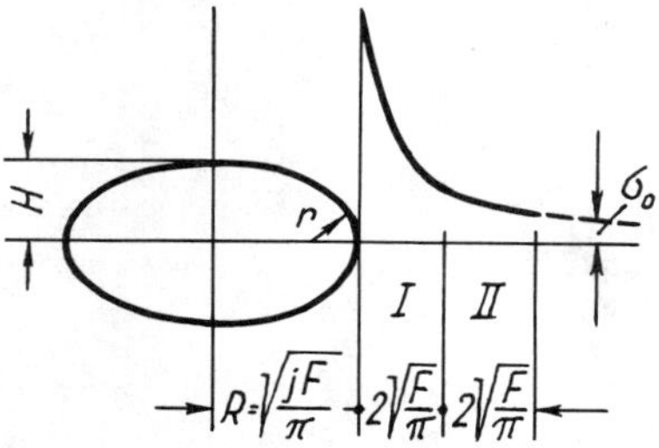

Figure 1 Stress Distribution at the Model Defect

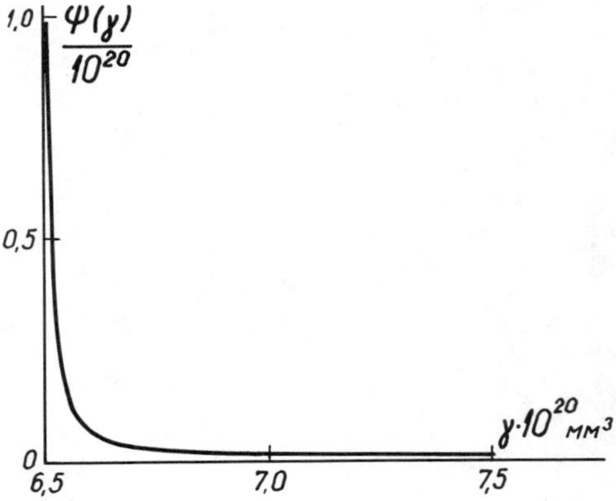

Figure 2 Distribution of Factor γ as Assumed in Calculation

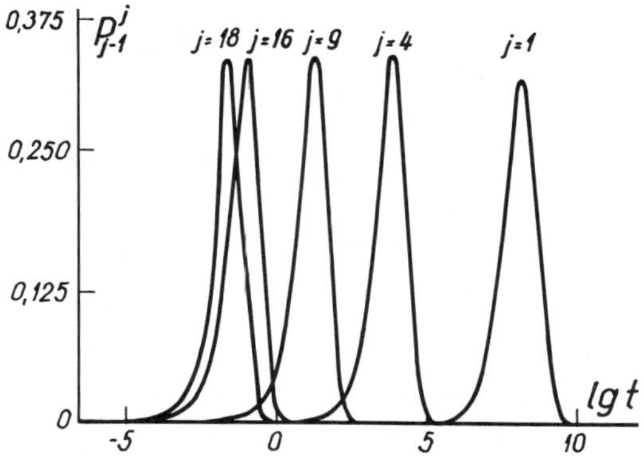

Figure 3 Probability Density of the Transition of a (j-1) - Defect into a j-Defect $\sigma_0 = 9.8 \cdot 10^8$ Pa

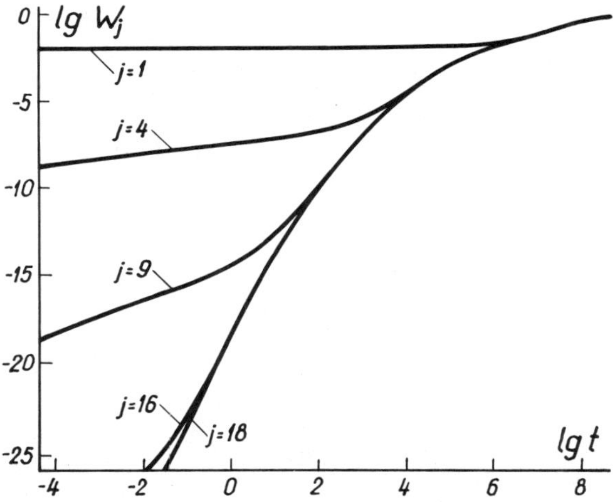

Figure 4 Probability that the Grain Face will Nucleate a Defect of Size of j $\sigma_0 = 9.8 \cdot 10^8$ Pa

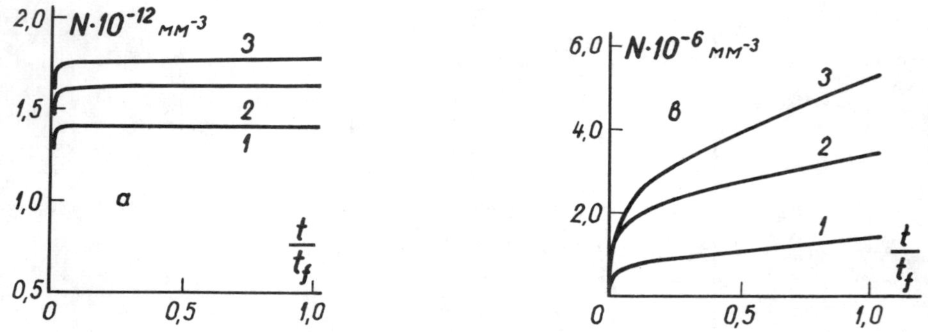

(a) (b)

Figure 5 (a) Accumulation of 1-Defects
(b) Accumulation of 4-Defects

t_f - time prior to fracture; 1 - σ_o = 4.9 . 10^8 Pa
2 - σ_o = 6.4 . 10^8 Pa; 3 - σ_o = 9.8 . 10^8 Pa

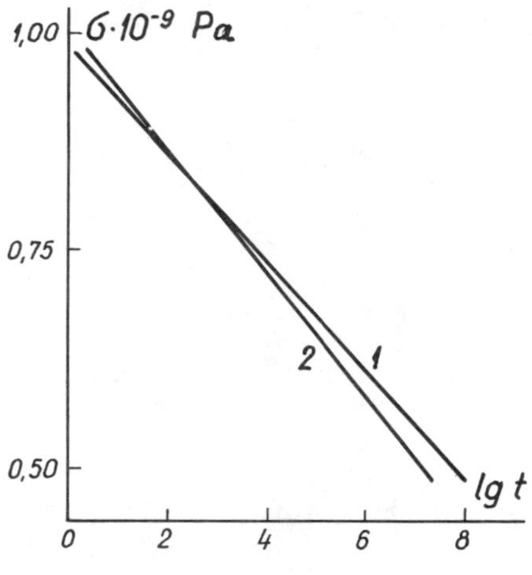

Figure 6 Curves of the Long-Term Strength of Oriented Capron
1 - Experimental Data [14], 2 - Calculation Data

Fracture 1977, Volume 3, ICF4, Waterloo, Canada, June 19 - 24, 1977

THE EFFECT OF LOAD BIAXIALITY ON THE FRACTURE TOUGHNESS PARAMETERS J AND G^{Δ}

A. P. Kfouri and K. J. Miller*

INTRODUCTION

It is well known that biaxiality of loading has no effect on the resistance to fracture of an ideal linear elastic material containing a flat sharp crack. Thus, in a centre-cracked plate of linear elastic material loaded in plane strain as shown in Figure 1, the critical stress intensity factor K_{IC} is independent of σ_Q. According to the Griffith fracture criterion, the energy release rate, G, at fracture is equal to the cohesive strength G_c of the material and

$$G_c = \frac{K_{IC}^2\,(1-\nu^2)}{E} \tag{1}$$

Further, for elastic materials Rice's path independent integral J is identical to G.

In elastic-plastic materials, crack tip plasticity plays an important role in enhancing the fracture toughness, i.e. the applied stress, σ_P required to cause the fracture of a ductile material with a sizeable crack tip plastic zone, is greater than the applied stress required to cause the fracture of a brittle material with similar values of E, ν and G_c, but having such a high yield stress that the crack tip plastic zone size at fracture is minimal. For an ideal, unyielding elastic material, with similar elastic properties, σ_P at fracture is approximately equal to that for the brittle materials.

With elastic-plastic materials, it is not possible to use either J or G as crack propagation or fracture instability parameters [1], although J can act as a characterizing parameter. Thus G, given by the right hand side of equation (1) with K_I replacing K_{IC}, can only pertain to an unyielding elastic material. However, a Griffith type criterion can be used for quasi-brittle materials in terms of the crack separation energy rate G^{Δ} associated with a small *finite* growth step Δa where Δa is assumed to be a characteristic property of the material. This approach is described in references [2, 3] and in a paper presented at this conference [4]. The crack separation energy rate G^{Δ} is defined as $\Delta W/\Delta a$ where ΔW is the work absorbed at constant applied stress during the proportional quasi-static release of the stresses holding the surfaces together at Δa. For small scale yielding, the ratio of Δa over the crack tip plastic zone size is given by

$$r = k(\sigma_y/K_I)^2\ \Delta a \tag{2}$$

*Department of Engineering, University of Cambridge, Cambridge, England.

MS294

where k is a constant which depends on the mode of loading and σ_y is the yield stress in uniaxial tension.

By combining elastic-plastic finite element analyses on quadrant 1 of Figure 1 with a crack tip release technique, it was possible to calculate G^{Δ} for different values of r. Material properties used were E = 207 GN m^{-2} $\nu = 0.3$, σ_y = 310 MN m^{-2} and a linear strain hardening tangent modulus of 4830 MN m^{-2}. The results show that when r is greater than three, G^{Δ} is approximately equal to J (or G for the unyielding elastic material). As r tends to zero at constant applied stress σ_p, G^{Δ} also tends to zero, thus confirming Rice's conclusions [1] already cited. Between these two values of r there is a strong dependence of G^{Δ} on r.

EFFECT OF BIAXIALITY OF LOADING ON G^{Δ}

It has been known for some time that the sizes of the crack tip plastic zones and the resistance to fracture of elastic-plastic materials depend on the biaxiality of the mode of loading [5, 6, 7, 8]. Using the load biaxiality parameter $\lambda = \sigma_Q/\sigma_p$, λ takes the values -1, 0 and 1 for the shear uniaxial and equibiaxial modes, respectively, with σ_p always positive, i.e. in tension

Figure 2 shows plastic zone sizes for different values of λ obtained by finite element analyses at approximately the same values of σ_p. The computer drawn crack profile is that for the shear mode with displacements being exaggerated by a factor of 50. It will be noted that the zone size corresponding to the shear mode is much the largest with the equibiaxial mode being the smallest. The respective ratios of the three zone sizes are approximately 6:1.5:1 corresponding to λ = -1, 0 and 1.

Figure 3 shows the principal stress, σ_1, perpendicular to the crack surface, and ahead of the crack tip, normalized with respect to σ_p for the loads given in Figure 2. The stress patterns differ appreciably with loading mode. The combined effects of hydrostatic tension and the elastic singularity are most pronounced in the equibiaxial loading mode, for which the plastic zone size is smallest, and are almost totally absent in the shear mode corresponding to the largest plastic zone.

Figure 4 summarizes the results of this study. For all loading modes the value of G^{Δ} decreases with increasing crack tip ductility. This effect can be attributed mainly to the increasing plastic work of deformation required to *extend* the plastic zone as the crack advances, with little or no recovery of energy from the wake region, in spite of the unloading which has occurred in the material now in the wake. Hence, at the same value of σ_p, G^{Δ} must be least when λ = -1 and the plastic zone is largest. By the same token G will be greatest when λ = 1 corresponding to the smallest plastic zone. The abscissa G/G_o of Figure 4 represents a normalized applied load. Here G_o is the value of G at incipient yielding under uniaxial loading of the plate of Figure 1. The value of G_o therefore depends on $\sigma_y^2\ \Delta a$ since Δa is the length of the side of the leading element at the crack tip. Hence G/G_o is also a measure of the plastic zone size at the crack tip, i.e. the ductility of the material. The ordinates give, in the case of the top three curves, the normalized values J/G for the three modes and in the case of the lower three curves, G^{Δ}/G. When G/G_o = 1, i.e. the truly brittle state, G, J and G^{Δ} are all equal since they correspond to an elastic solution. As G/G_o increases and becomes large, G^{Δ}/G tends to zero in all cases, but the paths are different for the three loading modes*.

In view of the increasing use of the path independent contour integral J as a characterizing parameter it is interesting to note in Figure 4 the values G/G_o at which J begins to diverge significantly from G and which indicate states in which small scale yielding assumptions are no longer justified. In the present study J was calculated along a path always running through elastic material.

ACKNOWLEDGEMENTS

The authors gratefully acknowledge the support of the Science Research Council.

REFERENCES

1. RICE, J. R., "An Examination of the Fracture Mechanics Energy Balance from the Point of View of Continuum Mechanics", Proc. First Int. Conf. on Fracture, Sendai, Japan, 1, (ed. T. Yokobori et al). The Japanese Soc. for Strength and Fracture of Materials, 1966.
2. KFOURI, A. P. and MILLER, K. J., "Stress Displacement, Line Integral and Closure Energy Determinations of Crack Tip Stress Intensity Factors", Int. J. Press. Vessels and Piping, 2 (3) 1974, 179-191.
3. KFOURI, A. P. and MILLER, K. J., "Crack Separation Energy Rates in Elastic-Plastic Fracture Mechanics", to be published in Proc. Inst. Mech. Eng. (London).
4. KFOURI, A. P. and RICE, J. R., "Elastic/Plastic Separation Energy Rate for Crack Advance in Finite Growth Steps", Fracture 1977, 1, (ed. D. M. R. Taplin), University of Waterloo Press, Canada.
5. HILTON, P. D., "Plastic Intensity Factors for Cracked Plates Subjected to Biaxial Loading", Int. J. Fracture, 9, No. 2, 1973, 143-156.
6. MILLER, K. J. and KFOURI, A. P., "An Elastic-Plastic Finite Element Analysis of Crack Tip Fields under Biaxial Loading Conditions", Int. J. of Fracture, 10, No. 3, 1974, 393-404.
7. LARSSON, S. G. and CARLSSON, A. J., "Influence of Non-Singular Stress Terms and Specimen Geometry on Small Scale Yielding at Crack Tips in Elastic-Plastic Materials", J. Mech. Phys. Solids, 21, 1973, 263-277.
8. RICE, J. R., "Limitations to the Small Scale Yielding Approximation for Crack Tip Plasticity", J. Mech. Phys. Solids, 22, 1974, 17-26.

---

*G/G_o and G^{Δ}/G_o are related to the ϕ, ψ parameters of reference [3] by equations

$$G/G_o = \psi$$

$$G^{\Delta}/G_o = \phi$$

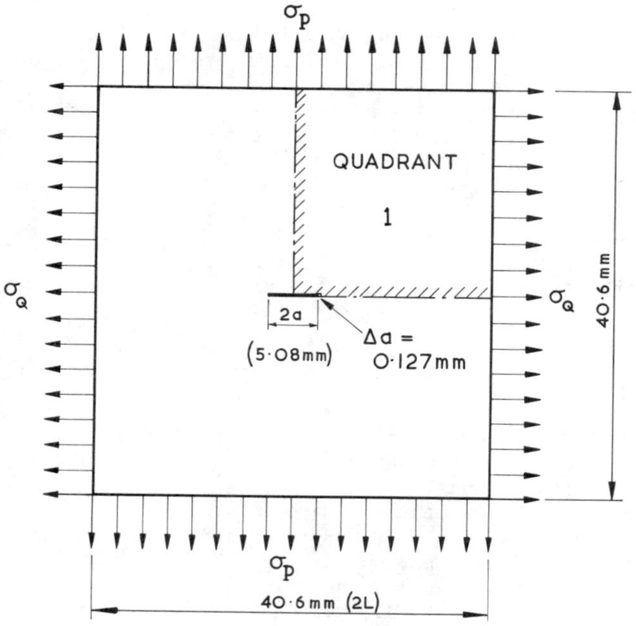

Figure 1 Centre-Cracked Plate in Plane Strain

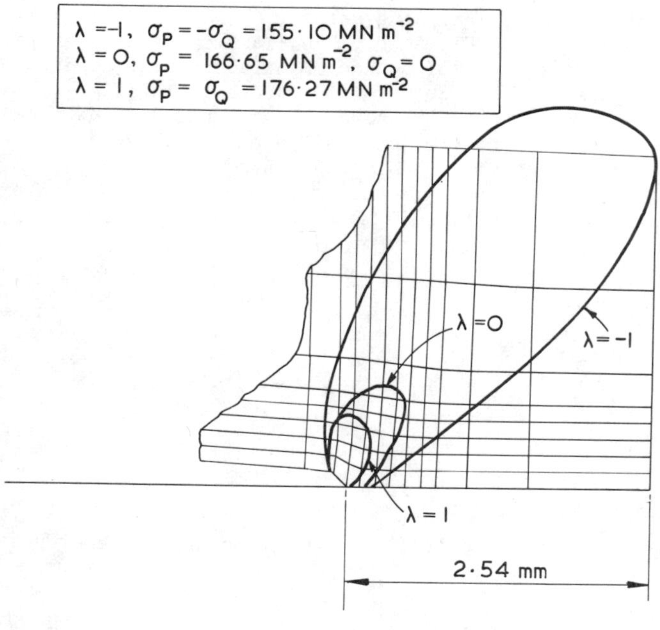

Figure 2 Crack Tip Plastic Zone Sizes for Different Biaxial Modes of Loading

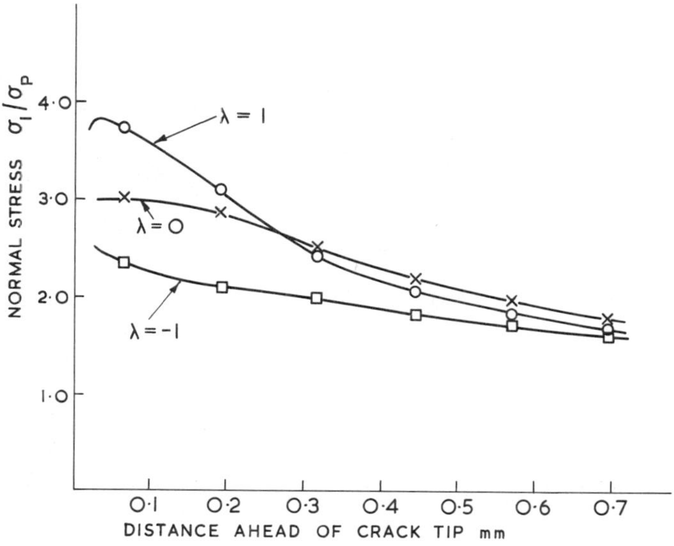

Figure 3 Normal Principal Stresses Ahead of the Crack Tip for Different Biaxial Modes of Loading

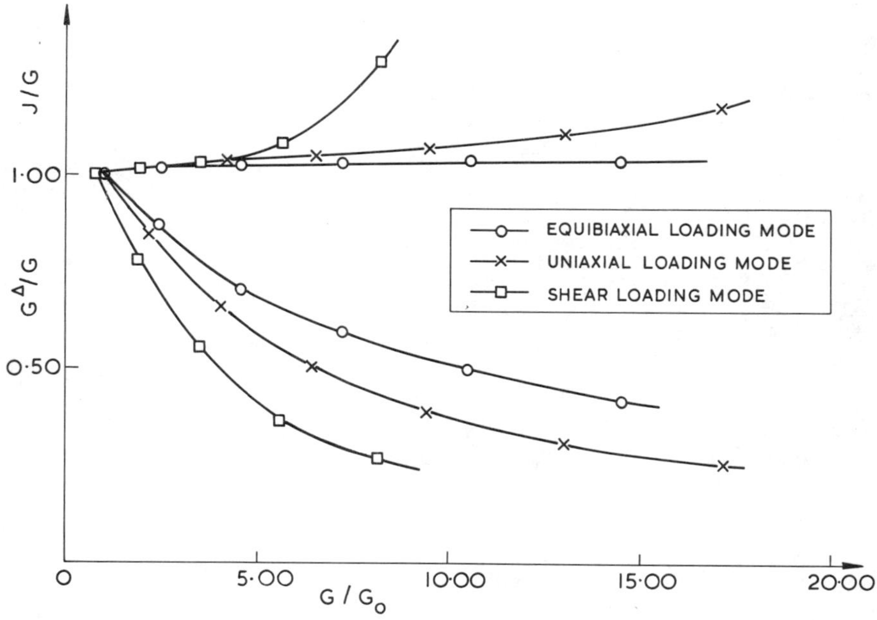

Figure 4 The Effect of Various Biaxial Modes of Loading on the Fracture Toughness Parameters J and G^{Δ}

Fracture 1977, Volume 3, ICF4, Waterloo, Canada, June 19 - 24, 1977

USE OF THE CALCULATION OF INTEGRAL J_1

R. L. Roche*

INTRODUCTION

The method of finite elements applied to the analysis of stresses and strains is of major interest in fracture mechanics. In particular, calculation of the integral J_1 or similar integrals by these methods leads to useful applications. Two types of application are presented here, one for the fast, relatively low cost determination of K_I in linear elastic fracture mechanics, and the second for the examination of certain conditions of validity of initiation criteria based of J for an elastic plastic material. These applications are illustrated for a number of calculation results obtained by means of the CEASEMT system developed at Saclay [1-2].

DETERMINATION OF K_I IN LEFM

A number of different methods are available to determine K_I by means of a calculation program by finite elements [3-4]. Most of these involve determination of certain values (displacement, stress, etc...) as a function of distance from the crack front (polar radius). Hence these are actually derivation methods. Consequently, they require a fine mesh and are therefore costly. This explains why they are rarely employed for industrial calculations.

Methods possessing an integration character are far more preferable. Furthermore, the method of finite elements gives better results on overall values such as energy, than on the detailed distriubtion of strains or stresses. The integral J_1 hence appears more suitable for the determination of K_I since it is a curvilinear integral resulting from an integration in the entire plane region enclosed by the integration path.

Calculations performed with the CEASEMT system showed that this method is, technically, fairly accurate and inexpensive. The PASTEL module enables calculation of J_I simultaneously on several contours.

A simple example is provided by the results obtained for a square plate exhibiting a crack whose length 2a is 1/4 of the side, subjected to a uniform tensile stress S on the sides parallel to the crack. The table one gives the reduced value F of $K_I = FS\sqrt{\pi a}$, the deviation in relation to known values [5] and the total cost of the calculation expressed in equivalent seconds of an IBM 360/91 computer for the different meshes employed. For reasons of symmetry, the calculation only covers a quarter of the plate. Some of these meshes are shown in Figure 1.

*DEMT, CEN Saclay, 91190 GIF, France.

It can be seen that satisfactory accuracy (2%) is achieved for low calculation cost. Moreover, the mesh is obtained automatically and requires very little labor.

This method is obviously limited to the area of validity of J_1 (homogeneity of the material, flatness of the crack). However, it is suitable for dealing with certain interesting cases, such as pseudo-cracks without sharp fronts.

As an example, a pseudo-crack with a thickness of 0.1 mm located at the center of a beam bent at three points was subjected to calculations (Figure 2). By definition, the reduced stress intensity factor is equal to:

$$F_J = \frac{1}{S} \sqrt{\frac{EJ}{(1 - \nu^2)\,\pi\, a}} \tag{1}$$

Where S is the reference stress (in this case equal to the maximum bending stress of the uncracked part) and J is calculated on four contours surrounding the entire crack front.

a(mm)	0.2	0.5	1	2
F_J	1.046	1.016	0.998	0.959

The standard deviation on the four contours is always less than 0.010. As for the calculation cost, it is less than 50 equivalent seconds for each case.

This method may be extended to three-dimensional cases, by calculation of the vectorial integral J on different surfaces enclosing the elements of the crack tip. Another development under way at Cadarache deals with thin shells, using both the integral J and the integral L of Knowles and Sternberg [6].

VALIDITY CONDITIONS OF THE CRITERION J_C

The integral J is employed as a crack propagation initiation criterion [7]. Two arguments may be employed to justify this view. The first relates to the energy available during crack propagation. The second is based on the property of independence of the contour, making J a characteristic of the crack front (like K_I in LEMF). For both arguments, it is necessary for J to be path-independent.

It has been shown that J is independent of the contour for non-linear elastic materials. It was suggested to extend this property to plastic materials by using the "strain energy" W, namely, the density of work received [8]. However, the validity of this extension is debatable.

A necessary condition for J to be independent of the contour may be written [9]. A defect vector W_k^* is defined, with a volume density of:

$$W_k^* = W_{,k} - \sigma_{ij}\varepsilon_{ij,\,k} \tag{2}$$

Where:

$$W = \int_0^{ij} \sigma_{ij} d\varepsilon_{ij}$$

($W,_k$ is the derivative with respect to x_k).

The surface and linear densities are similarly defined. This vector is null when the spatial variation in "strain energy" dW is equal to that which would result from the spatial variation in "strain", in other words:

$$dw = \sigma_{ij} d\varepsilon_{ij}$$

Applied to the surface of the body are the surface defect vectors $\bar{W}^{**}$ which, when no loads are applied, are equal to $W\bar{n}$ ($\bar{n}$ normal to the surface). It can easily be shown that the overall defect vectors applied have a null resultant (Figure 3).

It is easy to show that integrals J_1 and L_1 are the resultants of the defect vectors located in the volume V bounded by the integration surface:

$$\vec{J} = \int_V \vec{W}^* dv \tag{3}$$

$$\vec{L} = \int_V (\vec{OM} \wedge \vec{W}^*)\, dv \tag{4}$$

The conditions of independence of J and L are hence reduced to $\bar{W}^*{}_k = 0$.

In the plane case of a plane crack parallel to axis Ox (tip perpendicular to Oxy), the condition for J_1(component of J along x) to be independent of the contour is hence $W^*{}_1 = 0$, or:

$$\frac{\partial W}{\partial x} = \sigma_{ij} \frac{\partial \varepsilon_{ij}}{\partial x} \tag{5}$$

This condition does not require the use of the constitutive equation of the material.

A sufficient but not necessary condition may be suggested for materials with potential mechanical energy W. It is sufficient for W not to depend explicitly on the point in question, but only on the state of strain ε_{ij} (the material must be homogeneous).

This condition may be applied to non-linear elastic materials. It may also be applied to materials exhibiting deformation type plasticity. In effect, if unloading were not to occur, a relationship would exist in finite terms between strains and stresses, introducing a mechanical potential.

Unfortunately, the plastic behavior of materials is rather of the incremental type, and no mechanical energy potential exists. Consequently, it is not certain that J is independent of the contour, and that a criterion based on J is entirely valid [10].

NUMERICAL STUDIES OF THE INDEPENDENCE OF J

Calculation results obtained with the method of finite elements, employing an incremental plasticity model, can serve to evaluate the independence of J. Consequently, in the special cases calculated, it is possible to appreciate the validity of a criterion based on the value of J. Such calculations have been performed [11] in a number of cases. It does not appear that highly significant variations of J with the contour occur. Moreover, it is difficult to establish whether the variations observed are real or due to numerical appearances. However, some authors [12] believe that J is not path-dependent when the path crosses plastic zones.

A number of calculations of this type were performed with the CEASEMT system. The plasticity model employed was that of Von Mises normal flow and law (Prandtl-Reuss equations) [13].

The above plate (2 x 2 mm square with crack 2a = 0.5 mm) was analyzed (plane strain) (E = 206,800 MPa - ν = 0.3 - σ_y = 310 Pa). The load S consisted of a tensile force applied progressively to the sides parallel to the crack. The quarter plate mesh is represented in Figure 4. Figure 5 shows the variation of F_J with S for a material without strain hardening, and a material of which the tangent modulus is 1/10 of the Young's modulus.

For each of the 15 increasing values of stress S, over 20 contours, J was calculated, together with the portion due to elastic energy J_e, that due to plastic energy J_p and that due to the forces on the integration contour J_F. Table 2 gives, as an example, the results obtained for S = 260 MPa. It may be noted that J depends only slightly on the contour (unit N/m) despite a very broad plastic zone.

It should be observed that in all the cases dealt with, the loading was radial, in other words, all the forces applied increased proportionally. This procedure makes it possible to consider the behavior of the test sample like that of a material with deformation type plasticity. It is also interesting to analyse other types of loading and to compare them with the foregoing results. In effect, significant changes in strain may occur at certain points, and possibly local recovery in the elastic strain region. Analyses of this type are under way at Saclay on the plate previously investigated. Instead of increasing the applied stress S uniformly, it is progressively established at a selected value S_o, starting, for example, with the corners and moving towards the center of the side (Figure 6). Initial results obtained appear to differ from those obtained for radial loading up to S_o.

Consequently, calculations of J by the method of finite elements can serve to clarify the independence of the latter for material behavior which is not of deformation type plasticity, but incremental plasticity and visco-plasticity. They also make it possible to evaluate the effect of the loading procedure which, in actual structures, is not always of the radial type, but may be more complex.

REFERENCES

1. HOFFMANN, A., GOLDSTEIN, S. and ROCHE, R., Paper F4/4, SMIRT3, London, 1975.
2. GOLDSTEIN, S., et. al., Nuclear Energy Maturity, Pergamon Press, New York, 1976, 383.
3. CHAN, S. K., TUBA, I. S. and WILSON, W. K., Eng. Fract. Mech., 2, 1970.
4. TRACEY, D. M., Eng. Fract. Mech., 3, 1971, 255.
5. TADA, M., et. al., The Stress Analysis of Cracks Handbook, Dell Res. Corp., 1973.
6. BUDIANSKY, B. and RICE, J. R., J. of Appl. Mech., March 1973, 201.
7. BEGLEY, J. A. and LANDES, J. D., STP 514.
8. RICE, J. R., Fracture (Ed. H. Liebowitz), Academic Press, 1968, 191.
9. ROCHE, R. L., Note CEA-N-1829, CEN SACLAY, 1975.
10. TURNER, C. E. and BURDEKIN, F. M., Atom Ener. Rev., 12, 1974, 439.
11. SUMPTER, J. D. and TURNER, C. E., Second Int. Conf. on Pressure Vessel Tech., San Antonio, 1973.
12. NEALE, B. K., Report RD/B3253-CEGB, 1975.
13. SARTHE, M. H., Note CEA (to be published).

Table I

Mesh	nodes	elements	F	deviation	cost
A	25	32	0.855	- 17%	
B	81	128	0.962	- 8%	53
C	137	232	0.974	- 5%	61
D	289	512	1.009	- 2%	71
E	1089	2048	1.034	+ 0.3%	354

Table II

PATH	J_E	J_P	J_F	J	F_J
E1	26,33	235,57	249,42	511,32	1,479
E2	64,16	254,82	164,28	483,27	1,438
E3	72,76	308,19	108,31	489,27	1,447
E4	79,89	344,34	67,84	492,08	1,451
E5	87,70	300,01	102,69	490,40	1,448
E6	94,06	323,14	74,53	491,74	1,450
E7	100,09	336,24	56,41	492,76	1,452
E8	105,82	341,10	46,45	493,38	1,453
E9	111,39	341,10	41,26	493,77	1,453
E10	140,33	304,81	49,72	494,87	1,455
E11	171,86	271,49	53,25	496,61	1,458
E12	187,65	239,72	70,52	497,90	1,459
E13	203,93	179,33	116,81	500,08	1,463
E14	229,70	121,94	149,89	501,55	1,465
E15	236,29	69,80	196,16	502,26	1,466
E16	233,83	27,37	242,47	502,68	1,466
E17	206,94	0	296,11	503,06	1,467
E18	155,23	0	348,01	503,24	1,467
E19	127,17	0	375,99	503,16	1,467
E20	120,51	0	382,58	503,10	1,467

J unit N/m

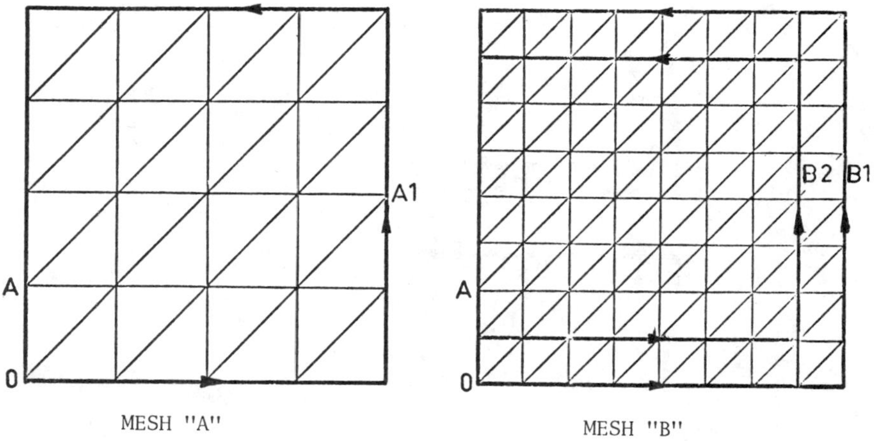

Figure 1 Meshs used in K_I computation

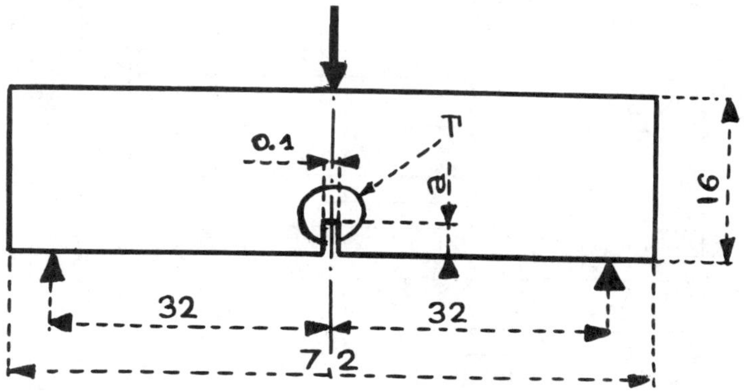

Figure 2 Open crack in a beam

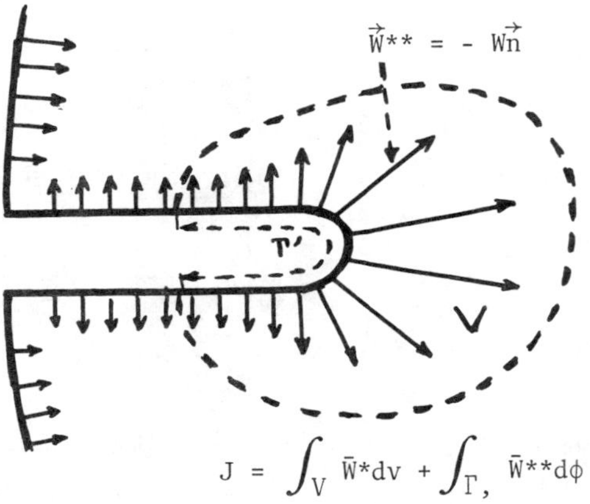

Figure 3 Defect vectors along a crack

Figure 4 Mesh used for plastic computation

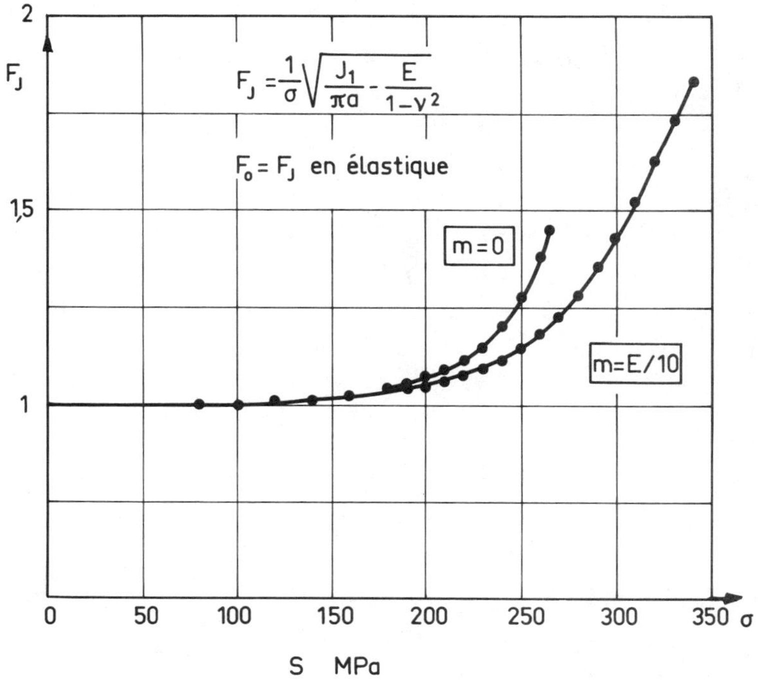

Figure 5 Reduced value of J_1 versus applied load

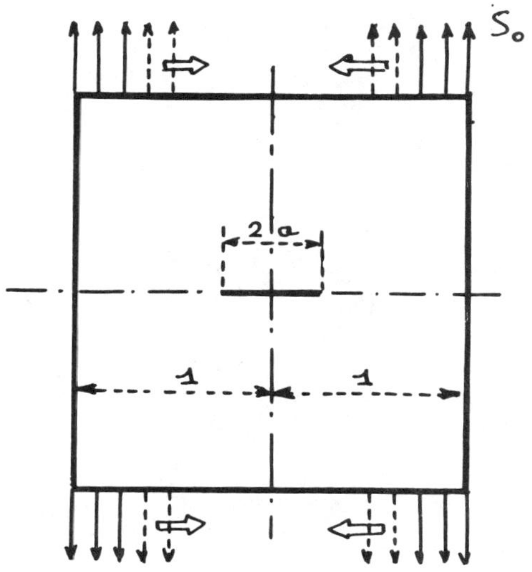

Figure 6 Non proportional loading process

Fracture 1977, Volume 3, ICF4, Waterloo, Canada, June 19 - 24, 1977

A PATH INDEPENDENT INTEGRAL FOR SYMMETRIC STRESS-DIFFUSION FIELDS SURROUNDING LINE CRACKS

E. C. Aifantis* and W. W. Gerberich**

INTRODUCTION

A careful thermomechanical analysis of crack propagation in continuous media led Cherepanov [1] to propose an integral form as a general fracture criterion. For slow crack growth, and in the absence of heat flux and body forces, this integral form is identified as the J-integral. This was independently discovered and popularized by Rice [2] as an outgrowth of the work of Eshelby [3], for calculating the path-independent fracture toughness of cracked metal sheets subjected to an elasto-plastic stress field. Knowles and Sternberg [4] discovered two more path independent integrals by applying Noether's theorem to the theory of linear elastostatics. Aifantis [5] extended the above results to generate conservation laws for linear isotropic stress fields in the presence of body forces derived from harmonic potentials.

The results of the present investigation may be considered as a partial answer to the question: Do path independent integrals exist when, in addition to the stress field, a diffusion field is present?

At first glance, it appears natural to attack the problem by modelling microscopically the change of the energetics of the solid-matrix, due to the motion of the diffusing species. But the uncertainty involved in specifying the details of the elastic interaction energy, suggests a continuum mechanics treatment. Thus, diffusion effects are taken into account by postulating the existence of an internal diffusion force which is properly introduced into the equation of motion to describe the exchange of momentum between the solute and solvent atoms. A steady state diffusion is considered and a simplified model for the diffusion force is adopted. The kinematics of the diffusing species are restricted by the principles of mass and momentum balance and their mechanical response is modelled with a constitutive law [6] for an elastic fluid. The mechanical response of the solid is determined within the theory of linear isotropic elastostatics.

We consider symmetric configurations for both the stress and solute density fields surrounding the line crack. The conservation law that we discover is independent of symmetry considerations but the path-independence is particularly sensitive to symmetry arguments.

Thus, for the cases under consideration, we derive a path-independent integral which includes terms due to diffusion. If we assume that the diffusion effects are negligible at infinity, then the first component of this integral, evaluated at infinity, is reduced to the familiar J-integral [2].

---

* Assistant Professor of Theoretical and Applied Mechanics, University of Illinois at Urbana-Champaign, Illinois, U. S. A.

** Professor of Materials Science, University of Minnesota, Minneapolis, Minnesota, U. S. A.

DIFFUSION MODELLING

We imagine a central crack in an infinite linear elastic isotropic medium, subjected to a symmetric tensile stress field at infinity, as shown in Figure 1. The crack tip serves as a source of solute atoms (e.g., dissociated hydrogen ions) which diffuse in the elastic medium symmetrically with respect to y-axis. The cloud of diffusing species is modelled to behave as a perfect fluid obeying the mechanical principles of mass and momentum balance. These principles, in local form, are expressed by the differential equations [6],

$$\frac{\partial \rho}{\partial t} + \mathrm{div}(\rho \underset{\sim}{v}) = 0 \tag{1}$$

and

$$\mathrm{div}\underset{\sim}{S} + \rho\underset{\sim}{\psi} = \dot{v} \tag{2}$$

respectively.

In the field equation (1) and (2), ρ, $\underset{\sim}{v}$, $\dot{\underset{\sim}{v}}$ and $\underset{\sim}{S}$ stand for the density, the velocity, the acceleration and the stress tensor of the gas and $\underset{\sim}{\psi}$ represents the diffusive force vector.

The following simple constitutive model is adopted for the stress tensor $\underset{\sim}{T}$ and diffusion force $\underset{\sim}{\psi}$:

$$\underset{\sim}{S} = -A\rho\underset{\sim}{1} \quad ; \quad \underset{\sim}{\psi} = B\underset{\sim}{v} \tag{3}$$

where A and B are constants, and $\underset{\sim}{1}$ represents the unit tensor. The above model is a special case of a general constitutive structure proposed in [6].

Next, we insert (3) into (2) and neglect the acceleration $\dot{\underset{\sim}{v}}$ in order to conform with classical diffusion theories [8]. The result is

$$\rho\underset{\sim}{v} = -\frac{A}{B}\nabla\rho \; . \tag{4}$$

Introducing (4) into (1) we obtain

$$\frac{\partial \rho}{\partial t} = \frac{A}{B}\nabla^2\rho \; , \tag{5}$$

which is the classical diffusion equation, if A/B is identified with the diffusivity D. We are interested in steady-state situations and in these cases (5) is reduced to

$$\nabla^2\rho = 0 \; . \tag{6}$$

STATIC EQUILIBRIUM OF THE SOLID

The infinite medium is modelled to behave as a linear elastic isotropic solid, obeying the familiar constitutive law

$$\underset{\sim}{T} = \lambda \, \mathrm{tr}\underset{\sim}{\varepsilon}\underset{\sim}{1} + 2\mu\underset{\sim}{\varepsilon} \tag{7}$$

where $\underset{\sim}{T}$ and $\underset{\sim}{\varepsilon}$ are the stress and strain tensors of the solid correspondingly; λ and μ are the Lamé constants; and $\mathrm{tr}\underset{\sim}{\varepsilon}$ represents the trace of $\underset{\sim}{\varepsilon}$. The strain tensor $\underset{\sim}{\varepsilon}$ is defined in the usual way by

$$\underset{\sim}{\varepsilon} = \frac{1}{2}\left[\nabla\underset{\sim}{u} + (\nabla\underset{\sim}{u})^T\right] \tag{8}$$

where $\underset{\sim}{u}$ is the displacement vector of the solid, the symbol "T" denotes transposition and ∇ is the gradient operator either for vector or scalar fields. The strain energy, W, of the solid is also given by the familiar relationship

$$W = \frac{1}{2}\,\mathrm{tr}(\underset{\sim}{T}\underset{\sim}{\varepsilon}) \; . \tag{9}$$

The solid is considered to be in static equilibrium, in the presence of the diffusive force field, $-\rho\underset{\sim}{\psi}$, which acts as a body force. Thus the equilibrium equations, in vector form, are given by

$$\mathrm{div}\,\underset{\sim}{T} - \rho\underset{\sim}{\psi} = 0 \; . \tag{10}$$

Using $(3)_2$ and (4) we can write (10) in the form

$$\mathrm{div}\,\underset{\sim}{T} + A\nabla\rho = 0 \; . \tag{11}$$

From now on we will consider the constant A to be given.

Substitution of (7) into (11) yields

$$\mathrm{div}\underset{\sim}{\varepsilon} = -\frac{A}{2\mu}\nabla\rho - \frac{\lambda}{2\mu}\nabla\,\mathrm{tr}\underset{\sim}{\varepsilon} \; . \tag{12}$$

Operating with the divergence in (12) and using a direct consequence of compatibility [8], we obtain

$$(\lambda + 2\mu)\nabla^2\mathrm{tr}\underset{\sim}{\varepsilon} = -A\nabla^2\rho \; . \tag{13}$$

We are interested in steady-state diffusion processes. Then, in light of (6), equation (13) results to

$$\nabla^2\mathrm{tr}\underset{\sim}{\varepsilon} = 0 \; . \tag{14}$$

Another relationship that will be useful in the subsequent analysis is the solution of (7) with respect to strain tensor $\underset{\sim}{\varepsilon}$. This solution is well known [9] and may be written as

$$\underset{\sim}{\varepsilon} = \frac{1 + v}{E} \underset{\sim}{T} - \frac{v}{E} \sigma \underset{\sim}{1} \tag{15}$$

where v and E are the Poisson's ratio and the modulus of elasticity and σ is the trace of the stress tensor of the solid, i.e.,

$$\sigma = \text{tr}\, \underset{\sim}{T} \,. \tag{16}$$

A CONSERVATION LAW

In this section, we establish a conservation law holding for any sub-region, free from singularities, of the infinite isotropic linear elastic medium under consideration. Towards this aim we prove the following theorem.

Theorem: If an isotropic linear elastic domain supports a diffusive force field, of the form $\underset{\sim}{f} = A\nabla\rho$, exerted by diffusing species of concentration ρ, then the following conservation law holds:

$$\underset{\sim}{J}^* = \oint_C \left\{ \left[W + \frac{1}{2\mu} \left(\rho^{*2} + 2\tau^*\rho^* + \frac{1}{2v} \tau^{*2} \right) \right] \underset{\sim}{n} + \left[2\rho^*\underset{\sim}{\varepsilon} + \tau^*\nabla\underset{\sim}{u} - \underset{\sim}{u} \otimes \nabla(\rho^* + \tau^*) - (\nabla\underset{\sim}{u})^T \underset{\sim}{T} \right] \underset{\sim}{n} \right\} d\ell = 0 \tag{17}$$

for every surface C that is the boundary of a finite regular closed sub-region of the elastic domain, provided $\underset{\sim}{n}$ is the unit outward normal vector of C. In the case of zero diffusion, i.e., $\nabla\rho = 0$, it may be shown that

$$\underset{\sim}{J}^*_{\nabla\rho=0} = \oint_C \left[W\underset{\sim}{n} - (\nabla\underset{\sim}{u})^T \underset{\sim}{T}\underset{\sim}{n} \right] d\ell = 0 \tag{18}$$

which is the familiar conservation law discussed in [4]. The scalar fields ρ* and τ* in (17) are defined by

$$\rho^* = A\rho \quad ; \quad \tau^* = \lambda\, \text{tr}\underset{\sim}{\varepsilon} \,. \tag{19}$$

Proof: For convenience we use the familiar indicial notation to define a vector $\underset{\sim}{J}$ by its components as

$$J_i = \int_{\partial\mathcal{D}} \left(Wn_i - t_j\, u_{j,i} \right) d\ell \,, \tag{20}$$

where $\underset{\sim}{n}$ is the outward normal of the surface $\partial\mathcal{D}$ that is the boundary of the finite regular closed subregion $\mathcal{D}$; and $\underset{\sim}{t}$ is the traction vector defined by

$$t_i = T_{ij} n_j \,. \tag{21}$$

The divergence theorem [10], the chain rule and the definitions (8), (9) and (21) allow for successive transformations of (20), as follows:

$$J_i = \int_{\mathcal{D}} \left\{ W,_i - \left[(\nabla \underset{\sim}{u})^T \underset{\sim}{T} \right]_{ij,j} \right\} dv = \int_{\mathcal{D}} \left\{ \frac{\partial W}{\partial \varepsilon_{mj}} \varepsilon_{mj,i} - \left(u_{m,i} T_{mj} \right),_j \right\} dv$$

$$= \int_{\mathcal{D}} \left\{ T_{mj}\varepsilon_{mj,i} - u_{m,ij}T_{mj} - u_{m,i}T_{mj,j} \right\} dv = - \int_{\mathcal{D}} u_{m,i}T_{mj,j} dv \tag{22}$$

where the identity

$$T_{mj}\varepsilon_{mj,i} = \frac{1}{2} T_{mj}u_{m,ji} + \frac{1}{2} T_{mj}u_{j,mi} = \frac{1}{2} T_{mj}u_{m,ji} + \frac{1}{2} T_{jm}u_{j,mi} = T_{mj}u_{m,ji} \tag{23}$$

was used.

Upon substitution of (11) into (22) we obtain

$$J_i = A \int_{\mathcal{D}} u_{m,i}\rho,_m \ . \tag{24}$$

With the aid of definition (8) and a trivial algebraic manipulation the last equation is written, in direct notation, as

$$\underset{\sim}{J} = 2A \int_{\mathcal{D}} \underset{\sim}{\varepsilon} \nabla\rho \, dv - A \int_{\mathcal{D}} (\nabla \underset{\sim}{u}) \nabla\rho \, dv \ . \tag{25}$$

It is convenient to introduce the definitions

$$\underset{\sim}{J_1} = \int_{\mathcal{D}} (\nabla \underset{\sim}{u}) \nabla\rho \, dv \quad ; \quad J_2 = \int_{\mathcal{D}} \underset{\sim}{\varepsilon} \nabla\rho \, dv \ . \tag{26}$$

Then using the easily shown identity,

$$\mathrm{div}(\underset{\sim}{u} \otimes \nabla\rho) = (\nabla \underset{\sim}{u}) \nabla\rho + \underset{\sim}{u} \, \mathrm{div}(\nabla\rho) \ , \tag{27}$$

when the symbol "$\otimes$" denotes the dyadic, we obtain

$$J_1 = \int_{\mathcal{D}} \mathrm{div}(u \otimes \nabla\rho) dv - \int_{\mathcal{D}} \underset{\sim}{u} \nabla^2 \rho dv \ , \tag{28}$$

which with the aid of (6) and the divergence theorem [10], gives

$$J_1 = \int_{\partial\mathcal{D}} (\underset{\sim}{u} \otimes \nabla\rho) \underset{\sim}{n} \, d\ell \ . \tag{29}$$

The divergence theorem and the identity

$$\mathrm{div}(\rho\underset{\sim}{\varepsilon}) = \underset{\sim}{\varepsilon}\nabla\rho + \rho\ \mathrm{div}\ \underset{\sim}{\varepsilon} \tag{30}$$

serve to write $(26)_2$ in the following form

$$J_2 = \int_{\partial D} \rho\underset{\sim}{\varepsilon}\underset{\sim}{n}d\ell - \int_{D} \rho\ \mathrm{div}\ \underset{\sim}{\varepsilon}\ dv\ . \tag{31}$$

Our efforts will be directed next in transforming the second term of the right hand side of (31) to a surface integral. Thus, the equilibrium equations (12) are used to write

$$\int_D \rho\ \mathrm{div}\ \underset{\sim}{\varepsilon}\ dv = -\frac{A}{4\mu}\int_D \nabla\rho^2 dv - \frac{\lambda}{2\mu}\int_D \nabla(\rho\ \mathrm{tr}\ \underset{\sim}{\varepsilon})dv + \frac{\lambda}{2\mu}\int_D \mathrm{tr}\underset{\sim}{\varepsilon}\ \nabla\rho\ dv\ , \tag{32}$$

which after the use of the divergence theorem is reduced to

$$\int_D \rho\ \mathrm{div}\ \underset{\sim}{\varepsilon}\ dv = -\frac{A}{4\mu}\int_{\partial D} \rho^2\underset{\sim}{n}d\ell - \frac{\lambda}{2\mu}\int_{\partial D} \rho\ \mathrm{tr}\underset{\sim}{\varepsilon}\ \underset{\sim}{n}\ d\ell + \frac{\lambda}{2\mu}\int_D \mathrm{tr}\varepsilon\ \nabla\underset{\sim}{\rho}\ dv\ . \tag{33}$$

Defining a new integral

$$\underset{\sim}{J}_3 = \int_D \mathrm{tr}\varepsilon\ \nabla\underset{\sim}{\rho}\ dv\ , \tag{34}$$

and expressing the equilibrium equations in terms of the displacement [9] we obtain

$$\underset{\sim}{J}_3 = -\frac{\mu}{A}\int_D \mathrm{tr}\underset{\sim}{\varepsilon}\ \nabla^2\ \underset{\sim}{u} - \frac{\lambda+\mu}{A}\int_D (\mathrm{tr}\underset{\sim}{\varepsilon})\nabla\mathrm{tr}\underset{\sim}{\varepsilon}\ dv\ . \tag{35}$$

Next, we use the identity

$$\int_D \mathrm{tr}\underset{\sim}{\varepsilon}\ \nabla^2\ \underset{\sim}{u} = \int_{\partial D} [\mathrm{tr}\underset{\sim}{\varepsilon}(\nabla\underset{\sim}{u})\underset{\sim}{n} - (\underset{\sim}{u}\otimes\nabla\ \mathrm{tr}\underset{\sim}{\varepsilon})\underset{\sim}{n}]d\ell\ , \tag{36}$$

a proof of which is given in [5]. Thus, with the aid of the divergence theorem again to transform the second term of the left hand side of (36) to a surface integral, equation (35) may be written as

$$\underset{\sim}{J}_3 = -\frac{\mu}{A}\int_{\partial D} [\mathrm{tr}\underset{\sim}{\varepsilon}(\nabla\underset{\sim}{u})\underset{\sim}{n} - (\underset{\sim}{u}\otimes\nabla\ \mathrm{tr}\underset{\sim}{\varepsilon})\underset{\sim}{n}]d\ell - \frac{\lambda+\mu}{2A}\int_{\partial D} (\mathrm{tr}\underset{\sim}{\varepsilon})^2 d\ell\ . \tag{37}$$

Combination of (33), (34) and (37) yields

$$\int_D \rho\ \mathrm{div}\underset{\sim}{\varepsilon}dv = -\frac{1}{2\mu}\int_{\partial D}\left\{\left[\frac{A}{2}\rho^2+\lambda\rho\ \mathrm{tr}\underset{\sim}{\varepsilon} + \frac{\lambda(\lambda+\mu)}{2A}(\mathrm{tr}\underset{\sim}{\varepsilon})^2\right]\underset{\sim}{n} + \frac{\lambda\mu}{A}[\mathrm{tr}\underset{\sim}{\varepsilon}\nabla\underset{\sim}{u}-\underset{\sim}{u}\otimes\nabla\mathrm{tr}\underset{\sim}{\varepsilon}]\underset{\sim}{n}\right\} \tag{38}$$

and this way the following expression for J_2 is derived

$$\underset{\sim}{J}_2 = \frac{1}{2\mu} \int_{\partial D} \left\{ \left[\frac{\lambda\mu}{A} \rho \underset{\sim}{\varepsilon} + \text{tr}\underset{\sim}{\varepsilon}\nabla\underset{\sim}{u} - \underset{\sim}{u}\otimes\nabla\text{tr}\underset{\sim}{\varepsilon} \right] \underset{\sim}{n} + \left[\frac{A}{2} \rho^2 + \lambda\rho\,\text{tr}\underset{\sim}{\varepsilon} + \frac{\lambda(\lambda+\mu)}{2A} (\text{tr}\underset{\sim}{\varepsilon})^2 \right] \underset{\sim}{n} \right\} . \quad (39)$$

Defining new scalar fields ρ^* and τ^* as in (19) and combining (25), (26), (29) and (39) we finally obtain

$$\underset{\sim}{J} = \int_{\partial D} \left\{ [2\rho^*\underset{\sim}{\varepsilon} + \tau^*\nabla\underset{\sim}{u} - \underset{\sim}{u}\otimes\nabla(\rho^*+\tau^*)]\underset{\sim}{n} + \frac{1}{2\mu}\left(\rho^{*2} + 2\rho^*\tau^* + \frac{1}{2v}\tau^{*2}\right)\underset{\sim}{n} \right\} d\ell . \quad (40)$$

Also equation (20) may be written in direct notation as

$$\underset{\sim}{J} = \int_{\partial D} \left[W\underset{\sim}{n} - (\nabla\underset{\sim}{u})^T \underset{\sim}{T}\underset{\sim}{n} \right] d\ell . \quad (41)$$

The results (40) and (41) establish the validity of the conservation law. (17), if the boundary ∂D is identifying with the closed surface C.

It is easily seen from (41) that the first component of the vector $\underset{\sim}{J}$ is the well-known J-integral. Equation (40) indicates that the value of J integral along a closed path is not zero when diffusion is considered. The appropriate form which replaces the J-integral, in the cases under consideration, is provided by the first component of the vector $\underset{\sim}{J}^*$ in equation (17).

It is natural to expect that the vector $\underset{\sim}{J}^*$ is identified to the vector $\underset{\sim}{J}$ when diffusion effects are neglected. In this case the density of the diffusing species is uniform, i.e.,

$$\nabla\rho^* = 0 \quad ; \quad \rho^* = \rho^*_o . \quad (42)$$

Under these conditions, equation (17) combined with (41) gives

$$\underset{\sim}{J}^*_{\nabla\rho=0} = \underset{\sim}{J} + \int_{\partial D} \left(2\rho^*_o\underset{\sim}{\varepsilon} + \frac{1}{2\mu}\rho^{*\,2}_o + \frac{1}{\mu}\rho^*_o\tau^* \right)\underset{\sim}{n}\,d\ell + \int_{\partial D} (\tau^*\nabla\underset{\sim}{u} - \underset{\sim}{u}\otimes\nabla\tau^*)\underset{\sim}{n}\,d\ell +$$

$$+ \frac{1}{4\mu v} \int_{\partial D} \tau^{*2}\underset{\sim}{n}\,d\ell . \quad (43)$$

First observe that use of the divergence theorem and $(19)_2$ yields

$$\int_{\partial D} \left(2\rho^*_o\underset{\sim}{\varepsilon} + \frac{1}{2\mu}\rho^{*\,2}_o + \frac{1}{\mu}\rho^*_o\tau^* \right)\underset{\sim}{n}\,d\ell = \int_{D} \left(2\rho^*_o\text{div}\underset{\sim}{\varepsilon} + \rho^*_o\frac{\lambda}{\mu}\nabla\,\text{tr}\underset{\sim}{\varepsilon} \right) dv = 0 . \quad (44)$$

This is true because the integrand in $(44)_2$ has a zero value, as it is easily seen by combining equations (12) and $(42)_1$. Next, we use the definition $(19)_2$, the identity (36) and the divergence theorem, to write

$$\int_{\partial D} \left(\underset{\sim}{\tau}^*\nabla \underset{\sim}{u} - \underset{\sim}{u}\nabla \tau^*\right)\underset{\sim}{n}d\ell + \frac{1}{4\mu v}\int_{\partial D} \tau^{*2}\underset{\sim}{n}d\ell = \lambda \int_{D} \left[(tr\underset{\sim}{\varepsilon})\nabla^2\underset{\sim}{u} + \frac{\lambda+\mu}{\mu}(tr\underset{\sim}{\varepsilon})\nabla tr\underset{\sim}{\varepsilon}\right]dv \ . \quad (45)$$

But the equilibrium equations (12) are expressed in terms of the displacement vector in the form

$$\nabla^2 u + \frac{\lambda+\mu}{\mu}\nabla tr\varepsilon = -\frac{A}{\mu}\nabla\rho \ , \quad (46)$$

which when is combined with $(42)_1$ and (45) results into

$$\int_{\partial D} (\tau^*\nabla \underset{\sim}{u} - \underset{\sim}{u}\nabla\tau^*)\underset{\sim}{n}d\ell + \frac{1}{4\mu v}\int_{\partial D} \tau^{*2}\underset{\sim}{n}d\ell = 0 \ . \quad (47)$$

Then combination of (41), (43), (44) and (47) yields

$$\underset{\sim}{J}^*_{\nabla\rho=0} = \oint_{C} \left[W\underset{\sim}{n} - (\nabla\underset{\sim}{u})^T \underset{\sim}{T}\underset{\sim}{n}\right]d\ell = 0 \quad (48)$$

and this way the result (18) is established.

A PATH INDEPENDENT INTEGRAL

In the case without diffusion, it is shown [2] that the first component of vector $\underset{\sim}{J}$ has the same value for any closed curve surrounding the singularity.

In the present investigation diffusion effects are introduced and we derive results analogous to those contained in [2]. We confine attention to a two dimensional symmetric configuration. Thus, we consider an infinite plate loaded symmetrically under plane strain conditions (Mode I) and containing a central crack acting as source of diffusing species symmetrically distributed with respect to x-axis. The scheme under examination is shown in Figure 1.

The objective is to show that the $\underset{\sim}{J}^*$ integral, defined in (17), has the same value for all paths surrounding the line crack. Towards this aim we consider the closed curve, $C'+\Gamma^++C+\Gamma^-$, as indicated in Figure 1. Then the conservation law (17) insures path independence if we show that

$$\int_{\Gamma^++\Gamma^-} \left\{[2\rho^*\underset{\sim}{\varepsilon}+\tau^*\nabla \underset{\sim}{u} - \underset{\sim}{u}\otimes\nabla(\rho^*+\tau^*)]\underset{\sim}{n} + \frac{1}{2\mu}\left[\rho^{*2}+2\rho^*\tau^* + \frac{1}{2v}\tau^{*2}\right]\underset{\sim}{n}\right\}d\ell = 0 \ . \quad (49)$$

Employing symmetry arguments, as well as traction free boundary condition, we deduce the following relations holding on the crack surface.

$$T_{xx} = T_{yy} = T_{xy} = 0 \ ; \quad u_x = 0 \ ; \quad u_y\Big|_{y=0^+} = -u_y\Big|_{y=0^-}$$
$$\rho^*\Big|_{y=0^+} = \rho^*\Big|_{y=0^-} \ ; \quad \frac{\partial\rho^*}{\partial y}\Big|_{y=0^+} = -\frac{\partial\rho^*}{\partial y}\Big|_{y=0^-} \ ; \quad \frac{\partial\tau^*}{\partial y}\Big|_{y=0^+} = -\frac{\partial\tau^*}{\partial y}\Big|_{y=0^-} \quad (50)$$

Then τ^* vanishes on the crack surface and condition (49) is equivalent to

$$\int_{\Gamma^+ + \Gamma^-} \left\{ [2\rho^* \underset{\sim}{\varepsilon} - \underset{\sim}{u} \otimes \nabla(\rho^* + \tau^*)]\underset{\sim}{n} + \frac{1}{2\mu} \rho^{*2} \underset{\sim}{n} \right\} d\ell = 0 \ . \tag{51}$$

With the aid of (15) we obtain

$$\int_{\Gamma^+ + \Gamma^-} 2\rho^* \underset{\sim}{\varepsilon} \underset{\sim}{n} d\ell = \int_{\Gamma^+ + \Gamma^-} 2\rho^* \left[\frac{1+\nu}{E} \underset{\sim}{t} - \frac{\nu}{E} \sigma \underset{\sim}{n} \right] d\ell = 0 \tag{52}$$

since $\underset{\sim}{t}$ and σ vanish on the crack surface. If $\hat{1}y$ is the base vector in the y-direction then

$$\int_{\Gamma^+ + \Gamma^-} \frac{1}{2\mu} \rho^{*2} n d\ell = \hat{1}y \frac{1}{2\mu} \left\{ \int_{\Gamma^+} \rho^{*2} dx + \int_{\Gamma^-} \rho^{*2} dx \right\} = 0 \tag{53}$$

since the integrand ρ^{*2} has the same value as it is integrated over the same interval in opposite directions. Also,

$$\begin{aligned} \int_{\Gamma^+ + \Gamma^-} [\underset{\sim}{u} \otimes \nabla(\rho^* + \tau^*)]\underset{\sim}{n} d\ell &= \int_{\Gamma^+ + \Gamma^-} \underset{\sim}{u}[\underset{\sim}{n} \cdot \nabla(\rho^* + \tau^*)] d\ell = \\ &= \hat{1}_x \int_{\Gamma^+ + \Gamma^-} u_x \left[\frac{\partial \rho^*}{\partial y} + \frac{\partial \tau^*}{\partial y} \right] dx + \hat{1}_y \int_{\Gamma^+ + \Gamma^-} u_y \left[\frac{\partial \rho^*}{\partial y} + \frac{\partial \tau^*}{\partial y} \right] dx \ . \end{aligned} \tag{54}$$

But $u_x = 0$ on the crack surface and the integrand $u_y \left[\frac{\partial \rho^*}{\partial y} + \frac{\partial \tau^*}{\partial y} \right]$ has the same value, because of the last two relations of (50), as it is integrated over Γ^+ and Γ^-. Thus

$$\int_{\Gamma^+ + \Gamma^-} [\underset{\sim}{u} \otimes \nabla(\rho^* + \tau^*)]\underset{\sim}{n} \, d\ell = 0 \ . \tag{55}$$

The results (51), (52) and (55) establish the validity of condition (49), and therefore the path independence of $\underset{\sim}{J}^*$ integral.

It has been shown that when J_x^* is integrated over paths away from the crack tip where the diffusion effects are neglected, then it has the same value as the familiar J-integral. This furnishes approximate knowledge of the value of J_x^* for these configurations in which the J-integral has been already evaluated. Then approximate estimates for the stress-diffusion field in the neighbourhood of the singularity may be attempted.

REFERENCES

1. CHEREPANOV, G. P., PPM, 31, No. 3, 1967, 476.
2. RICE, J. R., J. Appl. Mech., 35, 1968, 379.
3. ESHELBY, J. D., "Solid State Physics", Vol. 3, Academic Press, New York, 1956, 79.
4. KNOWLES, J. K. and STERNBERG, E., Arch. Rational Mech. Anal., 44, No. 3, 1972, 187.
5. AIFANTIS, E. C., J. Appl. Mech., (to be submitted).

6. AIFANTIS, E. C., Mech. Res. Communications, 3, No. 4, 1976, 245.
7. CHEREPANOV, G. P., Eng. Fract. Mechanics, 5, 1973, 1041.
8. AIFANTIS, E. C., Ph. D. Thesis, University of Minnesota, 1975.
9. SOKOLNIKOFF, I. S., "Mathematical Theory of Elasticity", McGraw Hill, New York, 1956, 73.
10. GURTIN, M. E., "The Linear Theory of Elasticity", Handbuch der Physik, VI a/2, Springer-Verlag, Berlin, Heidelberg, New York, 1972, 16.

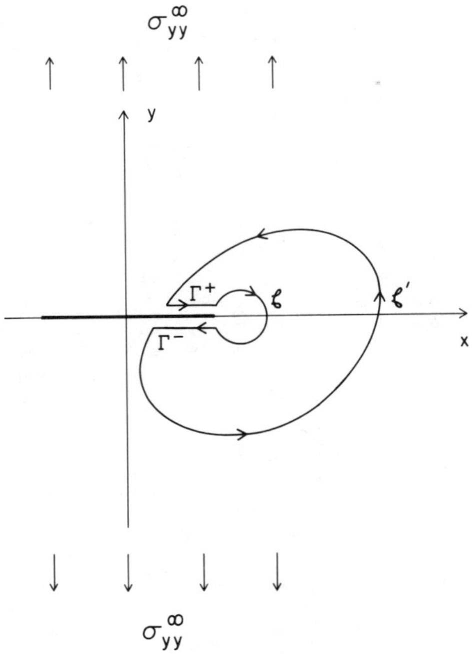

Figure 1

Fracture 1977, Volume 3, ICF4, Waterloo, Canada, June 19 - 24, 1977

SPECIMEN SIZE EFFECT ON J-INTEGRAL FRACTURE TOUGHNESS

D. Sunamoto, M. Satoh, T. Funada and M. Tomimatsu*

INTRODUCTION

In order to analyze quantitatively the safety of a structure against brittle fracture, current engineering level requires to measure the plane strain fracture toughness K_{IC} of the structural steel and to analyze conservatively with linear elastic fracture mechanics. This K_{IC} needs to satisfy ASTM E399 and requires very thick specimens. K_{IC} at room temperature, sometimes, requires a ten-inch thick specimen. Therefore it is not allowed to be taken as a routine work to measure a K_{IC} by ASTM E399. This specimen size requirement is a bottleneck in fracture safe analysis.

Today, J-integral developed by Rice is extended to elastic-plastic region and admitted as an acceptable measure of fracture analysis. With J-integral method, relatively small specimen can result in valid K_{IC} which, by ASTM E399, can be measured from very large specimens. There are many researches on J-integral test analysis (typical method listed in Table 1). The authors studied specimen size effect on J-integral fracture toughness K_J (fracture toughness measured by J-integral method) and defined valid specimen size experimentally.

TEST

240 mm thick A533 Gr.BCℓ.1 steel was used. Chemical compositions and mechanical properties were shown in Table 2. Plane strain fracture toughness K_{IC} of the test material which met ASTM E399 requirements were shown in Figure 1. They used a 240 mm thick compact tension specimen to get valid K_{IC} at -20°C and 4TCT specimen at -50°C.

First, they applied J-integral method to valid K_{IC} test data. They used Rice (1) method shown in Table 1. In plane strain fracture toughness test, notch root displacement was measured. They converted notch root displacement to load point displacement and measured J_c and J-integral fracture toughness K_J as shown in Figure 3 and equation (1).

$$K_J = \sqrt{EJ_c/(1-\nu^2)} \tag{1}$$

Where, E is Young's modulus, ν is Poisson's ratio, J_c is critical J-integral value. In Figure 3 rotational factor r of 1/3 was used considering the reports by Ingham [1] and Liebowitz [2]. As shown in Figure 3, K_J were almost equal to K_{IC} in elastic region and ASTM E399 requirements were satisfactory to J-integral method.

---

*Mitsubishi Heavy Industries, Limited, Takasago, Japan.

Next, they used 1/2TCT, 1TCT and 2TCT specimens with various crack length ratios shown in Figure 3 and conducted fracture test at -20°C and -50°C. They recorded load-displacement curves on a X-Y recorder and applied 4 J-integral methods shown in Table 1. At this time also notch root displacement was converted to load point displacement as before. Typical load-displacement curves at -20°C were shown in Figure 4. As shown in Figure 4, load-displacement curve of 1/2TCT specimen was not linear and indicated large plastic deformation. But the nonlinearity decreased as specimen size increased. The load-displacement curve of 2TCT specimen was almost linear and the curve of 240 mm specimen was linear.

Almost all specimens suffered from fibrous crack. Photograph 1 showed an example of fibrous crack and stretched zone of a 1TCT specimen (X 300). But this time, they didn't discuss on these micro-behaviour at crack tip. These matter will be reported at other chance.

Test results were listed in Table 4. Example of the comparison between 4 J-integral methods was shown in Figure 5. It was known that 4 methods gave almost same K_J.

The relation between K_J and specimen thickness was shown in Figure 6. K_J from 1/2TCT specimen varied very large. K_J data of 1TCT specimen varied less than 1/2TCT specimen and many of 1TCT specimens and all of 2TCT specimens, which number was limited, were almost equal to K_{IC} value. In Figure 6 also, K_{IC} estimated by another method were shown, such as Equivalent Energy (EE) method and COD. Fracture toughness by COD were greater than K_{IC} and those by EE method gave almost same value as K_J.

Considering the matter described above, they inquired the specimen size by which valid K_{IC} could be obtained. Necessary specimen size parameter α was shown in Figure 7. α was calculated by equation (2)

$$\alpha = \min\,(a,\ B,\ W-a)\cdot E\cdot\sigma_y\cdot/(1-\nu^2)\cdot K_J{}^2\ . \qquad (2)$$

Contrary to Begley's [7] and Griffis's [8] reports that required α to be greater than 50 and 100 respectively, α needed to be greater than 40 in order to get K_J as K_{IC}.

Figure 8 showed crack length ratio dependence of K_J, and it is known that crack length ratio should be greater than 0.4.

CONCLUSION

The authors conducted fracture test to a steel which plane strain fracture toughness was known and applied J-integral analysis. They obtained the following results.

1) No significant difference was recognized between J-integral methods.
2) Specimen size parameter α was to be greater than 40.

REFERENCES

1. INGHAM, T. et al, Proceedings of the Conf. Pract. Appl. Fracture Mech. Pressure-Vessel Technol., London, 1971.
2. LIEBOWITZ, H. et al, Symposium Mechanical Behaviour of Materials, Kyoto, 1974.
3. RICE, J. R. et al, ASTM STP 536, 1973, 231.
4. KANAZAWA, T. et al, IIW X-779-75, 1975.
5. RICE, J. R. et al, ASTM STP 514, 1972, 40.
6. MERKLE, J. G. et al, ASME Publication 74-PVP-33.
7. BEGLEY, J. A. et al, ASTM STP 514, 1972, 1.
8. GRIFFIS, C. A. et al, NRL Report 7676, 1974.

Photograph 1 Observation of Stretched Zone

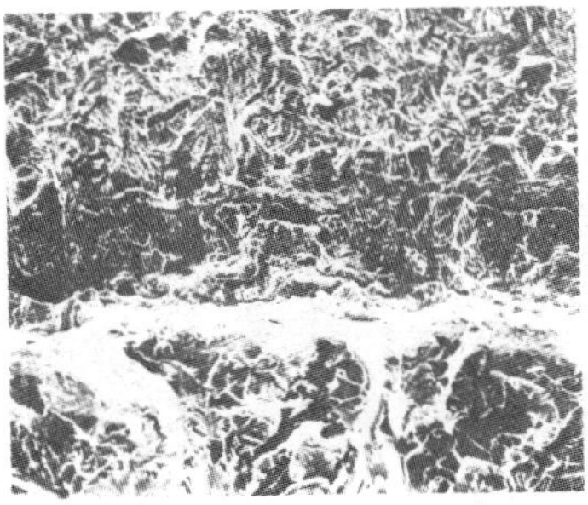

Table 1 J_C Method with Single Specimen

origin	equation
Rice (1) (3)	$J_C = \frac{2S}{B(W-a)}$
Kanazawa (4)	$J_C = -\left(\frac{1}{W-a} \cdot \frac{\alpha_{II}}{W}\right) P\delta_C + \left(\frac{2}{W-a} - \frac{\alpha_{II}}{W}\right) S$
Rice (2) (5)	$J_C = \frac{P\delta_C}{B(W-a)} \left\{1 + \frac{16D^2}{\beta\pi}\left(\frac{P}{P_L}\right)^2\right\}$ for $P < P_L$
Merkle (6)	$J_C = \frac{2}{B(W-a)} \cdot \frac{1+\alpha}{1+\alpha^2} \cdot S + \frac{2}{B(W-a)} \cdot \frac{\alpha(1-2\alpha-\alpha^2)}{(1+\alpha^2)^2} (P\delta_C - S)$

Table 2 Test Material

1) Chemical compositions (%)

	C	Si	Mn	P	S	Ni	Mo
Cheok analysis	0.19	0.23	1.40	0.008	0.007	0.63	0.55

2) Mechanical properties

Tensile properties				Charpy transition temperature		Drop weight test
σ_y (MPa)	σ_B (MPa)	δ (%)	φ (%)	${}_vT_{rE}$ (°C)	${}_vT_{rS}$ (°C)	NDT (°C)
483	607	25.0	47.0	-5	-5	-20

Table 3 Correlation Between K_{IC} and K_J by Rice (1) Method Obtained from Plane Strain Specimen

Temperature (°C)	Specimen	K_{IC} ($MPa \cdot m^{1/2}$)	K_J ($MPa \cdot m^{1/2}$)
-125	2TCT	74.4	73.2
-100	2TCT	70.1	70.4
-100	4TCT	67.0	66.0
-75	4TCT	98.6	101.4
-20	240mmCT	151.0	143.5

Table 4 Test Results

Temp. (°C)	Speci-men	No.	Specimen size a (mm)	B (mm)	W (mm)	a/W	max. load P_{max} (KN)	Fracture toughness ($MPa \cdot m^{1/2}$) K_{max}	$K_J^{(1)}$ (Rice 1)	$K_J^{(2)}$ (Kanazawa)	$K_J^{(3)}$ (Rice 2)	$K_J^{(4)}$ (Merkle)
-20°C	½TCT	JH-1	9.1	12.8	25.6	0.36	36.4	117	326	275	-	357
		-2	10.4	12.8	25.6	0.41	24.0	87.1	122	121	122	135
		-3	10.9	12.7	25.4	0.43	27.3	106	241	206	-	262
		-4	14.1	12.8	25.6	0.55	17.9	98.9	291	238	-	310
		-5	10.8	12.7	25.4	0.42	17.3	66.7	277	221	227	300
		-6	16.2	12.8	25.6	0.63	11.1	84.0	299	235	-	316
		-7	16.3	12.7	25.4	0.64	10.5	83.7	252	258	-	272
		-8	17.7	12.8	25.6	0.69	6.0	59.9	84	78	-	89
		-9	17.7	12.7	25.4	0.70	7.6	78.8	259	215	-	272
	1TCT	JO-1	15.7	25.4	50.8	0.31	142	148	313	283	-	344
		-2	20.9	25.4	50.8	0.41	93.2	123	157	162	162	174
		-3	20.7	25.4	50.8	0.41	93.7	122	156	160	160	172
		-4	25.9	25.4	50.8	0.51	71.3	123	162	129	-	173
		-5	25.8	25.4	50.8	0.51	62.4	107	147	127	136	159
		-6	30.0	25.4	50.8	0.59	42.8	97.4	120	122	129	130
		-7	30.1	25.4	50.8	0.59	49.4	114	193	172	-	206
		-8	38.1	25.4	50.8	0.75	16.5	80.6	155	135	152	162
		-9	41.1	25.4	50.8	0.81	9.6	66.4	157	126	144	163
	2TCT	JT-1	53.3	50.8	101.6	0.52	202	129	140	148	148	153
		-2	52.2	50.8	101.6	0.51	204	126	137	142	142	150
-50°C	½TCT	JH-11	8.5	12.7	25.4	0.33	31.0	96.1	155	182	-	174
		-12	8.7	12.7	25.4	0.34	32.8	103	195	180	-	214
		-13	10.5	12.7	25.4	0.41	27.2	102	184	166	-	200
		-14	10.9	12.7	25.4	0.43	24.5	95.5	134	130	137	147
		-15	10.3	12.7	25.4	0.41	17.7	64.8	116	114	-	128
		-16	11.1	12.7	25.4	0.44	15.1	60.2	125	115	111	136
		-17	16.6	12.7	25.4	0.65	6.9	58.0	65	68	72	72
		-18	15.8	12.7	25.4	0.62	10.7	78.1	155	133	-	165
		-19	18.2	12.7	25.4	0.72	5.3	61.7	136	114	-	143
		-20	18.4	12.7	25.4	0.72	5.0	60.2	82	77	87	86
	1TCT	JO-21	26.1	25.4	50.8	0.51	56.9	99.5	113	119	121	124
		-22	25.9	25.4	50.8	0.51	52.6	90.9	97	104	114	107
	2TCT	JT-21	52.7	50.8	101.6	0.52	182	114	117	127	125	129
		-22	53.2	50.8	101.6	0.52	131	83.7	84	91	87	92

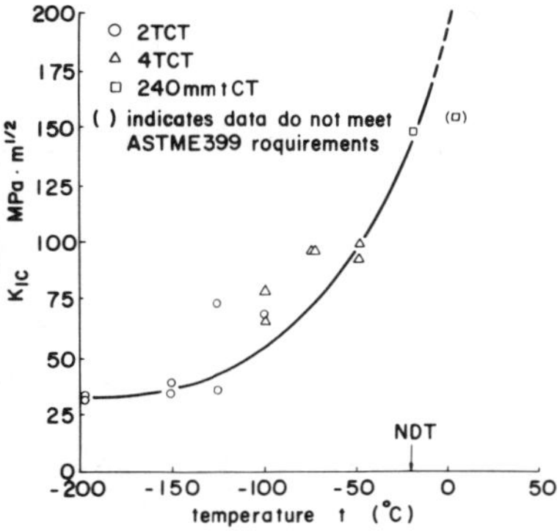

Figure 1 Plane Strain Fracture Toughness K_{IC} of Test Material

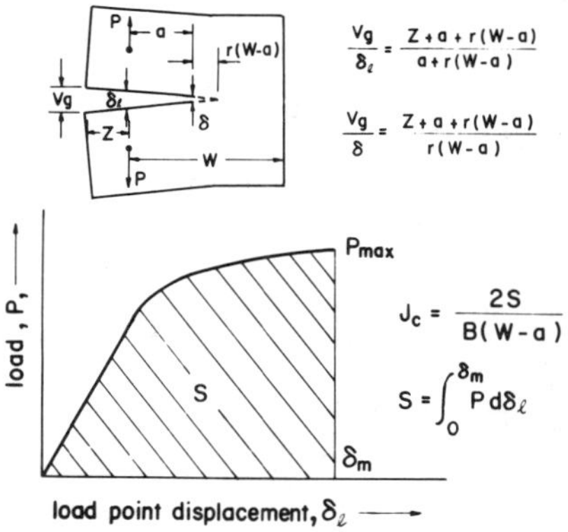

Figure 2 Experimental J-Integral Value

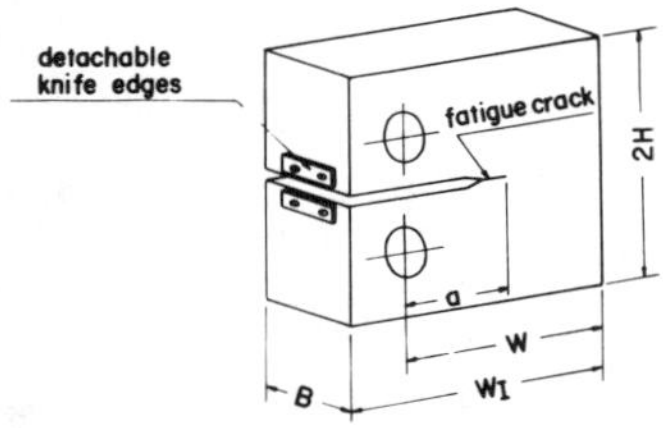

	B	W_I	2H	W	a/W
½TCT	12.7	31.8	30.4	25.4	0.3 ~ 0.8
1TCT	25.4	63.5	61.0	50.8	
2TCT	50.8	127.0	122.0	101.6	≈0.50

(mm)

Figure 3 Compact Tension Specimen

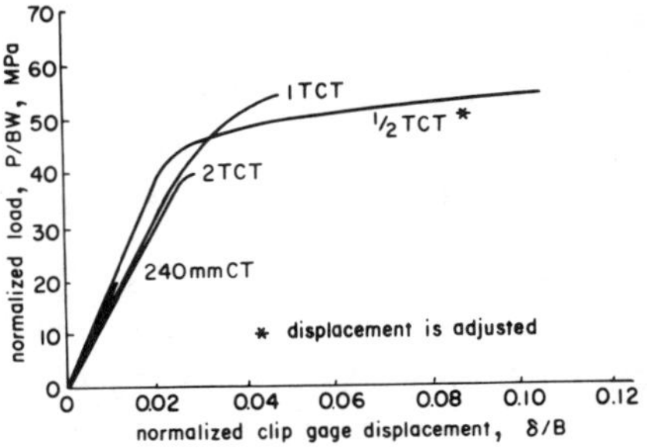

Figure 4 Typical Load-Displacement Curves (-20°C, a/w ≈ 0.5)

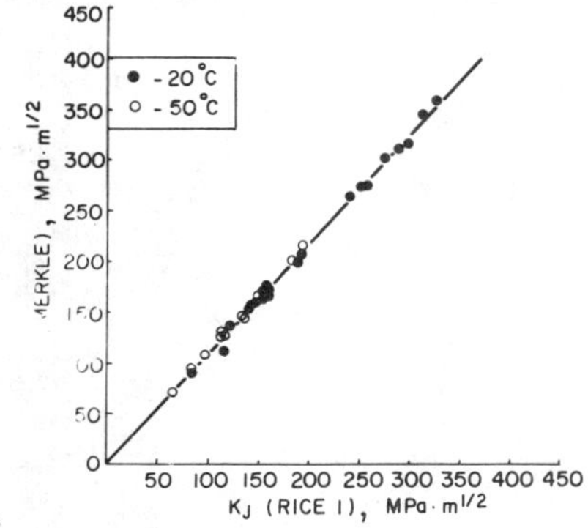

Figure 5 Comparison Between J-Integral Methods

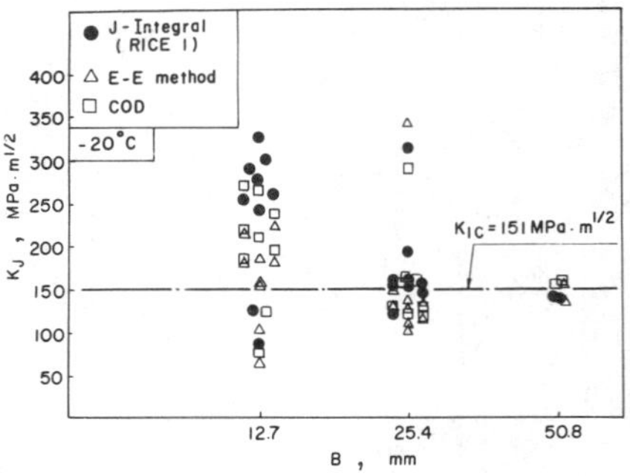

Figure 6 Variation of K_J on Specimen Thickness

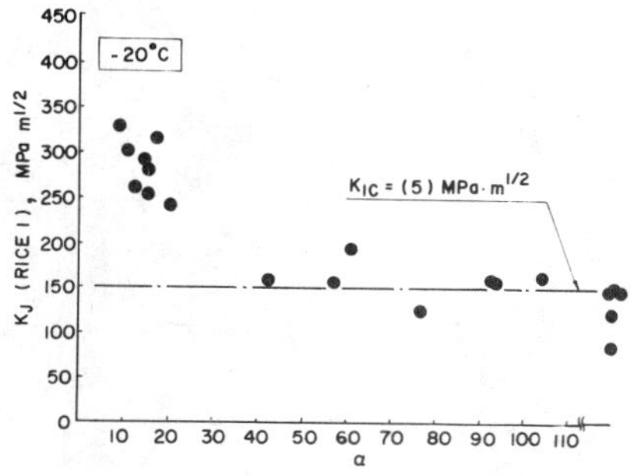

Figure 7 Necessary Specimen Size Parameter α to give Valid K_J

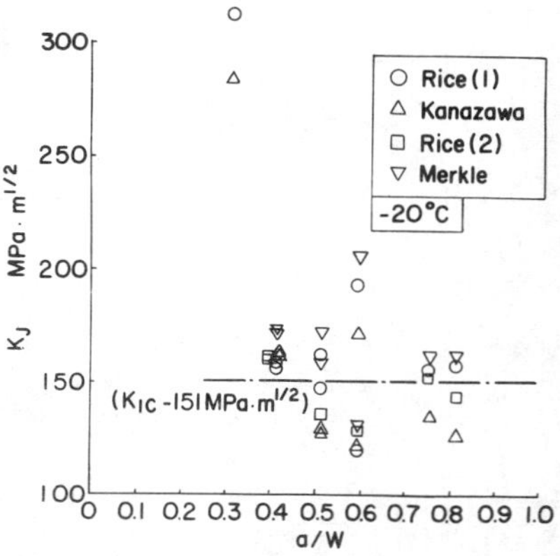

Figure 8 Correlation Between K_J and Crack Length Ratio

Fracture 1977, Volume 3, ICF4, Waterloo, Canada, June 19 - 24, 1977

THE J-INTEGRAL EVALUATION FOR CT SPECIMEN

T. Miyoshi* and M. Shiratori**

INTRODUCTION

Rice's method for the J-integral evaluation [1] is convenient for its simpleness, but the accuracy of Rice's method is not investigated sufficiently. The J-integral evaluation by Rice's method is compared to the one by the finite element method for the standard bend bar specimen (a/W=0.5). The result is that Rice's method gives the higher value than the finite element method by about 10%. For the compact tension specimen, Rice's method must be investigated on its accuracy, because it pays no consideration to the effect of axial force. In this paper, the J-integral for the compact tension specimen is evaluated by the finite element method and the accuracy of Rice's method is investigated based on the result by the finite element method.

RICE'S EQUATION AND MERKLE'S EQUATION [2]

Consider Rice's and Merkle's equations for the evaluation of J-integral. According to Rice's method, the J-integral is calculated by equation (1) based on the load-displacement curve as shown in Figure 1:

$$J = \frac{2A}{b} \qquad (1)$$

where b is the ligament length of the specimen.

Merkle et al. propose the equation which considers the axial force as well as the bending force. Considering that the axial force shifts the stress reversal point by αc as shown in Figure 2, they obtain equation (2) for the J-integral evaluation for CT specimen:

$$J = \frac{\eta_A A + \eta_C C}{b} \qquad (2)$$

where A is the strain energy, C is the complementary energy of the specimen and

$$\eta_A = \frac{2(1+\alpha)}{(1+\alpha^2)}, \qquad \eta_C = \frac{2\alpha(1-2\alpha+\alpha^2)}{(1+\alpha^2)^2} \qquad (3)$$

$$\alpha = \sqrt{\left(\frac{a}{c}\right)^3 + 2\left(\frac{a}{c}\right) + 2} - \left(\frac{a}{c} + 1\right) \qquad (4)$$

* University of Tokyo, Tokyo, Japan.
**Yokohama National University, Yokohama, Japan.

J-INTEGRAL EVALUATION BY FINITE ELEMENT METHOD

The J-integral evaluation is carried out for CT specimen shown in Figure 3. The mechanical properties of the material are represented in Table 1. The concentrated load is applied at the top of the loading pin hole shown in Figure 4, and the 318 elements and 194 nodes are used for the calculation of plane strain.

There are three ways to evaluate the J-integral based on the result by the finite element method:

1) J-integral is evaluated by equation (1) using the strain energy obtained by the finite element method.
2) J-integral is evaluated by equation (2) using the strain energy A and the complementary energy C obtained by the finite element method.
3) J-integral is evaluated by calculating the difference of the strain energy ΔU, as J-integral is given by equation (5):

$$J = -\frac{\partial U}{\partial a} = -\frac{U(a+\Delta a)-U(a)}{\Delta a} = \frac{\Delta U}{\Delta a} \tag{5}$$

J-integrals given by the methods (1), (2), and (3) are hereinafter referred to as J_R, J_R^*, and J_E, respectively.

As shown in the bend bar specimen, J and K_I are related by equation (6) for the elastic state. Therefore, K_I can be evaluated from J-integral of the elastic state:

$$J = \frac{1-\nu^2}{E} K_I^2 \quad \text{(for plane strain)} \tag{6}$$

In Table 2, the values of K_I obtained from J_E are compared to the analytical values, and the good coincidence is obtained. The values calculated from J_R and J_R^* do not agree with the analytical ones. This is because of the accuracy of J_R and J_R^* is getting worse when the deformation is small. The fact that K_I values calculated from J_E coincide with the analytical solutions is just one of the bases which take account of the validity of the J-integral evaluation by equation (5).

Numerical results of J_R, J_R^*, and J_E based on both the incremental theory and the deformation theory of the plasticity are presented in Table 3. It is shown from this table:

1) J_R, not taking account of the axial force effect, underestimates the J-integral value.
2) J_R^* is about 10% higher than J_E for the wide range of δ.

CONCLUSIONS

Rice's method, paying no consideration for the axial force effect, gives lower estimation for CT specimen. Therefore, when we evaluate the J-integral value by this type of equation, it is desirable to use Merkle's equation.

Rice's equation for the bend bar specimen and Merkle's equation for CT specimen seem to give higher values than equation (5) which, in the authors' opinion, gives most accurate values of J-integral. The dependence of J_{IC} values on the specimen geometry [3] seems to be based on the evaluation method of J-integral, partially.

REFERENCES

1. RICE, J. R., et al., ASTM STP 536, 1973, 231.
2. MERKLE, J. G. and CORTEN, H. T., Trans. ASME, Ser. J. 96-4, 1974, 286.
3. KANAZAWA, T., et al., IIW X-779-75, 1975.

Table 1 Mechanical Properties of A533B Steel

Young's Modulus (MPa)	Poisson's Ratio	Yield Stress (MPa)	Hardening Rate (MPa)
205800	0.3	480	2060

Table 2 $K_I{}^2B^2W/P^2$ for CT Specimen

a/W	Analysis	F.E.M.
0.50	92.16	91.59
0.52	104.24	102.47
0.54	118.59	117.31

Table 3 J-integral for CT Specimen (a/W=0.52)

Disp.	Incremental Theory			Deformation Theory		
δ	J_E	J_R	J_R^*	J_E	J_R	J_R^*
0.10	3.09	1.93	2.43	3.09	1.93	2.43
0.15	5.36	4.99	6.12	5.36	4.99	6.12
0.20	7.85	8.34	10.32	7.85	8.34	10.32
0.25	14.81	12.83	15.92	14.76	12.71	15.79
0.30	20.33	18.60	23.01	20.40	18.46	22.86
0.35	28.55	25.40	32.59	28.36	25.29	31.22
0.40	37.26	33.26	40.95	37.01	33.17	40.84
0.45	46.71	42.20	51.81	46.43	42.07	51.64
0.50	56.98	51.94	63.59	56.88	51.80	63.39

(δ:mm, J:kPa•m)

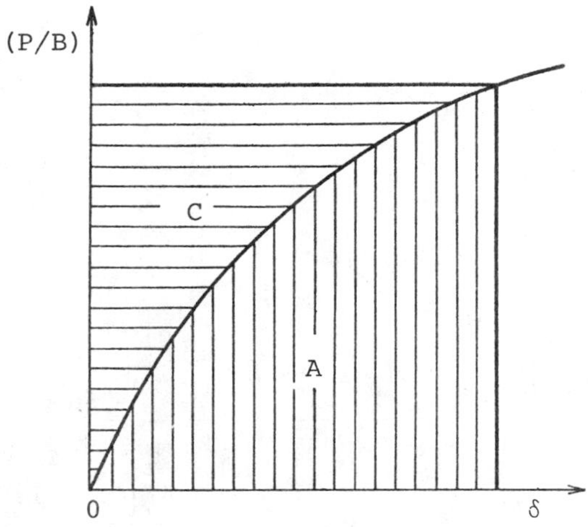

Figure 1 Load-Displacement Curve

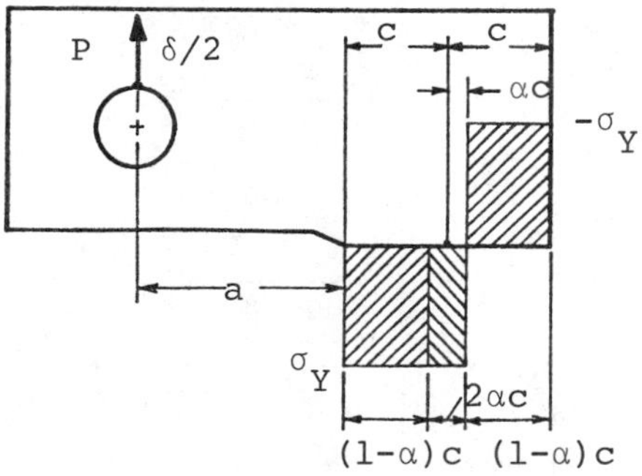

Figure 2 Reversals Point of Stress

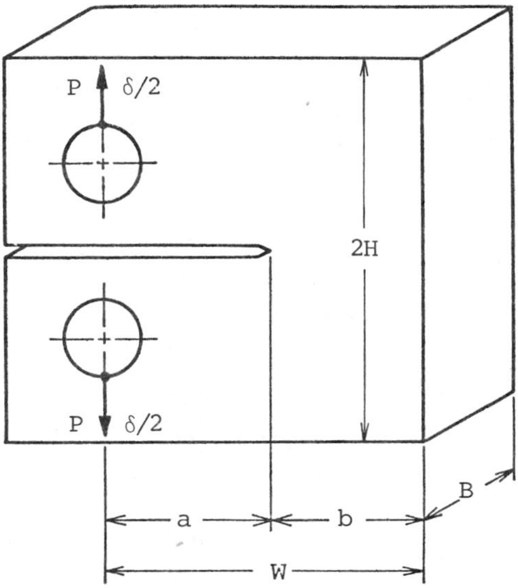

W=50mm, a/W=0.52
B/W=0.50, H/W=0.30

Figure 3 Geometry of CT Specimen

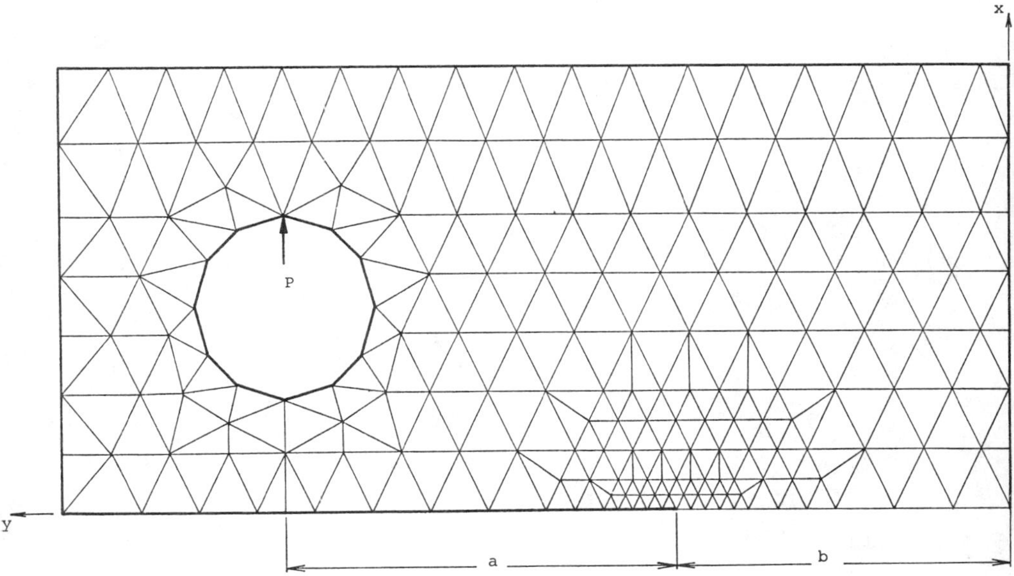

Figure 4 Nodal Breakdown of CT Specimen

Fracture 1977, Volume 3, ICF4, Waterloo, Canada, June 19 - 24, 1977

THE USE OF THE J INTEGRAL TO MEASURE THE RESISTANCE OF MILD STEEL TO SLOW STABLE CRACK GROWTH

S. J. Garwood and C. E. Turner*

INTRODUCTION

The development of yielding fracture mechanics allows the extension of the resistance (R) curve concept [1] to situations involving large scale plasti city. There are thee main advantages in this: firstly the possibility arises of predicting R curves for large structures from small specimens: secondly, an improvement in the accuracy of resistance curve measurements is attainable (since the build up of resistance is accompanied by an increase in plastic zone size, linear elastic fracture mechanics is inadequate): and finally, the measurements of "plane strain" R curves for low strength materials becomes feasible despite extensive yielding.

It is the purpose of this paper to examine the use of the J contour integral [2] as a measure of the resistance of mild steel to slow stable crack growth.

THEORY

From Rice [3] the area ΔU_p in Figure 1 can be related to J, for non-linear elastic material, by

$$\Delta U_p \simeq J_{1-2} \, B\Delta a \tag{1}$$

where opq represents the load deflection curve for an initial crack length a_1 and orq for a longer crack $a_2 = a_1 + \Delta a$. All the energy released is available to propagate the crack. J_{1-2} is the average value of J between the two crack lengths, and B is thickness. Begley and Landes [4] and Bucci et al [5] have extended this concept to elastic/plastic materials as a means of evaluating J, although the simple energetic meaning of energy available to drive the crack is not then maintained.

A technique for developing a J resistance (J_r) curve was derived for three point bending in [6], in which the three parameters of load, deflection and final crack length are all matched in the derivation of J. The basis of the method is the construction of a series of non-linear elastic curves to describe the history of the specimen as the crack extends (such as orq in Figure 1). This technique avoids the criticism that irreversible stress relaxation at the crack tip invalidates the J concept. Due, however, to the irreversible nature of plastic deformation the curves such as orq (Figure 1) are purely notional. Assuming a linear variation of the R curve for the increment of crack extension, and using the relationship between J and work done (U), as developed by Sumpter and Turner [7] for the three point bend geometry, the value of J at the new crack position can be

* Imperial College of Science and Technology, London, England.

found without the precise form of orq (Figure 1) being known. The expression is

$$J_2 = J_1 \frac{(W-a_2)}{(W-a_1)} + \frac{2\Delta U_d}{B(W-a_1)} \qquad (2)$$

where ΔU_d is the extra energy put into the specimen as the crack grows from a_1 to a_2 (see Figure 1) and W is the width of the specimen (Figure 2).

This equation can be applied for subsequent increments of crack growth to give the general form:

$$J_n = J_{(n-1)} \frac{(W-a_n)}{(W-a_{(n-n)})} + \frac{2(U_n-U_{n-1})}{B(W-a_{(n-1)})} \qquad (3)$$

Similar expressions may be obtained for all geometries where J : U estimates can be determined.

EXPERIMENTAL PROCEDURE

All tests were conducted at room temperature (~ 21°C). The properties of the mild steel (En32) used are listed in Table 1. Specimens were tested in three point bending. Both side grooved and plain types were used (Figure 2). All specimens were fatigue cracked at loads less than half the yield value. Initial crack lengths and subsequent amounts of crack growth were measured at ten points along the crack front and averaged to give mean values. Loading rate was constant at 1mm/min. A transducer was used to obtain load point displacement. As the technique relies on the load-displacement record, a correlation was made for the extraneous displacement due to roller indentation.

The three parameter technique is best suited to a one specimen test where crack extension is monitored together with load point displacement. However, due to the inadequacies of current techniques for detecting slow growth in the presence of large scale plasticity, a multi-specimen programme was adopted. Ten or so specimens were bent to give gradually increasing amounts of crack growth. Care was taken to ensure at least one specimen had crack growth $\leq$ 0.1mm. The specimens were then fractured in liquid nitrogen to facilitate measurement of the ductile crack extension. Crack growth determinations were made ignoring the stretch zone width (typically $\ll$ 0.1mm). Thus the initiation value of $J(J_i)$ is given by the intersection of the resistance curve with the Δa (the amount of slow growth measured) = 0 axis. Load-displacement areas were measured with a planimeter. The use of the formula:

$$J_o = \frac{2\,U_{total}}{B(W-a_o)} \qquad (4)$$

shown by Sumpter and Turner to be valid for a_o/W ranging from 0.4 to 0.7 precludes the need to subtract the energy attributable to the uncracked body recommended by Rice et al [8]. To begin with specimens of identical initial crack lengths were prepared but in later tests it was found sufficient to compensate for slight variations by making adjustments to the energy values.

The test method devised by Landes and Begley [9] was used to plot the $J_o : \Delta a$ curves (J_o is calculated using the initial crack length regardless of the amount of crack growth) in Figure 3. For larger increments of crack growth these curves overestimate the true J resistance (J_r) curve, but for small amounts the curves converge to give the same value for J_i. The $J_o : \Delta a$ curve was used as a check for any specimen behaving extraordinarily (due to blunted notch, preloading of bifurcation, for example). Any such point was omitted from the J_r calculation using the three parameter technique as it was found to affect subsequent points in the analysis. The first point on the J_o curve was taken at the J_1 value in equation (2) and the J_r curve was generated as in equation (3).

RESULTS

The effect of remaining ligament depth on the J_o curve for specimens with a side groove ratio (S.R. is defined in Figure 2) of 0.56 is shown in Figure 3. The variation of J_o with initial crack length translated into a J_r curve (Figure 4) which is independent of the remaining ligament for the values chosen. As expected J_o and J_r curves diverge after ~ 1mm crack growth. Both curves extrapolate to give a J_i for the material of 0.12 Jmm^{-2} (120 Nmm^{-1}). This value of J_i was found to be constant for all configurations of bend specimens tested. The side groove ratio of 0.56 was chosen to eliminate 3 mm deep shear lips present in the plain specimens. However when specimens of double the dimensions but with the same side groove depth were tested a steeper J_r curve was found (Figure 5). Essentially the same curve is produced for differing value of W. The curve for the smaller W however reaches a plateau as the crack moves into the plastic region of the loading roller. Figure 6 indicates that it is the S.R. ratio which is the over-riding factor, rather than the removal of shear lips, in the production of a lowest bound or "plane strain" resistance curve. The ratio of 0.56 appears sufficient to provide this curve whereas 0.25 is not.

Plain sided specimens (S.R. = 0) of varying thicknesses with constant W (Figure 7) show the expected dependence, i.e., the greater the amount of shear to flat rupture, the steeper the resistance curve.

Shear lip size was found to vary with specimen thickness and width in a rather complicated manner which is reflected in the resistance curves. Altering W with B constant gives very different J_r curves (Figure 8) attributable entirely to the variation of shear lip size in the two cases. Maintaining the W/B ratio constant (Figure 9) produces identical results.

CONCLUSIONS

By use of the three parameter technique, J_r curves of form similar to K_r curves are found. J_r curves are greatly affected by the presence of shear lips. Shear lip size is geometry dependent. This is reflected in the resistance curves of non side grooved specimens. It is not sufficient merely to eliminate the shear lips to produce the lowest bound curve. It would appear that the side groove ratio is itself a governing factor of the J_r curve obtained but by choice of a side groove ratio S.R. = 0.56 a "plane strain" curve is produced that is independent of initial crack length in three point-bending over the range of sizes tested.

REFERENCES

1. ASTM, STP 527, 1973.
2. RICE, J. R., J. Appl. Mech., 35, 1968, 379.
3. RICE, J. R., "Fracture", edited by H. Leibowitz, Vol. 2, 1968, 191.
4. BEGLEY, J. A. and LANDES, J. D., ASTM, STP 514, 1972, 7.
5. BUCCI, R. J., PARIS, P. C., LANDES, J. D. and RICE, J. R., ASTM, STP 514, 1972, 40.
6. GARWOOD, S. J., ROBINSON, J. N. and TURNER, C. E., Int. J. Fracture, 11, 1975, 528.
7. SUMPTER, J. D. G. and TURNER, C. E., ASTM, STP 601, 1976, 3.
8. RICE, J. R., PARIS, P. C. and MERKLE, J. G., ASTM, STP 536, 1973, 231.
9. LANDES, J. D. and BEGLEY, J. A., ASTM, STP 560, 1974, 115.

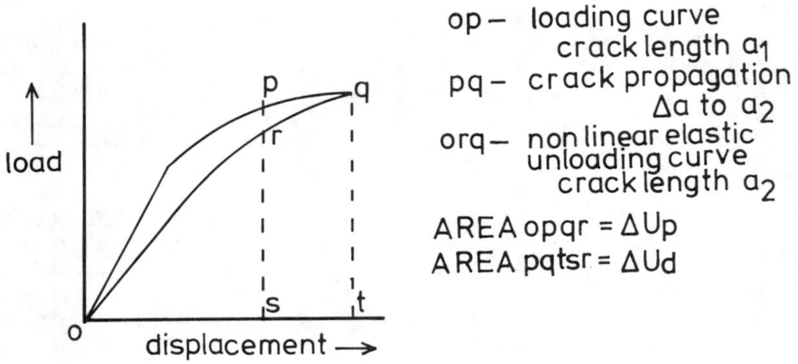

Figure 1

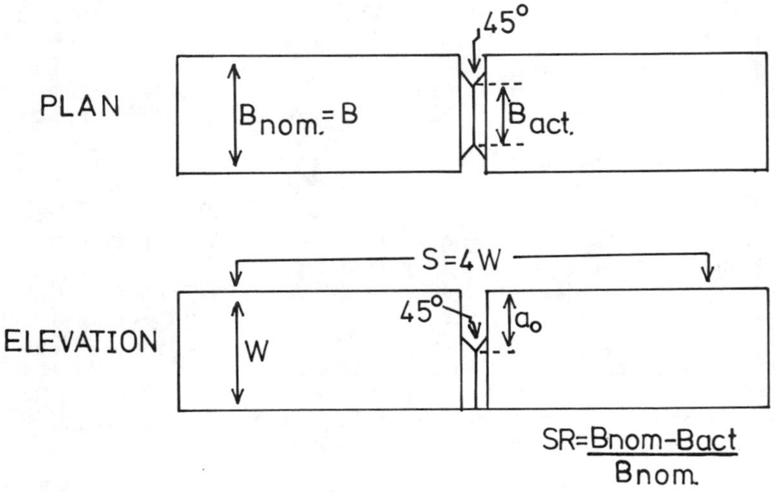

Figure 2

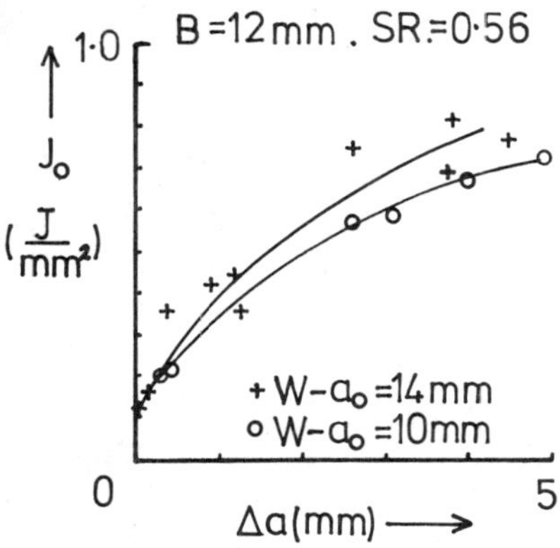

Figure 3

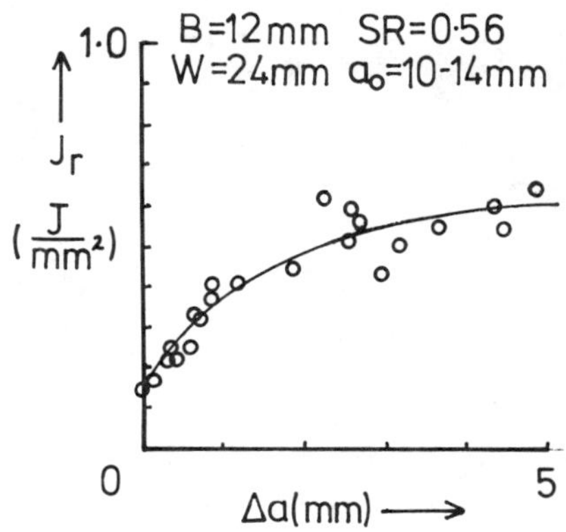

Figure 4

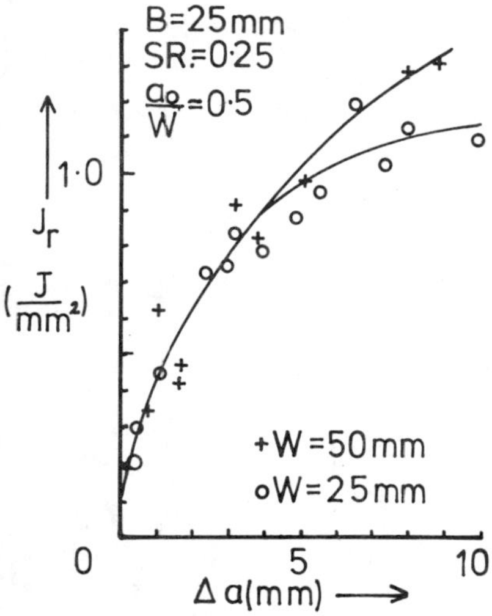

Figure 5

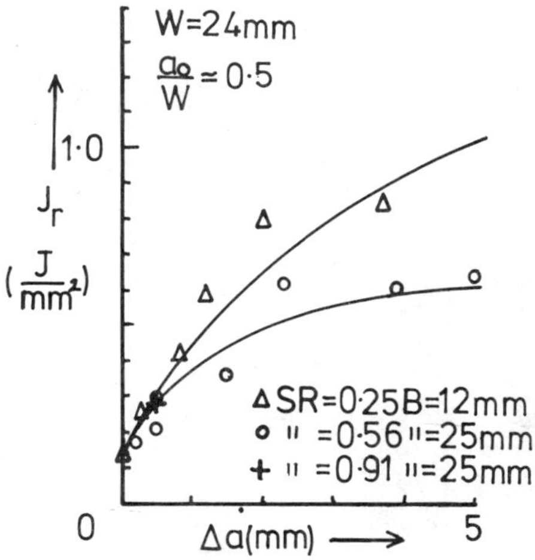

Figure 6

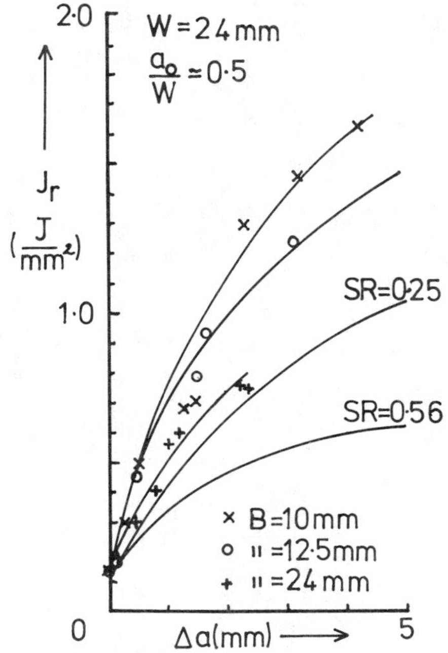

Figure 7

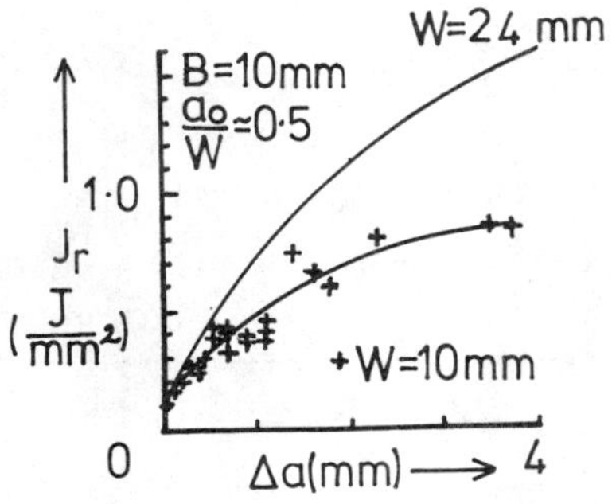

Figure 8

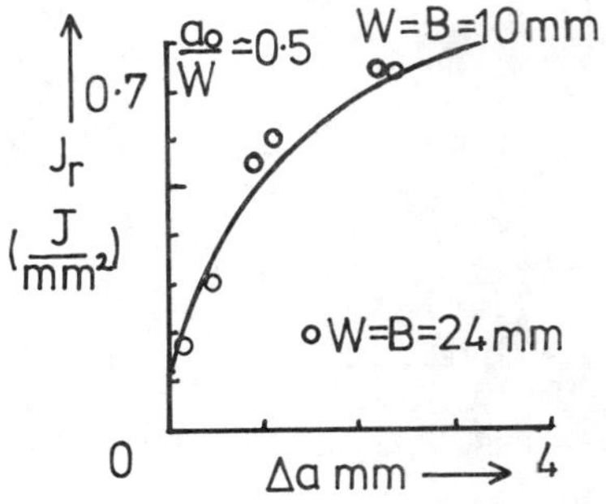

Figure 9

Table 1

MATERIAL	CONDITION	YIELD STRESS MN/m^2
En 32 steel (C-0·17% Si-0·23% Mn-0·74%)	Normalised (900°C, 1hr, air cooled)	290

Fracture 1977, Volume 3, ICF4, Waterloo, Canada, June 19 - 24, 1977

THE EFFECT OF GEOMETRY ON CRACK FORMATION

Terance V. Duggan[1] and Peter Sabin[2]

INTRODUCTION

In many components subjected to mechanical or thermal cycling it is frequently the case that, at a critical region, the maximum localized stress exceeds the yield strength. If continued cycling does not produce any further macroplasticity, the material will thereafter behave elastically although yielding on the first half cycle would establish a mean stress. In assessing the fatigue integrity of components, the local material behaviour must be understood, both from the point of view of the number of cycles required to initiate a crack (the crack formation life), and the subsequent crack growth whilst the crack is still within the influence of the geometry of a stress concentrator. A fundamental approach necessitates the ability to determine precisely the division between nucleation and Stages I and II crack propagation [1]. Whilst for a limited number of situations it might be possible from a diagnostic viewpoint to use such an approach, the prognostic situation is much more difficult, and the use of electron microscopy as a diagnostic tool is of only limited value to the designer. Consequently, efforts have been made to develop methods which enable crack formation in components to be predicted from data obtained from simple test pieces. An engineering crack is defined as one which can be detected using low power magnification (say X 25), and for a typical surface crack will be in the region of about 0.5mm long and 0.15mm deep.

FATIGUE LIFE OF PLAIN SPECIMENS

In order to estimate the crack formation life of a component it is necessary to relate the conditions at the critical region to known material behaviour and allow for influencing factors. If it is assumed that the conditions in a component can be represented by tests on plain specimens of the type used in obtaining fatigue data, this may provide the basis for assessing crack formation life.

There is considerable experimental support [2] to suggest that both the elastic and plastic strain range components, when plotted against cycles to failure, give approximately straight lines on logarithmic co-ordinates. Expressed mathematically,

$$\Delta\varepsilon_T = \Delta\varepsilon_p + \Delta\varepsilon_e = C_p N_f^{\alpha_1} + C_e N_f^{\alpha_2} \qquad (1)$$

where α_1 and α_2 represent the slopes of the plastic and elastic lines respectively on logarithmic co-ordinates, and C_p and C_e represent the strain range corresponding to the plastic and elastic intercept for one cycle. Equation (1) is based upon completely reversed strain cycling with zero mean stress. To allow for the effect of a mean strain the Sach's modification [3, 4, 5] is introduced into the plastic strain component, i.e.,

---

1. Reader in Mechanical Engineering, Portsmouth Polytechnic, England.
2. Research Engineer, Rolls-Royce (1971), Limited, Derby, England.

$$\Delta\varepsilon_p = (\varepsilon_f' - \varepsilon_m)N_f^{\alpha_1} \tag{2}$$

where ε_f' (= C_p) is defined as the fatigue ductility coefficient. If the elastic strain range corresponding to zero mean stress is $\Delta\varepsilon_{eo}$, then

$$\Delta\varepsilon_{eo} = C_e N_f^{\alpha_2} \tag{3}$$

To allow for the effect of a mean stress the Goodman relationship is used, i.e., in terms of strain range

$$\Delta\varepsilon_e = \Delta\varepsilon_{eo}\left(1 - \frac{\sigma_m}{\sigma_u}\right) \tag{4}$$

where σ_m is the mean stress and σ_u is the ultimate tensile strength. Combining equations (3) and (4) and substituting into equation (1),

$$\Delta\varepsilon_T = (\varepsilon_f' - \varepsilon_m)N_f^{\alpha_1} + C_e(1 - \sigma_m/\sigma_u)N_f^{\alpha_2} \tag{5}$$

If an endurance limit exists for the material, say $\Delta\varepsilon_{Lo}$ at N_e cycles, then

$$C_e = \Delta\varepsilon_{Lo} N_e^{-\alpha_2} \tag{6}$$

where $\Delta\varepsilon_{Lo}$ is the strain range equivalent of the endurance limit, for zero mean stress, i.e.,

$$\Delta\varepsilon_{Lo} = \frac{2\sigma_e}{E} \tag{7}$$

Now it will be observed that, using an iterative procedure, equation (5) enables an estimate of crack formation life in a component to be predicted, if the conditions at the crack region can be determined.

STRAIN DISTRIBUTION IN A COMPONENT

In most practical designs the local plastic strains will usually be sufficiently contained to limit the plastic zone to only a small region. The local behaviour will be dependent upon a number of factors, such as the relative magnitude of the plastically to elastically strained material, the strain distribution, the materials cyclic strain hardening or softening characteristics, and the effect of environment. Further even though the component may be subjected to constant amplitude cyclic loading, the materi in the vicinity of the concentration feature and which is locally plastic, will experience a variation of strain range with cycles [6]. It is the behaviour of this local material which governs crack formation.

Excluding finite element methods, two possible avenues are currently available for evaluating the strain at a concentration feature, namely the Neuber [7] and the Hardrath-Ohman or modified Stowell [8] methods. The total strain at a concentration feature can be estimated using an analysis suggested by Zwicky [9]. Based on the modified Stowell method:

$$\Delta\varepsilon_T = \frac{m\,\Delta\sigma_o(K_t - 1)\Delta\sigma}{E(\Delta\sigma - m\Delta\sigma_o)} \tag{8}$$

where K_t is the theoretical stress concentration factor; $\Delta\sigma_o$ is the nominal stress range; $\Delta\sigma$ is the local stress range and m is a factor which relates the effective stress to the principal stress. Using the Neuber method:

$$\Delta\varepsilon_T = \frac{m\ g\ K_t^{\ 2}\ \Delta\sigma_o^{\ 2}}{\Delta\sigma} \qquad (9)$$

where g is a factor which relates the effective strain to the principal strain.

For plane stress,

$$m = 1, \qquad g = 1.$$

For plane strain,

$$m = (1 - \nu' + \nu'^2)^{1/2} \qquad (10)$$

$$g = \frac{m}{1 - \nu'^2} \qquad (11)$$

where ν' is defined as the pseudo-Poisson's ratio, obtained from

$$\nu' = 0.5 - (0.5 - \nu')(E_s/E) \qquad (12)$$

and E_s is the secant modulus defining the local material behaviour.

Accepting the assumptions inherent in the derivation of equations (8) and (9) either the modified Stowell or the Neuber rule may be used to estimate the total strain range at a critical section in a component.

ESTIMATING CRACK FORMATION

If the behaviour of the material at a critical region in a component can be predicted from smooth specimen behaviour, then we have a ready method for assessing crack formation life. This consists essentially of using the known material cyclic behaviour, as expressed by equation (5) with the total strain range. The total strain range can be determined using an iterative procedure with either the modified Stowell or Neuber method, applied first to the loading half-cycle and then to the unloading half cycle. It would seem reasonable to assume, for the first half cycle, that the material will follow the monotonic stress-stress curve. On unloading various possibilities are likely, depending upon the magnitude of strain attained on the loading half cycle and the overall stress ratio. Thus the local material behaviour may be such that either (i) yielding in compression occurs; (ii) no yielding in compression but residual compressive stress is obtained; or (iii) residual tensile stress is achieved due to high normal stress ratio. In any real component the strain distribution will be such that strain gradients exist and the volume of plastically strained material and the strain exponent may be significantly different from that of the plain test piece where the strain gradient will usually be essentially zero. A recent study by Leiss *et al* [10], suggests that the theoretical stress concentration factor K_t, should be replaced by an experimentally determined value for K_f (termed the "fatigue notch factor"). Investigations conducted by the present authors suggest that if K_t is replaced by K_f in the relationship for estimating strain range, then closer predictions to crack formation lives are obtained. However, the values for K_f were obtained using the usual definition involving notch sensitivity index q, i.e.,

$$K_f = 1 + q\,(K_t - 1) \tag{13}$$

has been used to replace K_t in the method for calculating total strain range, and q was expressed in terms of the Neuber material constant [6]. Once this local material behaviour has been established, the mean strain may be determined from

$$\varepsilon_m = \frac{\varepsilon_{max}}{2}\,(1 + r_\varepsilon) \tag{14}$$

where r_ε is the local strain ratio. To extend the crack formation model to include the effect of bulk stress ratios, the nominal stress range in equations (8) and (9) has been modified by including the stress ratio for the unloading half cycle. Thus if σ_o is the nominal stress amplitude attained on the loading half cycle, then for the unloading half cycle

$$\Delta\sigma_o = \sigma_o\,(1 - R) \tag{15}$$

where R is the bulk stress ratio. This allowance for bulk stress ratio would not be expected to be applicable if time dependent influences, such as creep or stress relaxation are involved, without some modification.

CORRELATION OF EXPERIMENTAL DATA WITH MODEL

To investigate the validity of the proposed model, experimental data for fatigue crack formation has been obtained for a variety of materials, geometries, bulk stress ratios and elevated temperature. Laboratory fatigue tests were conducted on SEN bend specimens having different notch configurations and tension plates with central holes; the onset of fatigue crack propagation was located using the electric potential method [2]. Aero-engine model discs were tested by cyclic spinning, and the cracks detected by NDT methods. Values for K_t varied from less than 2 up to about 14. Table 1 summarizes the material properties, and for identification purposes the materials are designated, A, B and C. Predictions for crack formation life have been made using the model representated by equation (5). The total strain range $\Delta\varepsilon_T$ has been calculated using both the modified Stowell and the Neuber methods. For each situation it is necessary to know the service conditions (K_t and $\Delta\sigma_o$) and the appropriate material properties obtained from monotonic tests and fully reversed strain cycling fatigue tests on plain specimens (σ_u, E, ν, ε_f', α_1 and α_2). The material behaviour may also be predicted from a deformation model as suggested by Proctor and Ayers [11]. The iterative procedure required to solve equations (5) (8) and (9) make a computer solution desirable, and a suitable programme has been developed to facilitate the analysis. Strain measurements made using miniature electrical resistance strain gauges indicated good correlation with the predicted strains, except that for large plastic strains, (in excess of 1%) the assumption that the material behaviour during the unloading half cycle is similar to the loading half cycle was found to lead to an overestimate of the residual strain. The accuracy of the Neuber and modified Stowell methods is found to depend upon the shape of the cyclic stress strain curve and the degree of plastic strain. Strain conditions are dependent upon geometry and specimen or component dimensions, and strictly speaking a three dimensional finite element analysis is required to establish the precise conditions. If complete restraint exists at the notch plane strain will be obtained, and this condition is approached if the ratio of notch radius (ρ) to specimen on component thickness (B) is small. As (ρ/B) increases, so the conditions approached are those of plane stress. Figure 1 indicates a restraint factor (F), allowing for notch configuration, to be applied to the total strain range calculated on the basis of a plane stress analysis. Figure 2 shows the correlation between experimental and predicted results for crack formation lives obtained on the above basis, using the modified Stowell method of calculating local material behaviour. For the majority of results the scatter is within the tolerance band of ± 25%, thus suggesting that the predictive method presented has a most acceptable accuracy. Both Neuber and the modified Stowell methods were found to overestimate actual strain range but, Stowell method was generally more accurate. Further studies are continuing to extend the work to include creep-fatigue interactions.

ACKNOWLEDGEMENTS

This research has been undertaken in the Department of Mechanical Engineering, Portsmouth Polytechnic, in collaboration with Rolls-Royce (1971), Limited, Derby Engine Division, under Contract No. URC 104.

REFERENCES

1. DUGGAN, T. V., Chart. Mech. Engr., 17, 1970, 443.
2. KNOTT, J. F., "Fundamentals of Fracture Mechanics", Butterworths, London, 1973.
3. MANSON, S. S., Expt. Mechanics, 5, 1965, 193.
4. SESSLER, J. G. and WEISS, V., J. Basic Engrg., 85, 1953, 539.
5. SACHS, G., GERBERICH, W. W., WEISS, V. and LATTORRE, J. V., Proc. A.S.T.M., 60, 1960, 512.
6. DUGGAN, T. V., Ph. D. Thesis, Portsmouth, 1973.
7. NEUBER, H., J. Appl. Mech., 28, 1961, 544.
8. HARDRATH, H. F. and OHMAN, L., NACA Tech. Note 117, 1953.
9. ZWICKY, E. E., A.S.T.M. 67-WA-PVP6, 1967.
10. LEISS, B. N., GOWDA, C. V. B. and TOPPER, T. H., STP 519, 1973, 133.
11. PROCTOR, M. W. and AYRES, K. B., Second International Conference on Mechanical Behaviour of Materials, Boston, Massachusetts, 1976.

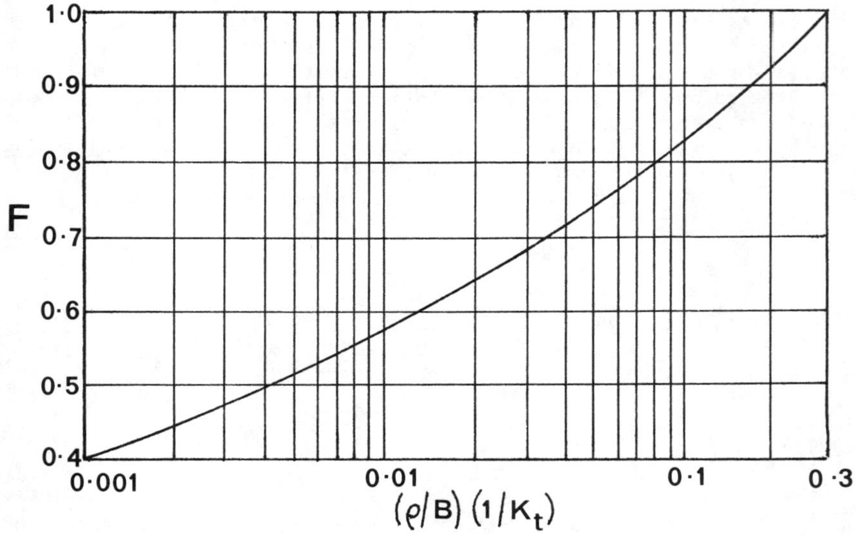

Figure 1 Variation of Material and Restraint Factor (F) with Geometric Parameter (ρ/b) $(1/K_t)$

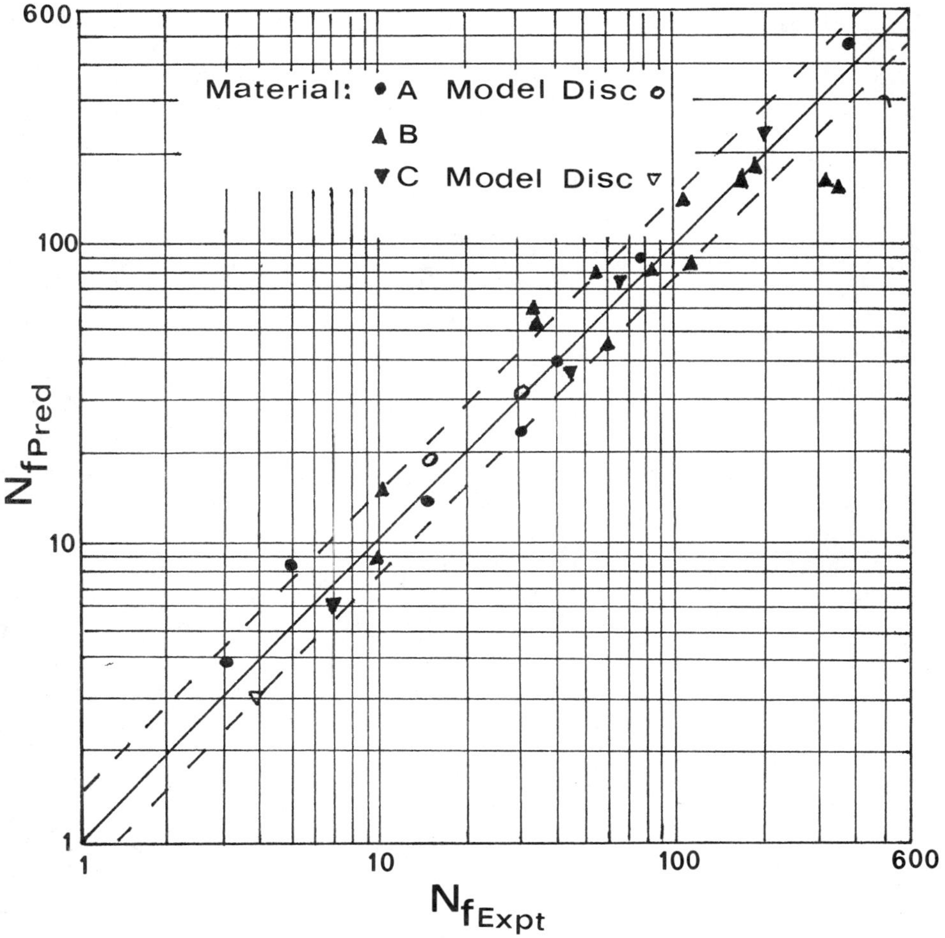

Figure 2 Correlation Between Predicted and Experimental Crack Formation Lives

Table 1 Summary of Mechanical Properties of Materials Studied

Material	A		B		C	
Temp °C	RT	500	RT	450	RT	500
σ_u MN/m^2	1095	817	900	693	1175	1027
σ_y MN/m^2	919	645	768	535	860	769
σ_e MN/m^2	600	-	464	-	310	277
E GN/m^2	214	173.9	205.4	182.0	179.6	166.2
α_1	-0.616	-0.617	-0.610	-0.596	-0.461	-0.439
ε_f'	1.046	1.138	1.127	1.076	0.214	0.194
α_2	-0.134	-0.138	-0.138	-0.137	-0.092	-0.092
C_e	0.0214	0.0207	0.0194	0.0167	0.0169	0.016

Fracture 1977, Volume 3, ICF4, Waterloo, Canada, June 19 - 24, 1977

THE K-COD RELATIONSHIP FOR PIN LOADED SINGLE EDGE NOTCHED TENSION SPECIMENS

S. A. Paranjpe* and S. Banerjee*

INTRODUCTION

The validity and usefulness of any fracture mechanics parameter as a suitable fracture criterion depends on the ease with which it can be calculated (like K or J), the ease and reproducibility of its measurement and its compatibility with K in the linear elastic range. The parameter can be the basis of a valid fracture criterion provided it is independent of specimen geometry and configuration. In this paper the well known K-δ (COD) relationships are examined for different widths, W and aspect ratios, a/W.

Well's analysis [1] and Burdekin et al [2] results based on Dugdales model indicate that K^2-δ relationship is linear.

$$\delta = \frac{K^2}{E\sigma_y} \qquad (1)$$

However, Begley and Lande's [3] results and those of Anderson's [4] results indicate a parabolic relationship between K^2/E (i.e. J) and δ in the linear elastic range

$$\delta \propto \sqrt{J} \propto K \qquad (2)$$

Apparently equations (1) and (2) do not agree in the linear range. Finite element observations [5] indicate a linear P (load)-δ relationship in the linear elastic range. Thus

$$P \propto \delta \qquad (3)$$

and $$K = \frac{Y P a^{1/2}}{B W} \qquad (4)$$

where $Y = f(a/W)$, B = thickness and a = crack length.

or $$P = \frac{B W K}{Y a^{1/2}} = \frac{B K \sqrt{W}}{Y (a/W)^{1/2}} \qquad (5)$$

Substituting equation (3) into (5) gives

$$\delta \propto \frac{B K \sqrt{W}}{Y (a/W)^{1/2}} \qquad (6)$$

Equation (6) indicates a linear K-δ relationship which is similar to that in equation (2). This equation implies that for a given K and a/W ratio,

* Department of Metallurgical Engineering, Indian Institute of Technology, Powai, Bombay 400 076, India

higher widths will give higher displacement. Similarly at a given W and K higher a/W ratio should give a lower displacement. However, Well's Dugdale type K-δ relationship does not include the effect of specimen geometry and configuragion where δ is only function of K, σ_y and E. Equation (1) has been used to support COD as a fracture criterion [6] indicating that as $K \to K_{IC}$, $\delta \to \delta_c$ and thus δ_c is a fracture characterizing parameter (for a given thickness of plate) which is independent of the width of the plate [7].

In steels, it is possible that low triaxiality induces fibrous fracture (a tough fracture and consequently a higher δ_c) while high triaxiality induces cleavage. Triaxiality is a function of specimen dimensions and therefore δ_c measurements on small specimens may not correspond to the δ_c at which crack initiates in a large structure (unless the constraints in the specimen and the structure are identical).

It is expected that the state of stress at the crack tip will depend on the extent of deformation at the tip and its proximity to unnotched free edge. Thus the state of stress at the crack tip will continuously change from plane strain to plane stress as loading progresses and the crack tip deformation increases. Though this phenomenon is appreciated, the continuous change of state of stress (which can be represented by the value of constraint) is not considered in any reported calculations. Instead, it is a common practice to assume a constant state of stress throughout the loading history. It has been suggested by Hayes and Turner [5] and Egan [6] that δ in a given state of stress can be obtained using

$$\delta = \frac{K^2}{mE\sigma_y} \tag{7}$$

where m is a measure of constraint at the crack tip [8]. It is further suggested [5, 6, 9] that a value of m = 2 represents plane strain situation and is equal to 1 in case of plane stress. In this paper the results of a simple analysis developed which takes into account the continuous change of the value of m is reported. Using this analysis COD values are computed and the various K-δ relationships are examined.

THEORETICAL CALCULATION OF COD

Dixon [10] has shown that a pin loaded SENT specimen can be represented by an axial force applied at the midpoint of a ligament and a bending moment. Richard and Ewing [11] using a similar representation have calculated yield point loads of SENT specimen, while Merkle et al [12] have used it for compact tension (CT) specimens. Dixon's work is limited to the elastic solution while the latter works do not refer to crack tip behaviour and strain hardening characteristics of the material. They assume a linear stress or strain distribution over the ligament. Liu et al [13, 14] have shown that the strain distribution ahead of the crack tip is of $1/\sqrt{r}$ type even in presence of considerable yielding. Use is made of this fact and a composite distribution comprising of $1/\sqrt{r}$ near the crack tip and linear strain distribution far away from the crack tip is assumed. For a smooth and continuous change over from the $1/\sqrt{r}$ to a -r type strain distribution, the magnitudes and slopes of the two strain distributions are matched at the change over point. The material is assumed to exhibit a linear strain hardening response. Figure 1 shows the general nature of strain distribution in the uncracked ligament with the various parameters used in the calculations. Applied axial load and moment (generated because of unsymmetric loading) was balanced with the reactive axial load and moment (generated because of the assumed strain distribution). The load and moment balance equations were simultaneously solved (using Newton-Raphsons iterative procedure) for various loads and specimen sizes. The output of this solution is R, the apparent plastic zone size ahead of the crack tip (which is defined as a point at which strain ε_{yy} is equal to ε_y), and X_2, the point at which strain ε_{yy} is zero, the rotation axis position.

The apparent plastic zone size was represented in the Irwin-McClintock [15] type of representation, i.e.,

$$R = \frac{1}{\pi\, m'}\left[\frac{K}{\sigma_y}\right]^2 \qquad (8)$$

m' calculated from equation (8) is a measure of the constraint. It indicates the average increase of local σ_{yy} stress at which yield occurs.

Three different δ values are calculated the procedures for which are indicated below.

(a) Compute δ using equation (1). This is termed as δ_D.
(b) Compute δ from V, the crack mouth opening displacement, using Boyles et al [9] δ-V relationship for SENT specimen. This is termed as δ_W.
(c) Compute δ at crack tip when V is joined linearly to the point of strain reversal (Figure 1). This is termed as δ.

V used in the calculation of δ_W **is obtained according to the following steps. On the other hand, V used for calculation of δ is obtained as reported in [3].**

(1) Compute K using equation (4), for a given specimen W, a/W and P [16].
(2) Compute R and r_y through load and moment balance equations.
(3) Compute $a_{eff}/W = (a+r_y)/W$.
(4) Compute the value of EVB/P for the a_{eff}/W [16].
(5) Compute V using the results of previous step and the values of E, B and P.

RESULTS AND DISCUSSION

Figure 2 shows the plot of m' versus K. The figure has two important features. Firstly, m' drops as K increases (i.e., loading and hence deformation at the crack tip progresses). Secondly, as the specimen width increases the m' decreases at a slower rate. This means that wider specimens maintain a higher constraint value than a narrower specimen of same a/W ratio at a given K possibly because of wider ligament. The same trends are exhibited by m'' which is obtained by dividing δ_D by δ_W. Even the two constraint values m' and m'' which are obtained independently compare quite well in magnitude.

Figure 3 shows K-δ_W plots for various widths. The figure shows an interesting trend that as W decreases δ_W value increases for a given K. This trend does not agree with equation (6), because the constraint decreases more rapidly in specimens with lower widths (Figure 2). The constraint dependence of δ_W is further evident from Table 1, in which δ_W was calculated for a constant constraint m' = 2 in equation (8). It is observed that for all widths studied, the δ is almost constant for a given K value if m' is constant.

The constraint dependence of δ_W probably originates from the definition of δ_W. δ_W has been defined as the resultant displacement at the crack tip when crack profiles are extended into the ligament. In this definition we approach from the crack mouth side and the overall stress-strain distribution in the ligament is not taken into account. Since the stress strain distribution in the ligament is ignored and the constraint value depends on these distributions, δ_W becomes a function of the constraint or specimen width.

Secondly, it is assumed in the definition that the crack faces open by a simple hinge mechanism about an apparent axis of rotation. The position of this rotation axis is assumed to be the intersection of the extrapolated crack profile with X axis. However, the "neutral axis" position (represented here as the *point of strain reversal*) determined in the present investigation for SENT and CT specimens, as well as Merkle's analysis for CT specimens [12] are quite different from the rotation axis positions as suggested by Wells and others [17, 18]. According to the theory of bending it seems unlikely that the specimen will rotate at the apparent rotation axis position where a finite positive (tensile) strain is present. The most likely position of rotation axis is expected to be the strain reversal point.

Moreover, the different formulae for V-δ conversion (based on crack profile extension technique) given in DD 19 [19] have been analytically and experimentally verified only for SEN bend and CT specimens and no comment is made upon its usefulness to SEN, centre notched and double edge notched tension specimens. In fact it has been pointed out [5, 9] that the crack profile extrapolation technique is not suitable for SENT specimens.

Taking all these observations into account it was decided to define the crack tip opening displacement by an alternative method.

AN ALTERNATIVE DEFINITION OF COD

The COD is defined as the resultant displacement at the crack tip when the crack mouth opening displacement is joined to the neutral axis. The results of the present investigation with this definition of δ are given below.

The $\delta E/W\sigma_y$ versus $VE/W\sigma_y$ relationship for various a/W ratios follow a trend similar to that indicated in experimental calibration given in CODA [19] and other equations [20]. However the $\delta E/W\sigma_y$ - $VE/W\sigma_y$ plot in the present study is a linear relationship while the finite element calculations report a nonlinear relationship at lower loads. Secondly for a given value of V, δ obtained here is more than the FEM δ [9] reported for SENT specimens. This is expected since the definition of the two δs are different. Secondly, the FEM results reported are valid for a constraint value of m = 2 where as m decreases continuously in the present results.

Figure 4 shows the K-δ relationship obtained in the present investigation. It is observed that K-δ obeys a linear relationship and is a strong function of specimen width and a/W ratio. Higher widths give higher δ values for a given K and a/W ratio. Similarly lower a/W ratios give higher δ values for a given K and W. These results naturally do not agree with the trends exhibited by δ_W as shown in Figure 3 but they are in agreement with the observations made in equation (6). Similar results are obtained for the CT specimens.

If the present definition of COD is adequate then according to equation (6) the plot of K and $\delta\sqrt{a}/W$ should yield a straight line independent of a/W and W. Figure 5 shows the plot of K versus $\delta\sqrt{a}/W$. It is observed that it does yield a unique straight line for all a/W ratios and widths studied. This implies that the present way of defining δ is in agreement with the proven relationships and observations of fracture mechanics. It must also be noted that this type of representation makes δ values independent of constraint.

SUMMARY

An alternative definition of COD is given based on strain reversal point as the rotation axis. The proposed parameter $\delta\sqrt{a}/W$ appears to be an unique function of K for the widths, a/W ratios and the constraints studied in linear elastic and small scale yielding situations for the SENT and CT specimens. However the applicability of this parameter needs to be examined for other specimen geometries where rotation axis may lie outside the specimen and also in the case of large scale yielding situations where the COD application is most appropriate.

REFERENCES

1. WELLS, A. A., Br. Weld. J., 10, 1963, 563.
2. BURDEKIN, F. A. and STONE, D. E. W., J. Strain. Anal., 1, 1966, 145.
3. BEGLEY, J. A. and LANDES, J. D., Fracture Toughness, ASTM STP 514, 1972, 1.
4. ANDERSON, H., J. Mech. Phys. Solids, 20, 1972, 33.
5. HAYES, D. J. and TURNER, C. E., Int. J. Fracture, 10, 1974, 17.
6. EGAN, G. R., Eng. Fract. Mech., 5, 1973, 167.
7. KNOTT, J. F., Material Sci. Engng., 7, 1971, 1.
8. LEVY, N., MARCAL, P. V., OSTERGREN, W. J. and RICE, J. R., Int. J. Fracture Mech., 7 1971, 143.
9. BOYLE, E. F. and WELLS, A. A., "A Finite Element Study of Plane Strain Fracture Criteria Under Elastic-Plastic Conditions", Dept. of Civil Engineering, Queens University, Belfast, Ireland, June 1973.
10. DIXON, J. R., STRANNIGAN, J. S. and McGREGOR, J., J. Strain Anal., 4, 1969, 27.
11. EWING, D. J. F. and RICHARDS, C. E., J. Mech. Phys. Solids., 22, 1974, 27.
12. MERKLE, J. G. and CORTEN, H. T., J. Press. Vessel Tech., Paper No. 74-PVP-33, 1.
13. LIU, H. W., GAVIGAN, W. J. and KE, J. S., Int. J. Fracture Mech., 6, 1970, 41.
14. HU, W. L., KUO, A. S. and LIU, H. W., Tech. Rep., Syracuse Univ., August 1975.
15. IRWIN, G. R. and McCLINTOCK, F. A., "Fracture Toughness Testing and Its Applications", ASTM STP 381, 1965, 133.
16. BROWN, W. F. and SRAWLEY, J. E., ASTM STP 410, 1967.
17. VEERMAN, C. C. and MULLER, T., Engng. Fracture Mech., 4, 1972, 25.
18. INGHAM, T., EGAN, G. R., ELLIOTT, D. and HARRISON, T. C., "Practical Application of Fracture Mechanics to Pressure Vessel Technology", Inst. Mech. Engrgs., London, 1971, 200.
19. "Methods for COD Testing", DD19:1972, British Standard Institution.
20. ARCHER, G. L., E/63/75, Res. Rep., Welding Inst., United Kingdom, March 1975.

Table 1 δ Obtained at Constant Constraint m' = 2 using FEM V-δ Relationship for a/W = 0.5 [9]

Width mms	K $MPa \cdot m^{1/2}$	δ mms
15.0	24.5	0.00432
30.0	25.0	0.00410
60.0	25.05	0.00393
120.0	25.20	0.00389
240.0	24.90	0.00376

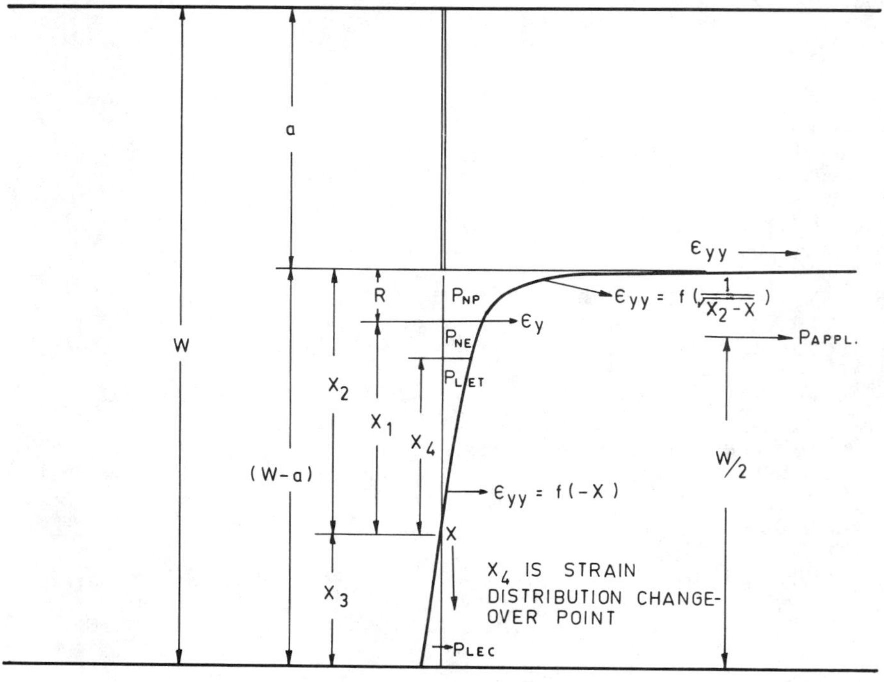

LOAD BALANCE EQUATION:

$P_{APPL} = P_{NP} + P_{NE} + P_{LET} - P_{LEC}$

MOMENT BALANCE EQUATION:

$P_{APPL}\,(A + X_2 - W/2)$ = SUM OF MOMENTS DUE TO P_{NP}, P_{NE}, P_{LET} AND P_{LEC} ABOUT STRAIN REVERSAL POINT

Figure 1 Strain Distribution in the Ligament of a SEN Tension Specimen

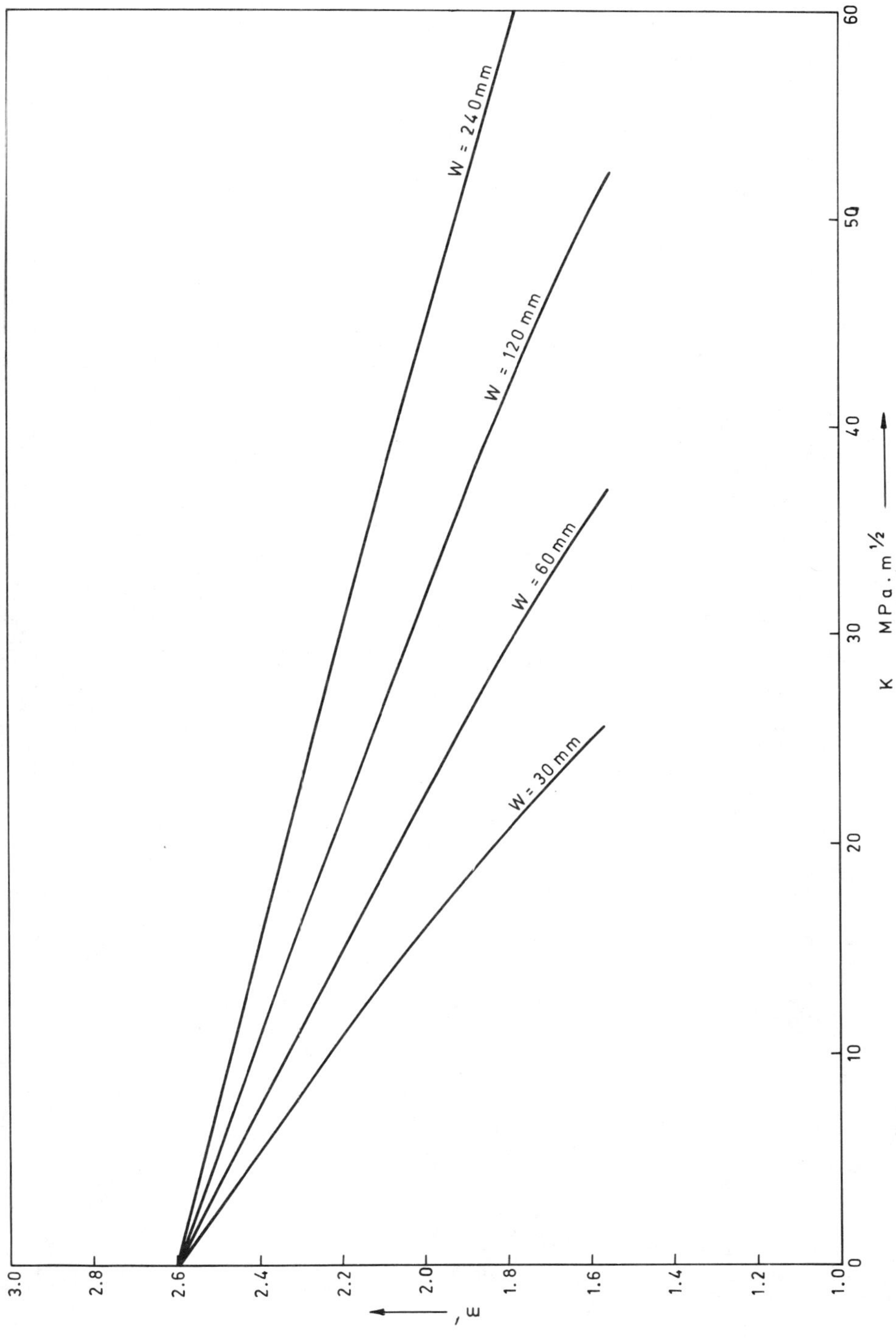

Figure 2 Constraint m' - K Plots for SENT Specimens a/W = 0.5

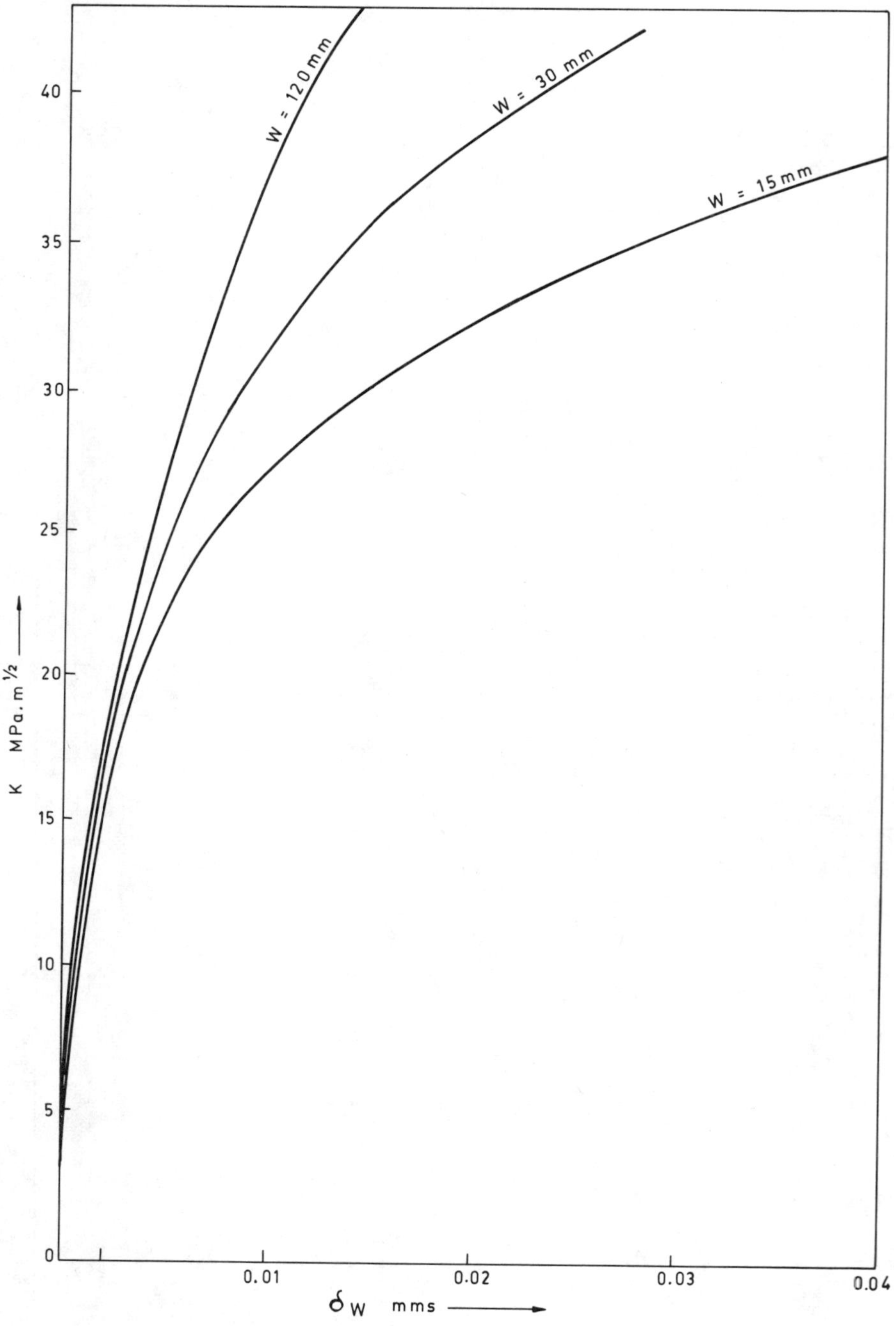

Figure 3 K versus δ_W Plots for SENT Specimens [9]a/W = 0.5

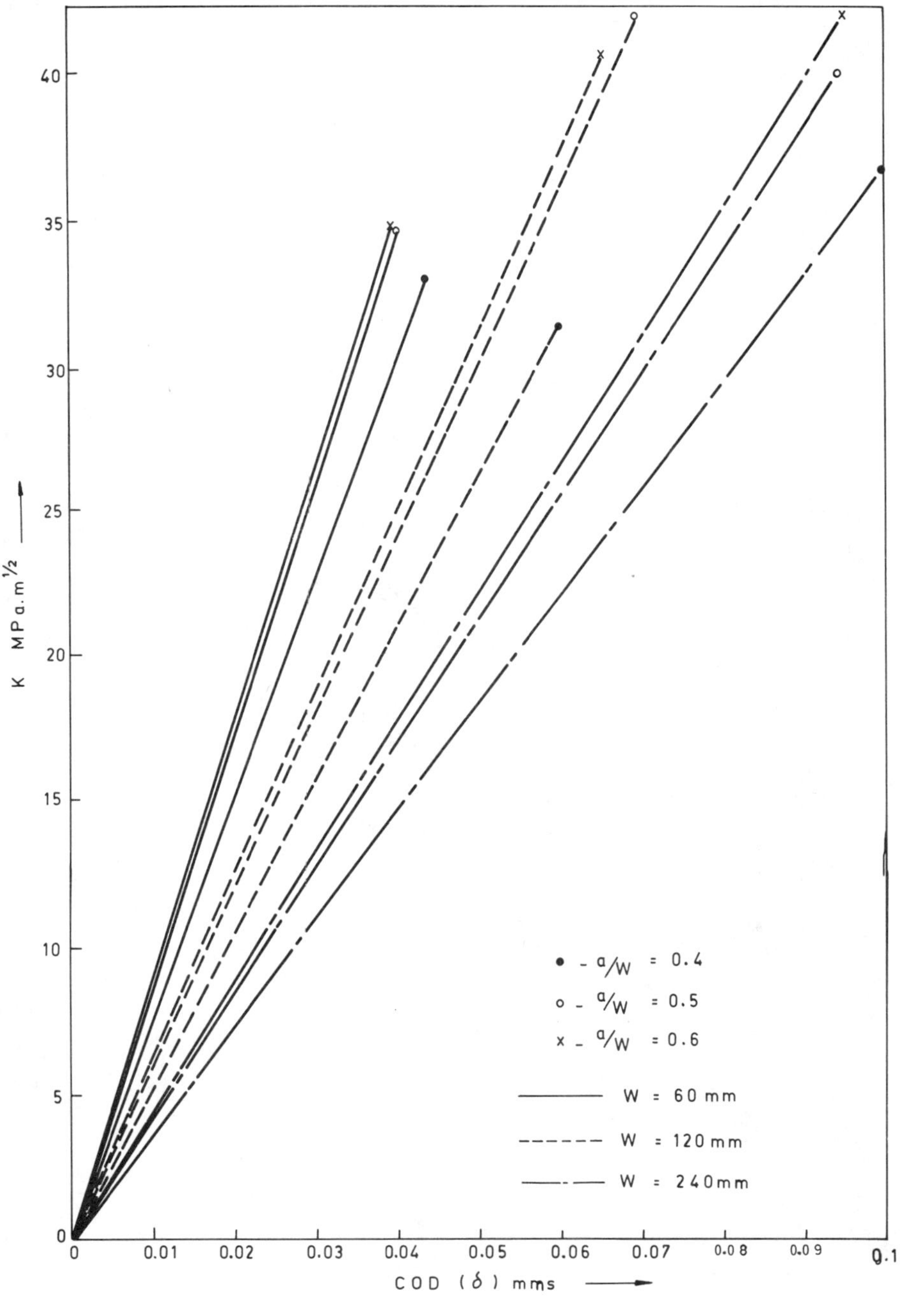

Figure 4 K versus δ Plot SEN Tension Specimen for W = 60,120,240 mm

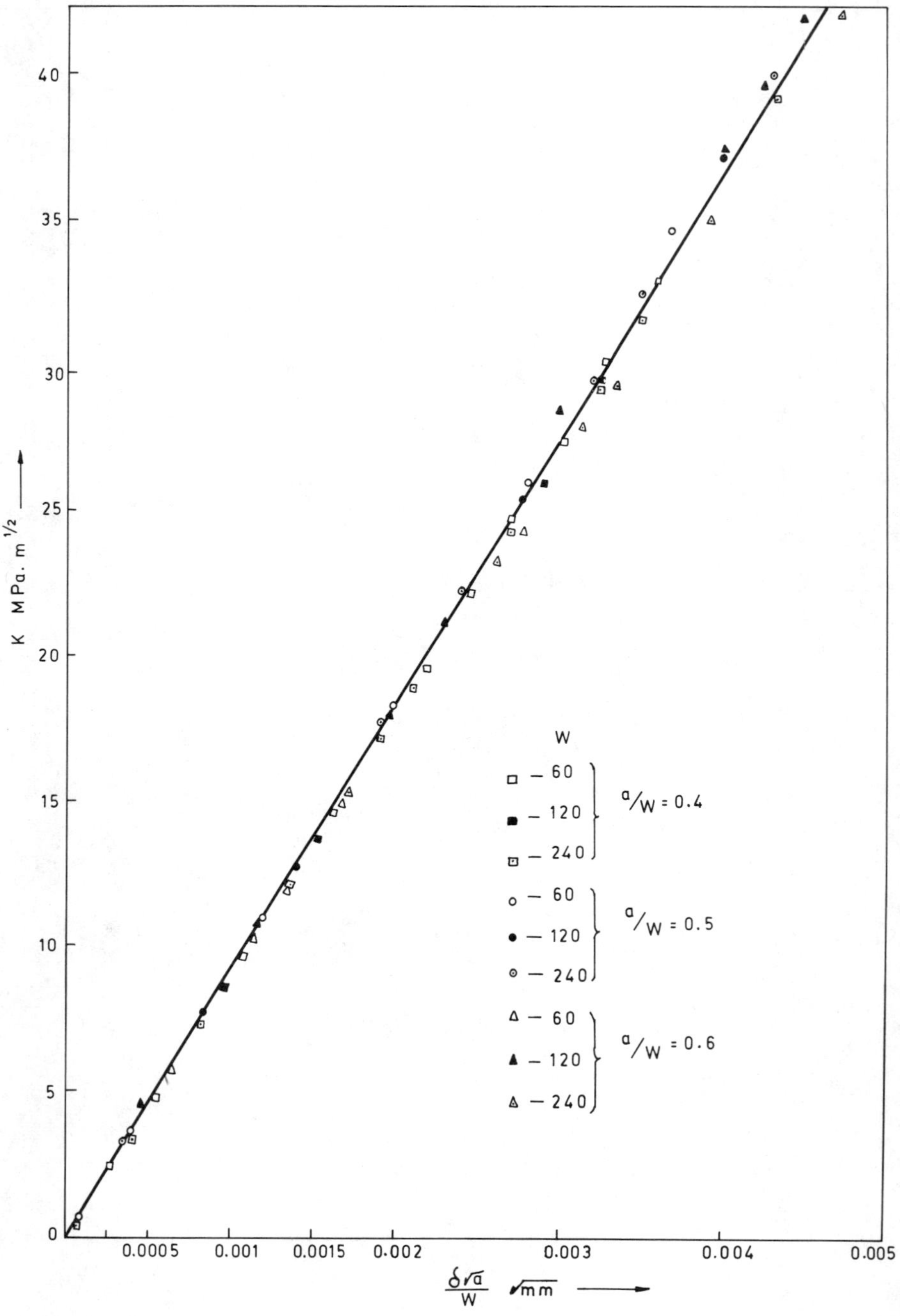

Figure 5 K - $\frac{\delta\sqrt{a}}{W}$ Plots for SENT Specimens with a/W = 0.4 , 0.5 and 0.6 and W = 60, 120 and 240 mm

Fracture 1977, Volume 3, ICF4, Waterloo, Canada, June 19 - 24, 1977

THE RELATION BETWEEN CRACK OPENING DISPLACEMENT AND FLOW STRESS

M. R. Piggott*

INTRODUCTION

The width of a crack in a material about to fracture, and in a material in which fracture is taking place is an important parameter in fracture analysis; in particular the change in width as the applied stress is raised from zero to the critical stress for fracture is related to the crack extension force (or fracture toughness) of the material [1]. Crack width has been estimated indirectly in the case of very brittle materials [2, 3] but the analysis of the results needs further clarification [4]. Crack opening displacement has been calculated using numerical techniques for an aluminum alloy [5], and has been measured for a number of steels [6]; however, care is required in the detailed interpretation of the results according to the position on the crack where the measurement is made [7].

It has been suggested that only limited flow at a crack tip (and hence limited widening of the crack) can take place without fracture being prevented altogether. Two approaches taking this into account have led to criteria to brittleness [8, 9] which depend in a simple way on the flow stress of the material. From these theories it follows that crack width should depend on flow stress, and indeed Wells obtained a relation between crack opening displacement and flow stress of the form [1]

$$\delta = \frac{\pi\sigma_c^2 a}{E\sigma_y} \tag{1}$$

for the C.O.D. occurring during the initial deformation of a material under an applied stress σ_c, with a flow stress σ_y, modulus E, and having a crack length 2a at its centre. Two materials were chosen to investigate this relation, a carbon steel which work hardness to a limited extent, and a brass which work hardness very considerably. Thin sheets of the material were used (i.e., plane stress conditions were maintained).

EXPERIMENTAL DETAILS

The steel used was a high carbon spring steel (SAE-1090). It was tested after various heat treatments to obtain different hardnesses. The brass was 70/30 alloy (SAE 70) and was work hardened and annealed to obtain the different hardnesses. The materials were in the form of strip 50 mm wide, and 0.38 mm thick in the case of the steel, and 0.50 mm thick in the case of the brass.

A crack 10 mm long and about 0.17 mm wide was formed in the strip by spark machining. In order to reduce the risk of micro cracks induced by the spark

* Department of Chemical Engineering, Centre for the Study of Materials, University of Toronto, Toronto, Ontario, Canada

machining, to retain a fine finish on the crack surfaces, and to minimize the thickness of the disturbed layer at the crack surface, a very low cutting rate was used. No microcracks were observed on the test specimens.

The specimens were tested on a Hounsfield Tensometer, the crack tip was observed using a microscope, and the crack size was recorded on film by a 35 mm camera attached to the microscope. C.O.D. was determined from prints of the photographs obtained, by measuring the distance between easily recognizable features on either side of the crack tip within 0.1 mm of the tip. The estimated error of the measurement was ± 1 micron (two prints of each photograph were made, and agreement between measurements on them was better than ± 1 micron). The applied stress was increased in steps of 53 $MPa \cdot m^{1/2}$ (100 kg) in the case of the steel, and 10 $MPa \cdot m^{1/2}$ (25 kg) in the case of the brass. The time at a given stress was kept as near to 30 s as possible, and the rate of applying stress was kept as constant as possible.

EXPERIMENTAL RESULTS

It was possible to obtain hardnesses ranging from about 150 to 250 VHN in samples of the steel by heat treatment. Stress-strain curves for the material are shown in Figure 1. C.O.D. is plotted as a function of stress for a number of hardnesses in Figure 2. The value of C.O.D. is given as a percentage of the undeformed crack length. There is a nearly linear region in the curves for all hardnesses; this ended at about 80% of the stress at which unstable propagation of the crack occurred. The results for hardnesses of 175, 200 and 250 VHN were not significantly different in the linear region.

In the case of the brass it was possible to vary the hardness over the range 40 to 100 VHN by rolling and annealing. Stress-strain curves for the material are shown in Figure 3. The observed C.O.D.s were much larger and more variable than in the case of the steel. Three specimens of each hardness were therefore tested and a typical set of results is shown in Figure 4. The mean results for each hardness were used for the plot shown in Figure 5. The curves relating C.O.D. to stress have two regions, an approximately linear one at low stress, and an approximately quadratic curve (C.O.D. $\propto$ $(\text{stress})^2$) at high stress.

DISCUSSION

C.O.D. in the steel and brass used in this investigation does not appear to depend on the square of stress at low stress as required by equation (1). (The effect of the finite specimen width [10] is neglected, since it is small (3%) compared with experimental error.) The relation appears to be much closer to a direct proportionality between C.O.D. and stress. This type of behaviour would be expected if the material were behaving elastically. However, the C.O.D. for the hardest brass and steel used is more than an order of magnitude greater than the elastic displacement at the region of the crack tip where the measurements were made. In addition, the dependence of C.O.D. on flow stress indicates that some plastic flow is taking place.

The linear dependence of δ on σ_c is in agreement with one of the theories for a brittleness criterion [9]. In the examples discussed in that paper, the crack opening displacement increases with stress, and is given by

$$\delta = \frac{\alpha\sigma_c a^2}{\beta a\sigma_y - \pi\sigma_c^2 a/E + 2\gamma} \tag{2}$$

where α and β are dimensionless constants, and γ is the surface energy of the material. The surface energy can be neglected compared with $\pi\sigma_c^2 a/E$ for moderate stresses, with crack lengths of 10 mm, so that curves exhibit the behaviour expected, i.e., a linear function of σ_c at low stresses, and a nonlinear deviation, above the line, at higher stresses.

At low stresses the expression reduced to

$$\delta = \frac{\alpha\sigma_c a}{\beta\sigma_y} \tag{3}$$

The experimental results only agree with this equation if $\alpha/\beta \propto (E/\sigma_y)^2$. Figure 6 shows the slopes, S, of the linear regions in Figure 2 and 4, plotted as a function of hardness. We assume hardness is proportional to flow stress, after Tabor [11]. The ordinate is $(S/E^2)^{1/3}$ and the abscissa is $(VHN)^{-1}$. The points for both steel and brass are reasonably close to a straight line going through the origin. Thus

$$\delta = c_1 a\sigma_c E^2/\sigma_y^3 \tag{4}$$

where c_1 is a dimensionless constant. Thus, comparing equations (3) and (4)

$$\alpha/\beta = c_1(E/\sigma_y)^2 \tag{5}$$

This suggests that the work in the plastic zone at the tip of a crack is a function of the yield strain. The theory attempted to separate the elastic and plastic components of the work at a crack tip. Equation (5) suggests that this may not be possible. However, the result is not at variance with the conclusion that metals obey a brittleness criterion of the form: flow stress/modulus > 7×10^{-3}.

CONCLUSION

Crack opening displacement does not appear to behave as indicated by simple theory. With relatively small plastic zones at the crack tip it obeys an expression of the form $\delta = c_1\sigma_c E^2/\sigma_y^3$ where c_1 has the same value for both brass and steel. At higher stresses, where the plastic zones is a significant fraction of the specimen width, the crack opening displacement increases more rapidly with applied stress.

ACKNOWLEDGEMENTS

Part of this work was done at Chalk River Nuclear Laboratories of Atomic Energy of Canada Limited, and the author is grateful to A. Hunton for help with the experimental work.

REFERENCES

1. WELLS, A. A., B.W.R.A. Report M13/63, 1963.
2. HULL, D., BEARDMORE, P. and VALINTINE, A., Phil. Mag., 12, 1965, 1021.
3. TETELMAN, A. and JOHNSON, T., Phil. Mag., 11, 1965, 389.
4. PIGGOTT, M., Phil. Mag., 15, 1967, 423.
5. WILLIAMS, M. and SWEDLOW, J., Trans. Met. Soc., AIME, 239, 1967, 162.
6. BURDEKIN, F. and STONE, D., J. Strain Analysis, 1, 1966,
7. SWEDLOW, J., Int. J. Fracture Mech., 3, 1967, 75.
8. KELLY, A., TYSON, W. and COTTRELL, A., Phil. Mag., 15, 1966, 567.
9. PIGGOTT, M., Int. J. Fracture, 11, 1975, 479.
10. PARIS, P. C. and SIH, G. C., ASTM STP 381, 1965, 41.
11. TABOR, D., "The Hardness of Metals, Oxford, 1951, 105.

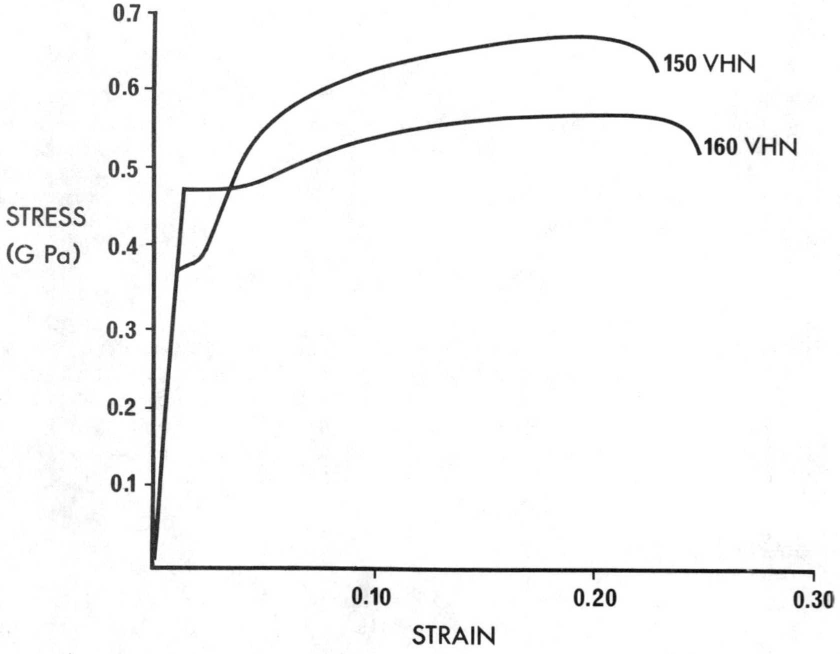

Figure 1 Stress-Strain Curves for Steel

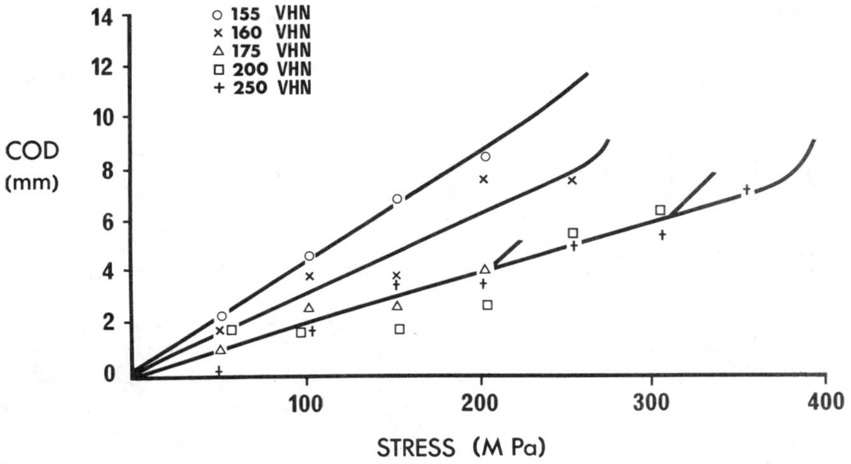

Figure 2 C.O.D. Measurements on Steel

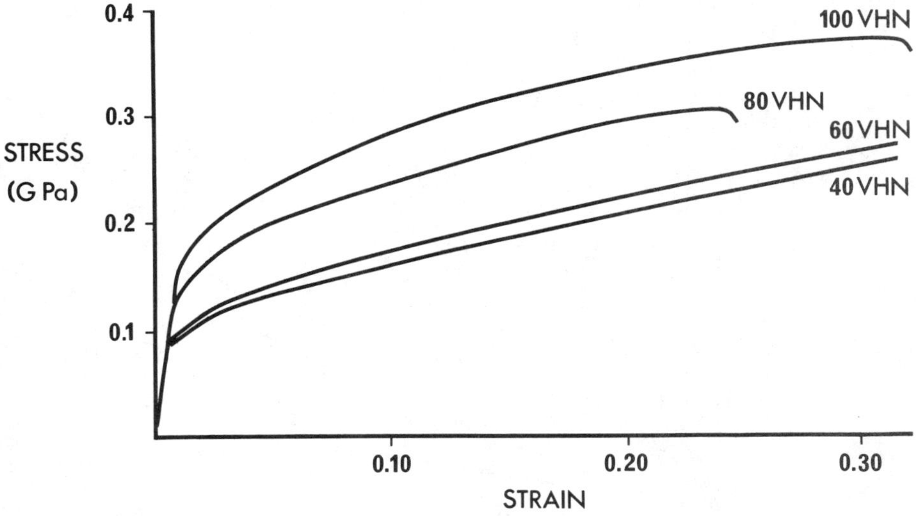

Figure 3 Stress-Strain Curves for Brass

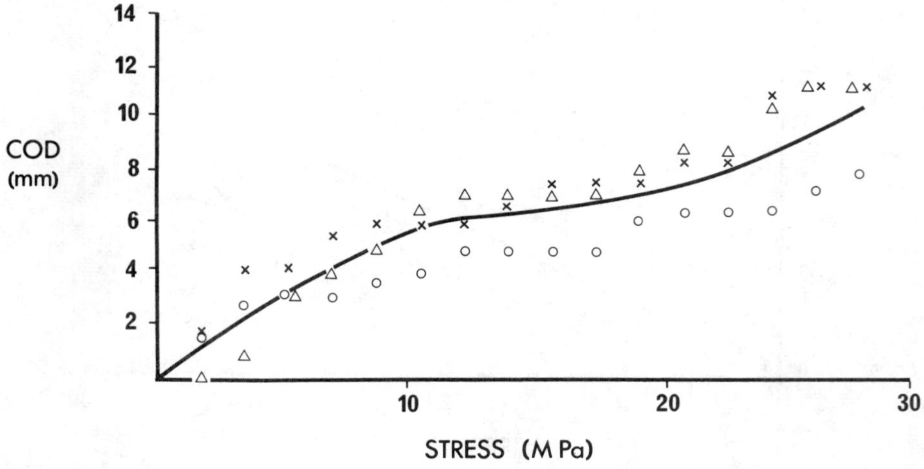

Figure 4 C.O.D. Measurements on Brass with Hardness 50 VHN

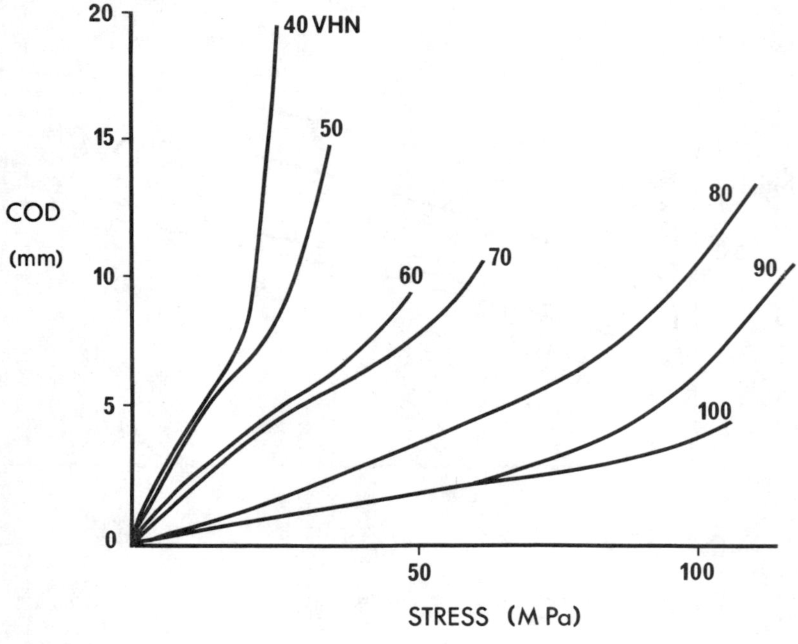

Figure 5 C.O.D. Measurements on Brass with Different Hardness Values

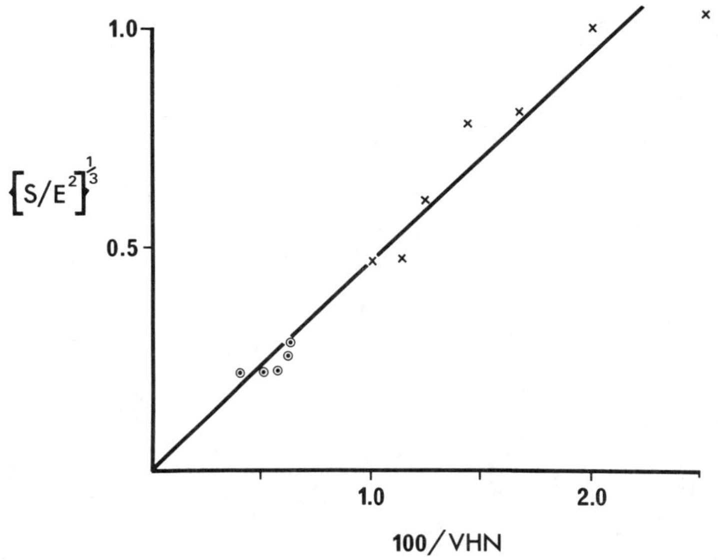

Figure 6 Data from the Approximately Linear Regions of the Curves in Figures 2 and 5

Fracture 1977, Volume 3, ICF4, Waterloo, Canada, June 19 - 24, 1977

AN EXPERIMENTAL AND FEM STUDY ON CRACK OPENING DISPLACEMENT

Albert S. Kuo and H. W. Liu*

INTRODUCTION

Wells [1] proposed crack tip opening displacement, δ_t, as a ductile fracture criterion at and beyond general yielding, and he verified the criterion experimentally with mild steel specimens. Recently, Green and Knott [2] have shown that δ_t, at the initiation of fibrous fracture, is a constant. McClintock [3] also proposed that fatigue crack growth per cycle is equal to half of the crack tip opening displacement. Ke and Liu [4] proposed near tip strain as a fracture criterion for ductile and tough materials, and they found a linear relationship between near tip strain and near tip crack opening displacement, δ. δ_t is related to the stress intensity factor K, in the case of small scale yielding, according to the strip yielding model [5 - 9]. Subsequently, it has been shown that δ is related to crack tip stresses and strains [10, 11]. All of these studies are two dimensional analyses which do not take into account the thickness consideration. In this study, the effect of thickness on crack opening displacement is studied experimentally. The results are further analyzed with the aid of the two-dimensional FEM calculations.

EXPERIMENTS AND RESULTS

Aluminum alloys, Al 2024-0, Al 2024-T3, and Al 2024-T351, were used in this study. The tensile yield strengths of these alloys were 53.8, 310, and 386 MPa, respectively. The specimens were 10.16 cm wide, with central cracks or central slots, $2a \simeq 1.78$ cm. All the specimens were centrally slotted, with the exception of specimens 4 and 5 which were centrally slotted with a jeweler's saw and then fatigue pre-cracked. The width of the slots was approximately 0.023 cm. Specimens of two different thicknesses, namely 0.041 cm and 0.625 cm, were tested.

Crack opening displacements, δ, between the upper and lower crack surfaces under static tensile load were measured with the moire method. The specimen surface was polished, cleaned, dried, and coated with photo-resist. A moire grille, of line density 5275 lines/cm, was printed onto the specimen surface with its grille lines perpendicular to the loading direction. The moire pattern was obtained by double exposures before and after deformation. A more detailed description of moire method was given by Schaefer, Liu and Ke [12]. The applied load and plate thickness were chosen to give a large plastic zone relative to plate thickness ($[K/\sigma_Y]^2/t \simeq 18$), and a small plastic zone relative to plate thickness ($[K/\sigma_Y]^2/t \simeq 1.12$).

The measured crack opening displacements will be compared with those calculated using the Dugdale model, the elastic model, and the elastic-plastic model. The crack opening displacements of an elastic crack in an infinite plate is given by Ang and Williams [13] as:

---

* Syracuse University, Syracuse, New York, 13210, U. S. A.

$$\delta = 4\ \sigma\ (a^2-x^2)^{1/2}/E \qquad \text{plane stress}$$
$$\delta = 4\ \sigma\ (1-\nu^2)\ (a^2-x^2)^{1/2}/E \qquad \text{plane strain,} \tag{1}$$

where 2a is the length of the crack laying along the x-axis with its tips at $x = \pm\, a$. ν, σ, and E are the Poisson's ratio, applied tensile stress, and Young's modulus, respectively. Goodier and Field [6] used the Dugdale model to calculate crack opening displacement, δ. For plane stress case,

$$\delta = \frac{2\sigma_Y\ (a+\ell)}{\pi E}\ [\cos\theta\ \ln\frac{\sin^2(\theta_2-\theta)}{\sin^2(\theta_2+\theta)} + \cos\theta_2\ \ln\frac{(\sin\theta_2+\sin\theta)^2}{(\sin\theta_2-\sin\theta)^2} \tag{2}$$

where σ_Y is the tensile yield strength, ℓ is the length of the strip yielding zone and $\ell = a(\sec\theta_2-1)$, $\theta_2 = \pi\sigma/2\ \sigma_Y$, $\cos\theta = x/(a+\ell)$ for $|x| < (a+\ell$ and $-\pi < \theta < \pi$. Similar results, using the continuous dislocation model, were obtained by Bilby, Cottrell, and Swinden [9].

Specimens 1 and 2 were made of the aluminum alloy, Al 2024-0, and both were loaded to the same K-value, 4.53 $MPa \cdot m^{1/2}$. Both specimens were slotted, but were of different thicknesses: 0.041 cm and 0.625 cm, respectively. The values of the quantity, $(K/\sigma_Y)^2/t$, were 17.5 and 1.12 for these two specimens, where t was specimen thickness. The results of the measurements are shown in Figure 1.

In Figures 1, 2, and 3, the solid lines indicate the elastic calculation, equation (1), and the dashed lines indicate the Dugdale calculation, equation (2). The dash-dotted line in Figure 1b indicates the elastic-plastic FEM calculation, which agrees well with the elastic calculation in the region $r > 7.6 \times 10^{-2}$ cm. Figure 1 indicates that the measurements of specimen 1 agreed with the plane stress Dugdale calculation, but the measurements of specimen 2 agreed with the plane strain elastic calculation. It should be noted that, in the region of measurements, the results of the FEM elastic-plastic calculation coincided with those of the elastic calculation. Unfortunately, no measurements were made in the region closer to the crack tip.

Specimens 3 and 4 were made of Al 2024-T3, and their thicknesses: 0.041 cm and 0.038 cm were nearly the same. Specimen 3 was slotted, and specimen 4 was fatigue pre-cracked. The applied K-values of these two specimens 26.2 and 28.2 $MPa \cdot m^{1/2}$. The values of $(K/\sigma_Y)^2/t$ for these two specimens were 17.5 and 18.8 respectively. The results in Figure 2 show that the measurements of both specimens agreed well with the plane stress Dugdale calculation in spite of the slot in specimen 3.

Specimens 5 and 6 were made of Al 2024-T351 and were of same thickness, 0.625 cm. Specimen 5 was fatigue pre-cracked, and specimen 6 was slotted. These two specimens were loaded to the nearly same K-values, 20.4 and 20.9 $MPa \cdot m^{1/2}$. The values of $(K/\sigma_Y)^2/t$ for these two specimens were 0.45 and 0.46, respectively. The results in Figure 3 show that the cracked specimen agreed well with the plane strain elastic calculation, but the slotted specimen fell in between the Dugdale and the elastic calculations. At the applied load level, the Dugdale and the elastic crack opening displacements were very close of each other. A thicker specimen loaded to a higher stres level was needed to make a distinction between these two models.

In an earlier work [12], crack opening displacements were measured in a very thin steel specimen, $(K/\sigma_Y)^2/t = 48$. In the thin specimen, extensive crack tip deformation caused localized crack tip strip necking. The strip necking zone was embedded in a diffused plastic zone which, in turn was embedded in the massive elastic plate as shown schematically in Figure 4. The crack opening displacement measurements in the steel specimen agreed well with the Dugdale calculation as shown in Figure 5. Even the opening displacements in the strip yielding zone agreed reasonably well with the Dugdale calculation.

Figure 6 is a picture of the typical moire patterns near a crack tip. It was taken from specimen 4 which was only 0.038 cm thick, and was loaded to $K = 28.2\ \text{MPa}\cdot\text{m}^{1/2}$. The heavily deformed plastic region appeared to be diffused. However, there was a slight indication in the picture that the plastic zone had the tendency to become strip-necked. The measurements of specimen 4 agreed with the Dugdale model as was mentioned earlier, even without a strip necking zone.

CALCULATION OF δ WITH THE FEM

Crack opening displacements were calculated with the finite element method under the conditions of plane strain and small scale yielding. In a region, r_e, near a crack tip, the elastic stresses are [15, 16]:

$$\sigma_{ij} = [K/(2\pi r)^{0.5}]f_{ij}(\theta) \tag{3}$$

where r and θ are the polar coordinates with the origin at the crack tip. The crack line lies along the negative x-axis. K is the mode I stress intensity factor. In the case of small scale yielding, a small plastic zone, r_p, is embedded in the elastic crack tip stress field. If $r_p \ll r_e$, according to Saint Venant's principle, it could be shown that a relaxation of the stresses in the plastic zone would not significantly change the stresses on the boundary of r_e. Therefore, for the case of small scale yielding, the crack tip deformation could be obtained from the calculation on a semi-circular region with the boundary stresses given by equation (3), [17].

The configuration of the elements is shown in Figure 7. There are 372 elements and 213 nodes. The smallest element at the crack tip is 5.33×10^{-3} cm with a crack length of 2.54 cm. Plane-strain and constant-strain elements are used. The computer programme is based on the elastic-plastic constitutive matrix obtained by Yamada and Yoshimura [18]. The values of yield strength, Young's modulus, strain-hardening exponent, and Poisson's ratio are 53.8 Pa, 68.9×10^3 Pa, 0.307, and 0.3, respectively.

The calculated crack opening displacements are shown in Figure 8. The different symbols in the figure denote the calculated values at different K-values. Close to the crack tip, δ could be expressed as,

$$\delta/r_p = \beta\,(r/r_p)^m \quad , \quad \beta = (\delta/r_p)_{r=r_p} \, . \tag{4}$$

The value of the slope, m, is 0.37. The value of m from HRR singularity analysis [10, 11] is 0.23. β is the value of δ/r_p at $r = r_p$; its value is 3.5×10^{-3} found from Figure 8. r_p is the plastic zone size, along the

$\theta = 60°$ line, and it is related to K,

$$r_p = \alpha\ (K/\sigma_Y)^2 = 0.112\ (K/\sigma_Y)^2\ . \tag{5}$$

The value of α, 0.157, was obtained by Levy et al [17] using the FEM for a non-hardening material, along the line $\theta = 70°$, where r_p is maximum.

Combining equation (4) and equation (5), one has the following relation for δ

$$\delta = \beta\ \alpha^{1-m}(K/\sigma_Y)^{2(1-m)}r^m = 8.74 \times 10^{-4}(K/\sigma_Y)^{1.27}r^{0.37}\ . \tag{6}$$

A comparison of the δ calculated by using the Dugdale model, the elastic model, and the elastic-plastic model by the FEM is shown in Figure 9. It is evident from the figure that near the crack tip, the Dugdale model gives a much higher value of δ than either the elastic model or the elastic-plastic model.

In the region $r/(K/\sigma_Y)^2 < 0.13$, the values given by FEM elastic-plastic calculation and the elastic results diverge as r approaches the crack tip. In the region $r/(K/\sigma_Y)^2 > 0.13$, the elastic-plastic calculation coincides with the elastic solution. This is also shown in Figure 1b. The moire measurements fell into the range where the FEM calculations and the elastic results coincide. In order to see the difference between the elastic and the elastic-plastic models, the measurements would have to be made in the region closer to the crack tip.

The Dugdale model is based on the physical model of strip yielding. In the experimental moire study, no strip yielding has been observed. The quantity $(K/\sigma_Y)^2/t$ is a measure of the size of plastic zone relative to thickness, and it is also an index of the deviation from the plane strain condition. When a specimen is very thin, for example, for the thin steel specimen at $(K/\sigma_Y)^2/t = 48$, strip necking takes place. The strip necking zone is embedded in a diffused plastic zone. In this case, the Dugdale calculation agrees well with the measurements. As the specimen thickness is increased, the strip necking zone disappears. However, the Dugdale model for the calculation of δ is still applicable in the region $(K/\sigma_Y)^2/t > 18$, even without a strip necking zone. As the plate thickness is further increased, and the applied K-value is reduced to the point $(K/\sigma_Y)^2/t < 1$, the Dugdale model δ differs considerably from the moire measurement. In this region, the measurements agree well with the elastic solution. In the region of measurements, the elastic solution coincides with the elastic-plastic FEM calculation. It is expected that in the region closer to a crack tip, the measurements would agree with the elastic-plastic calculation.

CRACK TIP OPENING DISPLACEMENT, δ_t, AND THE UNZIPPING MODEL

Based on the empirical results shown earlier, it is expected that δ_t calculated with the Dugdale model, is an appropriate measure of crack tip deformation only for thin specimens loaded to high K-values. The calculated δ_t with the Dugdale model by Goodier and Field [6] is

$$\delta_t = (8\ \sigma_Y\ a/\pi E)\ \ln\ [\sec(\pi\sigma/2\ \sigma_Y)]\ . \tag{7}$$

Based on the Dugdale model, in the case of small scale yielding, Rice obtained [7],

$$\delta_t = K^2/E\ \sigma_Y\ . \qquad (8)$$

Subsequently, δ_t was calculated by Levy et al using the FEM [17],

$$\delta_t = 0.425\ K^2/E\ \sigma_Y\ . \qquad (9)$$

δ_t given by these models could be used to characterize crack tip deformations and stresses in the analyses of brittle and ductile fractures.

McClintock and Pelloux [3, 19] suggested that fatigue crack growth rate, da/dN, is related to δ_t, for the decohesion plane inclined 45° to the crack plane. Recently, Kuo and Liu [21] made a calculation of the crack tip opening displacement and the crack tip advancing based on the unzipping model of shear decohesion taking place on two conjugate decohesion planes as shown schematically in Figure 10. As the applied stress on a cracked solid is increases, the decohesion processes take place along slip lines, α, b, β, c, γ, and d successively; while the "slabs' between the neighbouring slip lines move like the teeth of a zipper during the unzipping process, causing crack tip opening and advancing. The unzipping model separates the crack tip opening which contributes to crack tip advancing from those which cause crack tip blunting. Crack growth rate is related to δ_t, which in turn is related to ΔK. The final result is

$$da/dN = 0.019\ (1-\nu^2)\ K^2/E\ \sigma_{Y(c)} \qquad (10)$$

where $\sigma_{Y(c)}$ is the cyclic yield strength. If the ratio of the cyclic yield strength to the Young's modulus is 1/400, the agreement between the predicted crack growth rates and the empirical rates given by Barsom [22], Hahn et al [23] and Bates and Clark [24], is within a factor of 2; while the calculated growth rates, according to the classic Dugdale model, are more than 10 times higher than the empirical rates.

It can be concluded that the Dugdale model is applicable for analyzing δ and δ_t only in thin specimens loaded to high K-values. For a thick specimen at a low K-value, the general contour to the crack tip is, perhaps, better described by the elastic-plastic FEM calculations, such as those given by Levy et al. The crack tip opening which contributed to crack tip advancing is more realistically given by the unzipping model.

ACKNOWLEDGEMENTS

The authors wich to thank NSF (Grant No. GK-34047) for their support and Mrs. H. Turner and Mr. R. Ziemer for the preparation of the manuscript.

REFERENCES

1. WELLS, A. A., British Welding J., Nov., 1963, 563.
2. GREEN, G. and KNOTT, J. F., J. Eng. Mat. and Tech., Jan., 1976, 37.
3. McCLINTOCK, F. A., Fatigue Crack Propagation, ASTM STP 415, 1967, 170.
4. KE, J. S. and LIU, H. W., Int. J. of Fracture, 7, 1971, 487.
5. DUGDALE, D. S., J. of Mech. and Phys. of Solids, 8, 1960, 100.

6. GOODIER, J. N. and FIELD, F. A., Fracture of Solids, John Wiley, New York, 1963, 103.
7. RICE, J. R., Fatigue Crack Propagation, ASTM STP 415, 1967, 247.
8. RICE, J. R., J. of Appl. Mech., June, 1968, 379.
9. BILBY, B. A., COTTRELL, A. H. and SWINDEN, K. H., Proc. Roy. Soc., London, A285, 1965, 304.
10. HUTCHINSON, J. W., J. of Mech. and Phys. of Solids, 16, 1968, 13.
11. RICE, J. R. and ROSENGREN, G. F., J. of Mech. and Phys. of Solids, 16, 1968, 1.
12. SCHAEFFER, B. J., LIU, H. W. and KE, J. S., Exp. Mechanics, April, 1971, 172.
13. ANG, D. D. and WILLIAMS, M. L., J. of Appl. Mech., Sept., 1961, 372.
14. McCLINTOCK, F. A., ASME, J. of Basic Eng., 82, 1960, 423.
15. WILLIAMS, M. L., J. of Appl. Mech., March, 1957, 109.
16. IRWIN, G. R., J. of Appl. Mech., Sept., 1957, 361.
17. LEVY, N., MARCAL, P. V., OSTERGREN, W. J. and RICE, J. R., Int. J. of Fracture Mech., 7, 1971, 143.
18. YAMADA, Y. and YOSHIMURA, N., Int. J. of Mech. Sci., 10, 1968, 343.
19. PELLOUX, R. M. N., Trans. ASM, 62, 1969, 281.
20. HU, W. L. and LIU, H. W., Proceedings of ICM-II, published by ASM, 1976, 1058.
21. KUO, A. S. and LIU, H. W., Scripta Metallurgica, 10, 1976, August, 723.
22. BARSON, J. M., Damage Tolerance in Aircraft Structures, ASTM STP 486, 1971, 1.
23. HAHN, G. T., HOAGLAND, R. C. and ROSENFIELD, A. R., AF 33615-70-C-1630, Battelle Memorial Inst., Columbus, Ohio, August, 1971.
24. BATES, R. C. and CLARK, W. G., Jr., Trans. ASM., 62, 1969, 380.

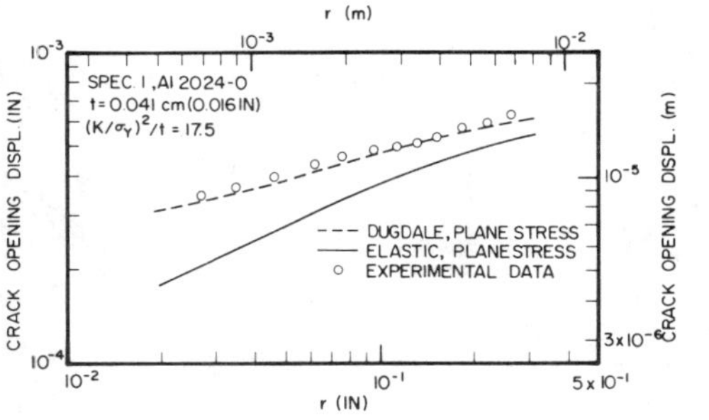

(a)

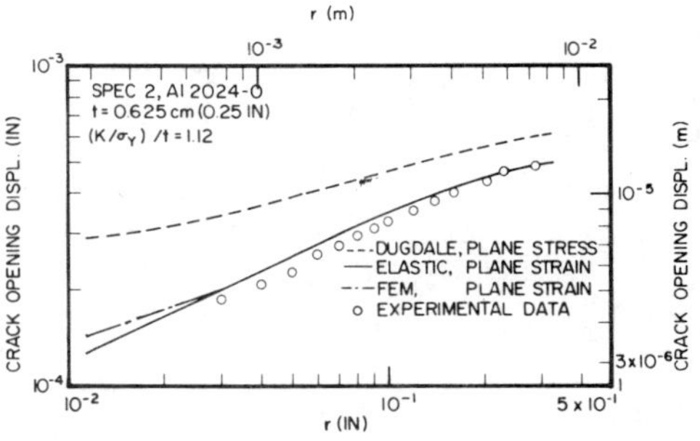

(b)

Figure 1 Effect of Thickness on δ, Al 2024-0 Aluminum Alloy

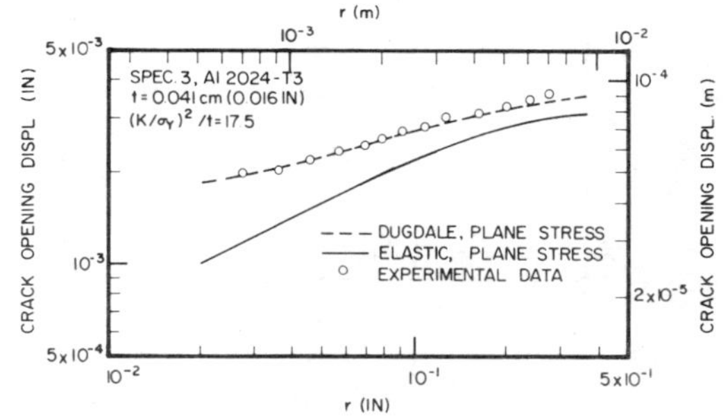

(a)

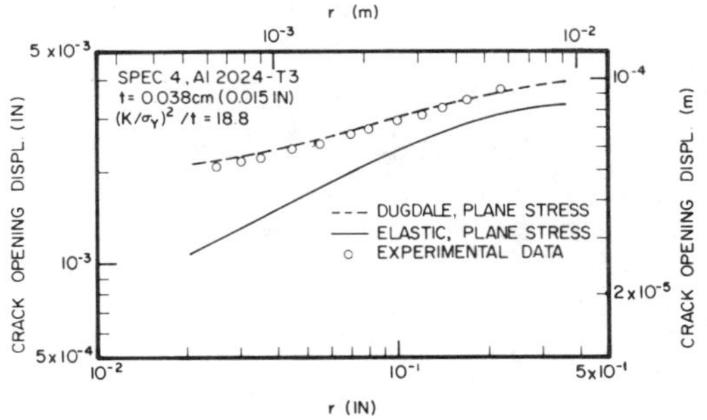

(b)

Figure 2 δ in Al 2024-T3 Aluminum Alloy

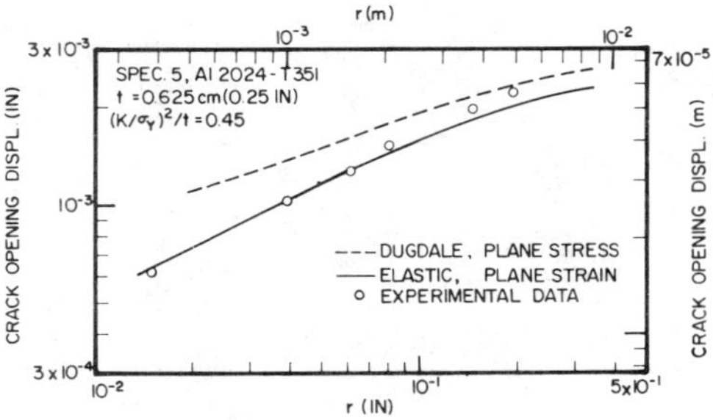

(a)

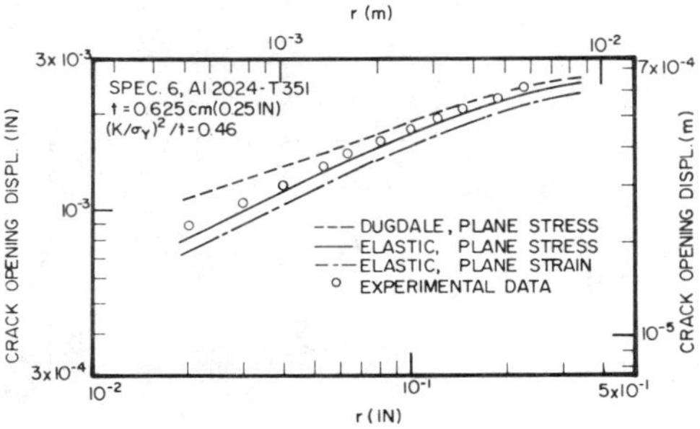

(b)

Figure 3 δ in Al 2024-T351 Aluminum Alloy

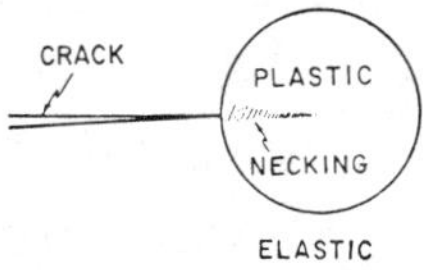

Figure 4 Schematic Diagram of Crack Tip Necking [14]

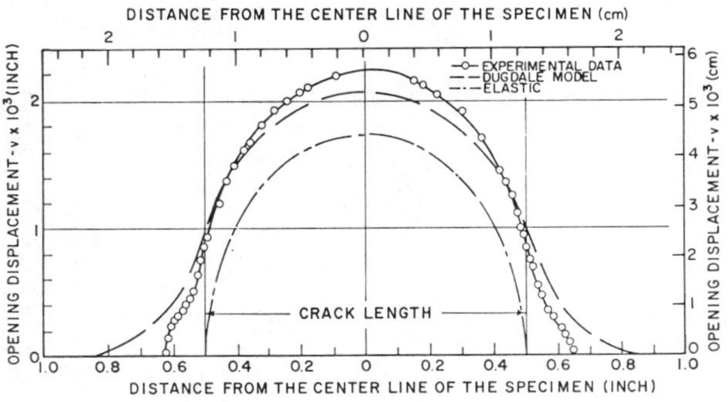

Figure 5 δ in Thin Steel Specimen [12]

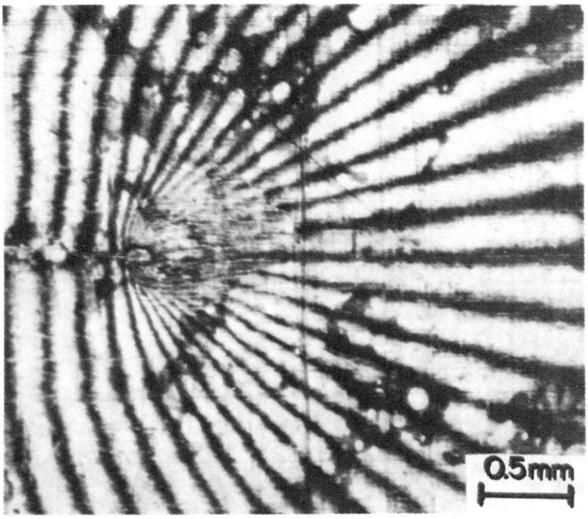

Figure 6 Moire Pattern at Crack Tip

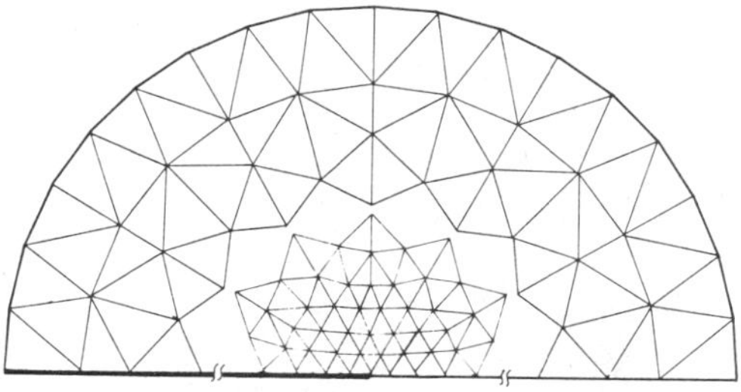

Figure 7 FEM Element Configuration

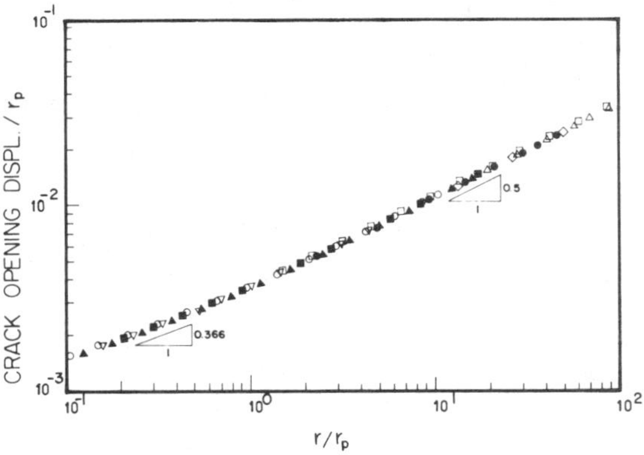

Figure 8 δ by Elastic-Plastic FEM Calculation - Small Scale Yielding

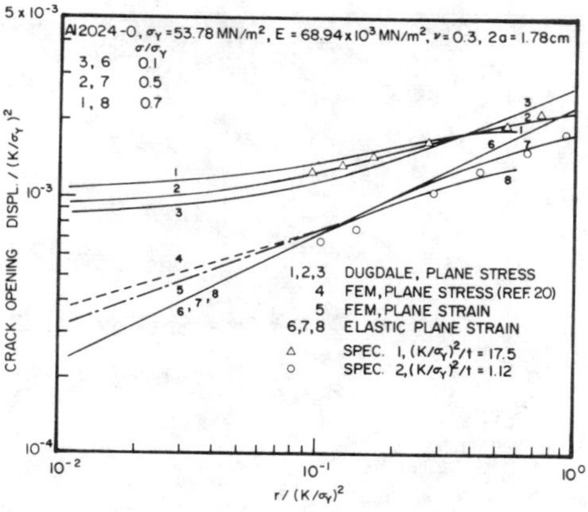

Figure 9 δ - A Comparison of Dugdale, Elastic and Elastic-Plastic Models

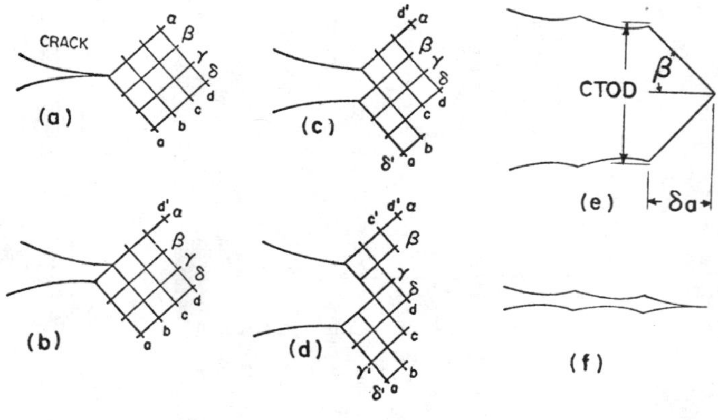

Figure 10 Unzipping Model for Crack Tip Opening and Advancing [21]

Fracture 1977, Volume 3, ICF4, Waterloo, Canada, June 19 - 24, 1977

STUDIES OF CRACK TIPS IN STEEL AND ALUMINUM ALLOYS

C. K. Clarke*

INTRODUCTION

This paper presents results of two studies: the nature and mechanism of formation of the stretched zone in steel and aluminum, and COD behaviour in aluminum as a function of specimen size and geometry. Some comparisons of COD and fractographically measured crack tip displacements were made in the second project.

Considerable interest has been shown in dimensional changes at crack tips for the past several years. Some of this interest has been in the area of fracture testing where investigators have attempted to relate fracture toughness in tough materials to crack tip dimensional changes in terms of Crack Opening Displacements (COD). Wells [1] postulated that a finite opening at the tip of a crack, δ, was developed as a consequence of the formation of the plastic zone at the crack tip. This should be on the order of:

$$\delta = \frac{4G}{\sigma_y} = \text{COD} \qquad (1)$$

where σ_y = yield stress, G = strain energy release rate. A critical δ_c similar to G_c and K_c was expected because of the relationship between G_c and δ_c.

COD has been observed to predict well nonplane-strain fracture in steels. (Reference [2] provides references in this area.) COD and crack tip strains measured at the onset of ductile tearing were observed to be material constants. Fractographic studies [3] of crack tips and finite element models [4, 5] of crack tips have provided a physical basis for Well's concept of a critical COD for fracture.

Fractographic studies of crack tips were initiated after the COD concept was established. A feature called the stretched zone was discovered at crack tips in fracture specimens. The size of this feature was observed to be proportional to the fracture toughness of the specimen. Reference [3] provides a review of stretch zone studies and efforts to explain its formation.

PROCEDURE

A 302D steel was tested in the form of standard ASTM 50.8 mm thick compact tension specimens at different temperatures [3]. Aluminum specimens of conventional and high toughness 7075-T651 alloy were tested at room temperature with compact tension and three-point bend specimens. Three point bend specimens had a B/W (thickness to width) ratio of 0.25 for specimens

---

* Department of Mechanical Engineering, University of South Alabama, Mobile, Alabama, U. S. A.

cut from 12.7 mm thick plate and a B/W ratio of 0.5 for specimens cut from 25.4 mm thick plate. All specimens were machined in the T-L orientation (crack propagation direction parallel to the rolling direction).

COD measurements were made on the aluminum specimens with crack mouth compliance gages using a technique described in Reference [6]. COD measurements were made at the first indication of crack instability or growth as determined from electric potential observations during testing.

All fractographic observations of the crack tip were performed using scanning electron microscopes (SEM's). Stereo pairs of matching fracture surfaces were used extensively in morphological studies of the actual crack tip blunting. Fractographic measurements of the amount of blunting or Crack Tip Opening Displacement (CTOD) were made using a technique in Reference [7]. This technique permitted crack tips to be measured *in situ* relative to the overall specimen geometry. Matching surfaces at the crack tip were measured to accurately determine CTOD as shown in Figure 1. (CTOD will refer to fractographically determined crack tip dimensions while COD will refer to mechanically derived crack tip dimensions).

RESULTS

Figures 2 and 3 are typical results for the steel specimens which fractured in a brittle fashion but displayed plasticity in terms of nonlinear load vs. deflection curves. An SEM picture of the steel crack tip is shown in Figure 2 with a cross section for orientation. The ripple marks at the crack tip are considered to be caused by alternating slip.

The metallographic cross section shown in Figure 3 shows the severity and localized nature of the crack tip deformation in the steel specimens. Fractographic examination of this specimen did not reveal evidence of ductile fracture before mounting for sectioning and polishing.

Blunted crack tips were observed only in steel specimens that displayed nonlinearity in load vs. deflection curves produced during the fracture test. Specimens which did not display nonlinearity in the load vs. deflection curves exhibited only localized crack tip deformation.

The aluminum specimens also exhibited crack tip blunting but in a fashion generally similar to the sharp crack tips predicted by Pelloux [8]. A complete crack tip similar to that predicted by Pelloux is preserved on one surface of a fracture specimen as shown in Figure 4. The included angle at the crack tip in this picture is approximately 70°. This feature was observed in several aluminum fracture specimens studied. Other variations on this morphology were observed in individual grains depending on the crystallographic orientation of the grain to the principal stress axis.

Several aluminum fracture specimens were tested to determine COD and CTOD. Results of the COD tests given in Figure 5 indicate that COD in 7075-T651 is affected by specimen geometry and the type of specimen used. Crack tips were fractographically measured in two aluminum specimens and CTOD results were observed to be approximately an order of magnitude smaller than COD results for the same specimen.

DISCUSSION

The fractographic study of the crack tips in steel and aluminum fracture specimens revealed that the stretch zone was the result of crack tip deformation. This crack tip deformation produced blunted crack tips in the steel and aluminum specimens and provides a physical basis for the COD approach to fracture.

Careful study of the blunt crack tips provides several clues to the nature of crack tip deformation:

a) Severe crack tip deformation in the steel specimens produced elongated grains at the crack tip (Figure 3).

b) Equiaxed cleavage facets were observed right up to the crack tip in the steel specimen.

c) The very sharp crack tip in the aluminum specimens which did not produce fracture suggests that linear elastic fracture mechanics cannot be used to describe the stress state at the crack tip.

These clues show that simple plastic slip line analyses can be applied to describe crack tip behaviour in tough materials. The ideal plastic slip line solution for a single edge notched tension specimen [8], or the solution for an edge notched specimen in bending [9] predicts finite constant stresses ahead of the crack tip. These solutions predict that plastic deformation can only occur on two mathematical shear planes emanating from the crack tip. This leaves the region ahead of the crack tip distortion free in agreement with the microstructures observed in the steel specimens.

The elongated grain structure at the crack tip in the steel specimens in combination with equiaxed cleavage facets fit the slip line description of the crack tip. In fact the highly deformed grains at the crack tip can be qualitatively reproduced by simulating Pelloux's proposed simultaneous slip on both shear planes (taking into account strain hardening). Finally, the sharp crack tips occasionally found remaining on the aluminum specimens support the concept of constant, finite stresses ahead of the crack tip.

The difference in COD measured in the two types of aluminum specimens probably reflects basic differences in the two specimen types. Compact tension specimens with both tension and moment forces present probably have different rotational factors whereas one factor was used for both specimen types in these tests to calculate COD. Other unpublished work at Boeing suggests that crack mouth displacements from compact tension specimens also contain a displacement component due to bending of the two loading arms (cantilever beams).

The disagreement between CTOD and COD measurements is not surprising when the empirical nature of the method of Elliott, Walker, and May [6] is considered. They assumed the specimen to rotate about a hinge point in the remaining ligament in a fashion predicted by the slip line solution for an edge notched bend specimen. The experimental determination of the location of the hinge point is essential to this analysis. Neither this approach or Well's analytical approach address whether crack tips actually deform.

This work reveals that crack tips do deform or blunt and single slip line plastic analysis can describe the blunting mechanism. This crack tip blunting process in fracture is also the mechanism by which cracks grow in fatigue. If CTOD critical for fracture can be analytically predicted, a method is available to determine fracture toughness in low strength, high toughness materials. A fatigue crack growth analysis would be available if CTOD per cycle could be predicted as a function of previous load history.

CONCLUSIONS

1. Blunted crack tips were observed in steel and aluminum alloys.
2. The occurrence and behaviour of the blunted crack tips are explained with simple plastic slip line fields.
3. COD was a function of the specimen type and geometry.
4. CTOD measurements made with scanning electron microscope techniques were about an order of magnitude smaller than corresponding COD measurements.

ACKNOWLEDGEMENTS

The Pressure Vessel Research Committee sponsored the fractographic studies on steel and aluminum at Lehigh University. The COD/CTOD studies in aluminum at Boeing-Wichita were conducted with Company IR & D funds.

REFERENCES

1. WELLS, A. A., Brit. Welding J., 1963, 563.
2. OTSUKA, A., MIYATA, T., NISHIMURA, S. and KASHIWAGI, Y., Eng. Frac. Mech., 1, 1975, 419.
3. CLARKE, C. K., "A Study of Microfracture Processes in Steel and Aluminum", dissertation, Lehigh University, Bethlehem, Pennsylvania, 1973.
4. NEWMAN, J. C. and ARMEN, H., AIAA Journal, 13, 1975, 1017.
5. HAYES, D. J. and TURNER, C. E., Int. J. of Frac., 10, 1974, 17.
6. ELLIOTT, D., WALKER, D. F. and MAY, M. F., "Practical Application of Fracture Mechanics to Pressure Vessel Technology", Inst. Mech. Eng., London, 1971, 217.
7. LANE, G. S., J. Sci. Instr., 2, 1970, 565.
8. PELLOUX, R. M. N., Eng. Frac. Mech., 1, 1970, 697.
9. GREEN, A. P. and HUNDY, B. B., J. Mech. and Phys. of Solids, 4, 1956, 128.

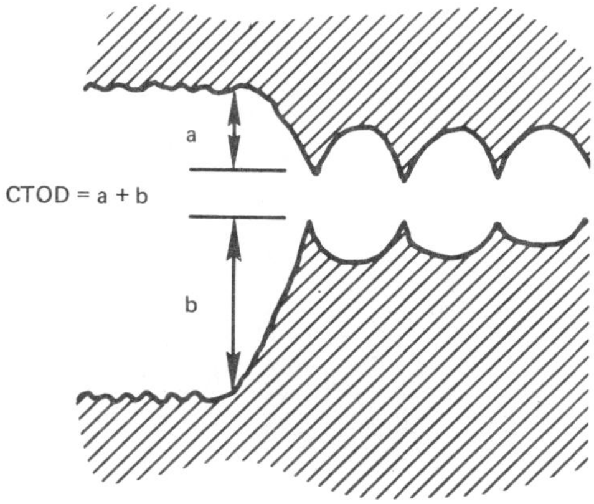

Figure 1 Fractographic Measurements of CTOD were made from Matching Fracture Surfaces

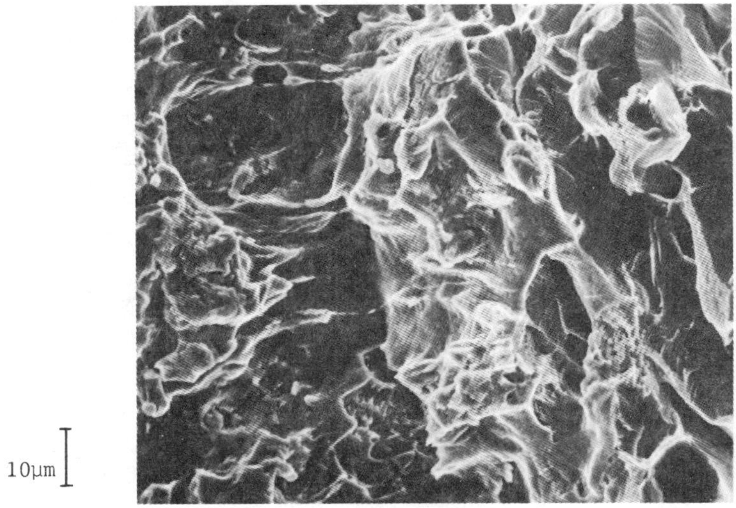

Figure 2A An SEM Picture of a Blunted Crack Tip (Stretch Zone) in a Steel Specimen

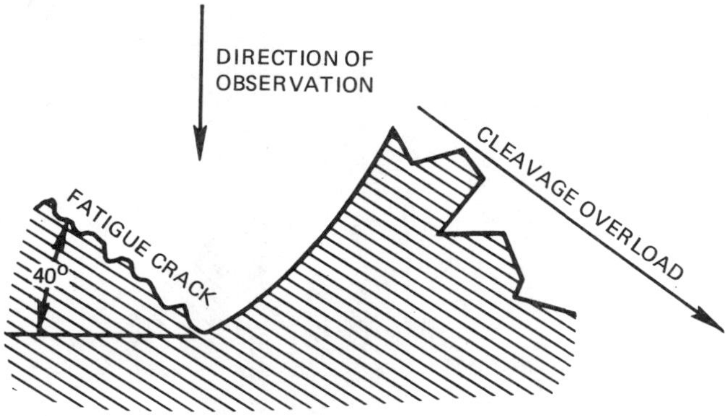

Figure 2B A Schematic Sketch of the Cross Section is Shown in this Figure to Provide Specimen Orientation in Figure 2A

Figure 3 Metallographic Cross Sections of Steel Specimens Revealed Elongated Grains at the Crack Tip

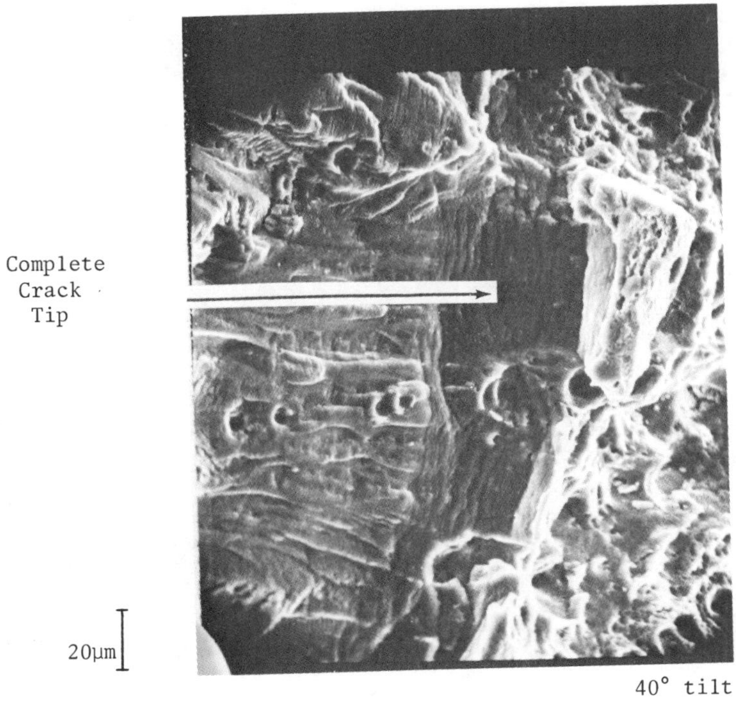

Figure 4 Sharp Crack Tips were Occasionally Found Intact on Aluminum Specimens

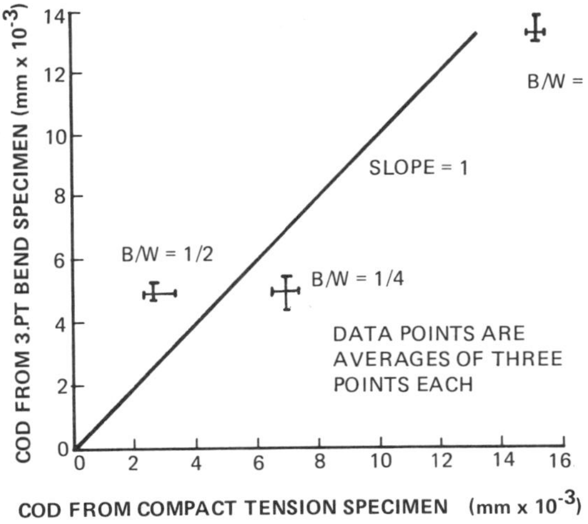

Figure 5 COD was Dependent on Specimen Type and Specimen Geometry for 7075-T651 Aluminum

MEASUREMENT OF STRETCH ZONE WIDTH AND δ_i IN A LOW ALLOY NAVAL STEEL

P. Hopkins and G. Jolley*

INTRODUCTION

The development of the low alloy high yield strength naval steels has taken place with great emphasis on the need for cleanliness and toughness. These steels sustain extensive plastic deformation before the onset of stable fracture at service temperatures and unstable fast fracture is not considered to be a problem in the thicknesses used in practice. From a design or material selection point of view the linear elastic fracture mechanics approach is therefore inapplicable and a greater interest has been shown in general yielding fracture mechanics parameters in particular the C.O.D. (crack opening displacement) concept. Safe defect size predictions for service components may be made on the basis of critical C.O.D. at the onset of fast fracture, (δ_c), measured on laboratory specimens [1]. The correlation is only accurate, however if fracture occurs before extensive plasticity [2]. Until recently when unstable fracture does not occur in the test, the critical C.O.D. has been taken to be the maximum load point on the load/C.O.D. curve (i.e. δ_{max}). This value has been shown to vary with specimen size and has proved difficult to use in the prediction of acceptable defect sizes in actual components.

Harrison and Fearnehough [3] have shown that the C.O.D. at which ductile crack initiation occurs, δ_i, is lower but gives a much more constant value of C.O.D. than δ_{max} over a range of specimen sizes. Work is taking place at several research establishments to quantify this parameter and establish its potential use in design. This paper describes the results to date of an ongoing project at Salford University to investigate the effects of metallurgical variables on δ_i in Q1(N) a low alloy high yield strength steel. The particular metallurgical variables investigated were testing temperature and orientation of the test specimens to the rolling direction of the plate.

EXPERIMENTAL

The material used was a 25mm thick Q1(N) steel the chemical composition of which is given in Table 1. The steel was in the 'as received condition' i.e. water quenched at 1203 K and tempered at 913 K for 1 1/2 hours. The microstructure of the material is shown in Figure 1.

The fracture toughness specimens used in the investigation were 10mm x 20mm x 100mm the dimensions conforming with recommended procedure [4]. The specimens were machine notched initially by milling and then by a rubber bonded slitting wheel to a depth of 7.7mm. The cracks were then extended to 10mm by fatigue pre-cracking. The test pieces were loaded in

*Department of Aeronautical and Mechanical Engineering, University of Salford, Salford M5 4WT, England.

3 point bending over a load span of 80mm at a loading rate of the order of 50MPa $m^{1/2}s^{-1}$ (cross head velocity of 8 x 10^{-5} ms^{-1}). Double cantilever beam clip gauges were used to measure the crack face displacement. The positive location of the clip gauge over the crack openings was achieved by the use of saddle type knife edges, clamped to the test piece by means of grub screws.

A number of techniques have been developed to detect the onset of slow crack growth during a fracture toughness test including potential drop (p.d.) techniques [5], acoustic emission [6] and ultrasonics [7]. In the present investigation a potential drop technique was used. The method monitors the potential change around the crack tip against the crack opening displacement at the tip. This is done by connecting a constant current supply (in this case capable of producing 50A) to the ends of the specimen and measuring the potential change around the crack tip as it opens and grows. Copper probes spot welded at either side of the crack measure this change while the C.O.D. is obtained by attaching a clip gauge above the crack. Both these readings can be monitored on an autographic XY plotter and a curve of the type shown in Figure 2 is obtained where δ_i is the value of C.O.D. at the point of gradient change on the graph. The curve in Figure 2 is idealised and in the present work it was often very difficult to detect the change in slope possibly because of the relatively small specimens used. To overcome this problem the potential change at the crack tip was plotted as a function of the applied load rather than C.O.D. Also by using an XYY plotter it was possible to monitor C.O.D. versus load simultaneously (load being the common abscissa). The potential change versus load curve shows a gradient change at a load Y between loads of X and Z (see Figure 3a), the two portions of the curve XY and YZ being linear. These three loads can be related to C.O.D. by referring to the C.O.D. versus load curve obtained simultaneously (Figure 3b). It will be shown later that XY represents the growth of the stretch zone and the point Y represents δ_i.

RESULTS

Fracture Mode

In general, a reasonably tough material after a C.O.D. test will have a fracture surface containing a region of fibrous crack growth at the base of the fatigue crack followed by a brittle cleavage fracture if the specimen has been broken open in liquid nitrogen. In a ductile material such as Q1(N) there is a stretch zone between the fatigue crack base and the beginning of the fibrous crack (Figure 4). This is due to conditions in the volume of material immediately ahead of the crack tip (i.e. the plastic zone) being such as to promote a 45° shear type mechanism of crack extension. As the C.O.D. test progesses the microcrack extension within the stretch zone is followed by macroscopic fibrous crack growth (microvoid coalescence) which continues until plastic collapse. The presence of stretch zones on the fracture surface of fracture toughness specimens has been the topic of several recent investigations [8 - 13] and correlation between fracture toughness parameters and the width of the zone have been suggested in all these investigations. It is now well established that the stretch zone is a result of blunting of the crack tip [14, 15]; strain and localised yielding at the crack tip enabling the stretch zone to form by alternating shear along slip bands originating from the tip.

Attempts to relate stretch zone width (S.Z.W.) to fracture toughness parameters have not yet produced a completely unified approach. Spitzig [17], proposed that S.Z.W. was equal to the critical crack tip opening displacement which was disputed by Gerberich and Hemmings [11], who concluded that the region was controlled by the fatigue pre-crack operation. Bates et al [10], have shown for both steels and aluminum alloys that the S.Z.W. can be correlated with the ratio of stress intensity to yield stress ($K_1/\sigma y$) giving the relationship

$$\text{S.Z.W.} = 10^{-3} \left\{\frac{K_1}{\sigma y}\right\}^{1.6} \qquad (1)$$

This was substantiated by Brothers et al [12] who showed additionally that the S.Z.W. could be numerically correlated with C.O.D. However the scatter in all the data is considerable, [12], and there are differences of opinion on many points, the most common being the definition of S.Z.W. and the angle of inclination of the zone to the fatigue crack plane, [18, 19]. Since the width of the zone varies along its length past workers have assumed a mean value between the maximum and minimum width.

Figure 5 shows the stretch zone length plotted against C.O.D. for the material used in the investigation. This figure was compiled from specimens loaded to increasing C.O.D. values and broken in liquid N_2. The length of the stretch zone was measured using a scanning electron microscope. It can be seen from Figure 5 that stretch zone formation does not commence upon initial loading. No stretch zone was evident on the fracture surface until a C.O.D. of .09mm was reached. Between C.O.D.s of .09mm and .17mm the length of the stretch zone increased from zero to .12mm. At C.O.D.s greater than .17mm fibrous crack growth was observed along the crack front, the amount of which increased with C.O.D.

Previous attempts to measure the angle of inclination of the stretch zone to the fatigue crack plane have involved stereo viewing of crack profile replicas, [14], which can be tedious. In these present investigations two further techniques have been used. The first employed the tilt stage of a scanning electron microscope. The fracture surface was placed on the stage in a horizontal position and any particular point on the stretch zone was chosen and tilted until its maximum length was realised. The angle through which it had been tilted was then the angle of inclination. The stretch zone formed in Q1(N) is very large and this technique might not be so successful in less ductile materials. Furthermore location of the corresponding point on the mating fracture surface can be difficult. The second method adopted was to trace the crack profile using a Talysurf which produced a trace similar to that shown in Figure 6. Where the stretch zone deviated from the crack plane is very clear enabling its angle of inclination to be measured very easily but it was impossible to identify the demarkation between the stretch zone and the initiation of fibrous fracture. The Talysurf instrument leaves a scratch along the crack profile which can be viewed using a scanning electron microscope where the stretch zone width can be measured and the information transferred to the Talysurf trace. Using the above techniques the angle of the stretch zone in Q1(N) was found to lie within the range 21 - 23°.

C.O.D. Results

Since the potential drop equipment used for this work had not been used on any previous investigations it was decided to calibrate the technique with the more laborious procedure of measuring fibrous crack growth on specimens tested to various values of C.O.D., the specimens being then broken in liquid nitrogen, and producing plots of fibrous crack growth versus C.O.D. [20]. Every specimen tested was also simultaneously tested by the p.d. method. Figure 7 shows the results obtained in this way for specimens tested at room temperature and at 193 K. Up to point A on the curve only a stretch zone was evident on the fracture surface and so the increase in crack length is solely due to the formation of the stretch zone. After point A fibrous crack growth is evident on the fracture surface along with the stretch zone. It can be seen from Figure 7 that the δ_i value of 0.17mm is the same for both testing temperatures. This value was also obtained by the potential drop technique when the point Y (Figure 3) gave a consistent value of 0.17mm in all specimens tested. The vertical line B in Figure 7 relates to point Z on the p.d. curves (Figure 3) and from scanning electron microscopy this point is the point at which the whole of the crack front moves by fibrous crack growth only. At levels beneath this figure there were still some regions of the crack front which were advancing solely by stretching. Both the points Y and Z on the p.d. curves gave consistent values of 0.17 and 0.31 respectively.

The effect of orientation to the plate rolling direction was examined on specimens machined 0°, 22½°, 45°, 77½° and 90° to the rolling direction. Table 2 gives the details of inclusion counts on sample specimens at each orientation which were obtained on a scanning X-ray image analyser. Figure 8 shows the room temperature values obtained for the specimens machined at 0° and 90° to the rolling direction which shows that δ_i is only slightly lowered by a change in specimen orientation in the rolling direction plane. Figure 9 shows details for all the orientations tested and also gives δ_{max} and charpy upper shelf figures. Further details of the charp tests are given in Figure 10. The differences in δ_i although small were consistently obtained by both methods used and the differences in fracture characteristics are further emphasised by the charpy and δ_{max} results.

DISCUSSION

The work reported in this paper deals with the development of techniques with which to study and determine the stretch zone width and δ_i values in Q1(N) as well as the effects of metallurgical variables on these values. It is felt that the combination of the scanning electron microscope and the Talysurf instruments described earlier offers a very good technique for studying the width and angle of inclination of stretch zones and will be used in the further work which is to be carried out. This technique as was previously mentioned, should be beneficial in other plastic materials.

Some problems were found with the p.d. equipment using the standard technique of plotting C.O.D. versus potential change at the crack tip, possibly because the samples were small compared with other published work but the revised technique described earlier does give consistent results which were much easier to interpret on the specimen tested.

The C.O.D. tests showed that stretch zone width and δ_i were unaffected by changes in temperature from 293K to 193K. Since the matrix flow stress of Q1(N) will undoubtedly increase with lower temperatures one might expect some difference in the strain to failure at inclusion/matrix interfaces. This change is obviously too small to be detected by the testing technique adopted.

The effect of rolling direction on δ_i was also rather small. As was mentioned earlier the plate material has been developed to have high cleanliness and therefore high ductility. The plate had also been cross-rolled to further remove orientation effects. The results in Table 2 show that the inclusion area etc. was not significantly affected by the change in orientation although the 90° sample did seem to possess an increased inclusion area. It can be seen that the testing technique consistently showed that this orientation had a lower value of δ_i than the other orientations which is substantiated by the δ_{max} and impact data. This was very reassuring and the technique can be used with confidence in future investigations. It is difficult to comment constructively on the higher values obtained for the 22½° orientation which were nevertheless consistently obtained. Further work will perhaps bring clarification on this point.

Further work is in progress using the techniques described in this paper, to further investigate the effects of metallurgical and geometric variables on δ_i in this material.

ACKNOWLEDGEMENTS

The authors would like to thank the Procurement Executive Ministry of Defence for the provision of a contract to carry out this work and in particular Mr. I. M. Kilpatrick of the Naval Construction Research Establishment for helpful discussions.

The opinions expressed in this paper are solely those of the authors and do not necessarily reflect current M.O.D. research policy.

REFERENCES

1. BURDEKIN, F. M. and DAWES, M. G., Conference on Practical Application of Fracture Mechanics to Pressure Vessel Technology, Institution of Mechanical Engineers, May 1971, 28-37.
2. FEARNEHOUGH, G. D. and WATKINS, B., Int. Journal of Fracture Mechanics 4, 1968, 233.
3. HARRISON, T. C. and FEARNEHOUGH, G. D., Int. Journal of Fracture Mechanics 5, 1969, 348.
4. Methods for C.O.D. Testing, Brit. Stand. Inst. DD19, 1972.
5. JOHNSON, H. H., Materials Research and Standards 5, 1965, 442.
6. HARTBLOWER, C. E., GERBERICH, W. W. and CRIMMINS, P. P., Final Contract Report NASA-4902 Aerojet General Corp. for Langley Research Centre, June 1968.
7. CLARKE, W. J. and CESCHINI, L. J., Materials Evaluation 27, 1969, 180.
8. SPITZIG, W. A., ASTM STP 436, 1968, 19.
9. SPITZIG, W. A., ASTM STP 453, 1969, 90.
10. BATES, R. C. et al., ASTM STP 453, 1969, 192.
11. GERBERICH, W. W. and HEMMINGS, P. L., Trans A.S.M. 62, 1969, 540.
12. BROTHERS, A. J. et al., ASTM STP 493, 1971, 3.
13. ELLIOTT, D. and STUART, H., Proc. 3rd Annual Symp. on Scanning Electron Microscopy, I.I.T., Chicago, 1970, 305.
14. WOLFF, W. E., ASTM STP 493, 1970, 20.
15. BROEK, D., Nat. Lucht. Ruimtevaarthals (T.R. 71021) 1971.
16. PELLOUX, R. M., Eng. Fract. Mech., 1, 1970, 697.
17. SPITZIG, W. A., Trans. A.S.M. 61, 1968, 340.
18. BROEK, D., 3rd Int. Congress on Fracture, Munich, 1973, paper 3, 422.

19. SCHWALDE, K., 3rd Int. Congress on Fracture, Munich, 1973, paper 2, 515.
20. SMITH, R. F. and KNOTT, J. F., Conference on Practical Application of Fracture Mechanics to Pressure Vessel Technology, Institution of Mechanical Engineers, May 1971, 65.

Table 1 Chemical Composition of Materials Used

Element	C	Si	Mn	S	P	Ni	Cr	Mo	Cu	V	Al
% Composition	0.17	0.31	0.31	0.005	0.010	2.80	1.47	0.39	0.09	<0.01	0.061

Table 2 Inclusion Counts on Material Used

Angle of specimen to rolling direction	No. of inclusions/mm^2			Average area %
	>3μ	>8μ	>15μ	
0°	9.3	1.1	0.13	0.0216
22½°	9.2	0.9	-	0.0218
45°	8.0	0.8	-	0.0223
77½°	9.5	0.9	0.1	0.0223
90°	8.7	1.1	0.1	0.0245

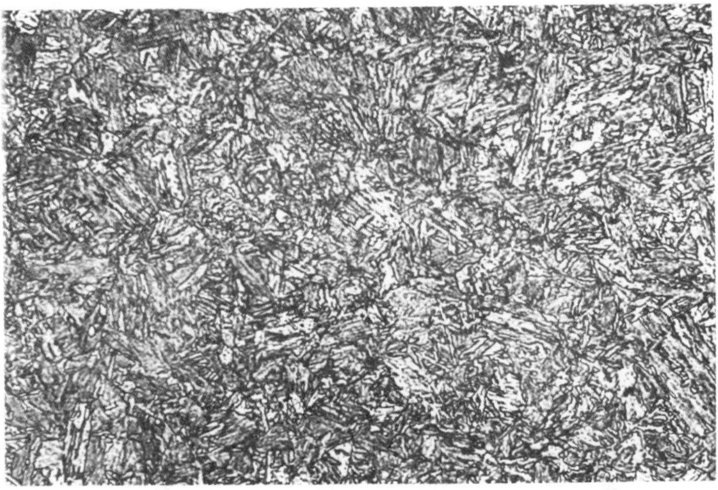

Figure 1 Microstructure of as received material x160.

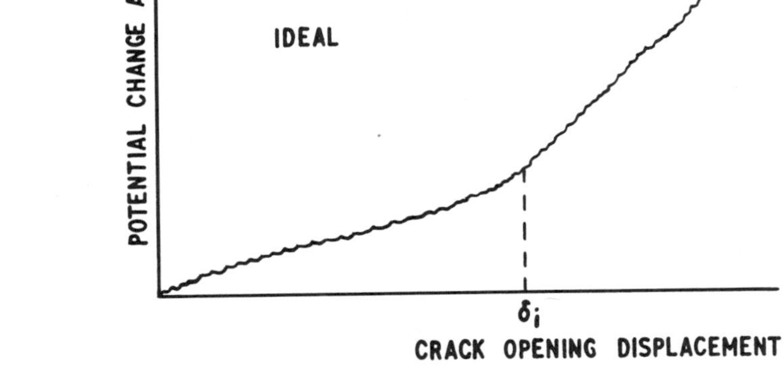

Figure 2 Ideal plot obtained by potential drop technique.

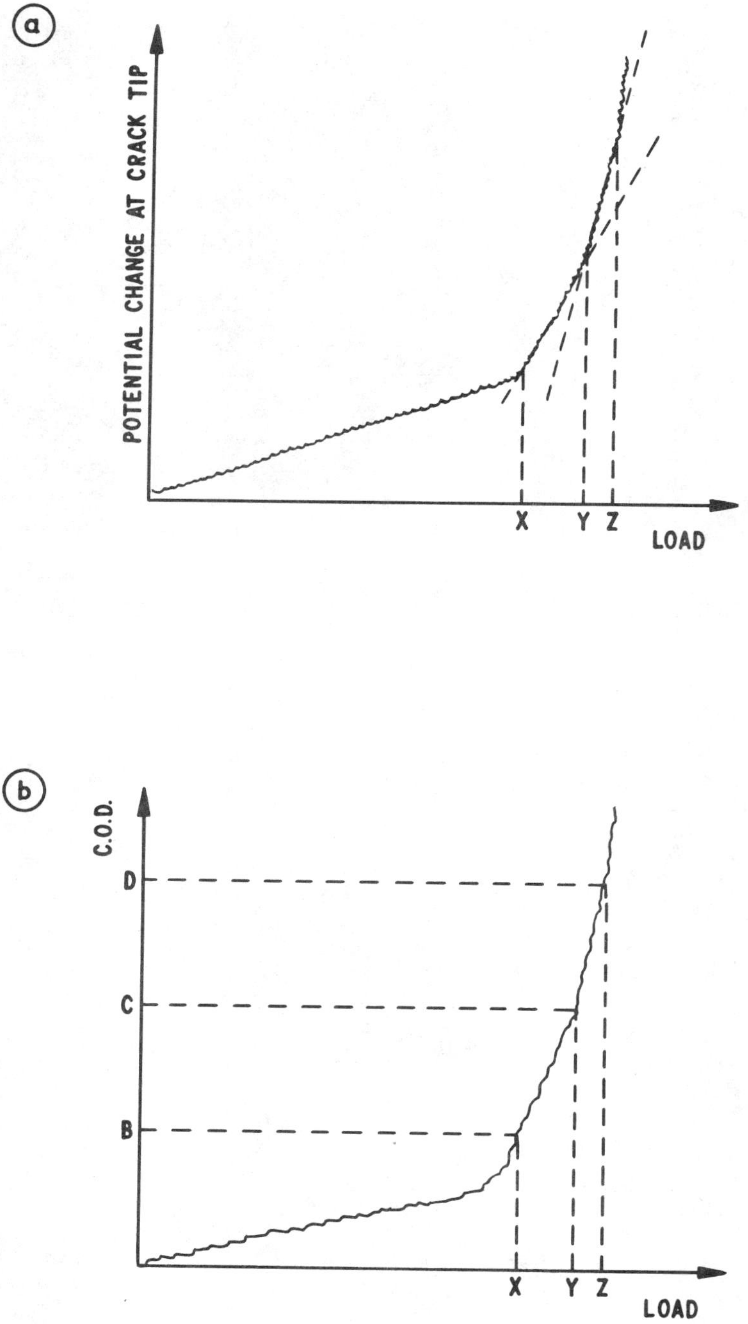

Figure 3 XY plots obtained using modified potential drop technique.

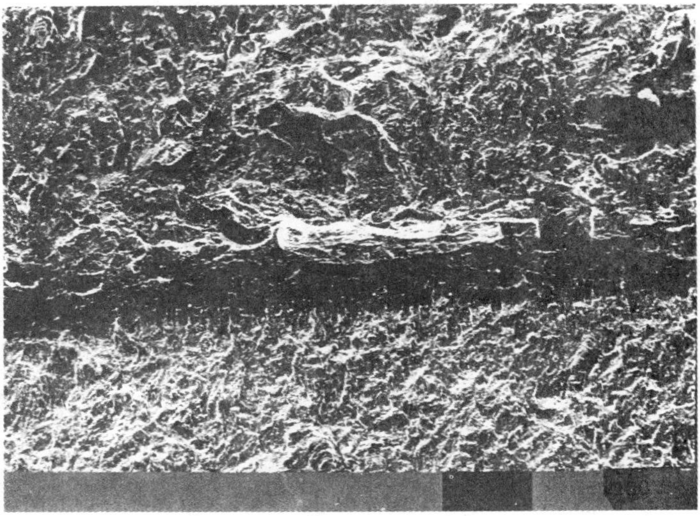

Figure 4 Presence of stretch zone on fractured specimen

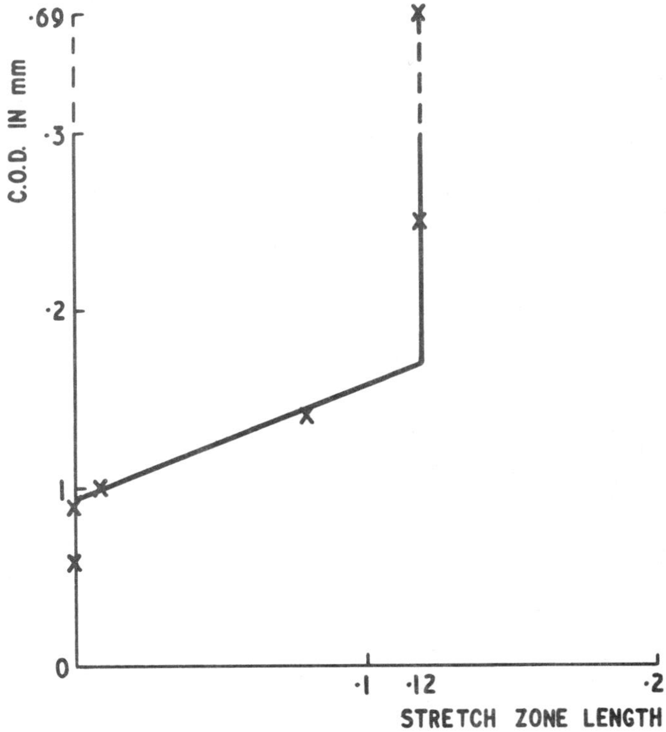

Figure 5 C.O.D. vs stretch zone length.

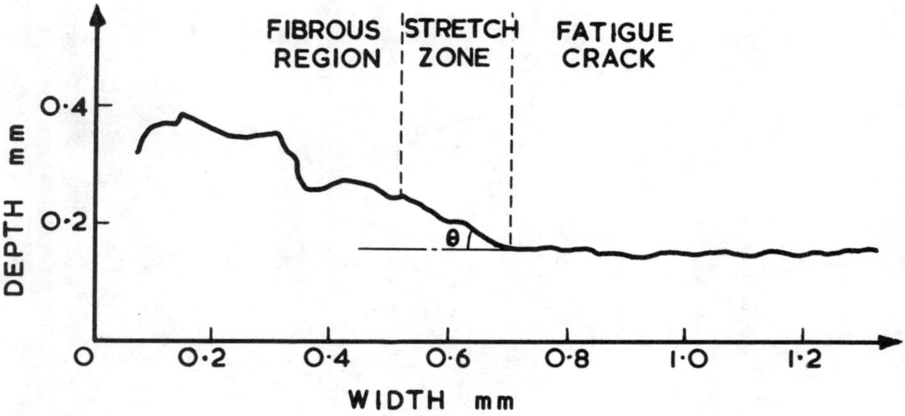

Figure 6 Talysurf trace across fracture initiation region on a broken specimen.

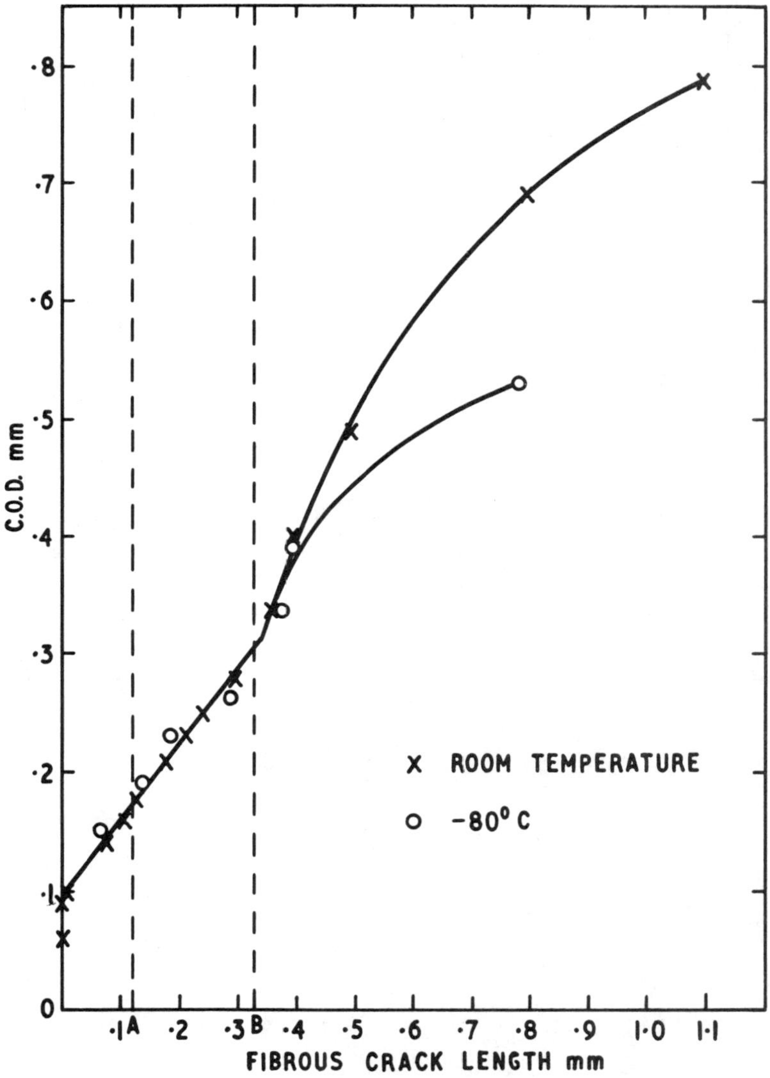

Figure 7 Effect of temperature on C.O.D. vs. fibrous crack growth

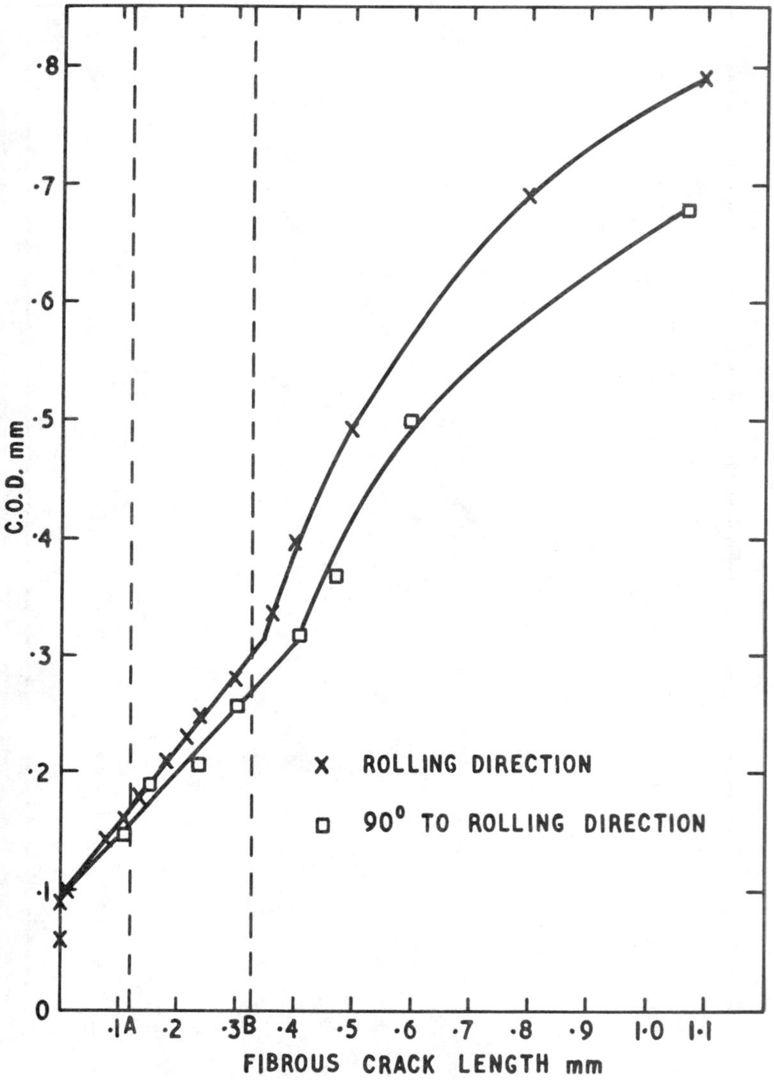

Figure 8 Effect of orientation to the rolling direction on C.O.D. vs. fibrous crack growth.

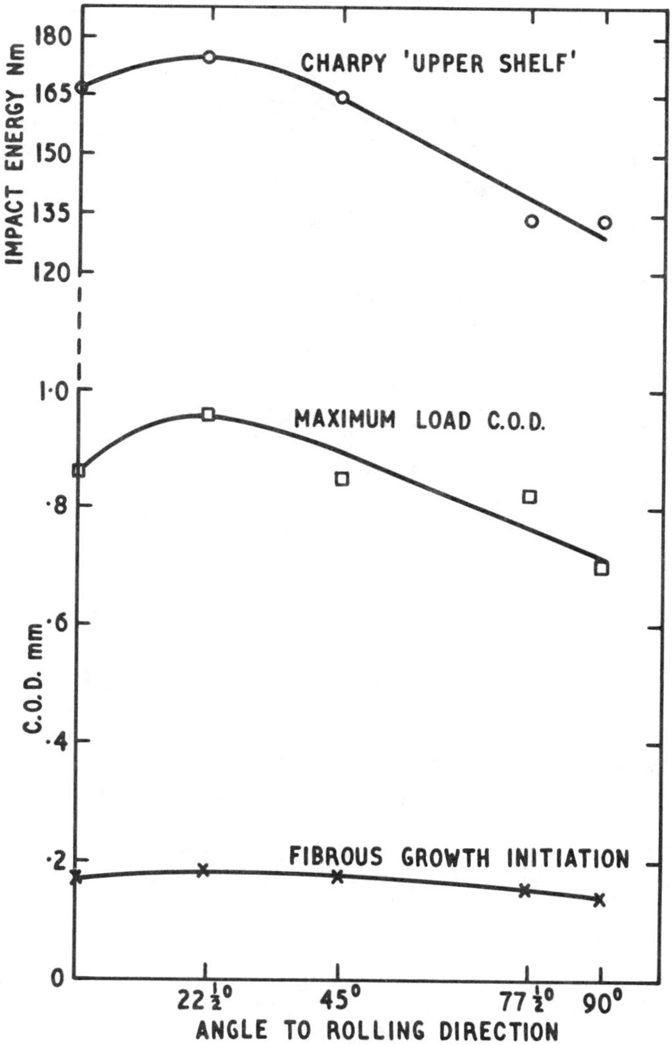

Figure 9 Effect of orientation to the rolling direction on fracture characteristics.

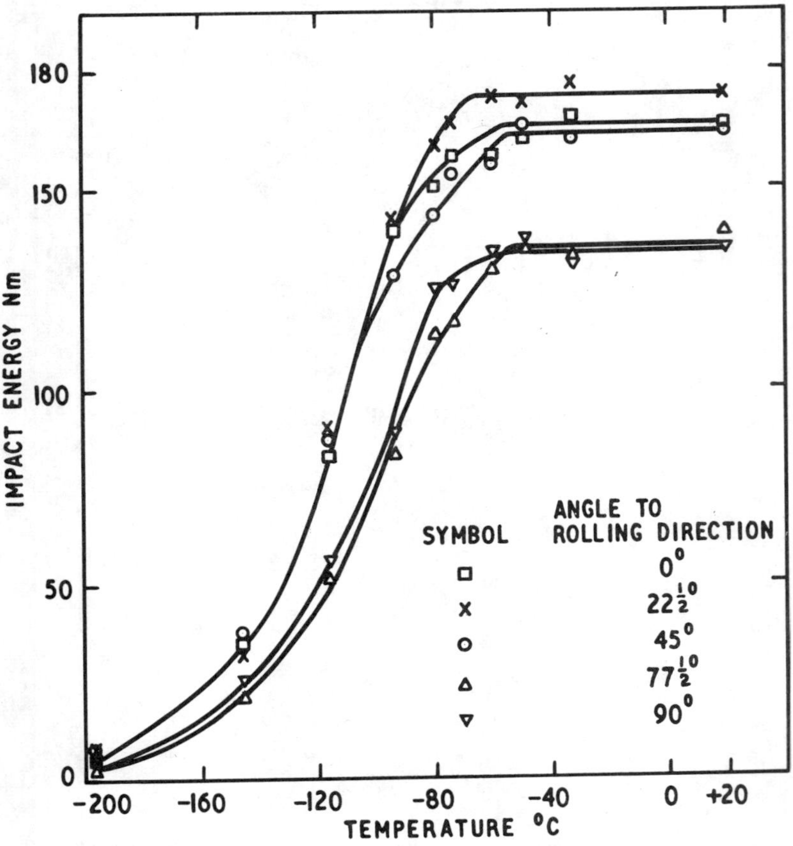

Figure 10 Effect of orientation to the rolling direction on charpy impact values

Fracture 1977, Volume 3, ICF4, Waterloo, Canada, June 19 - 24, 1977

BLUNTING EFFECTS ON FRACTURE TOUGHNESS OF LOW STRENGTH STEELS

R. K. Pandey* and S. Banerjee**

INTRODUCTION

The fracture toughness of low strength steel decreases considerably with decreasing temperature and increasing strain rate since the fracture micromechanism changes progressively from microvoid coalescence to cleavage. It is proposed [1] that crack initiation by cleavage occurs when σ_{yy}^{max} stress at some distance ahead of the crack tip reaches σ_f, the cleavage strength of the material at the crack tip. It is shown [2] that σ_f changes little with temperature and thus the temperature influences K_{IC} mainly due to the variation of σ_y with temperature wherein the crack tip radius ρ_0 remains unchanged [1]. However, K_I (K_{IC}) is an inadequate fracture mechanics parameter when elastic plastic or fully plastic loading situations are encountered such as in low strength steel specimens fractured at higher test temperatures and in such situations the parameter $J(J_{IC})$ should be used more appropriately. Accordingly K_{IC} calculated, from the experimentally measured J_{IC} (or J_Q, which is J_{cr} measured from specimens which do not satisfy requirements of valid J_{IC} measurements) value, is termed as K_{IC} (J) and is obtained from [3,4].

$$K_{IC}(J) = \left[\frac{E.\ J_{IC}}{1 - \nu^2}\right]^{1/2} \quad (1)$$

In the above loading situations, considerable crack tip blunting occurs. It is observed that [5] calculated K_{IC} values based on the achievable $\sigma_{yy}^{max}/\sigma_y$ value [6,7] in the plastic zone ahead of a sharp crack are significantly lower than the experimental K_{IC} (J) values. In addition the achievable $\sigma_{yy}^{max}/\sigma_y$ values were higher than σ_f/σ_y even at room temperature where cleavage micromechanism does not operate as confirmed by fractographic observation [5]. These observations support the contention [8] that the crack tip undergoes blunting. In fact the blunting of the crack tip is physically confirmed and supported by SZW measurement [9,10], COD measurements [11,12] and the results of crack tip profile measurements and observations [9,11,12,13].

RKR [14] used the stress distribution ahead of a crack [8] in conjunction with a two grain model [14] to calculate the K_{IC} values which were compared with the experimental K_{IC} values. However, the change in σ_y/E and n values with change in temperature and strain rate were not taken into account in their calculation.

The present investigation correlates the various elastic-plastic fracture toughness such as δ_c [15], J_{IC} [16,17] and $\bar{G}_{IC}$, the non-linear energy parameter proposed by Liebowitz and Eftis [18] of two different class of low strength steels in the temperature range -196°C to 28°C and crosshead

*Indian Institute of Technology, Delhi 110 029, India
**Indian Institute of Technology, Bombay 400 076, India

speed range 0.5 to 200 mm/minute. In particular the effect of crack tip blunting on K_{IC} (J) is assessed by an examination of the experimental data in the light of the well known theoretical results [6,7,8].

EXPERIMENTAL

The chemical analysis and the other details of the 12.5 mm thick plate materials are reported in Table 1. Three point bend fracture toughness specimens 12.5 x 15 x 75 mm cut with the long axis along the rolling direction were prepared as per ASTM E 399-72. The tests at low temperatures were carried out with the specimens immersed in a low temperature bath. The load, load line displacement and the crack opening displacement were measured as a function of time. Companion tensile test specimens of the two steels were pulled at various temperatures and strain rates.

The load-displacement [16,21] plots were suitably analysed to obtain δ_c [13,19,20], J_{IC} [16,21] and $\bar{G}_{IC}$ [18]. A measure of the crack tip strain rates was obtained, taking into account the tensile yield strength variation with temperature and strain rate [19,20]. The tensile test data were processed to obtain the n and σ_y/E values at various strain rates and temperatures.

RESULTS AND DISCUSSION

Correlation of the Fracture Mechanics Parameters

COD and J relation is given by

$$\text{COD} = \frac{J}{M \cdot \sigma_y}$$

where M takes into account the elevation of the local σ_{yy} stress at which yield occurs at the crack tip. The values of M have been theoretically determined to be 1.63 from deformation theory of plasticity [3], 2.32 using an incremental plasticity theory [24], 2 by finite element analysis [19,25]. Experimental values of M have been reported [26] for three point bend specimens and lie in the range of 0.83 to 1.75 depending upon a/W and W values.

Figure 1 shows the correlation between J_{cr}, δ_c and σ_y for the various test temperatures and crack tip strain rates for the two steels. The value of M is about 1.52 and is not unreasonable. It is interesting to note that M has a constant value in spite of the differing amounts of plasticity preceeding fracture in the specimens tested.

Figure 2 shows the relation between J_{IC} and $\bar{G}_{IC}$ for the whole range of small scale to extensive yielding situations. The correspondence between the two is reasonable up to the limits of valid J_{IC} measurements as specified in reference [16].

Stress Induced Fracture - Contribution of Progressive Crack Tip Blunting

An examination [5] of the K_{IC} (J) values in the light of Tetelman and Malkin's [1] anlaysis showed that sharp crack tip radius ρ_o of their analysis depends on temperature and strain rate and increases with increasing δ_c. It has also been shown by earlier investigators that considerable stretching occurs at the crack tip which indicates that crack tip undergoes blunting and the stretched zone width directly relates to δ_c.

It is now proposed that the crack tip undergoes progressive blunting with loading and the radius reaches a critical value ρ_L at the point of crack initiation and that

$$\rho_L = 0.5\ \delta_c \tag{2}$$

Thus the effect of temperature and strain rate on K_{IC} is reflected in two ways - firstly by virtue of their effect on yield strength and secondly through their effect on ρ.

The stress distribution ahead of a blunt crack [8] could not be used to examine experimental results since it is available only for a few σ_y/E values. On the other hand σ_y/E values change continuously in the experimental data obtained. The crack tip blunting effect is therefore assessed in an indirect manner.

The experimentally determined K_{IC} (J) values are suitably processed to calculate hypothetical K_{IC} (J) values as would be obtained if the crack tip were sharp. These hypothetical values are termed as K_{IC} (H). The K_{IC} (H) values are then compared with K_{IC} (NB), the fracture toughness values based on sharp crack stress distribution [6,7] and the two grain model [14]. At low σ_y/E values fracture of three grains is assumed to lead to fracture.

Calculation of $K_{IC}(H)$:

In the case of a blunt crack the stress σ_{yy} reaches a maximum value σ_{yy}^{max} at an approximate distance $X = 1.9\ \delta_t$ [8].

It has been shown [8] that crack opening displacement at the tip can be represented by

$$\delta_t = 0.717\ \frac{K^2}{E\ \sigma_y} \tag{3}$$

Therefore

$$(X)_{\sigma_{yy} = \sigma_{yy}^{max}} = 1.9\ \delta_t = 1.362\ \frac{K^2}{E\ \sigma_y} \tag{4}$$

It is shown [27]

$$(X)_{\sigma_{yy} = \sigma_{yy}^{max}} = \rho\ \left[\exp\left(\frac{\sigma_{yy}^{max}}{\sigma_y} - 1\right) - 1\right] \tag{5}$$

At crack initiation $\sigma_{yy}^{max} \rightarrow \sigma_f$, $\rho \rightarrow \rho_L$ and correspondingly $K_I \rightarrow K_{IC}$ (ρ). Combining equations (2), (4) and (5)

$$K_{IC}(\rho) = \frac{E\ \sigma_y\ \delta_c}{2.724}\ \left[\exp\left(\frac{\sigma_f}{\sigma_y} - 1\right) - 1\right]^{1/2} \tag{6}$$

where $K_{IC}(\rho)$ is the calculated fracture toughness for the different values of ρ_L and ρ_L value changes with strain rate and temperature.

$K_{IC}(\rho)$ can be calculated at various temperatures and strain rates since the experimental δ_c and σ_y are known and σ_f is calculated from the general yield and fracture initiation load as per a procedure reported in reference [1]. The calculated $K_{IC}(\rho)$ is plotted against the experimental $K_{IC}(J)$ values at various temperatures and strain rates in Figure 3. The experimental K_{IC} (J) values as reported in Figure 3 are obtained from equation (1). The $K_{IC}(J)$ values at intermediate strain rates and temperatures are obtained by graphical interpolation. The higher $K_{IC}(\rho)$ values could not be accommodated in the figure but they clearly obey the trend shown in the graph.

Figure 3 shows that $K_{IC}(\rho) = K_{IC}(J)$ up to $K_{IC}(J) = 60$ MPam$^{1/2}$ for the Mn-V steel and 50 MPam$^{1/2}$ for the ship building steel. Above these, the experimental $K_{IC}(J)$ changes linearly with $K_{IC}(\rho)$ and obeys an equation of the form

$$K_{IC}(J) = A+B \cdot K_{IC}(\rho) \quad (7)$$

where the constants A and B can be evaluated from Figure 3.

If ρ_L for a sharp crack is known, $K_{IC}(\rho)$ values for a sharp crack can be calculated from equation (6). In these calculations ρ_L corresponding to a sharp crack is assumed to be 2.7 µm since the fatigue crack width and the minimum δ_c value (at -196°C) is 5 to 6 µm. Once $K_{IC}(\rho)$ for a sharp crack is known the corresponding $K_{IC}(J)$ values for a sharp crack can be calculated from equation (7) and these $K_{IC}(J)$ values are termed as $K_{IC}(H)$, the hypothetical $K_{IC}(J)$ value for a sharp crack. The $K_{IC}(H)$ values calculated in this manner are plotted in Figure 4. These $K_{IC}(H)$ values can be compared with $K_{IC}(NB)$ values calculated at the different temperatures and strain rates since both these parameters refer to a sharp crack.

It should be noted that equation (6) and Figure 3 are not based on any microstructural features such as grain size. The change in slope of the lines at various points occur due to various reasons such as: 1) the crack tip becomes too blunt (of the order of the grain size) at higher $K_{IC}(J)$ values, to obey equation (3); 2) the fracture micromechanism changes from a stress induced to a strain induced one; 3) the higher $K_{IC}(J)$ values (>66 MPam$^{1/2}$ for Mn-V steel and >45 MPam$^{1/2}$ for ship building steel) are not derived from valid J_{IC} values and 4) there is a progressive loss of constraint at higher temperatures (i.e. higher $K_{IC}(J)$ values) and lower strain rates. It is interesting however to note that in spite of the above factors, the relation between $K_{IC}(J)$ and $K_{IC}(\rho)$ obeys a linear relationship as given by equation (7).

Calculation of $K_{IC}(NB)$:

The stress distribution ahead of the crack is influenced by n [8]. The result of the previous investigators [28,29] that n depends only on σ_y irrespective of temperature and strain rate was confirmed and the relations between σ_y and n for the two steels investigated were found out [5]. It was also confirmed from the examination [5] of the unnotched and notched tensile test data reported in reference [30,31] that n values are not significantly influenced by triaxiality. Thus n values of the material at the crack tip can be calculated from the σ_y value at a given temperature and crack tip strain rate.

Figure 4 in reference [8] plots σ_{yy}/σ_y versus $X/(K/\sigma_y)^2$ in case of a sharp crack for n values 0, 0.1 and 0.2. An approximate plot for n = 0.15 was generated by graphical interpolation. According to the two grain model [14], X = 2 grain diameter. With σ_f/σ_y and n values known for a given temperature and crack tip strain rate, the value of $X/(K_{IC}/\sigma_y)^2$ is known from the above figure. Since X is known the corresponding K_{IC} can be calculated. This K_{IC} is $K_{IC}(NB)$. The $K_{IC}(NB)$ values are also plotted in Figure 4.

It may be noted that at higher temperatures when σ_y values are low, σ_f is not reached in the very first grain near the crack tip. In these cases based on R_{max} quantity, a value of X=3 grain diameter is assumed and $K_{IC}(NB)$ values are computed.

Comparison of $K_{IC}(H)$ and $K_{IC}(NB)$:

The $K_{IC}(NB)$ values exhibit a somewhat irregular trend since the intermediate n values are appropriately approximated to either 0, 0.1, 0.15 or 0.2 in the calculations. If one were to ignore this irregularity, $K_{IC}(NB)$ values fall reasonably close to $K_{IC}(H)$ values up to a temperature where cleavage mode of fracture operates for a hypothetical sharp crack. The agreement between $K_{IC}(NB)$ and $K_{IC}(H)$ has an interesting consequence. The experimental $K_{IC}(J)$ values contain a contribution due to the crack tip blunting in addition to the contribution due to yield strength. It may be written as

$$K_{IC}(\text{Blunting}) = K_{IC}(J) - K'_{IC}(H)$$

where $K_{IC}(H) = K_{IC}(NB)$.

CONCLUSIONS

1. In the specimen geometry investigated, the critical COD is observed to relate to J_{cr} by the relation

$$\text{COD} = \frac{J_{cr}}{1.53}\ \sigma_y .$$

2. Experimental J_{IC} values show a reasonably good agreement with $\bar{G}_{IC}$ as long as J_{IC} measurements are valid.

3. Significant amount of crack tip blunting occurs during elastic-plastic loading situation and this contributes considerably to the toughness. The contribution can be evaluated from a relation of the type

$$K_{IC}(\text{Blunting}) = K_{IC}(J) - A - B\sqrt{\rho_s\ E\ \sigma_y}\left[\exp\left(\frac{\sigma_f}{\sigma_y} - 1\right) - 1\right]^{1/2}$$

where the value of ρ_s corresponds to the critical crack tip radius at the initiation of a sharp crack and A and B are numerical constants.

REFERENCES

1. MALKIN, J. and TETELMAN, A. S., Engng. Fracture Mech., 3, 1971, 151.
2. KNOTT, J. F., JISI, 204, 1966, 104.
3. RICE, J. R., J. Appl. Mech., 35, 1968, 379.
4. BEGLEY, J. A. and LANDES, J. D., ASTM STP 514 (II), 1972, 1.
5. PANDEY, R. K., Ph.D. Thesis, IIT-Bombay, 1976.
6. HUTCHINSON, J. W., J. Mech. and Phys. Solids, 16, 1968, 13.
7. RICE, J. R. and ROSENGREN, G. F., J. Mech. and Phys. Solids, 16, 1968, 1.
8. RICE, J. R. and JOHNSON, M. A., "Inelastic Behavior of Solids" (ed. Kanien, et al), McGraw Hill, New York, 1970.

9. TANAKA, J. P., PAMBILLO, C. A. and LOW, J. R., ASTM STP 463, 1970, 191.
10. PANDEY, R. K. and BANERJEE, S., Engng. Fract. Mech., 5, 1973, 965.
11. SPITZIG, W. A., ASTM STP 453, 1969, 90.
12. BROTHERS, A. J., et. al., ASTM STP 493, 1971, 3.
13. BATES, R. C., CLARK, W. G. and MOON, D. M., ASTM STP 453, 1969, 148.
14. RITCHIE, R. O., RICE, J. R. and KNOTT, J. F., J. Mech. and Phys. Solids, 21, 1973, 395.
15. Methods for COD Testing, DD19:1972, British Standards Institution, London.
16. BEGLEY, J. A. and LANDES, J. D., Westinghouse Scientific Paper 73-1E7-FMPWR-P3, October 1973.
17. BEGLEY, J. A. and LANDES, J. D., ASTM STP 514, 1972, 1.
18. EFTIS, J. and LIEBOWITZ, H., Engng. Fract. Mech., 7, 1975, 101.
19. HAYES, D. J., and TURNER, C. E., Int. J. Fract., 10, 1974, 10.
20. VEERMAN, C. C. and MULLER, T., Engng. Fract. Mech., 4, 1972, 25.
21. LANDES, J. D. and BEGLEY, J. A., "Guidelines for J_{IC} Tests", Report of ASTM Task Group E.24.01.09, 1974.
22. SHOEMAKER, A. K. and ROLFE, S. T., Trans. ASME, 91D, 1969, 512.
23. PANDEY, R. K., HARIDAS, J. D. and BANERJEE, S., Engng. Fract. Mech., 6, 1974, 105.
24. LEVY, N., et al., Int. J. Fract., 7, 1971, 143.
25. BOYLE, E. F. and WELLS, A. A., Report on 'Finite Element Study of Plane Strain Fracture Criteria Under Elastic Plastic Condition', June 1973, Queens Univ., Belfast.
26. DAWES, M. G., The Welding Institute Research Report, E/65/75, March 1975.
27. HILL, R., Mathematical Theory of Plasticity, Oxford Press, London 1950.
28. ROSENFIELD, A. R. and HAHN, G. T., Trans. ASM, 59, 1966, 962.
29. HOLLOMAN, J. H. and ZENER, C., Trans. AIME, 158, 1944, 283.
30. HAYDEN, H. W. and FLOREEN, S., Acta Met., 17, 1969, 213.
31. COX, T. B. and LOW, J. R., Met. Trans., 5, 1974, 1457.

Table 1 Chemical Analyses, Tensile Properties and Grain Size of Steels Investigated

Steel	C	Si	P	S	Mn	V	YS MPa	UTS MPa	E%	Average Grain Size in μm
Mn-V (TISCO)	0.20	0.235	0.032	0.027	1.60	0.12	434	600	38	18
Ship Building (Lloyds Grade A)	0.16	0.03	0.009	0.019	0.80	-	234	365	36	27

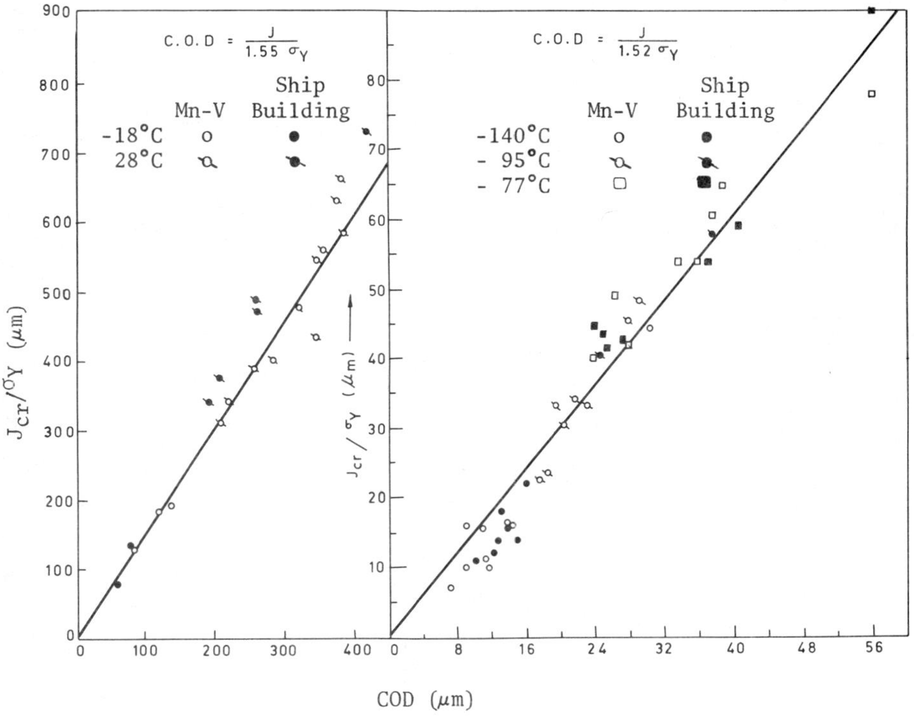

Figure 1 Relation Between Critical COD and J_{cr}/σ_Y

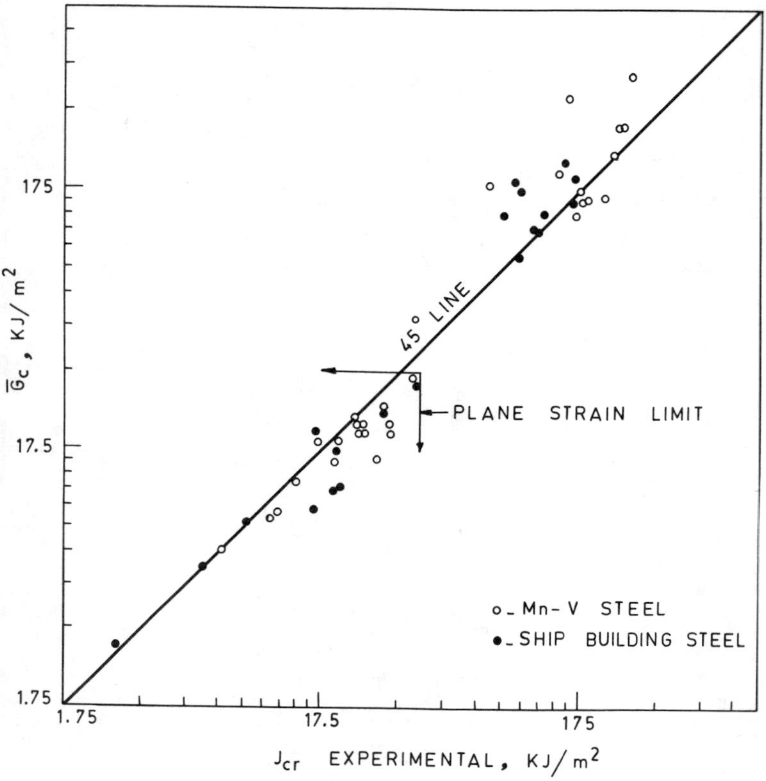

Figure 2 Comparison Between Experimental J_{cr} Values and $\overline{G}_c$ Values of Liebowitz and Eftis

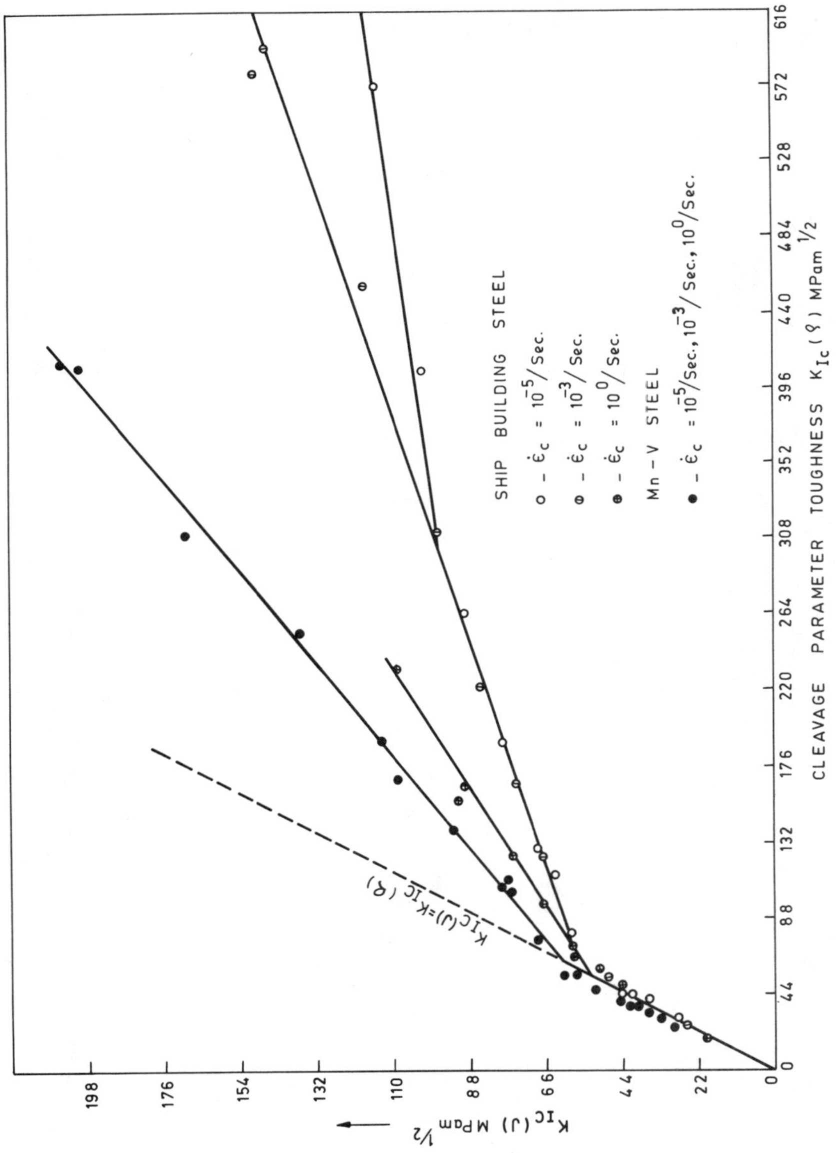

Figure 3 $K_{IC}(J)$ as a Function of Cleavage Parameter Toughness

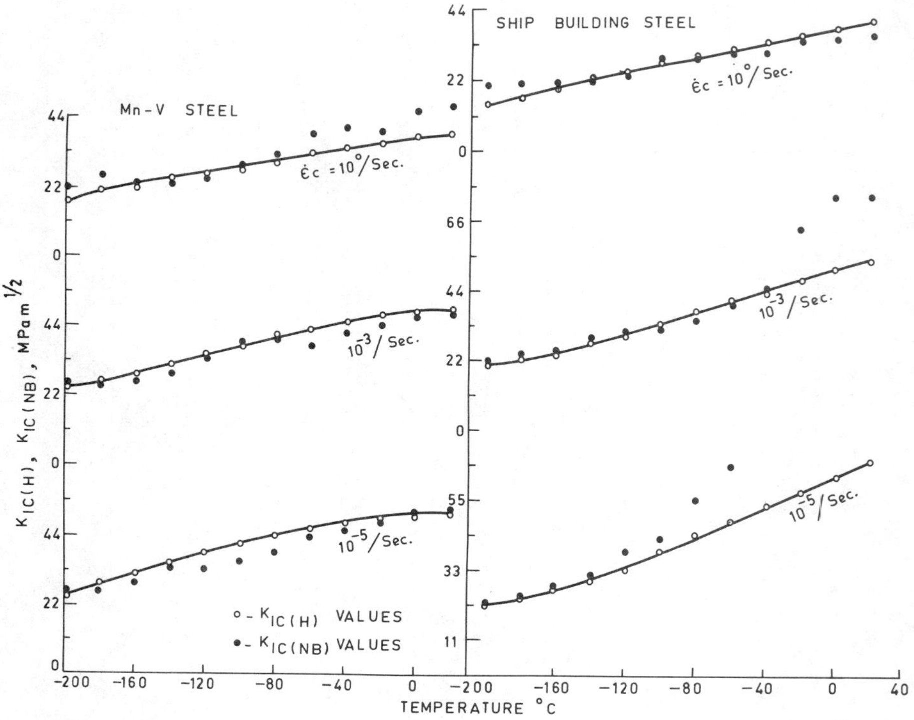

Figure 4 Temperature Variation of $K_{IC(H)}$ and $K_{IC(NB)}$

Fracture 1977, Volume 3, ICF4, Waterloo, Canada, June 19 - 24, 1977

VARIATIONAL BOUNDS AND QUALITATIVE METHODS IN FRACTURE MECHANICS

V. M. Entov* and R. V. Goldstein*

INTRODUCTION

According to modern fracture mechanics, to determine the conditions of subsequent growth of a crack of given geometry, it is necessary to know the stress intensity factor in the points of the initial crack contour as well as in the points of all the subsequent positions of the crack contour. This is of minor importance in plane and axisymmetrical problems but gives rise to great difficulties in three-dimensional problems such as the problem of growth of an opening mode crack in the plane of symmetry of an elastic body.

The paper is concerned with some methods of determination of conditions sufficient for a crack to be dangerous or safe in the principal three-dimensional case mentioned above. The key to the problem is the notion of *positivity* which is introduced here. The problem is said to be positive if the application of arbitrary positive (wedging) tractions to crack surfaces gives rise to positive normal displacements of the surfaces and positive normal stresses in the plane of symmetry outside the crack. For positive problems there is the following *comparison principle:* the stress intensity factor at a given point of the crack contour grows as the crack extends outside some arbitrary small region around the point. The stress intensity factor grows also if the additional wedging forces are applied to the crack surfaces. It follows that for positive problems a given crack is more dangerous (i.e., gives rise to fracture during a shorter period of time) than any crack it contains and is less dangerous than any crack that contains it. This makes it possible to take into consideration the conditions of growth of cracks with comparatively simple ("standard") contours only.

DISCUSSION

The principle of comparison makes it possible to construct a two-sided estimate of the stress intensity factor at a given point on the arbitrary smooth contour of a crack considering contours of simpler form enveloping the given crack and enveloped by it and having a common tangent in the point under consideration.

It has been shown [1] that the problem of a plane crack in an infinite three-dimensional body is positive. In [1] it was demonstrated for the example of an elliptical crack in an infinite elastic body that the comparison principle is a highly efficient means of construction of two-sided estimates of stress-intensity factors. The theorem may be extended to include the case of an opening mode crack in a bounded body as soon as

*Institute of Mechanical Problems USSR Academy of Science, Moscow, USSR.

the boundary is sufficiently far from the crack. It has been proved that the problem of an opening - mode crack situated in the central plane of an elastic layer with traction-free faces is positive as soon as $(d/h) < 0.7$, d - being the diameter of the plane domain occupied by the crack, 2h - the thickness of the layer. The crack opening and stress intensity factor at any point of the crack contour diminish as the thickness of the layer grows, the crack goemetry and tractions on its surfaces being fixed.

The stress intensity factor is a local differential quantity, so that its evaluation involves enormous difficulties. Nevertheless, the load distribution being prescribed there exists a family of specific so-called "extremal" contours with the following two properties: 1) The stress intensity factor is constant along the crack contour. 2) The elastic energy of a body with a crack of given area attains its maximum value for cracks with contours which belong to the family of extremal contours. The respective values of energy and stress intensity factor are functions of area bounded by extremal contours, so the stress intensity factor may be expressed through the derivative of the elastic energy on the crack area. The extremal contours may be used as "barriers" for crack growth in the sense of the introduction. As a result it becomes possible to express the notion of a dangerous or a safe crack in terms of integral quantities such as the crack area and elastic energy of the body with a crack. If the comparison principle is valid for the body under consideration and there is an extremal contour such that the corresponding stress intensity factor is equal to its critical value, then all the cracks contained inside the contour are safe and all the cracks containing the contour are dangerous. Consider an extremal contour Γ which bounds a domain G of area S. It is assumed that there is a supporting domain G_o of area S_o, $G_o < G$. Then the extremal contour generally has two parts one of which $\Gamma'' = \Gamma \cap \partial G_o$ and the second Γ' is free, i.e. lies outside G_o. The stress intensity factor on the free part of Γ is expressed through the corresponding values of elastic energy W as a function of area S: $W = W(S)$ by the formula

$$N^2/_{\Gamma'} = \frac{\mu}{(1-\nu)\pi}\frac{dW}{dS} \tag{1}$$

The true displacements in the points of the surface of a crack with a contour of given form minimize the elastic energy W of the cracked body. So through the definition of an extremal contour is devised a solution of a "maximin" problem:

$$W_o = \max_{\text{mes } G = S} \min_{G} W \tag{2}$$

Consider an example of an extremal contour. Let a crack occupy a domain G_o in the plane $x_3 = 0$ in elastic space with surfaces act normal tractions acting on its surfaces:

$$\sigma_{33} = -p(1+\varepsilon x_1^2) \;, \; p=\text{const} \;, \; \varepsilon=\text{const} \tag{3}$$

The constant ε is assumed to be small. If $\varepsilon = 0$ the extremal contour is evidently a circle containing G_o. For small ε and with symmetry assumed it seems reasonable to seek the extremal contour as an elliptical contour of form:

$$x_1 = a_1\cos\theta,\; x_2 = b_1\cos\theta,\; a_1 = a(1+\delta_1),\; b_1 = a(1+\delta_2),\; \delta_{1,2} \ll 1 \tag{4}$$

The unknown $\delta_{1,2}$ may be determined using the condition N=const along the contour. Some algebra gives:

$$\delta_1 = \varepsilon a^2 \frac{29}{345} \quad , \quad \delta_2 = - \varepsilon a^2 \frac{3}{115} \tag{5}$$

Now the equations (4), (5) give a family of extremal contours corresponding to load of form (2), a being a parameter. The family may be used to estimate the conditions of limiting equilibrium of plane cracks of arbitrary geometry under loading of form (2) as outlined before.

To apply the approach presented here we must have some effective solutions of three-dimensional crack problems for cracks bounded by etalon contours. Such solutions may be constructed by straightforward use of variational and variational-difference methods. Some numerical results are presented. The results may be obtained using medium range digital computers.

The conditions for a crack to be dangerous or safe having been expressed in terms of an energy criterion there is a possibility of further simplification of the elasticity problem under consideration. The simplification is based on the following statement. Let the external loads $\vec{\sigma}_n = \vec{f}(\vec{x})$ be prescribed on a part S' of the surface S of an elastic body D, on the rest of S being prescribed the displacements, $\vec{u} = \vec{g}(\vec{x})$. The elastic constants of the material are considered to be functions of coordinates: $\lambda = \lambda(\vec{x})$, $\mu = \mu(\vec{x})$. The quantity Q:

$$Q = \iint_{S'} \vec{f}\,\vec{u}\, d\sigma - \iint_{S''} \vec{\sigma}_n\, \vec{g}\, d\sigma \tag{6}$$

may be considered as a functional of λ and μ; $Q = Q(\lambda,\mu)$. It may be shown that the functional is *monotonic:* $\lambda'(\vec{x}) \geq \lambda(\vec{x})$, $\mu'(\vec{x}) \geq \mu(\vec{x})$, $\nabla \vec{x} \in D$, then $Q' = Q(\lambda',\mu') \leq Q(\lambda,\mu)$. In particular, for $\vec{g}\equiv 0$ (the part S" of S is clamped) Q is equal to the work done by the external loads. So it follows that the work increases as the material rigidity in some subdomain of S decreases and vice versa. Thus it is possible to estimate strain energy for a given cracked body in terms of the energies of bodies of simpler geometry with cracks bounded by extremal (at the prescribed crack area) contours. Consider, for example, infinite space with a crack under uniform tension normal to the crack plane. Now the free extremal contours are, evidently, circles; the elastic energy is equal to half the crack volume multiplied by the applied stress. This assumes that the volume V of a plane crack of arbitrary geometry and area S, under internal pressure P is not greater than the volume of a penny-shaped crack of the same area, so:

$$V \leq \frac{P}{E} \frac{16(1-\nu^2)}{3\pi^{3/2}} S^{3/2} \tag{7}$$

The inequality is an analogue of a well-known inequality for the capacity of a plane domain [2].

In certain cases the energetic bounds may be applied directly to estimate the stress intensity factors. For example, consider a circular crack of radius ℓ around a spherical cavity of radius ρ. The crack and the cavity are under an internal pressure P. The potential energy is, by using linearity and dimensional arguments:

$$W = \frac{P^2 \ell^3}{\mu} \phi \left(\frac{\rho}{\ell}\right) \tag{8}$$

ϕ being a dimensionless function. Through the statements of this section we have $\phi' \geq 0$. Thus:

$$\frac{\partial W}{\partial \ell} = \frac{3P^2\ell^2}{\mu}\,\phi\left(\frac{\rho}{\ell}\right) - \frac{P^2\ell\rho}{\mu}\,\phi'\left(\frac{\rho}{\ell}\right) \leq \frac{3P^2\ell^2}{\mu}\,\phi\left(\frac{\rho}{\ell}\right) = \frac{3W(\rho,\ell)}{\ell} \leq \frac{3W(\ell,\ell)}{\ell} \tag{9}$$

It implies, using Irwin's formula:

$$N^2 \leq N_o^2 = \frac{3P^2\ell}{4\pi(1-\nu)} \tag{10}$$

For $\nu = 0.25$, $N \leq 0.56\ P\sqrt{\ell}$. For a penny-shaped crack of radius

$$r_o\ ,\ N = \frac{P\sqrt{2r_o}}{\pi} \approx 0.45\ P\sqrt{r_o}\,.$$

So a crack of radius ℓ surrounding a spherical cavity of radius ρ is less dangerous than a penny-shaped crack of radius

$$r_{eff} = \frac{\pi\ell}{8(1-\nu)} \quad (\text{for } \nu = 0.25,\ r_{eff} = \frac{\pi\ell}{2}).$$

In elastic contact problems this makes it possible to construct bounds for the displacement of a die and/or the force acting on the die, with the complex geometries of contact areas and elastic body and various types of contact (sliding contact, frictional contact, etc.), on the basis of a solution of the respective problems for dies and/or bodies of simple geometries.

REFERENCES

1. GOLDSTEIN, R. V. and ENTOV, V. M., Int. J. Fracture, 6, 1975, 11.
2. POLYA, G. and SZEGO, G., Isoperimetric inequalities in mathematical physics, Princeton University Press, 1951.

Fracture 1977, Volume 3, ICF4, Waterloo, Canada, June 19 - 24, 1977

AN APPROXIMATE THREE-DIMENSIONAL STATE OF STRESS IN THE VICINITY OF A CRACK

Yu Chen*

INTRODUCTION

The classical linear fracture mechanics solutions of the crack-tip stresses are based on two-dimensional formulation of the stress field under either plane stress or plane strain conditions. For contained plasticity the slip-line field is not enough for predicting the plastic zone size. Besides numerical treatment, several papers presented asymptotic analysis of the stress singularity at the crack tip [1, 2, 3]. Three-dimensional solutions are difficult and not available at the present time. The object of this paper is to study the effect of the plate thickness on the stress distribution in the plastic zone of a through crack with the assumption of an ideally plastic material.

MATHEMATICAL ANALYSIS

Consider the stress function defined by

$$F = \frac{1}{2} r^2 f(\theta) \ . \tag{1}$$

The stress components defined by equation (1) are [4]

$$\left.\begin{aligned} \sigma_r &= \frac{1}{r}\frac{\partial F}{\partial r} + \frac{1}{r^2}\frac{\partial^2 F}{\partial \theta^2} = f + \frac{1}{2} f'' \ , \\ \sigma_\theta &= \frac{\partial^2 F}{\partial r^2} = f \ , \\ \tau_{r\theta} &= -\frac{\partial^2}{\partial r \partial \theta}\left(\frac{F}{r}\right) = -\frac{1}{2} f' \ . \end{aligned}\right\} \tag{2}$$

If one specifies that the octahedral shearing stress τ_o defined by [4]

$$9\tau_o{}^2 = (\sigma_1-\sigma_2)^2 + (\sigma_2-\sigma_3)^2 + (\sigma_3-\sigma_1)^2 \tag{3}$$

to be constant, one will get the classical solution of the plastic stress field at the crack tip. In the above expression σ_1, σ_2, σ_3 are principal.

To construct an approximate solution in three dimensions let us assume a set of two Maxwell function [5]

* Department of Mechanics and Materials Science, Rutgers University, The State University of New Jersey, New Brunswick, New Jersey, U.S.A. 08903

$$X = Y = \psi H \quad \text{and} \quad Z = \phi F \tag{4}$$

where H and F are functions of x and y only, ψ and ϕ are functions of x, y and z except that these functions vary so slowly with x and y that their derivatives with respect to x and y can be ignored. The resulting stress components are given by the following:

$$\left.\begin{array}{ll} \sigma_x = \phi F_{yy} + \psi'' H\,, & \tau_{yz} = -\psi' H_y \\ \sigma_y = \phi F_{xx} + \psi'' H\,, & \tau_{zx} = -\psi' H_x \\ \sigma_z = \psi \nabla^2 H\,, & \tau_{xy} = -\phi F_{xy} \end{array}\right\} \tag{5}$$

where the subscripts represent partial differentiation and the primes denote differentiations with respect to z.

If we further introduce the assumption that the functions F and H are identical, then the above stress components will be reduced to four remaining components; namely,

$$\left.\begin{array}{ll} \sigma_x = \phi\, \bar{\sigma}_x\,, & \sigma_y = \phi\, \bar{\sigma}_y\,, \\ \sigma_z = \psi(\bar{\sigma}_x + \bar{\sigma}_y)\,, & \tau_{xy} = \phi\, \bar{\tau}_{xy}\,, \end{array}\right\} \tag{6}$$

where $\bar{\sigma}_x = F_{yy}$, $\bar{\sigma}_y = F_{xx}$, $\bar{\sigma}_{xy} = -F_{xy}$. (7)

It is important to note that the above set of stresses do not satisfy the equations of equilibrium exactly, but the quantities neglected will be small by assumption.

The task now is to suggest a function F that will yield a fan-shaped plastic region in the xy-plane. In this region the octahedral stress τ_0 will be constant. This function F can be taken to be the same classical solution that solves the plane-strain case in the region $0 \leq |\theta| \leq \pi/4$ and a new solution outside the region. (See Figure).

Thus, for $0 \leq |\theta| \leq \frac{\pi}{4}$,

$$f(\theta) = \frac{p+q}{2} - \frac{p-q}{2} \cos 2\theta \tag{8}$$

and for $0 \leq |\alpha| \leq \frac{3\pi}{4}$, $(\alpha = \pi - \theta)$

$$f(\alpha) = \frac{p+q}{2} - \frac{3q}{4} \cos \frac{2\alpha}{3} + \frac{p-2p}{4} \cos 2\alpha \tag{9}$$

where p and q are the stresses at $\theta = 0$.

It can be verified that these solutions satisfy all the boundary conditions and at $\theta = \pi/4$ all the stress conditions are matched.

By using equation (6) we can calculate τ_o as

$$9\tau_o^{\ 2} = (\sigma_1-\sigma_2)^2 + (\sigma_2-\sigma_3)^2 + (\sigma_3-\sigma_1)^2 = \frac{1}{2}(\phi-2\psi)^2(\bar{\sigma}_x+\bar{\sigma}_y)^2 + \frac{3}{2}\phi^2 V \qquad (10)$$

where V is defined by

$$V = (\bar{\sigma}_x-\bar{\sigma}_y)^2 + 4\,\bar{\tau}_{xy}^{\ 2}. \qquad (11)$$

For the case of plane strain $\phi = 2\psi$, we have

$$9\tau_o^{\ 2} = \frac{3}{2}\phi^2 V \qquad (\phi = 1)\ , \qquad (12)$$

and for plane stress, $\psi = 0$, thus

$$9\tau_o^{\ 2} = 2\phi^2\,(\bar{\sigma}_x^{\ 2} - \bar{\sigma}_x\bar{\sigma}_y + \bar{\sigma}_y^{\ 2} + 3\bar{\tau}_{xy}^{\ 2})\ . \qquad (13)$$

If we denote the value of ϕ at $z = 0$ by ϕ_m and that at $z = \pm c$ by ϕ_s and equate the two τ_o values, we obtain from equations (12) and (13) the ratio of the two ϕ-values as

$$\left(\frac{\phi_m}{\phi_s}\right)^2 = \frac{4}{3}\left[1 + \frac{p/q}{(1-p/q)^2}\right] \qquad (14)$$

The ϕ-function is further restricted by

$$\int_{-c}^{c} \phi\, dz = 2c \qquad (15)$$

The ψ-variation can also be estimated based on the condition τ_o being constant. The more specific form of ϕ and ψ can be suggested based on experimental results available in the literature, but will not be the subject of this paper.

CONCLUSION

An approximate three-dimensional solution of the state of stress in the vicinity of a crack has been suggested. Although these stresses do not satisfy the equilibrium equations exactly, the error is expected to be small for moderately thick plate. Variations of the stresses in the thickness are defined by two functions which can be estimated.

REFERENCES

1. RICE, J. A. and ROSENGREN, A. F., "Plane Strain Deformation Near a Crack Tip in a Power Law Hardening Material", J. Mech. and Phys. of Solids, 6, 1968, 1.
2. HUTCHINSON, J. W., "Plastic Stress and Strain Fields at a Crack Tip", J. Mech. and Phys. of Solids, 16, 1968, 337.
3. HUTCHINSON, J. W., "Singular Behavior at the End of a Tensile Crack in a Hardening Material", J. Mech. Phys. of Solids, 16, 1968, 13.

4. NADAI, A., "Theory of Flow and Fracture of Solids", 1, McGraw-Hill, New York, 1950, 531, 201.
5. LOVE, A. E. H., "Mathematical Theory of Elasticity", Dover Publications, New York, 1944, 84.

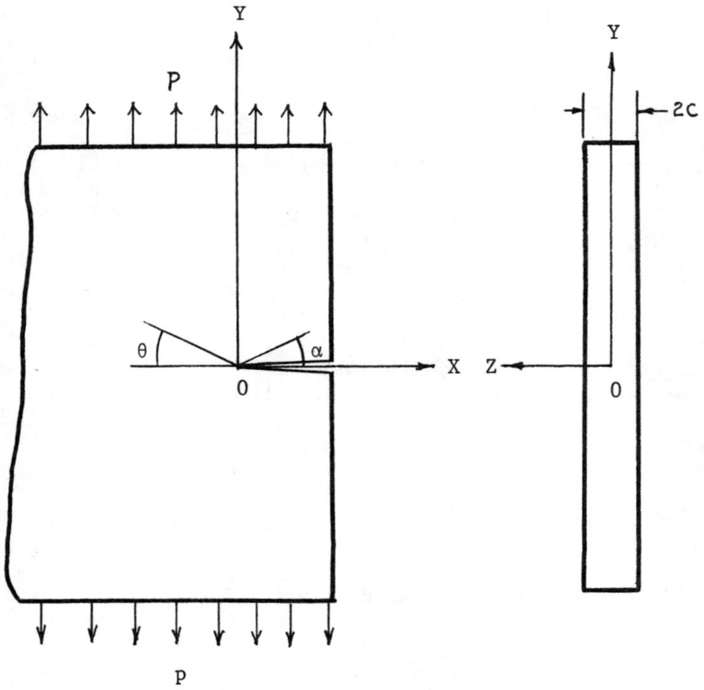

Figure 1 Plate with Finite Thickness with Crack Under Tension

Fracture 1977, Volume 3, ICF4, Waterloo, Canada, June 19 - 24, 1977

ON THE THREE DIMENSIONAL THEORIES OF CRACKED PLATES

S. K. Bhandari*, B. Barrachin** and J. L. Picou*

INTRODUCTION

It is well known since the beginning of Fracture Mechanics that the Strain-Energy-Release-Rate G_c depends on the thickness B of the specimen used; the value of G_c at smaller thicknesses could sometimes be seven times that for very thick specimens ((see for example Figure 3.36) in reference [1]). It is evident that the plasticity correction used after Irwin and based on plane-stress plane-strain argument could not help account for this variation and it was generally believed that a three-dimensional theory was needed to explain the triaxiality effect. A three-dimensional elasto-plastic theory is beyond the present reach. Even in the elastic domain, the problem seems to be extremely complicated. Nevertheless, recently two theories of elastic cracked plates have been proposed which reveal the variation of the stress-intensity factor K in the thickness direction. One would therefore be tempted to explain at least partially the thickness effect noting that G and K are related through material constants.

The aim of the present communication is:
1) to discuss two theories: one due to Hartranft and Sih [2] (we shall call it H-S theory) which starts with a certain proposed variation of stresses through the thickness, the other due to Folias using a certain integral representation for the displacements;
2) to present some numerical results obtained on a cracked plate of moderate thickness. These will be critically examined to evaluate the available theories;
3) finally, to look into some of the fundamental hypotheses of the present day fracture theory in the light of available 3-D fracture results.

THEORY PROPOSED BY HARTRANFT AND SIH

In a series of papers [2,3,4], Hartranft and Sih have developed an approximate 3-D theory of plates as applied to crack problems. In [3], they have shown that the singular part of the normal stress σ_{yy}, for the case of a plate subjected to uniform stress σ_o (Figure 1) can be written as

$$\sigma_{yy} = \frac{k_s(z)}{\sqrt{2r}} \cos\frac{\theta}{2}\left(1 + \sin\frac{\theta}{2} \cdot \sin\frac{3\theta}{2}\right)$$

where the stress-intensity factor is

$$k_s(z) = g(p)\ \phi(1)\ \sigma_o\sqrt{a}\ f''(\zeta)$$

*G.A.A.A., 20 Avenue Edouward Herriot, Le Plessis Robinson, 92350-France.
**D.S.N., C.E.N.-Saclay, C.E.A., Gif-sur-Yvette, 92150-France.

where $\zeta = z/B$, g(p) is a function of a constant parameter p and $\phi(1)$ is given as a function of the parameter p and the ratio B/a (Figure 2).

The mean value of $k_s(z)$

$$\overline{K}_s = \frac{1}{2B}\int_{-B}^{+B} k_s(z)\, dz = \frac{4}{p}\,\frac{\sin^2 p}{(2p+\sin 2p)}\,\phi(1)\,\sigma_o\sqrt{a}$$

which using p = 0,4 as suggested in [4] becomes

$$\overline{K}_s = \phi(1)\,\sigma_o\sqrt{a}$$

It is interesting to note that $\overline{K}_S$ varies with B/a through $\phi(1)$, a variation similar to the one implied by the experimental data. Thus one might expect to account for a partial thickness effect with the help of this theory.

THEORY PROPOSED BY FOLIAS [5]

Folias has treated the same problem using the method of Lure [6]. The analysis is sufficiently complex. The final result for the stress σ_{yy} in the crack-tip vicinity is

$$\sigma_{yy} = \Lambda\, F(\zeta)\,\frac{\sigma_o\sqrt{a}}{\sqrt{2r}}\cos\frac{\theta}{2}\left(1 + \sin\frac{\theta}{2}\,.\,\sin\frac{3\theta}{2}\right),\quad 0 < \zeta < 1,$$

which leads to the stress-intensity factor

$$k_F(z) = \Lambda\, F(\zeta)\,\sigma_o\sqrt{a},\quad F(\zeta) = \frac{1}{2}\left\{\frac{1}{(1-\zeta)^{2\nu}} + \frac{1}{(1+\zeta)^{2\nu}}\right\}$$

and $\Lambda = \Lambda(B/a)$ is shown in Figure 3 for $\nu = 1/3$.

The average value of $k_F(z)$ is given by

$$\overline{K}_F = \frac{\Lambda\,\sigma_o\sqrt{a}}{(1-2\nu)(2)^{2\nu}}$$

It should be noted that the mean value of K_F is greater than the two-dimensional value except for $\nu = 0$ when the plane-stress result is recovered. Moreover, one finds that contrary to the results of H-S theory, Folias' theory indicates an increase of stresses as z/B is varied from 0 to the value of 1. On the other hand, the variation in the mean value of $\overline{K}_F$ ($\nu \neq 0$) with respect to B/a (Figure 3) is of the order of 11% and one would tend to believe that a 3-D elastic theory would not give even a partial answer to the experimental variation in G_c with respect to thickness.

NUMERICAL RESULTS [7]

We have carried out certain numerical computations for the value of K in a moderately thick centrally through the thickness cracked plate.

The configuration used is that of Figure 1 with a width of 60 mm and two values for the thickness and for the crack length to obtain finally three different values of B/a (1.0, 0.75 and 0.25). The computations were performed using two different methods:

The Finite Element Method

We shall denote

$$K_\sigma = \lim_{r \to 0} \sqrt{2r}\ \sigma_{yy} \quad \text{(using stress data)}$$

$$K_u = \lim_{r \to 0} \frac{E\ U_y}{2\sqrt{2r}\ (1-\nu^2)} \quad \text{(using displacement data and plane strain assumption)}$$

Note that, although both K_σ and K_u values are given in the text, K_u is the solution to be compared because the computer codes we used assure the displacement compatibility as most of the F.E. computer programmes. This point has already been stressed in the literature [8] and we shall not go into more details:

The F.E. programmes used are:

1) SAFE 2D for a two dimensional analysis in order to compare with the three-dimensional analysis
2) SAP IV for the analysis of the finitely thick plate. In this case 16-node brick elements were used. The mesh is shown in Figure 4 for 1/8th of the plate considered. We have used 336 elements. Two values of B/a = 0.75 (Case I) and B/a = 0.25 (Case II) were run with a uniform tensile stress of 147 MPa applied at the edges of the plate parallel to the crack.

The results concerning σ_{yy}- distribution through the thickness for the plane y = 0 are shown in Figure 5. The computed mean K-values are given in Table 1. These results will be discussed later.

The Boundary Integral Equation Method

This method has been developed due to the efforts of Cruse [9], Lachat and Watson [10] and others. The results we present here were obtained with a EITD programme developed at CETIM, using a surface discretisation shown on Figure 6. This programme gives the value of the J-integral for any given closed contour around the crack-tip. We calculated the J-value using three contours in order to verify that this value is indeed independent of the contour. The K_J value is

$$K_J = \left[\frac{E\ J}{\pi(1-\nu^2)}\right]^{1/2} \quad \text{(cf. Table 1)}$$

Note that the stress distribution obtained with the EITD programme was practically identical to the one calculated with SAP programme.

Before comparing the different K values, we would like to discuss the precision on numerical results. Firstly, before carrying out the costly 3-D analysis, we check that the mesh used was sufficiently refined comparing 2-D calculations with available theoretical value. ($K_{Th} = \alpha\ \sigma \sqrt{a}$, α being the finite-width-correction factor (11)). The comparison was extremely good (cf. Table 1). After this step, we proceed to 3-D calculations. The use of two different methods (F.E. and B.I.E.M.) gives us another cross-check on 3-D results. The comparison was once again extremely good (cf. Table 1). This confirms our confidence in the results and we feel that their precision is certainly better than 10%.

COMPARISON BETWEEN THEORETICAL AND NUMERICAL RESULTS

Stress Distribution

From Figure 5 we note that the numerically calculated stress σ_{yy} falls down as we go from the centre of the plate towards the free surfaces. This result is in contradiction to Folias' theory. The H-S theory, though shows the same tendencies as the numerical results, predicts that the stress distribution for different values of B/a should be sufficiently different. For example, the ratio of stress in cases I and II (same crack-length), in the vicinity of crack-tip

$$\left(\sigma_{yy}\right)_{I} / \left(\sigma_{yy}\right)_{II} = \left[\phi(1)\right]_{I} / \left[\phi(1)\right]_{II} = 1.42$$

instead of nearly 1 as given by the numerical analysis.

K-Values

From Table 1, for the 3-D geometry, we shall make two remarks on the mean value of K:

- firstly, concerning the values of K: we find that the 3-D numerical results ($\bar{K}_u$) are close to 2-D values while the predictions of the H-S theory are very much lower and that of Folias very much higher.
- secondly, concerning the variation in K-values with respect to thickness: The numerical results indicate that there is hardly any effect of thickness on the K-values. This result is similar to that given by Folias' theory whereas the H-S theory predicts a variation of 42% in going from Case I to Case II.

It might be of interest to point out here that the analysis of Sternberg and Sadowsky [12], though for the case of a circular hole in a plate of arbitrary thickness, also showed little dependence of the stress distribution on the thickness of the plate.

Figure 7, indicating the distribution of the S.I.F. along the crack front for Cases I and II, was drawn to visualize still better the comparison between different methods. Note that the H-S theory predicts that for moderately thick plates, the results are quite different from the plane solutions, which does not seem to be the case. The Folias' theory though predicts that the thickness effect is about 10%, gives a K-distribution along the crack front through the thickness which is entirely different from that of numerical results. Moreover, the mean value of K from the Folias' theory ($\bar{K}_F$), even for large thicknesses is much higher than the theoretical 2-D plane-strain solution.

CONCLUSION

It is evident that the three dimensional problem of cracked plates is not yet fully resolved even in the linear elastic domain. But the important conclusion to which the present study leads us is that one would be incapable to predict the experimentally observed variation of G_c with respect to thickness, even through an exact 3-D elastic theory. This comes from the fact that the 3-D numerical results are close to the 2-D ones.

On the other hand, looking at the parameter G_c as obtained through the Griffith-Irwin theory, we find that its value remains nearly the same for the cases of plane-stress and plane-strain, although the plastic flow at the crack tip is entirely different in both the cases. This leads us to believe that certain hypothesis of this theory in formulating G_c may have to be re-examined.

In particular, the plastic energy dissipation rate, considered constant in this theory was shown to vary with the geometry and the applied loading (see [13] and the references given there). This result might be of significance in formulating a realistic fracture governing parameter.

REFERENCES

1. TETELMAN, A. S. and McEVILY, A. J., "Fracture of Structural Materials", John Wiley, New York, 1967, 137.
2. HARTRANFT, R. J. and SIH, G. C., Int. J. Engng. Sci., 8, 1970, 711.
3. SIH, G. C. and HARTRANFT, R. J., Int. J. Fracture, 9, 1973, 75.
4. VILLARREAL, G., SIH, G. C. and HARTRANFT, R. J., ASME Paper No. 75-APMW-29, 1975.
5. FOLIAS, E. S., ASME Paper No. 75-APM-24, 1975.
6. LURE, A. I., "Three Dimensional Problems of the Theory of Elasticity", Interscience Publishers, New York, 1964.
7. BHANDARI, S. K. and BARRACHIN, B., unpublished G.A.A.A. Report MP.D310, May 1976.
8. TRACEY, D. M., Nuclear Engng. and Design, 26, 1974.
9. CRUSE, T. A., Computers and Structures, 3, 1973, 509.
10. LACHAT, J. C. and WATSON, J. O., App. Mech. Div. Nat. Conf., A.S.M.E., New York, 1975.
11. BHANDARI, S. K., GATEAU, M. and SERTOUR, G., Int. J. Fracture 12, 1976, 533.
12. STERNBERG, E. and SADOWSKY, M. A., J. App. Mech. 1949, 27.
13. BHANDARI, S. K., Int. J. of Fracture, 9, 1973, 345.

Table 1 Mean Value of S.I.F. as Given by Different Methods (MPa. $\sqrt{m}$)

			CASE I	CASE II	CASE III
CRACK LENGTH (a,mm)			10.0	10.0	7.5
RATIO B/a			0.75	0.25	1.0
2 - D VALUES	FINITE ELEMENTS	(K_σ)	18.2		
		K_u	15.65		
	THEORY*	K_{Th}	15.55		13.11
3 - D VALUES	FINITE ELEMENTS	$(\overline{K}_\sigma)$	18.44	18.35	
		$\overline{K}_u$	15.52	15.52	
	INTEGRAL EQUATIONS	K_J	15.53	16.2	12.83
	H-S THEORY	$\overline{K}_s$	13.21	9.27	12.11
	FOLIAS THEORY	$\overline{K}_F$	23.64	23.13	18.78

$^*K_{Th} = \alpha\ \sigma\sqrt{a}$

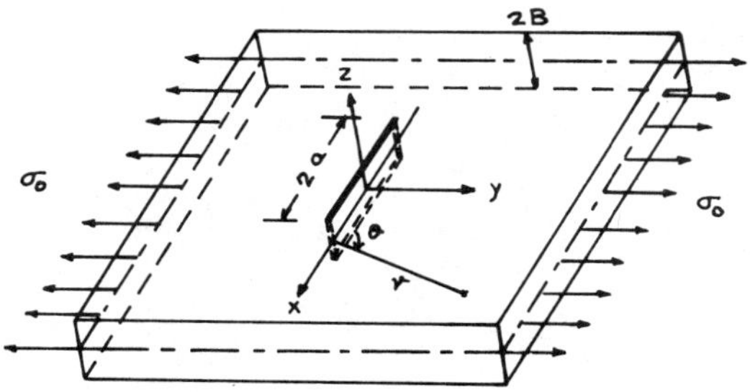

Figure 1 Plate with a Central Crack

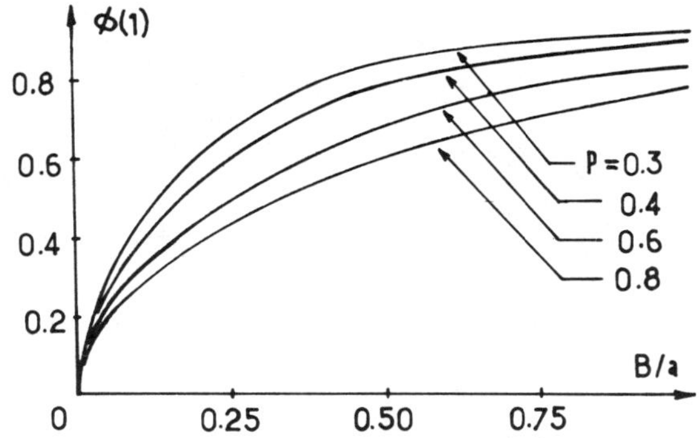

Figure 2 ∅ (1) as a Function of B/a and p
(from reference [1])

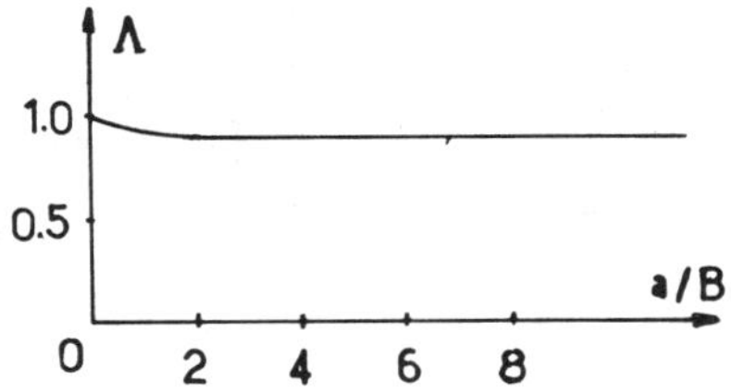

Figure 3 Λ as a Function of a/B
(from reference [5])

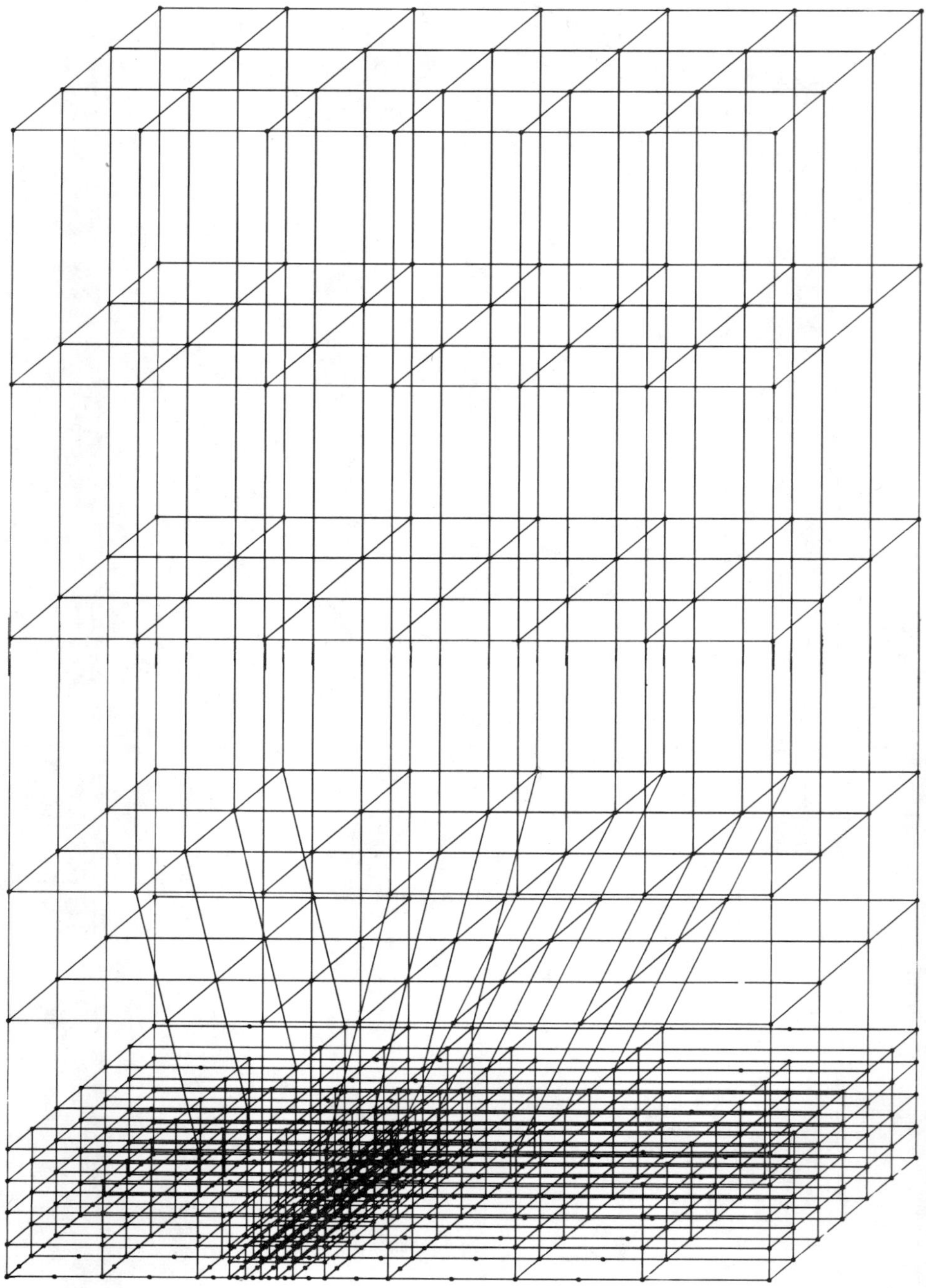

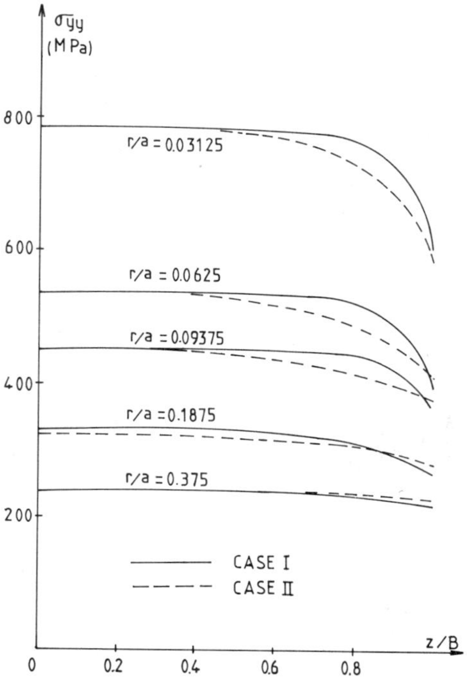

Figure 5 Normal Stress Distribution Through the Thickness of the Plate in the Plane y = 0

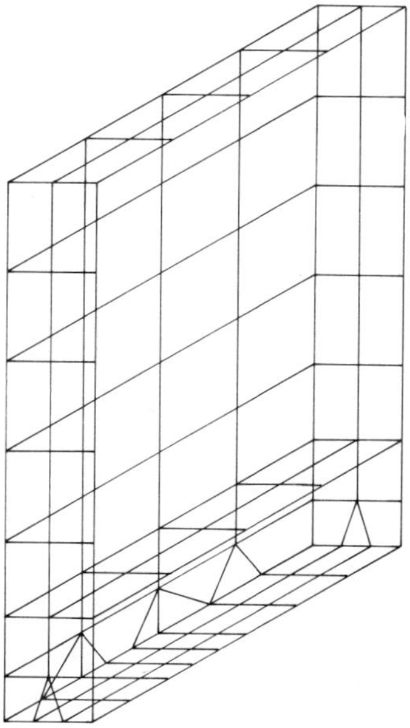

Figure 6 Surface Discretisation for Integral-Equations Analysis

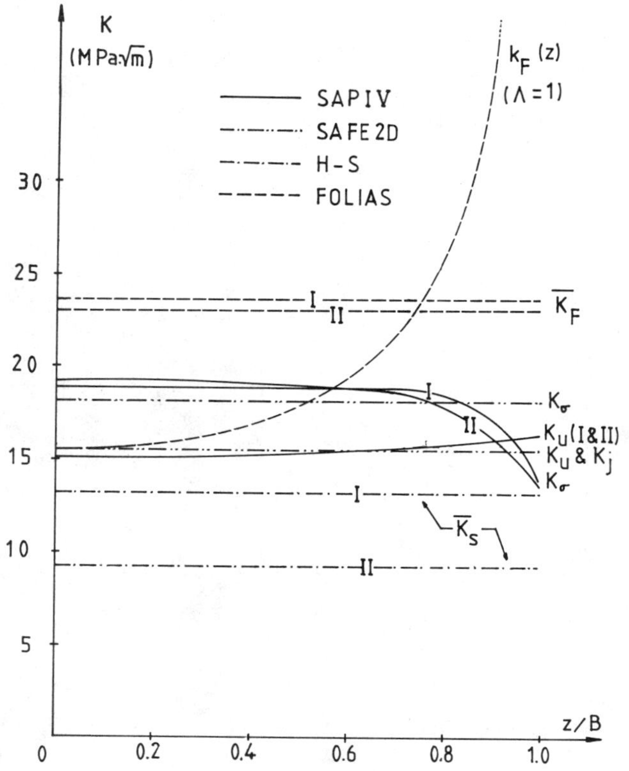

Figure 7 Distribution of Stress-Intensity Factors for Cases I and II

Fracture 1977, Volume 3, ICF4, Waterloo, Canada, June 19 - 24, 1977

GENERAL NUMERICAL METHOD FOR THREE-DIMENSIONAL SINGULARITIES IN CRACKED OR NOTCHED ELASTIC SOLIDS

Z. P. Bažant* and L. F. Estenssoro**

INTRODUCTION

The objective of this paper is to report preliminary results of a numerical finite element study concerned with the elastic deformation field near the point where a crack front edge intersects the surface of an elastic body. The crack plane as well as the front edge are assumed to be normal to the surface. However, the numerical method, briefly outlined herein, has a general applicability to three-dimensional elastic singularities involving singular points located on stress singularity lines, such as crack edges, corners, notches, inclusion edges, etc.

The problem is of fundamental interest for the propagation of cracks intersecting a surface, and a solution is needed to assess the effect of thickness of thin sheets, plates and layers upon crack propagation. A solution of this problem has been attempted many times without success. Recently, Benthem [1] presented an analytical solution.

Problems of similar type also arise in potential theory, where they are, of course, much easier to treat. Very accurate analytical solutions of certain three-dimensional singularities in potential theory have been recently obtained by Morrison and Lewis [2], and by Keer and Parihar [3].

A general numerical method which is capable of handling any three-dimensional singularity in potential theory has been developed in reference [4]. The basic ideas of the present solution, involving the separation of variables postulated here in equations (1) and (11), and the use of finite difference or finite element method to formulate and solve a large nonlinear generalized eigenvalue problem (equation (17) in the sequel), are the same as those in reference [4].

VARIATIONAL EQUATION FOR THE EIGENSTATES

Consider a singular point, 0, located at a smooth singularity line 00' which terminates at point 0 (e.g., Figure 1). Let r, θ, ϕ be a spherical coordinate system centred at point 0, such that ray $\theta = 0$ coincides with the singularity line. It will be assumed that in the vicinity of point 0 the displacement in r, θ and ϕ directions can be expressed in the form

$$u = r^{\lambda} F(\theta,\phi) \tag{1a}$$

* Professor, Technological Institute, Northwestern University, Evanston, Ill., USA.
**Graduate Research Assistant, Northwestern University, Evanston, Ill., USA.

$$v = r^{\lambda} G(\theta,\phi) \tag{1b}$$

$$w = r^{\lambda} H(\theta,\phi) \tag{1c}$$

Substituting these expressions into the well-known differential equations of equilibrium in terms of u, v, w, it is found that the radius coordinate, r, cancels out of the equations, and the following differential equations of equilibrium in r, θ, φ directions in terms of functions F, G, H result:

$$X_r = (Q+2)(\lambda-1)\left[\lambda F+F+G_\theta+G\cot\theta+\frac{1}{\sin\theta}H_\phi\right]-\left[(\lambda+1)G_\theta-F_{\theta\theta}\right]$$

$$-\cot\theta\left[(\lambda+1)G-F_\theta\right]+\frac{1}{\sin\theta}\left[\frac{1}{\sin\theta}F_{\phi\phi}-H_\phi-\lambda H_\phi\right] = 0 \tag{2a}$$

$$X_\theta = (Q+2)\left[\lambda F_\theta+2F_\theta+G_{\theta\theta}+G_\theta\cot\theta-\frac{1}{\sin^2\theta}G+\frac{1}{\sin\theta}H_{\theta\phi}-\frac{\cos\theta}{\sin^2\theta}H_\phi\right]$$

$$-\frac{1}{\sin\theta}\left[H_{\theta\phi}+H_\phi\cot\theta-\frac{1}{\sin\theta}G_{\phi\phi}\right]+\lambda\left[(\lambda+1)G-F_\theta\right] = 0 \tag{2b}$$

$$X_\phi = \frac{1}{\sin\theta}(Q+2)\left[\lambda F_\phi+2F_\phi+G_{\theta\phi}+G_\phi\cot\theta+\frac{1}{\sin\theta}H_{\phi\phi}\right]$$

$$-\lambda\left[\frac{1}{\sin\theta}F_\phi-H-\lambda H\right]+\left[H_{\theta\theta}+H_\theta\cot\theta-\frac{1}{\sin^2\theta}H+\frac{\cos\theta}{\sin^2\theta}G_\phi\right.$$

$$\left.-\frac{1}{\sin\theta}G_{\theta\phi}\right] = 0 \tag{2c}$$

where ν = Poisson ratio, $Q = 2\nu(1-2\nu)$, and subscripts of F, G, and H denote partial derivatives; e.g., $F_{\theta\theta} = \partial^2F/\partial\theta^2$. Furthermore, substituting equations (1a - c) into the well-known expressions for spherical stress components $\sigma_{r\theta},\ldots,\sigma_{\phi\phi}$ terms of u, v, w, it is found that

$$s_{r\theta} = \frac{1}{2Gr^{\lambda-1}}\sigma_{r\theta} = \lambda G - G + F_\theta \tag{3a}$$

$$s_{\theta\theta} = \frac{1}{Gr^{\lambda-1}}\sigma_{\theta\theta} = Q\left[\lambda F+2F+G_\theta+G\cot\theta+\frac{1}{\sin\theta}H_\phi\right]+2(G_\theta+F) \tag{3b}$$

$$s_{\theta\phi} = \frac{1}{2Gr^{\lambda-1}}\sigma_{\theta\phi} = H_\theta - H\cot\theta+\frac{1}{\sin\theta}G_\phi \tag{3c}$$

$$s_{r\phi} = \frac{1}{2Gr^{\lambda-1}}\sigma_{r\phi} = \frac{1}{\sin\theta}F_\phi+\lambda H - H \tag{3d}$$

$$s_{\phi\phi} = \frac{1}{Gr^{\lambda-1}}\sigma_{\phi\phi} = Q\left[\lambda F+2F+G_\theta+G\cot\theta+\frac{1}{\sin\theta}H_\phi\right]$$

$$+2\left[\frac{1}{\sin\theta}H_\phi+G\cot\theta+F\right] \tag{3e}$$

in which G = elastic shear modulus.

Point 0 is assumed to lie at the surface of the body. Expressions (3a - e) may then be used to express surface conditions at surfaces consisting of radial rays emanating from point 0. Let $\underset{\sim}{n} = (n_\theta, n_\phi)$ represent the unit normal to the surface of the body when plotted in the (θ,ϕ)-plane, with θ and ϕ being regarded as the cartesian coordinates in such a fictitious plane; thus, $\underset{\sim}{n} \sim (d\phi/ds, -d\theta/ds)$ where s = length of boundary curve, or $n_\theta/n_\phi = -d\phi/d\theta$ where $d\phi$, $d\theta$ are increments along the boundary. The boundary condition of a free surface may be written in the (θ,ϕ)-plane in the form

$$p_r = s_{r\theta} n_\theta \sin\theta + s_{r\phi} n_\phi = 0 \tag{4a}$$

$$p_\theta = s_{\theta\theta} n_\theta \sin\theta + s_{\theta\phi} n_\phi = 0 \tag{4b}$$

$$p_\phi = s_{\theta\phi} n_\theta \sin\theta + s_{\phi\phi} n_\phi = 0 \tag{4c}$$

The differential equations (2a - c) together with the boundary conditions (4a - c) may be combined to form the following variational statement

$$\iint_A \left[X_r \delta F + X_\theta \delta G + X_\phi \delta H\right] \sin\theta \, d\theta d\phi - \int_s \left[p_r \delta F + p_\theta \delta G + p_\phi \delta H\right] ds = 0 \tag{5}$$

in which s = length of the boundary of the region in the (θ,ϕ)-plane; $ds^2 = d\theta^2 + d\phi^2$; A = area of this region; and variations δF, δG, δH are arbitrary continuous functions of θ and ϕ which have piece-wise continuous derivatives and satisfy all displacement boundary conditions (if any). Conversely from the fact that equation (5) must hold for any kinematically admissible functions δF, δG, δH it follows that equations (2) and (4) must be satisfied. Thus, equation (5) is equivalent to equations (2) and (4).

Equation (5) involves second derivatives of F, G, H (which are contained in the expressions for $s_{r\theta},\ldots,s_{\phi\phi}$). To be able to apply the finite element method, it is necessary to transform equation (5) to a form which involves no higher than first-order derivatives of F, G, H and of δF, δG, δH. At the same time, it is necessary that during this transformation the boundary integral in equation (5) be eliminated (or else natural boundary conditions would not be automatically satisfied when the finite element technique is used). Indeed, a transformation by Green's integral theorem, applied in the Cartesian (θ,ϕ)-plane, has been found, such that both objectives are reached simultaneously. The resulting variational equation is

$$\iint_A \Big[\Phi_F \delta F + \Phi_{F_\theta} \delta F_\theta + \Phi_{F_\phi} \delta F_\phi + \Phi_G \delta G + \Phi_{G_\theta} \delta G_\theta + \Phi_{G_\phi} \delta G_\phi + \Phi_H \delta H$$
$$+ \Phi_{H_\theta} \delta H_\theta + \Phi_{H_\phi} \delta H_\phi\Big] \sin\theta \, d\theta d\phi = 0 \tag{6}$$

in which the following notations are made

$$\Phi_F = [Q(1-\lambda)+2]\left[(\lambda+2)F+G_\theta+G\cot\theta+\frac{1}{\sin\theta}H_\phi\right]-2\lambda(\lambda+2)$$

$$\Phi_{F_\theta} = (\lambda-1)G+F_\theta; \quad \Phi_{F_\theta} = \frac{1}{\sin\theta}\left[\frac{1}{\sin\theta}F_\phi+(\lambda-1)H\right]$$

$$\Phi_G = \left\{(Q+2)\left[(\lambda+2)F+G_\theta+G\cot\theta+\frac{1}{\sin\theta}H_\phi\right]-2(G_\theta+F)-2\lambda F\right\}\cot\theta$$

$$-2(F_\theta-G)-\lambda(\lambda+1)G-\lambda F_\theta$$

$$\Phi_{G_\theta} = Q\left[(\lambda+2)F+G_\theta+G\cot\theta+\frac{1}{\sin\theta}H_\phi\right]+2(G_\theta+F)$$

$$\Phi_{G_\phi} = \frac{1}{\sin\theta}\left[H_\theta-H\cot\theta+\frac{1}{\sin\theta}G_\phi\right] \tag{7}$$

$$\Phi_H = -\left\{\left[H_\theta-H\cot\theta+\frac{1}{\sin\theta}G_\phi\right]\cot\theta+2\left[\frac{1}{\sin\theta}F_\phi-H\right]+\lambda(\lambda+1)H\right.$$

$$\left.+\frac{\lambda}{\sin\theta}F_\phi\right\}$$

$$\Phi_{H_\theta} = H_\theta-H\cot\theta+\frac{1}{\sin\theta}G_\phi$$

$$\Phi_{H_\phi} = \frac{1}{\sin\theta}\left\{Q\left[(\lambda+2)F+G_\theta+G\cot\theta+\frac{1}{\sin\theta}H_\phi\right]+2\left[\frac{1}{\sin\theta}H_\phi\right.\right.$$

$$\left.\left.+G\cot\theta+F\right]\right\}$$

Alternatively, it is possible to derive equation (6) from equation (5) by means of Stokes theorem applied on a unit sphere $r = 1$, domain A being considered as a domain on a unit sphere. It has been checked that this gives the same result. It may be also checked that equation (6) can be transformed by means of Green's theorem (or Stokes theorem) back to equation (5).

The variational statement of the problem is: Functions F, G and H are the solution of the problem if and only if they satisfy equation (6) for any kinematically admissible variations δF, δG, δH.

Existence of variational equation (6) which contains no boundary integral indicates that natural boundary conditions (4) will be automatically fulfilled when the finite element method is used.

It is particularly noteworthy that the integrand of equation (6) is non-symmetric (and that $\Phi_F,\ldots,\Phi_{H_\phi}$ are not partial derivatives of some function Φ). This means that the variational principle cannot be written in the form of a stationary principle, $\delta W = 0$ (or minimum principle, $W = \min.$). At first this might seem surprising for an elastic material. However, a deeper analysis indicates that it must be so. To clarify it, assume that the integrand of equation (6) is symmetric with regard to F, G, H. Then the discrete eigenvalue problem for λ resulting from equation (6) as indicated in the sequel would have a symmetric matrix, and this would imply that all roots λ would have to be real. This is not possible because the same variational equation (equation (6)) must hold also for problems with two-material interfaces, which are known to exhibit oscillating singularities for which λ is complex. Hence, equation (6) cannot be symmetric. This contrasts with the analogous potential theory problem, for which a minimum variational principle in the (θ,ϕ)-plane does exist (see reference [4]), with the consequence that in potential theory the eigenvalues λ are always real.

The basic variational equation (equation (6)) can be also derived from the principle of strain energy, in a similar way as a minimum principle has been derived for the potential theory (equation (18) of reference [4]). The derivation is more direct but it involves certain steps which are difficult to justify without recourse to the derivation just presented. These steps involve the facts that the factor r^λ must be treated at first as an unknown function, $R(r)$, even though r^λ is known in advance to satisfy all governing equations, and that the integral must be integrated by parts with respect to dr, leaving the question of the meaning of the boundary terms arising from integration by parts when actually no boundary intersecting the radial rays is specified in the eigenstate problem.

FINITE ELEMENT FORMULATION IN (θ,ϕ) PLANE

Compared to finite difference solutions, a finite element solution of the variational equations (6) - (7) has the tremendous advantage that stress boundary conditions are automatically implied whenever a free boundary is considered. Therefore, the finite element technique has been selected to approach the problem.

Functions F, G and H must exhibit gradient singularities at the point where they intersect the gradient singularity line (crack edge) emanating from point 0. Such functions are not suitable for numerical solution, since it is known that the rate of convergence of the finite element method with piece-wise polynomial distribution functions is $O(\sqrt{h})$ when there is crack-type singularity, while without singularity it is $O(h^2)$, h being the maximum element size. This difficulty could be avoided, e.g., by using singular finite elements near the singularity line. But a more convenient method has been proposed and used with success in reference [4]. In this method the displacements in r, θ, ϕ directions are expressed as

$$u = f^n r_1^p \; f(\theta,\phi) = r^\lambda \rho^p \; f(\theta,\phi) \tag{8a}$$

$$v = r^n r_1^p \; g(\theta,\phi) = r^\lambda \rho^p \; g(\theta,\phi) \tag{8b}$$

$$w = r^n r_1^p \; h(\theta,\phi) = r^\lambda \rho^p \; h(\theta,\phi) \tag{8c}$$

in which p = exponent for the field near the singularity line; $\lambda = n+p$; $r_1 = r\rho$; ρ = any chosen smooth continuous function of θ and ϕ which is non-zero everywhere except on the singularity ray $\theta = 0$ and in the vicinity of this ray (i.e., for $\theta \to 0$) represents the distance from the ray measured on a unit sphere. Possible choices are $\rho = \theta$ or $\rho = \sin\theta$. The latter choice will be made here, and ρ will then represent the exact distance from the ray not only for $\theta \to 0$ but everywhere in the domain.

(It must be noted, however, that $\rho = \sin\theta$ cannot be used when $\theta = \pi$ is part of the domain and no line of singularity exists at $\theta = \pi$.) For crack edge, $p = 1/2$, which has been considered in all calculations presented here. (However, exponents $p = 0,1,\ldots$ are also possible [1]. It will be convenient to introduce the notations:

$$F(\theta,\phi) = \rho^p\ f(\theta,\phi) \qquad \left[\rho^p = (\sin\theta)^p\right] \tag{9a}$$

$$G(\theta,\phi) = \rho^p\ g(\theta,\phi) \tag{9b}$$

$$H(\theta,\phi) = \rho^p\ h(\theta,\phi) \tag{9c}$$

If the field near the singularity line varies as $\rho^{1/2}$ and p is set equal to 1/2, functions f, g, h may not exhibit any singularity at $\theta = 0$. This would make the convergence rate quadratic, $O(h^2)$. On the other hand, if components of types $\rho^{1/2}$ and ρ^0 (possibly with components of other exponents) were both present in the solution, as is indicated by Benthem's solution [1], the rate of convergence would not be quadratic, but slower than quadratic.

Choosing a finite element grid in (θ,ϕ)-plane, functions F, G, H within each finite element may be represented as

$$F = \textstyle\sum_i X_i F^i, \qquad F^i = \rho^p f^i \tag{10a}$$

$$G = \textstyle\sum_i X_i G^i, \qquad G^i = \rho^p g^i \tag{10b}$$

$$H = \textstyle\sum_i X_i H^i, \qquad H^i = \rho^p h^i \tag{10c}$$

in which X $(i = 1,2,\ldots,M)$ are the nodal values of f, g and h such that $f_k = f^{3k-2} = X_{3k-2}$, $g_k = g^{3k-1} = X_{3k-1}$, $h_k = h^{3k} = X_{3k}$, k being the mode number; and f^i, g^i, h^i are the corresponding distribution functions within the finite elements, normally chosen as polynomials in θ and ϕ. The variations of functions F, G, H and their derivatives may now be expressed as follows:

$$\delta F = \textstyle\sum_j F^j \delta X_j, \quad \delta G = \sum_j G^j \delta X_j, \quad \delta H = \sum_i H^j \delta X_j \tag{11a}$$

$$\delta F_\theta = \textstyle\sum_j F_\theta^j \delta X_j, \quad \delta G_\theta = \sum_j G_\theta^j \delta X_j, \quad \delta H_\theta = \sum_j H_\theta^j \delta X_j \tag{11b}$$

$$\delta F_\phi = \textstyle\sum_j F_\phi^j \delta X_j, \quad \delta G_\phi = \sum_j G_\phi^j \delta X_j, \quad \delta H_\phi = \sum_j H_\phi^j \delta X_j \tag{11c}$$

Substituting equations (10a - c) and (11a - c) into equation (7), it follows that

$$\Phi_F = \sum_i \Phi_F^i X_i, \quad \Phi_{F_\theta} = \sum_i \Phi_{F_\theta}^i X_i, \ldots, \Phi_{H_\phi} = \sum_i \Phi_{H_\phi}^i X_i \tag{12}$$

in which

$$\Phi_F^i = [Q(1-\lambda)+2]\left[(\lambda+2)\rho^p f^i+(\rho^p)_\theta g^i+\rho^p g_\theta^i+\rho^p g^i \cot\theta + \frac{\rho^p}{\sin\theta} h_\phi^i\right]-2\lambda(\lambda+2)\rho^p f^i, \quad \Phi_{F_\theta}^i = \ldots, \ldots$$

$$\Phi_{H_\phi}^i = \frac{1}{\sin\theta}\left\{\left[Q(\lambda+2)\rho^p f^i+(\rho^p)_\theta g^i+\rho^p g_\theta^i+\rho^p g^i \cot\theta + \frac{\rho^p}{\sin\theta} h_\phi^i\right]+2\left[\frac{\rho^p}{\sin\theta} h_\phi^i+\rho^p g^i \cot\theta + \rho^p f^i\right]\right\} \tag{13}$$

Finally, substitution of equations (11) - (13) into variational (6) yields a discrete variational equation of the form

$$\sum_{j=1}^{M}\left[\sum_{i=1}^{M} k_{ij}X_j\right]\delta X_i = 0 \tag{14}$$

in which k_{ij} are stiffness coefficients expressed as follows

$$k_{ij} = \iint_A \left\{\Phi_F^i F^j+\Phi_{F_\theta}^i F_\theta^j+\Phi_{F_\phi}^i F_\phi^j+\Phi_G^i G^j+\Phi_{G_\theta}^i G_\theta^j+\Phi_{G_\phi}^i G_\phi^j+\Phi_H^i H^j + \Phi_{H_\theta}^i H_\theta^j+\Phi_{H_\phi}^i H_\phi^j\right\}\sin\theta\, d\theta d\phi \tag{15}$$

Note that the stiffness matrix $[k_{ij}]$ is non-symmetric; i.e., $k_{ij} \neq k_{ji}$ in general. The variational equation (14) must hold for any choice of δX_i ($i = 1,\ldots,M$), and this requires that

$$\sum_{i=1}^{M} k_{ij}X_j = 0 \qquad (i = 1,\ldots,M). \tag{16}$$

This is a system of M linear homogeneous algebraic equations, representing an eigenvalue problem. All stiffness coeficients k_{ij}, not just the diagonal ones, depend on singularity exponent λ, and so the eigenvalue problem is of the generalized type. Furthermore, it is easy to see that k_{ij} are polynomials in λ, as well as in Poisson ratio ν (when multiplied by $1-2\nu$);

$$k_{ij} = k_{ij}(\lambda,\nu). \tag{17}$$

So, the generalized eigenvalue problem is a non-linear one. Various methods of numerical solution of this problem have been discussed in detail in reference [4], and method B from page 230 of reference [4] has been used here to search for the root λ. The root of smallest value (or of smallest $\text{Re}(\lambda)$, in case of complex root) is of main practical interest. A method of solution when root λ is complex has been described in more detail in reference [5] in connection with another problem.

NUMERICAL RESULTS FOR CRACK EDGE TERMINATING PERPENDICULARLY AT SURFACE

The method of solution just outlined has been programmed in FORTRAN IV. The finite elements were chosen as simple four-node quadrilaterals (with 12 degrees of freedom), obtained by the mapping of a rectangle on a general quadrilateral in the (θ,ϕ)-plane. The distribution functions for F, G and H on the original rectangle have been considered as bilinear in θ and ϕ, i.e., as $a+b\theta+c\phi+d\theta\phi$. The stiffness coefficients k_{ij} were calculated by Gaussian numerical integration, using 9 integration points.

The programme is general and capable of handling various situations, such as intersections of crack edge with body surface of any orientation at an arbitrary angle, corners of any angle on the crack edge, intersection of a line notch with a surface, pyramidal notches, possibly intersecting with cracks, etc. However, so far only the case when λ is a real number has been programmed. The programme will be also capable of handling cases when complex λ must be expected, such as intersections of crack edges with two-material interfaces. But this would require conversion of the FORTRAN programme to complex arithmetic and a generalization of the root search subroutine; this has not yet been done. The results presented in the sequel are all obtained under the restriction that root λ is real.

The correctness of the programme has been checked by a number of cases of known solution. First, elementary solutions of various special cases for the domain $0 \leq \theta \leq \pi/2$, $0 \leq \phi \leq \pi$ have been substituted in equation (16). These were: (a) three rigid body rotation fields, for which $\lambda = 1$, $p = 0$; (b) the field of homogeneous uniaxial stress in the direction $\theta = \pi/2$, $\phi = 0$, for which $\lambda = p = 0$; (c) the near tip plane strain field for a mode I crack with $\nu = 0$ ($\lambda = p = 1/2$); and (d) the same for mode II crack ($\lambda = p = 1/2$). In all cases the right-hand sides of equations (16) for all i were negligibly small (compared to $\Sigma_i|k_{ij}||X_j|$). Also substituted were: (e) homogeneous strain fields, with any of the six strain components being constant ($\lambda = 1$, $p = 0$); (f) plane strain mode I and mode II near tip fields for various ν ($\lambda = p = 1/2$); (g) antiplane mode III near-tip field ($\lambda = p = 1/2$); these fields cannot satisfy equations (16) for the nodes on the body surface, but they must satisfy them for all other nodes, and this was found to be true.

The programme was then applied to analyzing the field near the terminal point 0 of a crack whose plane and edge are normal to halfspace surface. Because of symmetry, it is sufficient to consider the domain $0 \leq \theta \leq \pi/2$, $0 \leq \phi \leq \pi$ (Figure 1), which has a rectangular shape in the (θ,ϕ) plane. The stress boundary conditions on crack surface ($\phi = 0$) and on half-space surface ($\theta = \pi/2$) are automatically satisfied, the boundary condition at $\theta = 0$ (pole) are irrelevant and were considered also as a free boundary, and the boundary conditions on the side $\phi = \pi$ (symmetry plane) must ensure a statically determinate support of the body and at the same time properly reflect the symmetry and antisymmetry properties of displacement field in the plane $\phi = 0$ that is normal to crack front edge. These conditions are achieved, in case of mode I crack, by imposing at the nodes with $\theta = \pi$ the condition $w = 0$ or $h = 0$, and only this case has been considered thus far in the solution of non-trivial cases. In case of mode II crack, antisymmetry of displacements in the plane $\phi = 0$ with respect to crack front edge requires that $u \sin\theta = v \cos\theta = 0$ for $f \sin\theta = g \cos\theta = 0$ at $\phi = \pi$; and in case of mode III crack, antisymmetry of displacements in the plane $\phi = 0$ with respect to the ray $\theta = \phi = \pi/2$ requires that $u \cos\theta - v \sin\theta = 0$ or $f \cos\theta - g \sin\theta = 0$ at $\theta = \pi$.

To obtain a picture of accuracy and convergence, root λ was first solved for $\nu = 0$, in which case the solution is known to be $\lambda = 0.5$. Grids of increasing numbers of finite elements, with N = 18, 32, 72 and 128 elements (and 84, 135, 273 and 459 degrees of freedom), were used. In the (θ,ϕ)-plane all elements were rectangular and identical; the subdivisions of the region in the θ- and ϕ-directions were 3 x 6, 4 x 8, 6 x 12, and 8 x 16. The results of these calculations are given in Figure 3; see line $\nu = 0$. For the finest grid used (128 elements, 459 simultaneous equations), the computed value of the root was 0.5097, which is still 1.9% in error. This indicates that for accurate calculation of λ a finer grid and more complicated finite elements will be required. Work in this direction is in progress.

Nevertheless, even from the results for the relatively crude grids used thus far, interesting results can be extracted if the practical convergence is studied more carefully. It is well known that ordinary finite element method exhibits quadratic convergence, i.e., it has error of the order $O(h^2)$, h being the maximum size of the finite element, provided that there are no singularities within the domain. Functions f, g, h and their gradients are nonsingular, and so the convergence should be also quadratic in the present case. Noting that $h^2 \sim 1/N$, it follows that error $\approx k/N$ where k = constant and N = number of finite elements. This relation should hold accurately when N is sufficiently large. Hence, $\log(\text{error}) = \log(\lambda - 0.5) = \log k - \log N = \log k - 2 \log\sqrt{N}$, which indicates that the plot of log(error) versus $\log\sqrt{N}$ must become a straight line of slope -2 when N is sufficiently large. This plot is shown for $\nu = 0$ in Figure 2, and it is seen that the plot is indeed a straight line, and that the slope of this line is exactly -2.0. Thus, for $\nu = 0$ the present formulation seems to follow a systematic pattern of quadratic convergence already for rather crude grids. This can be used to advantage in extrapolating the convergence pattern and estimating the results for $N \to \infty$.

Thus, expecting that $\lambda - \lambda_{exact} = k/N$, the numerical results for various values of ν obtained with various numbers of finite elements may be used to construct a plot of λ versus 1000/N (Figure 3). Again, for quadratic convergence these plots would have to be straight lines for a sufficiently large N. According to Figure 3 this seems indeed to be true. Therefore, straight lines (regression lines) have been passed in Figure 3 to obtain estimates of the values for $N \to \infty$, i.e., estimates of the exact solution. In case of $\nu = 0$, the point $N \to \infty$ falls exactly in 0.5. However, calculations with much finer grids will be required to make definite conclusions about the values for $N \to \infty$. Especially, caution is necessary in view of the fact that the estimates for $N \to \infty$ significantly deviate from Benthem's solution [1] (Figure 3). According to Benthem, the field in equation (8) with $p = 1/2$ is not the complete solution, unless $\nu = 0$, and components of the form of equation (8) but with p = 0 and p = 1 also significantly participate in the exact solution. If this were indeed true, the convergence of the present method could not be quadratic, $O(h^2)$, but slower; then, for high N the points in Figure 3 would have to begin deviating from a straight line. Based on the crude grids used thus far, this possibility cannot be discounted. If p were set equal to 0 rather than 1/2, the convergence rate would then be $O(\sqrt{h})$; accordingly, the results would have to give a straight line in a plot of λ versus $N^{-1/4}$, regardless of whether components with $p \geq 1$ are present. This case must be examined when results for very fine grids become available.

The solutions of λ for $N \to \infty$ obtained in this manner for various ν are shown in Figure 4, along with the results for various grids. Also shown in Figure 4 is the recent approximate analytical solution by Benthem [1].

For values of ν which exceed 0.4, the root search subroutine converged poorly or not at all. In this regard, it is noteworthy that the lines for a chosen number N of elements turn sharply upwards as ν exceeds 0.4. The search for root λ may be gometrically interpreted as intersection of the line of solution for constant N with the vertical line ν = const. For $\nu > 0.4$ either the intersections occur at very small angles or (for low N) no intersection seems to exist. To circumvent this difficulty, equation (16) may be considered as an eigenvalue problem for ν at a fixed λ. Then the solution represents an intersection of the line of constant N with the horizontal line λ = const. This intersection is at large angle and appears to exist for ν-values well over 0.4. So, the convergence should be rapid and, indeed, this was found to be the case. The convergence should again be quadratic, and so the plots of ν versus 1000/N at constant λ should be straight lines. Numerical results have confirmed it. Passing straight regression lines, similarly to Figure 3 (but for λ = const.), the extrapolated values for $N \to \infty$ have been determined. These values are also plotted in Figure 4, and the solution is extended to ν-values beyond 0.4. However, when ν becomes very close to 0.5, the present formulation breaks down because the value of Q increases without bounds. A special programme would have to be written for ν close to 0.5.

However, in view of the Benthem's solution, the same cautious view as expressed earlier must be adopted with regard to extrapolations to ∞. It may also be noted that the line for $N \to \infty$ in Figure 4 seems to be aiming into the point $\lambda = 1$ and $\nu = 0.5$. Although this point has been given by Benthem [1] as a point of exact solution, this would mean that there would be no singularity for $\nu = 0.5$, and this would be in disagreement with Benthem's experimental study which indicated that for $\nu = 0.5$ the singularity exponent λ should be much less than 1 and closer to 0.5.

CONCLUSION

A general numerical method for determination of three-dimensional singular fields in elasticity has been presented and verified. However, it would be premature to make conclusions on the basis of the numerical examples presented here.

ACKNOWLEDGEMENT

Thanks are due to Leon M. Keer, Professor at Northwestern University and co-director of the project, for his stimulating discussions and valuable suggestions during the progress of this work. The writers are also obliged to Jan P. Benthem, Professor at Delft University of Technology, for pointing out the possibility of simultaneous appearance of components of various p. S. S. Kim, R. J. Krizek and W. Thonguthai are thanked for helping to check some of the calculations, alternative approaches, and subroutines.

Support by the U.S. Air Force Office of Scientific Research under Grant No. AFOSR 75-2859 is gratefully acknowledged.

APPENDIX - NOTE ON THE ANGLE OF PROPAGATING CRACK

From the practical point of view, the case of a propagating crack is of main interest. There is no reason why the angle β of the crack front edge with the surface should have the value of $\pi/2$ which has been considered in the preceding analysis.

There exist certain simple physical restrictions for the solution of a propagating crack: (a) the flux E_0 of energy into the moving crack front edge per unit length of edge must be finite and non-zero because the surface energy γ is finite and non-zero, and (b) the flux E_1 of energy into any point on crack front, including the surface point 0, must be zero, because the trace of the surface point 0 as it moves is a line, and a line can be associated only with a negligible amount of additional surface energy.

The first condition requires that $E_0 = \oint_L \sigma_{ij}(\partial u_i/\partial x) r_1 d\phi$ where r_1, ϕ is a polar coordinate system in a plane normal to crack front edge, L is a circle of radius r_1 in this plane centred around the edge, x is the direction of crack propagation, σ_{ij} is the cartesian stress tensor, and u_i are cartesian displacements. Noting that $u_j \sim r_1^p$, $\partial u_j/\partial x \sim r_1^{p-1}$, $\sigma_{ij} \sim r_1^{p-1}$ it follows that $E_0 \sim r_1^{2p-1}$, and for this to be finite as $r_1 \rightarrow 0$, it is necessary that $2\,\text{Re}(p) - 1 = 0$ or $\text{Re}(p) = 1/2$, as is well known.

The second condition (b) requires that $E_1 = \oint_\Omega \sigma_{ij}(\partial u_j/\partial x) d\Omega = 0$ where σ_{ij} = cartesian stress tensor, u_j = cartesian displacements, x = coordinate in the direction of crack extension, Ω = surface of a sufficiently small sphere with centre at point 0. Noting that $u_j \sim r^\lambda$, $\partial u_j/\partial x \sim r^{\lambda-1}$, $\sigma_{ij} \sim r^{\lambda-1}$ and $d\Omega = r^2 \sin\theta\, d\theta d\phi$, it follows that $E_1 \sim r^{2\lambda}$, and for this to be zero as $r \rightarrow 0$ it is necessary that

$$\text{Re}(\lambda) > 0. \tag{18}$$

Furthermore, consider condition (a) and assume that the value of surface energy of crack extension, γ, is constant. Then the value of the stress intensity factor, K, must be also constant along the crack front edge, $\theta = 0$. Factor K is proportional to the displacements at a chosen fixed distance r_1 from the edge $\theta = 0$; i.e., $K \sim u \sim r^{\lambda-p}\, r_1^p\, f(\theta,\phi)$. Along the edge $\theta = 0$, only the value of r varies while r_1 and f do not. Thus, a constancy of K along the crack front edge requires that $\lambda - p = 0$ ($\text{Re}(p) = 1/2$). So, it has to be concluded that

$$\text{Re}(\lambda) = 1/2 \tag{19}$$

must hold for the terminal surface point of a crack that propagates.

In consequence, the most relevant problem is to determine the value of angle β which the crack front edge must form with the surface in order to yield $\lambda = 1/2$. This should be the main objective of further investigations.

REFERENCES

1. BENTHEM, J. P., "Three-Dimensional State of Stress at the Vertex of a Quarter-Infinite Crack in a Half-Space", Report No. 563, Laboratory of Engng. Mechanics, Delft University of Technology, Netherlands, September, 1975.

2. MORRISON, J. A. and LEWIS, J. A., "Charge Singularity at the Corner of a Flat Plate", SIAM J. Appl. Math., 31, 1976, 233-250.
3. KEER, L. M. and PARIHAR, K. S., "Singularity at the Corner of a Wedge-Shaped Punch or Crack", Northwestern University, August 1976 (private communication).
4. BAŽANT, Z. P., "Three-Dimensional Harmonic Functions Near Termination or Intersection of Gradient Singularity Lines: A General Numerical Method", Int. J. Engng. Sci., 12, 1974, 221-243.
5. ACHENBACH, J. D., BAŽANT, Z. P. and KHETAN, R. P., "Elastodynamic Near Tip Fields for a Crack Propagating Along the Interface of Two Orthotropic Solids", Int. J. Engng. Sci., 14, 1976, 811-818.

ADDENDUM

In subsequent work, the question of the proper value of the exponent p of distance $\rho = \sin\theta$ from the singularity line has been studied more carefully. Let $(r\rho)^p$ be the term of lowest exponent in the field near the singularity line ($\rho = \theta = 0$). For crack edge singularity, the lowest p corresponding to deformed states is $p = 1/2$. However, for rigid body rotations in the neighborhood of the singularity line, one has $p = 0$. For $p = 0$, the term $(r\rho)^p$ does not cause any singularity as $p \to 0$ (or $\theta \to 0$) at finite fixed r. However, this may cause gradient singularity of the type θ^{p-1} or θ^{-1} for $r \to 0$, which is more severe than the singularity $\theta^{-1/2}$ associated with the planar near tip field of a crack. That terms of θ^{-1} should indeed be present is indicated by Benthem's solution [1].

Therefore, all finite element solutions were rerun with $p = 0$. The resulting values of λ were plotted versus $1000/N^{m/2}$ for various chosen values of m, and the convergence rate exponent m which gives the best straight-line fit, as indicated by the sum of square deviations, was selected. This exponent varied between $m = 1.7$ and $m = 1.9$ for all ν. Furthermore, plotting $\log(\lambda - 0.5)$ versus $\log\sqrt{N}$ for $\nu = 0$ (and $p = 0$), the convergence rate was obtained as $m = 1.9$. Obviously, the convergence should not have been quadratic ($m \neq 2$) (since the gradients of F, G and H exhibit singularities at $\theta = 0$), but it is of interest that the exponent m is so close to 2, giving still quite a rapid convergence. Using the plots of λ versus $1000/N^{m/2}$ for the optimum value of m, the λ-values have been extrapolated for $N \to \infty$. In these plots the points fell on straight lines just about as closely as the points in Figure 3, but the lines were more steeply inclined than those in Figure 3. The extrapolated values for $N \to \infty$ values agreed within about 0.4% with Benthem's values for all ν between 0 and 0.48. This *confirms the Benthem's solution* as well as the present one as sufficiently accurate.

The choice $p = 1/2$ in previous computations (Figures 2-4) was motivated by the fact that the term $(\rho\theta)^{1/2}$ is dominant at $\theta \to 0$ and finite r. However, the results just reported show the choice $p = 1/2$ was inappropriate. This can be also deduced as follows. Let $u \sim r^{\lambda}\,\theta^p\,F(\theta,\phi)$ be the term of lowest p present at $r \to 0$ for the exact solution ($\rho = \sin\theta \approx \theta$), and assume that a different exponent, $p^* \neq p$, is considered instead of p for the numerical solution, i.e., $u \sim r^{\lambda}\,\theta^{p^*}\,F^*(\theta,\phi)$. Then, for the stress components, $\sigma_{ij} \sim \partial u/\partial\theta \sim r^{\lambda}\,\theta^{p-1}\,F(\theta,\phi)$ for the exact solution and $\sigma_{ij} \sim r^{\lambda}\theta^{p^*-1}F(\theta,\phi)$ for the numerical solution. Equating these two expressions for σ_{ij}, one has

$$F^*(\theta,\phi) = \theta^{p-p^*}\,F(\theta,\phi). \tag{20}$$

In previous computations (Figures 2-4), $p^*=1/2$ while $p = 0$ exists, giving $F^*(\theta,\phi) = \theta^{1/2}\,F(\theta,\phi)$. Obviously, $F^*(\theta,\phi) \to \infty$ as $\theta \to 0$, and so $F^*(\theta,\phi)$ can in no way be adequately represented numerically. If $p^* = 0$, then $p^*-p = 0$ for the term which prevails at $\theta \to 0$, giving $F^*(\theta,\phi) = F(\theta,\phi)$ which ought to be a bounded smooth function that can be adequately represented numerically.

A less severe singular term with $p = 1/2$ is always present at the same time, of course. For this term, $F^*(\theta,\phi) = \theta^{1/2} F(\theta,\phi)$; $F^*(\theta,\phi)$ is still bounded and acceptable for numerical representation, but because $\partial F^*(\theta,\phi)/\partial\theta$ tends to infinity as $\theta \to 0$, the accuracy of representation will be worse, causing the convergence rate to become less than quadratic.

Therefore, a quadratic convergence cannot be achieved with the present method of analysis.

Because the slope of $F^*(\theta,\phi)$ in the θ direction tends to infinity as $\theta \to 0$, it seems appropriate to refine the grid step $\Delta\theta$ as θ decreases. Irregular rectangular networks in which $\Delta\phi$ was constant and in which $\Delta\theta$ was refined so as to keep $\Delta\theta$ roughly equal $(\sin\theta)\,\Delta\phi$, have been tried, using same numbers of divisions in both θ and ϕ directions. Curiously, however, the results were not any better than those for regular grids; the plots of λ versus $1000/N^{m/2}$ had about the same inclination. But the extrapolated λ-values for $N \to \infty$ agreed again with simple check cases and with Benthem's solution within a 0.4% error ($N=121$ being the finest grid used).

As an additional check, the case of a right-angle corner on the front edge of a plane crack was solved. The solution for this case was obtained in equation (39) of reference [4] as $\lambda = 0.296$ for any ν and more accurately as $\lambda = 0.2966$ in references [2] and [3]. The extrapolated value of λ for $N \to \infty$ agreed with this within 0.2% error for $\nu = 0$ and $\nu = 0.3$, using $N = 128$ as the finest grid.

Presently, computations of λ are in progress for cracks whose plane is normal to the half-space surface but the front edge forms angle $\beta \neq \pi/2$ with the surface. Preliminary results indicate that λ decreases (below Benthem's values for $\beta = \pi/2$) as β exceeds $\pi/2$. The practically most important case $\lambda = 1/2$ is obtained for about $\beta = 101°$ if $\nu = 0.3$. This solution will be reported separately.

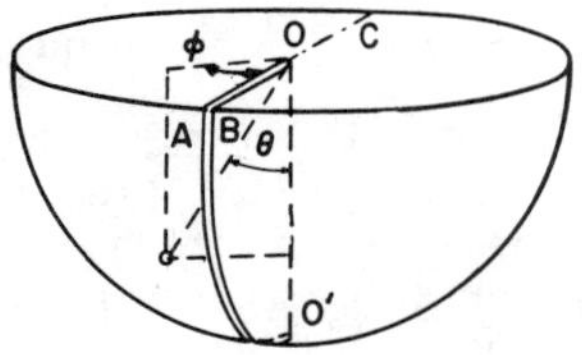

Figure 1 Geometry of the Crack Intersecting a Surface, in Spherical Coordinates

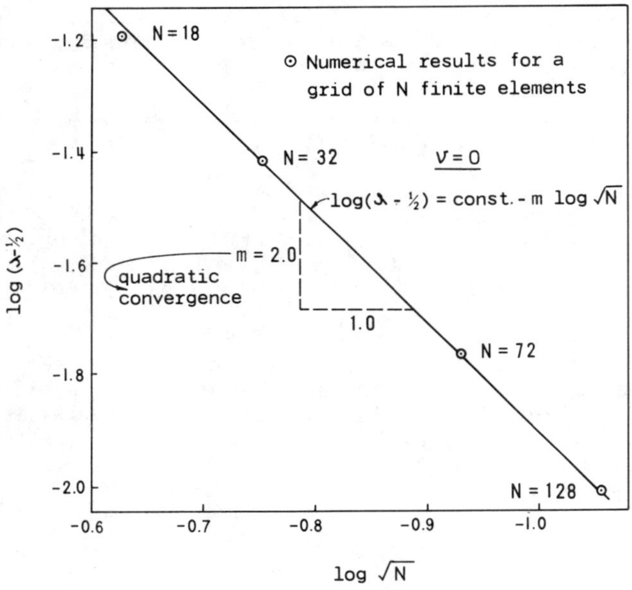

Figure 2 Determination of the Rate of Convergence with Increasing Number of Elements

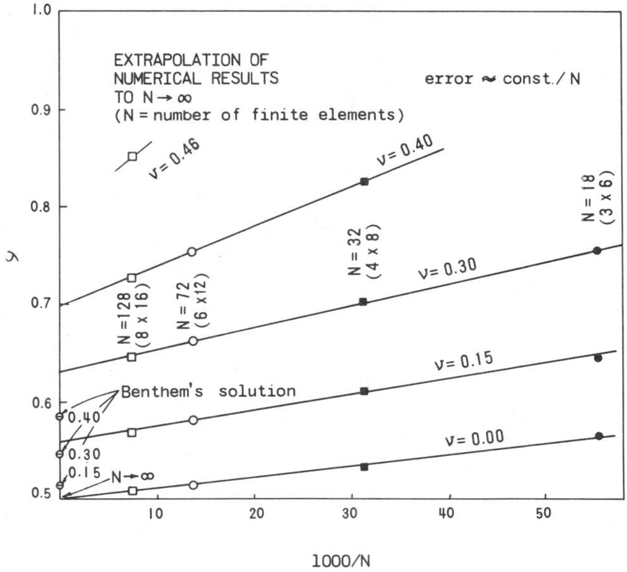

Figure 3 Tentative Extrapolation of Numerical Results to Infinit Number of Elements

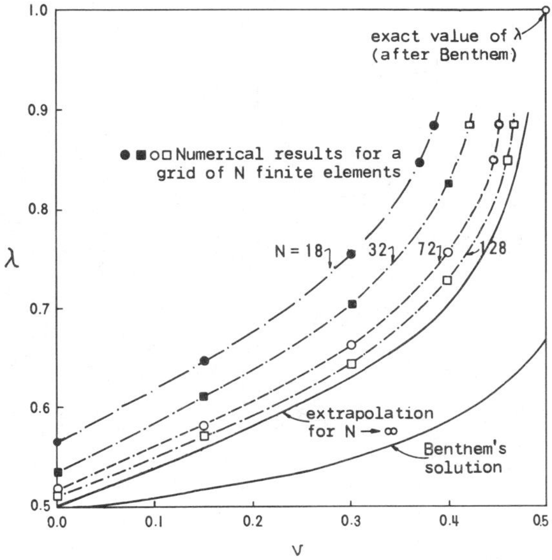

Figure 4 Singularity Exponent λ for Various Values of Poisson Ratio

Fracture 1977, Volume 3, ICF4, Waterloo, Canada, June 19 - 24, 1977

NONLINEAR LATTICE THEORY OF FRACTURE*

E. R. Fuller, Jr. and R. Thomson**

INTRODUCTION

In recent theoretical literature [1, 2, 3] regarding fracture in atomic lattices some questions have arisen regarding the fundamental role of the surface energy in brittle fracture. This paper represents an attempt to clarify some of these basic ideas. Hsieh and Thomson [1] have shown that in a two-dimensional lattice containing a crack there is a range of applied stress for which the crack is mechanically stable, and accordingly, is "trapped" by the lattice. This lattice trapping regime is bounded by a regime of fast fracture for stresses larger than an upper critical stress, σ_+, and by a spontaneous crack healing regime for stresses smaller than a lower stress, σ_-. These stress boundaries were found to vary as the inverse square root of the crack length, a,

$$\sigma_{\pm} = \sqrt{Y\gamma_{\pm}/a} \ , \qquad (1)$$

in precisely the same manner as in the continuum theory of Griffith [4]. Y is a constant of proportionality. The effective surface energy densities, γ_+ and γ_-, in this Griffith-like expression define the limits of the lattice trapping regime. When the macroscopic surface energy, γ_0, as defined by one-half the area under the cohesive force law curve, is compared with γ_+ and γ_-, it is found that γ_0 lies between these two limits. This result is then used as a basis for constructing a theory of subcritical crack growth, or healing, when the stress is either above, or below, the Griffith stress $\sigma_0 = \sqrt{Y\gamma_0/a}$. These general ideas have been picked up and expanded by Lawn [2] in a self-consistent and straightforward manner to form a basis for subcritical crack growth in the presence of an external atmosphere.

However, a recent paper by Esterling [3] has indicated that when a more realistic cohesive force law is used in a lattice theory, the macroscopic surface energy, as defined above, no longer lies within the lattice regime. A Griffith thermodynamic surface energy can be defined by the condition where thermal fluctuations cause a crack to advance and recede at equal rates. If this definition of surface energy corresponds to γ_0, the subcritical crack growth theory loses its basis since thermodynamic equilibrium occurs in a regime where the crack is mechanically unstable towards spontaneous healing. Since these ideas are considered to be basic to the fracture process in general, and to subcritical crack growth in particular, this paper will reinvestigate the relationship between the surface energy and fracture with particular attention to the subcritical crack growth regime.

---

* Contribution of U. S. National Bureau of Standards, not subject to copyright.

** Physical Properties Section, National Bureau of Standards, Washington, D. C. 20234, U. S. A.

 MS280

ONE-DIMENSIONAL LATTICE MODEL

The quasi-one-dimensional lattice model of a crack to be considered here is similar to that of Thomson et al [5]. The model consists of two semi-infinite chains of atoms that are bonded with two types of interactions, Figure 1. These interactions are modeled as bendable (horizontal) spring elements and stretchable (transverse) spring elements. The free ends of the chains are subjected to equal and opposite vertical opening forces P at the zeroth atoms. All displacements are assumed to be vertical with the displacement of the jth atom from its equilibrium separation, c, being denoted by u_j. The stretchable elements up to the nth atom are considered to be stretched beyond their range of interaction, or "broken", thus forming a crack of finite length.

The total potential energy of this system consists of three contributions: the change in potential energy of the external loading system (the negative of the work done by the external force); the strain energy of the bendable bonds; and the strain and/or surface energy of the stretchable bonds across the crack plane. The potential energy of the external loading system is given simply by $U_{ext} = -W_{ext} = -2\,P\,u_0$. The interaction of the bendable spring element is modeled as a second-neighbour interaction between atoms at j-1 and j+1 that resists flexure about their common nearest neighbour at j. The strain energy of this interaction about atom j is given by

$$\frac{1}{2}\,\beta\,[(u_{j-1}-u_j)-(u_j-u_{j+1})]^2 = \frac{1}{2}\,\beta\,[u_{j+1}-2u_j+u_{j-1}]^2 \,, \tag{2}$$

where β is the spring constant for this interaction. The total strain energy of the bendable spring elements, U_{bend}, is twice the summation of these contributions for atoms j = 1, 2, ... (one contribution for each side of the crack). The strain energy contained in the interaction of the jth stretchable spring element across the crack plane can be written as $2\gamma(u_j)c$, where $\gamma(u_j)$ is defined as the density of the surface energy assigned to each surface of the chains of atoms [6]. This surface energy per unit length of surface is given by

$$\gamma(u_j) = \frac{1}{2c}\,[2\int_0^{u_j} f_j(u)du] \,, \tag{3}$$

where $f_j(u_j)$ is the cohesive force of the jth stretchable bond which has been extended a distance $2\,u_j$ from its equilibrium separation. A finite range of interaction is assumed for this nonlinear cohesive force, so that elements which are stretched beyond a critical separation, $c + 2\,u_c$, are taken to be "broken". The surface energy for the "broken" spring element (j = 0 to n - 1) is given by one-half the area under the cohesive force law curve, $\gamma_0 c$. The strain (and/or surface) energy contained in the non-linear element at the "crack tip" is $2\gamma(u_n)c$, and represents the non-linear *elastic* energy of that bond. All stretchable elements beyond j = n are assumed to be linear elastic, $f_j(u_j) = \alpha(2\,u_j)$ for j = n + 1, ... The spring constant, α, is the linear part of the nonlinear force law, $\alpha = [df_j/d(2\,u_j)]$ at $u_j = 0$. The total energy of the stretchable bond elements, $U_{stretch}$, is given by the summation of these contributions for j = 0, 1, ...

Combining these potential energy terms, the total potential energy of the system is given by

$$U = -2Pu_o + \beta \sum_{j=1}^{\infty} (u_{j+1} - 2u_j + u_{j-1})^2 + 2\gamma_o nc + 2\gamma(u_n)c + 2\alpha \sum_{j=n+1}^{\infty} u_j^2. \tag{4}$$

For a given applied force P and crack length $a = nc$, necessary conditions for equilibrium configurations of the crack are $(\partial U/\partial u_j) = 0$, for $j = 0, 1, \ldots$. These equations of stability give an infinite set of fourth-order difference equations which can be solved analytically for the displacements u_j. The solutions for $j = 0, 1, \ldots, n - 1$ are

$$u_j = [\xi + (n-j)c]u_n/\xi + P(n-j)[2n^2 + 3n(\xi/c) + 1 - j(j+n)]/6\beta \tag{5}$$

and for $j = n+1, n+2, \ldots$ $(i = 1, 2, \ldots)$ are

$$u_{n+i} = [u_n \cos(\eta i) - (Pn/2\beta \sinh\lambda) \frac{\sin(\eta i)}{\sin(\eta)}]e^{-\lambda i} \tag{6}$$

where

$$\cosh(\lambda) = 1/\cos(\eta) = \sqrt{1+(\alpha/8\beta)} + \sqrt{\alpha/8\beta}$$

and ξ is a length defined by the spring constant ratio, β/α, according to $\xi = c/\tanh(\lambda)$, or equivalently, $(2\beta/\alpha) = \xi^2(\xi^2 - c^2)/c^4$. The displacement of the nth atoms, which interact through the nonlinear cohesive force $f(u_n)$, is determined from the nonlinear coupling equation, $(\partial U/\partial u_n) = 0$

$$\frac{P(nc+\xi)}{2\alpha u_c \xi} = \Psi(u_n) = \frac{f(u_n)}{\alpha(2u_c)} + \left(\frac{\xi - c}{2c}\right)\left(\frac{u_n}{u_c}\right). \tag{7}$$

The solution for u_n, from which the other displacements can be determined by equations (5) and (6), can be illustrated graphically. Consider an idealized nonlinear atomic force law and its corresponding surface energy density, as plotted in Figures 2a and 2b, respectively. A graphic solution of equation (7) for this nonlinear stretchable force law is shown in Figure 3 for three ratios of bendable to stretchable force constants, β/α. For given elastic properties (i.e., β/α) and a given position of the nonlinear, stretchable spring element (i.e., n), there exist a range of applied loads P, over which equation (7) has three solutions for u_n. The first and third solutions, denoted on the figure by u_n (1) and u_n (3), respectively, correspond to stable equilibrium configurations for cracks of length $a = nc$ and $a = (n+1)c$, respectively. For the first solution, the crack-tip bond is just beginning to see the influence of the nonlinear elastic region; whereas, for the third solution, the crack-tip bond is linear elastic by assumption, and the last broken bond has started to heal nonlinearly. Viewed in multi-dimensional configuration space, the total potential energy as a function of displacements, $U = U[n, u_o, u_1, \ldots u_j, \ldots]$, has a relative minimum at configurations corresponding to both u_n (1) and u_n (3). Topological arguments require that at least one saddle point exists between these two minima. Since the configuration corresponding to u_n (2) is the only possible candidate for an extremum, this configuration is the required saddle point.

Unstable bond rupture occurs when the applied force P is increased to a critical value P_+, so that solutions u_n (1) and u_n (2) coalesce to give $\Psi_+ = \Psi[u_n\ (1)] = \Psi[u_n\ (2)]$. Spontaneous bond healing occurs when the applied force is decreased to a critical value P_- so that solutions u_n (2) and u_n (3) coalesce. In the intermediate regime of applied force, $P_- < P < P_+$, the crack is lattice trapped. As noted by Thomson et al [5], increasing the ratio of β/α (that is, "stiffer" bendable spring elements and/or "softer" stretchable spring elements) results in a decrease in the lattice trapping regime. In constast to their model, however, lattice trapping will vanish for some critical *finite* ratio of spring constants, β/α (the upper curve in Figure 3). Thus, the existence of lattice trapping depends on the elastic properties of the solid. In general, the nonlinear nature of the crack-tip bond tends to decrease the range of lattice trapping in comparison to their "bond-snapping" model.

ONE-DIMENSIONAL CONTINUUM MODEL

In order to compare these results with the macroscopic surface energy γ_o, it is necessary to obtain a continuum model for the same type of crack. The simplest approach is to take the linear-elastic continuum limit of the total potential energy of the system, and use the Griffith approach. The potential energy of the system can be calculated by substitution of equations (5) and (6) into equation (4). The energy contained in the bendable and stretchable bonds is given by

$$U_{bend} + U_{stretch} = Pu_o + 2\gamma_o nc + [2\gamma(u_n)c - u_n f(u_n)] \ . \tag{8}$$

For linear elasticity, the term in square brackets vanishes, yielding the usual fracture mechanics relationship that the strain energy in the bendable and unbroken stretchable bonds is equal to one-half the work done by the external force, $W_{ext}/2$, or $-U_{ext}/2$.

To obtain the one-dimensional continuum model, the limits as $n \to \infty$ and $c \to 0$ is taken in such a manner that $nc \to a$, and βc^3 and α/c remain constant. [The bendable spring constant β must scale as c^{-3} (stiffen) in order to maintain a finite displacement at the zeroth atom for an infinite number of small spring elements. For this stiffening, the stretchable elements must soften proportional to c to maintain a non-zero crack-tip displacement.] Taking this limit and setting $(\partial U/\partial a) = 0$, gives the continuum relationship,

$$P(a+\xi) = \sqrt{2\beta c^3 \gamma_o} \ , \tag{9}$$

which is analogous to the force-crack length relationship for the double cantilever beam. Since equation (7) has the same form as equation (9), effective surface energy densities $\gamma_\pm$ can be defined from equation (7) for the discrete lattics model, analogous to equation (1).

DISCUSSION

When a crack is lattice trapped, thermally activated subcritical crack propagation, or crack healing, is possible [1, 2, 7]. Previous treatments used a modified continuum model to predict the character of the thermally activated crack growth. Since the present model predicts the configuration of the saddle point, it is possible to calculate the forward and backward activation energy barriers within the framework of the model. The forward activation barrier is given by $\Delta U_+ = U[\ldots u_n\ (2)\ldots] - U[\ldots u_n\ (1)\ \ldots]$; and the backward barrier is given by $\Delta U_- = U[\ldots u_n\ (2)\ldots] - U[\ldots u_n\ (3)\ldots]$. When the applied force is P_+, the forward barrier vanishes and catastrophic rapid fracture ensues. Similarily, spontaneous crack healing occurs at P_-. The equilibrium thermodynamic (Griffith) condition corresponds to an applied force P_G at which the forward and backward energy barriers are equal. Since P_G is always bounded by P_+ and P_-, the Griffith thermodynamic condition must always lie within the rapid fracture and spontaneous healing limits. The regime of applied forces for thermally activated subcritical crack growth is between P_G and P_+, where rapid fracture occurs.

The result obtained by Esterling [3], $\gamma_o < \gamma_-$, implies, therefore, that the macroscopic thermodynamic surface energy, γ_o, is not related to the microscopic thermodynamic surface energy, γ_G. This apparent "paradox" is best illustrated by two examples: for the cohesive force law plotted in Figure 2, $f(u_n) = \alpha(2\ u_n)(1-u_n/u_c)^2$, equation (7) with $\xi = 3c/2$ (or $\beta/\alpha = 45/32$) gives the central curve of Figure 3. The lattice trapping regime is given by

$$\gamma_+/\gamma_o = 6/5 > 1 \qquad \text{and} \qquad \gamma_-/\gamma_o = 250/243 > 1\ .$$

Thus, for this choice of nonlinear cohesive force law the macroscopic surface energy density γ_o is not bounded by the lattice trapping limits γ_+ and γ_-, similar to the findings of Esterling [3]. However, the present model is not self-consistent for this force law and choice of elastic constants. This inconsistency is easily seen from Figure 3. As previously mentioned, u_n (3) is the solution for the displacement of the nonlinear atom at n which is one atomic spacing behind the crack tip at n + 1. Using u_n (3) the crack-tip displacement can be calculated from equation (6). For self-consistency this displacement should correspond to solution 1 of equation (7), u_{n+1} (1), when the nonlinear spring element is assumed to be at n + 1. This is not possible in general, since there is no "weakly interacting" nonlinear spring element one lattice spacing behind the non-linear atom at n + 1. That spring element has been assumed to be "broken". In order to reduce this inconsistency for an arbitrary nonlinear cohesive force law, additional nonlinear interactions must be included (i.e., a larger crack-tip cohesive region). The feasibility of this extension is presently under investigation.

A second example more clearly illustrates that this inconsistency in the present model is the probable cause of the macroscopic surface energy density lying outside the lattice trapping regime. A cohesive force law (see Figure 4) is chosen so that the conditions previously mentioned are self-consistent. For this nonlinear cohesive force law,

$$P/P_o = \Psi/\sqrt{\Psi_+\Psi_-}$$

where

$$\Psi_+ = \left(\frac{\xi+c}{2c}\right)\left(\frac{u_m}{u_c}\right) \qquad \text{and} \qquad \Psi_- = \left(\frac{\xi-c}{2c}\right).$$

Thus,

$$\gamma_+/\gamma_o = \Psi_+/\Psi_- > 1 \qquad \text{and} \qquad \gamma_-/\gamma_o = \Psi_-/\Psi_+ < 1 \; .$$

For a given nonlinear bond at n, calculation of the applied force P_G when the forward and backward activation barrier are equal gives $\gamma_G/\gamma_o = (P_G/P_o)^2 = 1$. Thus, in this case, not only does γ_o lie within the lattice trapping range, but it is also equal to γ_G.

CONCLUSIONS

It appears that the result obtained by Esterling, $\gamma_o < \gamma_- \leq \gamma_G \leq \gamma_+$, for some nonlinear cohesive forces might be due to an assumed linearity of atomic interactions beyond the crack-tip bond, which is not necessarily a self-consistent assumption. In one case where the assumption of linearity was forced to be satisfied in the present model, not only was the macroscopic surface energy density bounded by the lattice trapping limits, but it also was equal to the Griffith thermodynamic value. This explanation, however, requires further investigation, since the possibility exists that the microscopic and macroscopic thermodynamic surface energies are not equivalent.

It is interesting to note that the model is self-consistent as a stress corrosion model. A two step process is required to advance the crack by one atomic spacing. The crack-tip bond is first broken by a thermally activated process. The activation energy barrier for this process, ΔU_+, is easily calculated within the framework of the present model. Now, the strong linear bond which was originally one lattice spacing ahead of crack-tip is partially exposed to the corrosion environment and can be corroded by a chemical activation process to return the crack-tip status to its original configuration with the crack length advanced by one lattice spacing.

REFERENCES

1. HSIEH, C. and THOMSON, R., J. Appl. Phys., 44, No. 5, 1973, 2051.
2. LAWN, B. R., J. Mater. Sci., 10, 1975, 469.
3. ESTERLING, D. M., J. Appl. Phys., 47, No. 2, 1976, 486.
4. GRIFFITH, A. A., Phil. Trans. Roy. Soc., (London), A221, 1920, 163.
5. THOMSON, R., HSIEH, C. and RANA, V., J. Appl. Phys. 42, No. 8, 1971, 3154.
6. BLEKHERMAN, M. Kh. and INDENBOM, V. L., Phys. Stat. Sol. (a), 23, 1974, 729.
7. TYSON, W. R., CEKIRGE, H. M. and KRAUSZ, A. S., J. Mater. Sci., 11, 1976, 780.

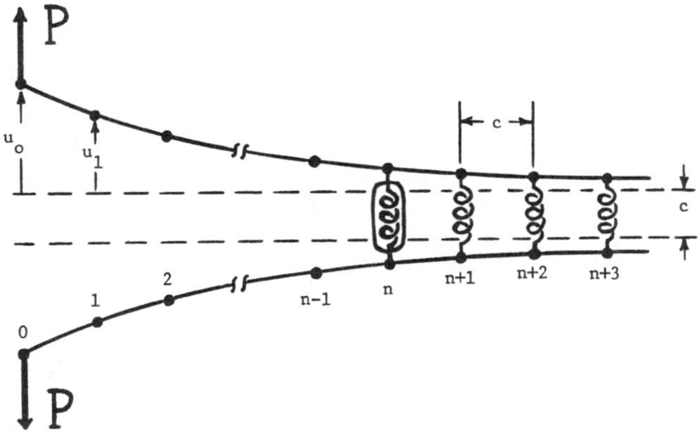

Figure 1 Quasi-One-Dimensional Lattice Model of a Crack

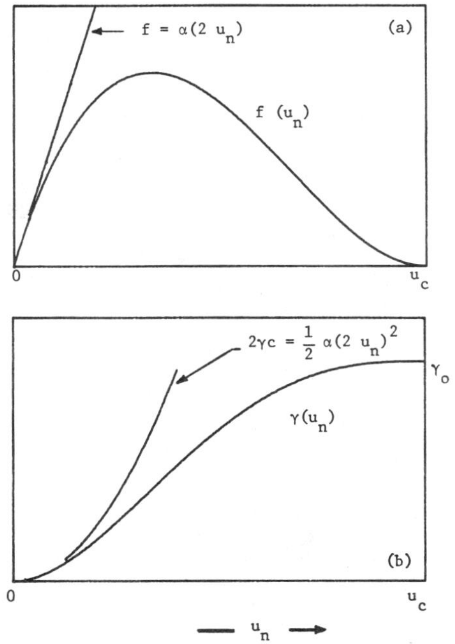

Figure 2 (a) Idealized Atomic Force Law

(b) Corresponding Surface Energy Density

Also Plotted are the Linear Elastic Relations

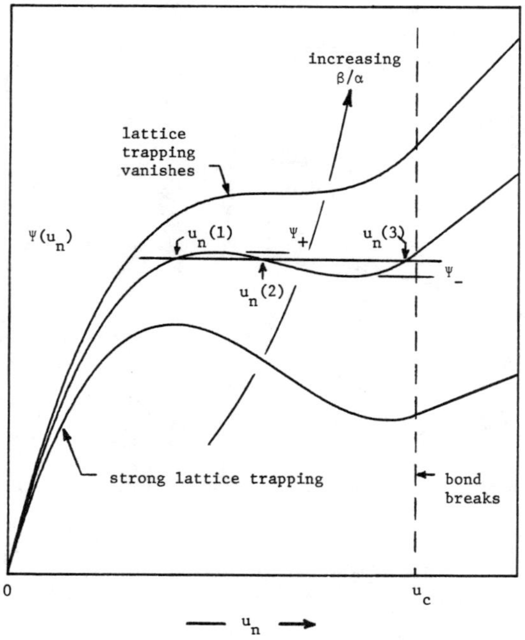

Figure 3 Graphic Solution of equation (7) for the Cohesive Force Law in Figure 2. The Three Ratios of β/α Vary Between Strong and Vanishing Lattice Trapping

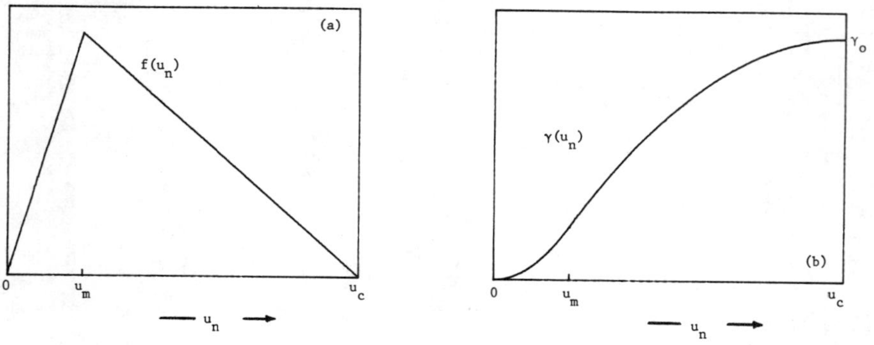

Figure 4 (a) Cohesive Force Law

(b) Corresponding Surface Energy Density

Fracture 1977, Volume 3, ICF4, Waterloo, Canada, June 19 - 24, 1977

ON ENERGY RELEASE RATES IN AXISYMMETRICAL PROBLEMS

M. A. Astiz, M. Elices and V. Sánchez Galvez*

INTRODUCTION

The problem of calculating energy release rates from a given stress and strain field in an elastic cracked body has been solved by means of the J integral [1]. This integral has been extensively applied to plane stress and plane strain problems. In three dimensional problems such a powerful tool does not exist. Three kinds of energy release rates have been defined [2, 3] corresponding to a movement of translation, rotation or expansion of the crack border, but it seems very difficult to find an energy release rate which will be adequate to describe the real movement of the crack border. In axisymmetrical problems this question seems easier to solve because a crack has only one degree of freedom.

INTEGRAL EXPRESSIONS OF THE ENERGY RELEASE RATE

Statement of the Problem

We will consider an axisymmetrical homogeneous and elastic solid with an internal circular crack, as shown in Figure 1 (the solid has not necessarily to be a cylinder). The problem of an external circular crack can be treated in the same way. The purpose of this paper is to find an integral expression for the energy release rate during crack extension.

To solve this problem we will make use of the integrals [2, 3]:

$$J_k = - \frac{\partial U}{\partial x_k} = \int_S (w\, n_k - T_i\, u_{i,k})\, dS \qquad (1)$$

$$M = - l\, \frac{\partial U}{\partial l} = \int_S (w\, x_i\, n_i - T_j\, u_{j,i}\, x_i - \frac{1}{2} T_i\, u_i)\, dS \qquad (2)$$

where w is the energy density, S is a closed surface surrounding the crack border, $\underline{n}$ is the outer normal, $\underline{T}$ is the stress vector acting on the outer side of $\underline{S}$, $\underline{u}$ is the displacement vector and l is any characteristic length of the crack.

We will now evaluate these integrals in an axisymmetrical problem.

* Escuela de Ingenieros de Caminos, Universidad Politècnica de Madrid, Ciudad Universitaria - Madrid 3, Spain

M. Integral

The surface S will be axisymmetric, as shown in Figure 1. Since the integrand is a scalar quantity, it is independent of the angular cylindrical coordinate, θ. This integrand can be evaluated on the plane $\theta = 0(x_2 = 0)$. Considering that the element of area is:

$$dS = \rho \; d\theta \; dl \tag{3}$$

where dl is the element of arc length of the curve C, which is the section of S by the plane $\theta = 0$, the M integral can be expressed as:

$$M = 2\pi \int_C \rho(wx_i n_i - T_j u_{j,i} x_i - \frac{1}{2} T_i u_i) \; dl \; . \qquad i,j = 1,2,3 \tag{4}$$

But along the curve C we have $x_2 = u_2 = 0$; then the indexes i and j of equation (4) may have only the values 1 and 3.

M is the energy release rate with respect to relative scale change da/a, then:

$$M = - \frac{dU}{\frac{da}{a}} = - \frac{2\pi \; a^2 \; dU}{2\pi \; a \; da} = 2\pi \; a^2 \; G \; . \tag{5}$$

As a consequence the energy release rate with respect to cracked area becomes:

$$G = \frac{1}{a^2} \int_C \rho(wx_i n_i - T_j \; u_{j,i} x_i - \frac{1}{2} T_i \; u_i) \; dl \; . \quad i,j = 1,3 \tag{6}$$

This expression is valid only for linear elastic materials as in equation (2) [4].

J. Integral

The application of the M integral to an axisymmetrical problem has been easy because the symmetric movement of the crack is an expansion equivalent to a relative scale change. It is more difficult to apply the J integral to this problem because this integral is the energy release rate with respect to a translation movement of the crack border.

To circumvent this problem it will be necessary to consider a sector of angular amplitude $d\theta$, as shown in Figure 2, because we are dealing with an axisymmetrical problem and the energy release rate is computed per unit of crack area. The movement of the element of crack border inside this sector can be considered as a translation parallel to the x_1 axis. The hypothesis can be justified by considering that the energy release caused by an increment da of the crack radius will be $Ga \; da \; d\theta$ and the difference between this energy release and the actual energy release would be $G \; da^2 \; d\theta$, which can be neglected. The surface of integration is composed of a surface of revolution, S_1, and two bases, S_2 and S_3.

To perform the integration over S_1 it must be taken into account that the components of all the vectorial quantities, $\underline{v}$, of equation (1) ($\underline{v}$ can be the normal, the displacement vector or the stress vector) are:

$$\begin{aligned} v_1 &= v_\rho \cos\alpha \\ v_2 &= v_\rho \sin\alpha \\ v_3 &= v_z \end{aligned} \tag{7}$$

where v_ρ and v_z are the components of $\underline{v}$ in the cylindrical system, and θ is the angular coordinate as shown in Figure 3 and varies between

$$-\frac{d\theta}{2} \text{ and } \frac{d\theta}{2} .$$

It can also be shown [5] that the derivatives of the displacement vector are:

$$u_{1,1} = \cos^2\alpha \frac{\partial u_\rho}{\partial \rho} + \sin^2\alpha \frac{u_\rho}{\rho} \tag{8}$$

$$u_{2,1} = \left(\frac{\partial u_\rho}{\partial \rho} - \frac{u_\rho}{\rho}\right) \sin\alpha \cos\alpha \tag{9}$$

$$u_{3,1} = \frac{\partial u_z}{\partial \rho} . \tag{10}$$

The integral over S_1 can be expressed as a sum of integrals of the type

$$\int_{S_1} f(\rho,z) g(\alpha) \rho d\alpha \, dl = \int_{-\frac{d\theta}{2}}^{+\frac{d\theta}{2}} g(\alpha) d\alpha \int_C f(\rho,z) \rho dl . \tag{11}$$

This integral must be proportional to $d\theta$ because the higher order terms ($d\theta^2$, $d\theta^3$, etc.) would not be significant to the integration in θ. Then only the zero order term will be taken into account in the function $g(\alpha)$. This is equivalent to supposing that:

$$\begin{aligned} \sin\alpha &\simeq 0 , \\ \cos\alpha &\simeq 1 . \end{aligned} \tag{12}$$

Applying equations (7, 8, 9, 10, 11, 12) to equation (1) we obtain:

$$\int_{S_1} \left(wn_1 - T_i \, u_{i,1}\right) dS = d\theta \int_C \left(wn_\rho - T_\rho \frac{\partial u_\rho}{\partial \rho} - T_z \frac{\partial u_\rho}{\partial \rho}\right) \rho \, dl \tag{13}$$

Along the surfaces S_2 and S_3 only the component T_θ of the stress vector and the component n_θ of the normal will be non-zero. After having neglected again the terms of higher order in $d\theta$ the J_1 integral over S_2 and S_3 becomes:

$$\int_{S_2} \left(w n_1 - T_i\, u_{i,1} \right) dS = \frac{d\theta}{2} \int_{S_2} \left(-w + T_\theta \frac{u_\rho}{\rho} \right) dS \,, \tag{14}$$

$$\int_{S_3} \left(w n_1 - T_i\, u_{i,1} \right) dS = \frac{d\theta}{2} \int_{S_3} \left(-w + T_\theta \frac{u_\rho}{\rho} \right) dS \,. \tag{15}$$

Then the J_1 integral will be:

$$J_1 = d\theta \int_C \left(w\, n_\rho - T_\rho \frac{\partial u_\rho}{\partial \rho} - T_z \frac{\partial u_z}{\partial \rho} \right) \rho\, dl - \int_{S_2} \left(w - T_\theta \frac{u_\rho}{\rho} \right) dS \,. \tag{16}$$

The energy release rate per unit angle would be $J_1/d\theta$ and, considering the whole solid, the energy release rate with respect to an increment of the crack radius in axisymmetrical problems, J_A, will be:

$$J_A = 2\pi \int_C \left(w n_\rho - T_\rho \frac{\partial u_\rho}{\partial \rho} - T_z \frac{\partial u_z}{\partial \rho} \right) \rho\, dl - 2\pi \int_{S_2} \left(w - T_\theta \frac{u_\rho}{\rho} \right) dS \tag{17}$$

and the energy release rate, G, will become:

$$G = \frac{J_A}{2\pi a} = \frac{1}{a} \int_C \left(w n_\rho - T_\rho \frac{\partial u_\rho}{\partial \rho} - T_z \frac{\partial u_\rho}{\partial \rho} \right) \rho\, dl - \frac{1}{a} \int_{S_2} \left(w - T_\theta \frac{u_\rho}{\rho} \right) dS \,. \tag{18}$$

A condensed expression for J_A can be obtained by applying Green's theorem:

$$J_A = 2\pi \int_{S_2} \left(\rho \frac{\partial w}{\partial \rho} + T_\theta \frac{u_\rho}{\rho} \right) dS - 2\pi \int_C \left(T_\rho \frac{\partial u_\rho}{\partial \rho} + T_z \frac{\partial u_z}{\partial \rho} \right) \rho\, dl \,. \tag{19}$$

These expressions are valid for nonlinear elastic materials as is equation (1) [4].

The relation between J_A and G can also be found by performing the integration (17) along a circle centred at the tip of the crack with a very small radius (5), and by supposing the stress and strain fields are the same as in a plane strain problem [6].

CONCLUSIONS

Two integral expressions have been derived for energy release rate in axisymmetrical problems. Expression (6) is valid for linear elastic materials and expression (17) is valid for nonlinear elastic materials. Then this expression can be applied to plasticity problems if one can suppose that the deformation theory of plasticity with no unloading gives a good description of the behaviour of this material. Then one can repeat the reasoning of Rice and Rosengren [7] to show that the energy density exhibits a 1/r singularity at the crack tip. For a strain hardening material the stress, strain and displacement field associated with the near-tip dominant singularity must have the form [7, 8]

$$\sigma_{ij} = K_\sigma \, r^{-\frac{1}{n+1}} \, \tilde{\sigma}_{ij}(\theta)$$

$$\varepsilon^p_{ij} = K_\varepsilon \, r^{-\frac{n}{n+1}} \, \tilde{\varepsilon}_{ij}(\Theta) \qquad (20)$$

$$u_i = K_\varepsilon \, r^{\frac{1}{n+1}} \, \tilde{u}_i(\theta)$$

where n is the strain hardening exponent and $\tilde{\sigma}_{ij}$, $\tilde{\varepsilon}_{ij}$ and $\tilde{u}_i$ are a dimensional functions of θ as in [8].

REFERENCES

1. RICE, J. R., Fracture, An Advance Treatise, Vol. II, Academic Press, 1968, 191.
2. KNOWLES, J. K. and STERNBERG, E., Archive for Rational Mechanics and Analysis, 44, 1972, 187.
3. BUDIANSKY, B. and RICE, J. R., Journal of Applied Mechanics, 40, 1973, 40.
4. ESHELBY, J. D., Prospects on Fracture Mechanics, Noordhoff, 1974, 69.
5. ASTIZ, M. A., Thesis Doctoral, Universidad Politècnica de Madrid, 1976.
6. KASSIR, M. K. and SIH, G. C., Three Dimensional Crack Problems, Noordhoff, 1975.
7. RICE, J. R. and ROSENGREN, G. F., Journal of the Mechanics and Physics of Solids, 16, 1968, 1.
8. HUTCHINSON, J. W., Journal of the Mechanics and Physics of Solids, 16, 1968, 13.

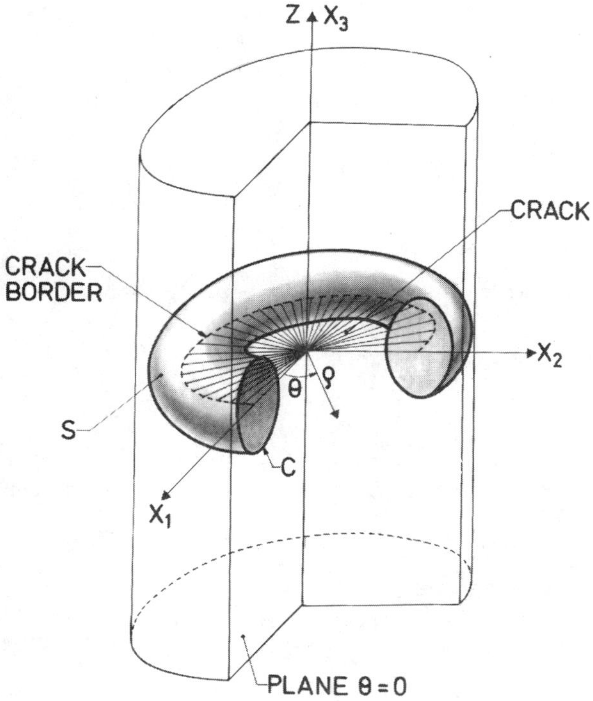

Figure 1

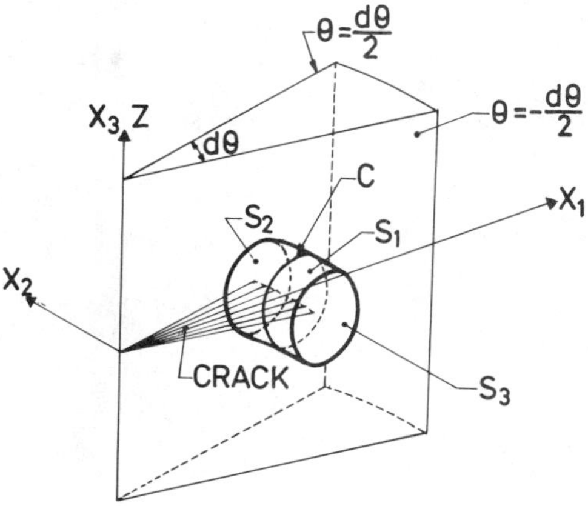

Figure 2

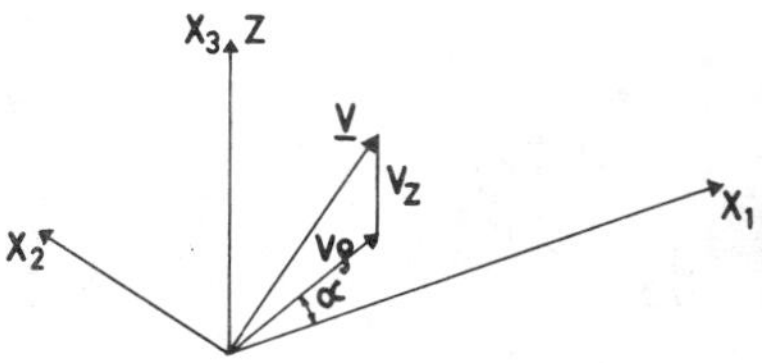

Figure 3

Fracture 1977, Volume 3, ICF4, Waterloo, Canada, June 19 - 24, 1977

STATISTICAL CONSIDERATION ON INHOMOGENEITY OF MECHANICAL PROPERTIES OF MATERIALS

Y. Kishino*

INTRODUCTION

Recently, various theories of continuum mechanics have been published to take the microscopical inhomogeneity of materials into account [1 - 4]. Mechanical quantities appearing in such theories are macroscopic usually and they are considered as averages of the actual quantities distributed inhomogeneously in materials. However, it does not seem that the deriving processes of such average quantities are sufficiently discussed. This paper is concerned with a fundamental consideration of the principle for averaging the inhomogeneous mechanical quantities distributed statistically in materials. A method to obtain mean quantities is introduced by using the internal work done by the inhomogeneous stress acting upon a closed surface of a small region in the material body. Some properties of mechanical quantities derived through the above process are discussed, and it is attempted to argue the relation between the inhomogeneity of stress and the yielding of materials. It is assumed here that the statistical characteristics of distribution of mechanical inhomogeneity are uniform at any place in a material and that the material is considered homogeneous from the macroscopic viewpoint.

AVERAGING OF MECHANICAL QUANTITIES

Stress and strain in continuum mechanics are regarded as mean quantities determined (explicitly or implicitly) through certain averaging processes, as is indicated by their definitions. A mean field quantity is averaged over a certain region in the material body. As far as the mean quantity is expected to be expressed in tensorial form, the averaging process is to be such that it maintains tensorial significance. It is known that this requirement is satisfied when the averaging is carried out isotropically.[1] In this paper, the sphere R with a radius ρ (surface area: A, volume: V) is adopted for the region upon which the mechanical quantities are averaged.

The increment of the internal work per unit volume of R due to the increment of the displacement $D\underline{u}$ is given by

$$Dw = \frac{1}{V} \oint_{\partial R} \underline{tn} \cdot D\underline{u} \, da \qquad (1)^{2}$$

where $\underline{tn}$ is the stress acting on a surface element of ∂R with unit normal

---

* Tohoku University, Sendai, Japan

1. It is considered that averaging of a tensor over open surface which appears in [1, 2] does not fully satisfy this requirement.
2. In this paper the body force is ignored for brevity.

$\underline{n}$. Quantities $\underline{tn}$ and $D\underline{u}$ are generally regarded as inhomogeneous microscopically. In terms of the stress tensor $\underline{\sigma}$ and the increment of the strain tensor (including the rotation) $D\underline{\gamma}$, quantities $\underline{tn}$ and $D\underline{u}$ are given respectively by

$$\underline{tn} = \underline{n} \cdot \underline{\sigma} + \Delta\underline{tn} \tag{2}$$

$$D\underline{u} = \underline{r} \cdot D\underline{\gamma} + \Delta(D\underline{u}) \tag{3}$$

where quantities to which the symbol Δ is attached denote residuals from mean values and $\underline{r}$ is the position vector from the center of R. Now the mean quantities $\underline{\sigma}$ and $D\underline{\gamma}$ are regarded as constant in the following averaging process. It is considered that the expected values of the components of $\underline{\sigma}$ and $D\underline{\gamma}$ are obtained at the vicinity of the center of R when the work done by the residual parts

$$\Delta(Dw) = \frac{1}{V} \oint_{\partial R} \Delta\underline{tn} \cdot \Delta(D\underline{u})da \tag{4}$$

is not dependent on the mean quantities $\underline{\sigma}$ and $D\underline{\gamma}$. This equation is transformed as

$$\Delta(Dw) = \frac{1}{V} \oint_{\partial R} (\underline{tn}-\underline{n} \cdot \underline{\sigma}) \cdot (D\underline{u}-\underline{r} \cdot D\underline{\gamma})da \ , \tag{5}$$

using equations (2) and (3). By differentiating the right hand side of equation (5) with respect to each component of $\underline{\sigma}$ and $D\underline{\gamma}$, and by equating these expressions to zero, we obtain

$$\underline{\sigma} = \frac{1}{V} \oint_{\partial R} \underline{r}\ \underline{tn}\ da \tag{6}$$

$$D\underline{\gamma} = \frac{1}{V} \oint_{\partial R} \underline{n}\ D\underline{u}\ da \tag{7}$$

where products of two vectors without intermediate symbol denote dyad. In the derivation of the above two equations, the following relation is used:

$$\oint_{\partial R} \underline{n}\ \underline{r}\ da = V\underline{I} \tag{8}$$

where $\underline{I}$ is the unit tensor. Equations (6) and (7) are also obtained by the least squares method applied for vectors $\underline{tn}$ and $D\underline{u}$. Namely, $\underline{\sigma}$ and $D\underline{\gamma}$ expressed by these equations give the minimum values of

$$I_t = \oint_{\partial R} \Delta\underline{tn} \cdot \Delta\underline{tn}\ da \tag{9}$$

and

$$I_u = \oint_{\partial R} \Delta(D\underline{u}) \cdot \Delta(D\underline{u})da \tag{10}$$

respectively. Furthermore, for these quantities the equation

$$Dw = \underline{\sigma} \cdot\cdot D\underline{\gamma} + \Delta(Dw) \tag{11}$$

holds, where the symbol $\cdot\cdot$ denotes the double inner product. Equation (11) represents the independence of the work done by the mean quantities and the work done by the residuals.

The magnitude of $\Delta(Dw)$ is affected by the size of R. As it is assumed that statistical characteristics of inhomogeneous distribution of $\underline{tn}$ and $D\underline{u}$ are uniform at any place in material, the value of integral

$$\oint_{\partial R} \Delta\underline{tn} \cdot \Delta(D\underline{u})da = \Delta(Dw)\ V \tag{12}$$

may be regarded as proportional to A in the case where the distributions of the stress and the strain are macroscopically uniform. Thus the magnitude of $\Delta(Dw)$ is inversely proportional to ρ. On the other hand, when the macroscopic fields of stress and strain have gradients in the material body, the errors involved in the approximations for $\underline{tn}$ and $D\underline{u}$ in terms of linear coefficients $\underline{\sigma}$ and $D\underline{\gamma}$ increase as the size of $\overline{R}$ increases. From the above consideration, the most reasonable tensor quantities maybe obtained by use of a sphere with a certain radius and regarded as the macroscopic quantities at the center of the sphere. In the following, ρ is assumed to be constant throughout the material body.

FIELDS OF MECHANICAL QUANTITIES

If a stress field is given by microscopically differentiable continuous stress tensor $\underline{\sigma}^*$ and the equilibrium condition

$$\underline{\nabla} \cdot \underline{\sigma}^* = \underline{0} \tag{13}$$

is fulfilled, equation (6) is transformed as

$$\begin{aligned} \underline{\sigma} &= \frac{1}{V}\oint_{\partial R} \underline{r}(\underline{n} \cdot \underline{\sigma}^*)da \\ &= \frac{1}{V}\int_R \underline{\sigma}^*\ dv + \frac{1}{V}\int_R \underline{r}\ \underline{\nabla} \cdot \underline{\sigma}^*\ dv \\ &= \frac{1}{V}\int_R \underline{\sigma}^*\ dv\ . \end{aligned} \tag{14}$$

Thus it is seen that $\underline{\sigma}$ is a volume average of $\underline{\sigma}^*$. Then $\underline{\sigma}$ becomes a tensor which satisfies the equilibrium condition as shown below.

$$\underline{\nabla} \cdot \underline{\sigma} = \underline{\nabla} \cdot (\frac{1}{V}\int_R \underline{\sigma}^*\ dv) = \frac{1}{V}\int_R \underline{\nabla} \cdot \underline{\sigma}^*\ dv = \underline{0}\ . \tag{15}$$ [3]

If $D\underline{u}$ is differentiable and continuous, equation (7) is transformed as

$$D\underline{\gamma} = \frac{1}{V}\oint_{\partial R} \underline{n}\ D\underline{u}\ da = \frac{1}{V}\int_R \underline{\nabla}\ D\underline{u}\ dv = \underline{\nabla}(\frac{1}{V}\int_R D\underline{u}\ dv)\ . \tag{16}$$ [3]

3. The commutability of the nabla $\underline{\nabla}$ and the averaging operator $\frac{1}{V}\int_R dv$ is assured because the region R should be considered to move together with the concerning material point while the shape of R itself is kept constant for all points.

This equation indicates that $D\underline{\gamma}$ is regarded as the strain tensor derived by differentiation of a displacement vector which is obtained as a volume average of $D\underline{u}$. Thus $D\underline{\gamma}$ satisfies the following compatibility equation[4]:

$$\underline{\nabla} \times D\underline{\gamma} = \underline{0} \ . \tag{17}$$

Even if the mean quantities satisfy the macroscopic field equations as stated above, the residual part of work

$$\Delta(Dw) = Dw - \underline{\sigma} \cdot\cdot D\underline{\gamma} \tag{18}$$

takes a finite value generally, and it may be said that the work done by microdeformations defined in the generalized continuum theories [4] is corresponding to this part of internal work.

Now let a plastic deformation takes place in a material body in which magnitude and direction of the stress vector acting on each point of the spherical surface are assumed as constant. Then the following moment is produced:

$$D\underline{m} = \frac{1}{V}\oint_{\partial R} \underline{t}n \times D\underline{u}\, da$$

$$= \underline{\sigma} \cdot \times D\underline{\gamma} + \Delta(D\underline{m}) \tag{19}$$

where

$$\Delta(D\underline{m}) = \frac{1}{V}\oint_{\partial R} \Delta\underline{t}n \times \Delta(D\underline{u})da \ . \tag{20}$$

As $D\underline{m}$ is the total moment acting on R, it can be equated to zero and we obtain

$$\underline{\sigma} \cdot \times D\underline{\gamma} = -\Delta(D\underline{m}) \ . \tag{21}$$

The above equation indicates that the moment related to mean quantities does not vanish unless the residual part of moment equals to zero. This conclusion seems to be significant, because the plastic deformation of an inhomogeneous material needs not to satisfy the St. Venant's assumption stating that principal directions of stress tensor and increment of strain tensor coinside. It may be possible to consider that $\Delta(D\underline{m})$ is produced by the couple stress [4]. Using such a theoretical model we can derive and eigen equation [6] which is very similar to the equation appearing in the Kondo's theory of yielding [3] in which the yielding is analyzed as analogus phenomenon to the buckling of plates.

It is noted that $\Delta(Dw)$ and $\Delta(D\underline{m})$ are; given respectively, by the trace and the antisymmetric part of the following tensor

---

4. When differentiability (and/or continuity) of $D\underline{u}$ is not guaranteed, $D\underline{\gamma}$ does not generally satisfy equation (17), and a material space after such an incompatible deformation becomes a material manifold with the teleparallelism [5].

$$\underline{T} = \frac{1}{V} \oint_{\partial R} \Delta \underline{tn}\ \Delta(D\underline{u})da \ . \tag{22}$$

This tensor denotes covariances between random variables in the statistical terminology.

INFLUENCE OF INHOMOGENEITY ON YIELD CRITERION

Let us begin with the consideration on von Mises' yield criterion

$$\underline{\sigma}' \cdot\cdot\ \underline{\sigma}' = k^2 \tag{23}$$

where k^2 is a constant and $\underline{\sigma}'$ is the stress deviation defined in terms of the mean stress p and unit tensor $\underline{I}$ as

$$\underline{\sigma}' = \underline{\sigma} - p\ \underline{I} \ . \tag{24}$$

The expression $\underline{\sigma}' \cdot\cdot\ \underline{\sigma}'$ is considered as a measure of variation from the mean stress. This measure may be acceptable particularly when the distribution of stress is assumed to be homogeneous. However, when the inhomogeneity of stress distribution is taken into account, the extended yield criterion may be written as

$$\frac{1}{A} \oint_{\partial R} (\underline{tn} - p\underline{n}) \cdot (\underline{tn} - p\underline{n})da = K^2 \tag{25}$$

where K^2 is a constant. The intensity of stress in yield criteria is usually given in terms of relative values to the initial state where the external forces are absent, and further it is generally considered that the initial stress exists in inhomogeneous materials. Thus $\underline{tn}$ given by equation (2) and the mean stress p are expressed as

$$\underline{tn} = \underline{n} \cdot (\underline{\sigma}_0 + \underline{\sigma}) + \Delta\underline{tn}_0 + \Delta\underline{tn} \tag{26}$$

$$p = \frac{1}{3}\underline{I} \cdot\cdot\ (\underline{\sigma}_0 + \underline{\sigma}) \tag{27}$$

where quantities to which a suffix o is added represent initial quantities. Substitution of equations (26) and (27) into equation (25) gives

$$K^2 = \frac{1}{3}(\underline{\sigma}'_0 \cdot\cdot\ \underline{\sigma}'_0 + 2\underline{\sigma}'_0 \cdot\cdot\ \underline{\sigma}' + \underline{\sigma}' \cdot\cdot\ \underline{\sigma}') +$$

$$+ \frac{1}{A} \oint_{\partial R} (\Delta\underline{tn}_0 \cdot \Delta\underline{tn}_0 + 2\Delta\underline{tn}_0 \cdot \Delta\underline{tn} + \Delta\underline{tn} \cdot \Delta\underline{tn})da. \tag{28}$$

Equation (28) is generalized yield criterion in which the initial stress and the microscopical inhomogeneity of stress are taken into account.

Now a simple case where $\underline{\sigma}'_0$ is zero while $\Delta\underline{tn}_0$ is not equal to zero is considered. Further the following two assumptions may be laid down from the physical consideration.

(Assumption 1) Statistically, the residual stress $\Delta\underline{t}n$ acts in such a way that the inhomogeneity of initial stress increases when $p > 0$ (tensile state) and decreases when $p < 0$ (compressive state).

(Assumption 2) Mean value of the inner produce $\Delta\underline{t}n \cdot \Delta\underline{t}n$ increases together with $\sigma \cdot\cdot \sigma$.

For brevity, replacing these assumptions by linear relations as

$$\frac{1}{A}\oint_{\partial R} \Delta\underline{t}n_0 \cdot \Delta\underline{t}n \, da = c_1 p \tag{29}$$

$$\frac{1}{A}\oint_{\partial R} \Delta\underline{t}n \cdot \Delta\underline{t}n \, da = c_2 \, \sigma \cdot\cdot \sigma \, . \tag{30}$$

Then we obtain

$$\sigma' \cdot\cdot \sigma' = C_1 - C_2 \, p - C_3 \, p^2 \tag{31}$$

where

$$\left. \begin{aligned} C_1 &= \frac{3}{1+3c_2} \left(K^2 - \frac{1}{A}\oint_{\partial R} \Delta\underline{t}n_0 \cdot \Delta\underline{t}n_0 \, da\right) \\ C_2 &= \frac{6c_1}{1+3c_2} \, , \qquad C_3 = \frac{c_2}{1+3c_2} \, . \end{aligned} \right\} \tag{32}$$

These coefficients are considered as material constants determined by the initial stress distribution. It is noticed that C_1 takes the smaller value as the amount of inhomogeneous initial stress increases. If we put as $C_2 = C_3 = 0$, we get von Mises' equation (23), and on the other hand, equation (31) reduces to Griffith type equation for two axial stress state in the case where $C_3 = 0$.

CONCLUDING REMARKS

The macroscopic field quantities appearing in ordinary continuum mechanics are considered to be derived through such an averaging process as stated here. In this paper, the examples of measures of deviation from these mean quantities are given in the forms of the right hand sides of equations (9), (10), (22) and (28). As indicated in the preceding sections, such quantities seem particularly important to explain nonelastic behaviours of materials.

ACKNOWLEDGEMENT

The author wishes to express sincere gratitude to Professor Masao Satake for his valuable advice and constant encouragement in the preparation of this paper.

REFERENCES

1. VOLKOV, S. D., Statistical Strength Theory, Gordon and Breach, New York, 1962.
2. ERINGEN, A. C. and SUHUBI, E. S., Int. J. Eng. Sci., 2, 1964, 189.
3. KONDO, K., RAAG Memoirs, 1, 2, 3, 4, Division D, 1955, 1958, 1962, 1968.
4. KRÖNER, E., (editor), Mechanics of Generalized Continua, IUTAM Symposium, Stuttgart, 1967.
5. SCHOUTEN, J. A., Ricci-Calculus, Second Edition, Springer-Verlag, 1954.
6. KISHINO, Y., Tech. Rep. Tohoku Univ., 37, 1972, 61.

Fracture 1977, Volume 3, ICF4, Waterloo, Canada, June 19 - 24, 1977

STABILITY CONSIDERATIONS IN THE GENERALIZED THREE DIMENSIONAL 'WORK OF FRACTURE' SPECIMEN

J. I. Bluhm*

INTRODUCTION

The determination of fracture toughness in brittle metals is generally accomplished by the mechanical testing of pre-cracked specimens. In these tests the load, at spontaneous fracture, and the associated crack length are recorded and fed into appropriate energy release rate relations to determine the critical energy release rate. This latter value is the fracture toughness. Essential to the adequacy of the test is the presumption of a sharp crack. In metallic specimens this 'crack' is obtained artificially by cyclically loading a specimen with a machined starter notch until a crack of adequate length has grown from the base of the machined (and generally relatively blunt) notch.

However, in extremely brittle materials such as ceramics for example, the techniques for introducing the sharp crack are not so simple. In the usual situation a ceramic specimen containing an initial blunt starter notch will, when loaded, sustain an inordinately high load and then fail catastrophically. Attempts to introduce a sharp crack by fatiguing are generally not successful. Either loads are so small as to not initiate or propagate a crack or attempts to increase the fatigue load merely lead again to inordinately large loads at which spontaneous fracture then occurs. If this load at which fracture actually occurs is fed into the energy release rate relations a fictitiously high fracture toughness is obtained.

One promising alternative approach which provided the motivation of the present study is based upon the use by Tattersall and Tappin [1] of a relatively simple 3-point beam bending specimen having a nominally square cross section and inclined notches such that the remaining ligament is an isosceles triangle with the apex on the tension side. Those authors suggested that that specimen (with an initially blunt machined notch) tended to behave in a stable fashion during crack initiation and extension. The area under the load-deformation curve (i.e., the work) can then be related to the fracture toughness. This specimen has become identified as the 'work of fracture' (WOF) specimen.

The initial optimism with respect to the potential of the WOF specimen has been somewhat dampened in light of the observations that its stability appeared to be dependent upon crack depth and other unspecified geometric parameters. Therefore it seemed appropriate to analytically explore the stability characteristics of such a specimen with the eventual objective of identifying and optimizing a specimen configuration which might be stable throughout the possible ranges of crack depth.

*Army Materials & Mechanics Research Center, Watertown, MA, U.S.A.

PRIOR ANALYSIS

In an earlier paper, Bluhm [2] using a slice synthesis technique developed the essential compliance relations to a generalized form of the WOF specimen. In the present paper, the form of the shear transfer coefficient which is used to compensate for the three-dimensional effects of the specimen is more fully determined; stability criteria are formulated in dimensionless form; and these stability criteria are used in a parametric optimization study.

Figure 1 shows the generalized notch/crack specimen configuration treated in the prior analysis. Figure 2 provides additional nomenclature. Note the heavy line outlining the notch/crack front.

Use of this generalized configuration permits ease of applicability to other specimen types. The original Tattersall-Tappin specimen configuration for example can be approximated from this generalized form by letting the central span $l_2 \to 0$ (3 pt. vs. 4 pt. loading)*, setting $C_o = 0$ and $\omega = W$. On the other hand by letting C_o be arbitrary and $\omega = W$ the Simpson [3] specimen configuration is obtained. Additionally by letting $C_o = \omega < W$, the conventional 'straight through' crack specimen configuration is obtained.

In the earlier work by Bluhm, the specimen compliance λ_s of this generalized work of fracture specimen was developed. The reader is referred to that paper for the form of λ_s and the related definition of terms.

THE SHEAR TRANSFER FUNCTION, $k(\phi_1\ \omega/W)$

In the earlier representation of the shear transfer function [Bluhm 2] it was recognized that k was in fact potentially a function of both ϕ, and ω/W; nevertheless because of limited experimental data it was tentatively presumed to be a function only of ϕ. That data (for $\omega/W = 1$) is shown as the circled points in Figure 3. However subsequently data for $\omega/W = 0.4$ was obtained and is indicated by the circled crosses $\oplus$. It is immediately obvious from the data that the influence of the ratio ω/W is significant indeed. In order to provide a guide for formulating a more general form of $k = k\ (\phi_1\ \omega/W)$ additional data were then obtained at $\phi = 22.5°$ (0.39 rad). That data is shown as circled $\otimes$'s. Although some experimental error is obvious, the trends are quite clear and it was possible to describe the significant effects of both ϕ and ω/W in the following form.

$$k(\phi,\omega/W) = \begin{cases} 1 + (\omega/W)^{3.12}\left(\sum_{n=1}^{4} A_n\phi^n\right) & 0 \le \phi \le 1 \\ 1 + (\omega/W)^{3.12}\left(\sum_{n=1}^{4} A_n\right) & 1 \le \phi \le \tan^{-1}\left(\frac{W-C_o}{H}\right) \end{cases} \qquad (1)$$

*In the previous compliance derivation by Bluhm, only conditions of 4-pt. loading are generally valid since state-of-the-art analysis of cracked specimens do not yet provide satisfactory treatment of the interaction of the crack tip with the bearing load. For 3-pt. loading with deep cracks, this interaction effect may be significant.

where the following constants are for ϕ expressed in radians.

where $A_1 = +2.263$, $A_2 = -4.744$, $A_3 = +4.699$, $A_4 = -1.774$.

The solid lines of Figure 3 represent the prediction of k using this relation. It is obvious that reasonable correlation with experiment is thus enforced.

STABILITY CONSIDERATIONS

The energy release rate G for a loaded specimen containing a crack is given by the familiar relation

$$G = \frac{1}{2} P^2 \, d\lambda_t/dA \tag{2}$$

where P is the applied load, A is the crack area and λ_t is the total system compliance, i.e. that of the testing machine λ_m as well as that of the specimen λ_s; then if for convenience the machine compliance λ_m is expressed as a factor n times the uncracked specimen compliance, one may write

$$\lambda_t = \lambda_s + \lambda_m = \lambda_s + n\,\lambda_{so} \tag{3}$$

where λ_{so} is the compliance of the notched but uncracked specimens.

Stability of crack extension is presumed if

$$dG/dA \leq dG_{cr}/dA \tag{4}$$

i.e. if the rate of energy release rate available for propagation is equal or less than the required energy absorption rate. In the present paper where the emphasis is principally on very brittle materials and where growth of shear lips is not a factor, it is assumed that the fracture toughness is independent of crack growth, i.e.

$$dG_{cr}/dA = 0 \tag{5}$$

Hence, equation 4 becomes merely

$$dG/dA \leq 0 \tag{6}$$

It is anticipated however that stability may alternatively be dictated not merely by the criteria that dG/dA be negative, i.e. equation 6, but more resolutely by the condition that dG/dA be *sufficiently* negative particularly if the initial 'crack' does not in fact simulate a theoretically sharp crack. Under this latter condition the overload to initiate crack propagation supplies a surplus of available energy which accelerates the crack after initiation. Hence only by an excessively negative value of dG/dA can one ever hope to stabilize such a specimen. Accordingly, in the present paper we assume a theoretical crack and utilize a dimensionless form of dG/dA, $\bar{S}'s$ as defined later.

Inserting now equation 2 into equation 6, we obtain

$$\frac{dG}{dA} = \frac{1}{2}\left[P^2 \frac{d^2\lambda_t}{dA^2} + 2P \frac{dP}{dA}\frac{d\lambda_t}{dA}\right] \leq 0 \tag{7}$$

If crack extension occurs under a constant critical load P_{cr} condition, this reduces to

$$\left(\frac{dG}{dA}\right)_P = G_{cr}\left[\frac{d^2\lambda_t}{dA^2} \Big/ \frac{d\lambda_t}{dA}\right]_{A = A_{cr}} \tag{8}$$

where G_{cr} is the critical value of G (determined by equation 2) at which crack propagation starts and A_{cr} is the corresponding value of A. On the other hand, if crack extension occurs under fixed grip or constant deflection conditions, then equation 7 reduces (in lieu of the definition of compliance), i.e. $\lambda_t = \delta_t/P$ to

$$\left(\frac{dG}{dA}\right)_\delta = G_{cr}\left\{\left[\frac{d^2\lambda_t}{dA^2} - \frac{2}{\lambda_t}\left(\frac{d\lambda_t}{dA}\right)^2\right] \Big/ \frac{d\lambda_t}{dA}\right\}_{A = A_{cr}} \tag{9}$$

When expressed in terms of the dimensionless crack length $\alpha = a/W$, equations 8 and 9 can be reduced to the following forms respectively.

$$-\bar{S}_p = d(G_p/G_{cr})/d(A/A_o) =$$

$$\begin{cases} \text{not defined} \ldots. & 0 \leq \alpha \leq C_o/W \\ \left[\dfrac{\omega/W - C_o/W}{\alpha - C_o/W}\right]\left\{\dfrac{d^2\lambda_t/d\alpha^2 - [1/(\alpha - C_o/W)]d\lambda_t/d\alpha}{d\lambda_t/d\alpha}\right. & C_o/W \leq \alpha \leq \omega/W \\ (d^2\lambda_t/d\alpha^2)/(d\lambda_t/d\alpha) & \omega/W \leq \alpha \leq 1 \end{cases} \tag{10}$$

$$\bar{S}_\delta = \bar{S}_p + (2/\lambda_t)(d\lambda_t/d\alpha)\begin{cases} \text{not defined} \ldots & 0 \leq \alpha \leq C_o/W \\ \left[\dfrac{\omega/W - C_o/W}{\alpha - C_o/W}\right] & C_o/W \leq \alpha \leq \omega/W \\ 1 & \omega/W \leq \alpha \leq 1 \end{cases} \tag{11}$$

where $\bar{S} = -\dfrac{dG/dA}{G_{cr}/2HW} = -\dfrac{d(G/G_{cr})}{d(A/A_o)}$ and the subscript P or δ on the $\bar{S}$ implies constant load and displacement conditions respectively.

On the other hand for the constant deflection condition of equation 11 when one replaces λ_t therein by its equivalent form, the presence of the λ_t term in a non-derivative term implies a machine compliance effect.

Hence in equation 11 the λ_t's can be replaced by λ_s only in the derivative terms and as would then be expected the machine compliance plays a significant role in the stability of a constant deflection test condition.

PARAMETRIC STUDIES

Reference to λ as defined in Bluhm [2] and to equations 10 and 11 herein suggest the following dimensionless parameters as defining the stability characteristics of those specimens falling within the scope of the generalized configuration of Figures 1 and 2:

$$\ell_1/\ell_2,\ \ell_1/W,\ C_o/W,\ \omega/W,\ W/H \text{ and } \nu$$

The stability characteristics $\bar{S}$ of any given specimen are then determined by these parameters. However, if the compliance λ per se or energy rate G per se is desired then in addition, the value of Young's modulus, E and one dimension must also be specified.

A limited parametric study was undertaken to explore the potential of defining a stable specimen.

Crack depths α were taken in increments $\alpha = 0.05$; it was evident early in the evaluation of the $\bar{S}_p$ data (constant load case) that satisfactory configurations were highly unlikely under this condition.

SUMMARY AND CONCLUSIONS

1) Representative stability curves for specific cases are indicated schematically in Figure 4. Typical computer outputs for $\bar{S}_p$ and $\bar{S}_\delta$ are shown in Figures 5 and 6 respectively. It was noted that in no instance was the constant load stability tendency $\bar{S}_p$ positive throughout the range of α's. Stability was indicated only for short crack lengths. (See Figures 5 and 6). (Note H is given in inches and E in pounds/inch^2).
2) On the other hand as anticipated a somewhat greater tendency for stability was indicated using the $\bar{S}_\delta$ criteria. Figure 6 shows typical outputs. It was also evident early in the evaluation of parametric studies that the ratio W/H did not significantly influence the results; this was subsequently substantiated by re-examination of equations 10 and 11 in which it was evident that the ratio W/H shows up only in its influence on the shear transfer factor k. Figure 7 shows more clearly the effects of the machine compliance $\lambda_m = n\lambda_{so}$, i.e., equation 3. Note that for $n = 0$, i.e., a very stiff machine, the system tends to be stable but as n increases the tendency is toward *instability* (corresponding to a constant load system).
3) Based upon the analysis and typical computational results of the present paper, it appears that even with sharply cracked notches the derivative forms of the Tattersall-Tappin 'Work of Fracture' specimen are incapable of providing, under constant load conditions, the slow stable growth necessary for a valid determination of fracture toughness.
4) On the other hand, for constant deflection conditions our parametric study suggests several modifications of the WOF specimen which do tend to be stable. The deeply notched Simpson [3] type of specimen as well as the deeply notched through crack specimen for example do

tend to be stable PROVIDED of course that the initial blunt starter notch does not lead to such an excessive over load that castastrophic failure occurs directly from the blunt notch.

5) Critical experimental verification of the utility of such modified WOF specimens is now called for.

ACKNOWLEDGEMENT

The author is indebted to Mrs. Athena Harvey of the Management Science Office, AMMRC for all the programming and computational effort reflected in this paper.

REFERENCES

1. TATTERSALL, H. G. and TAPPIN, G., J. Mat. Sciences, 1, 1966, 296.
2. BLUHM, J. I., Eng. Frac. Mech., 7, 1975, 593-604.
3. SIMPSON, L. A., J. Am. Ceramic Society, 56, 1973, (1), 7.

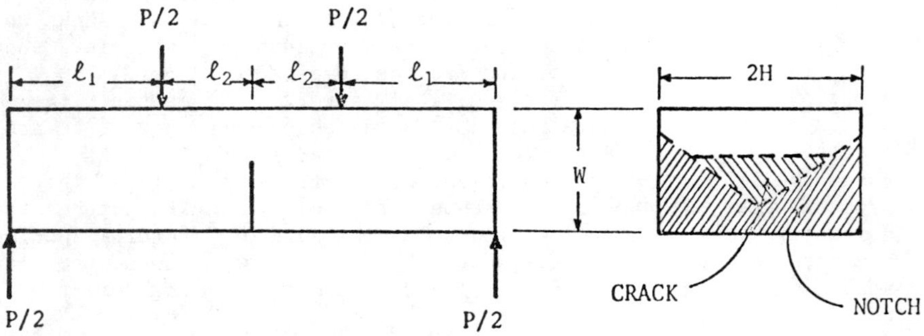

Figure 1 Generalized Notch/Crack Configuration

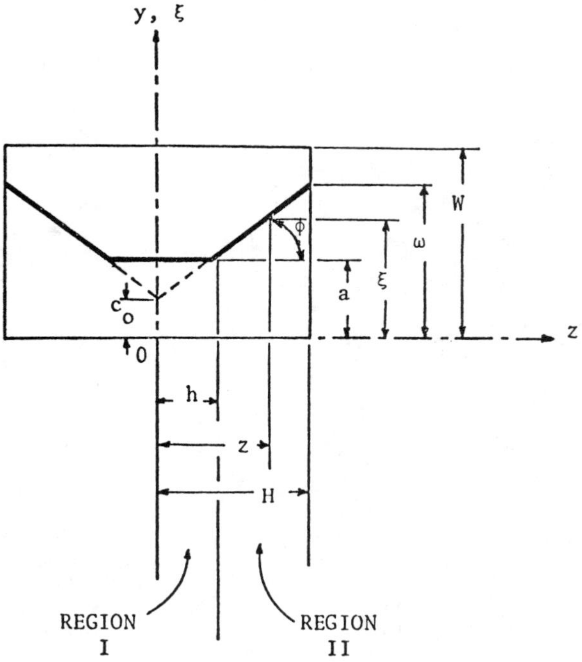

Figure 2 Nomenclature of Notch/Crack Cross Section

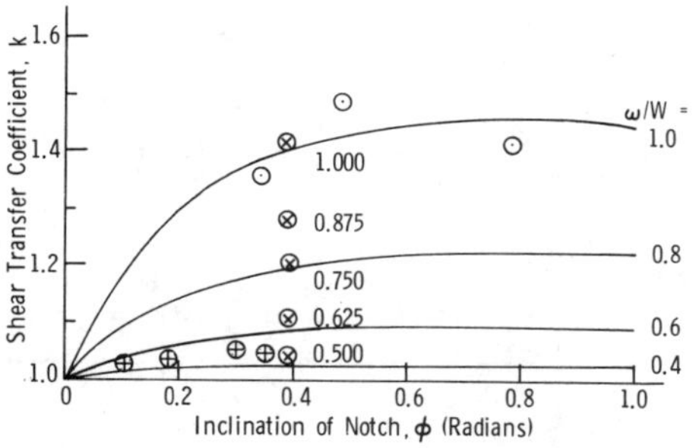

Figure 3 The Shear Transfer Coefficient k

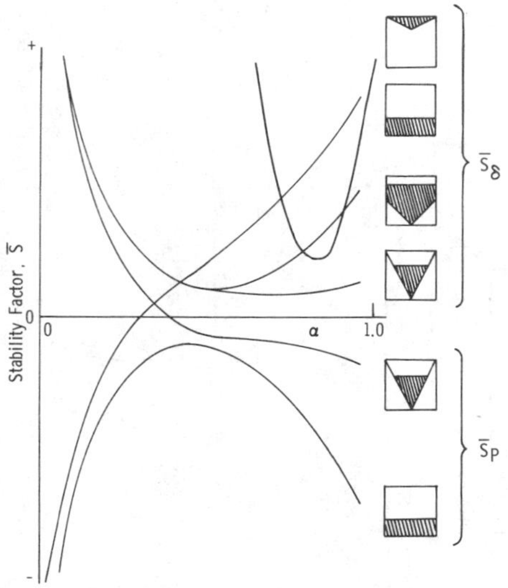

Figure 4 Schematic Showing Stability Tendency As a Function of Crack Length for Various Notch/Crack Configurations and Load Conditions

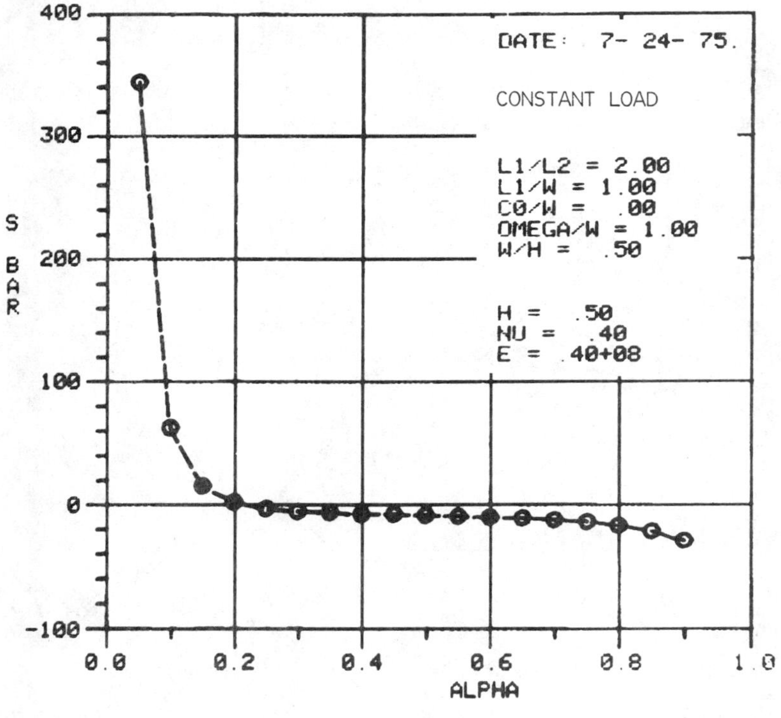

Figure 5 Typical Computer Output Showing Stability Factor vs. Crack Length (at constant load)

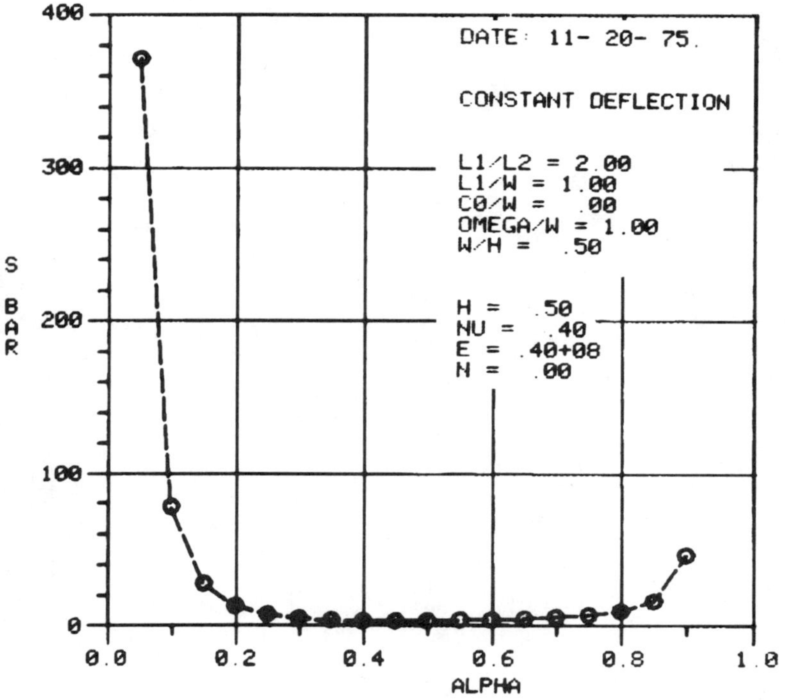

Figure 6 Typical Computer Output Showing Stability Factor vs. Crack Length (at constant deflection)

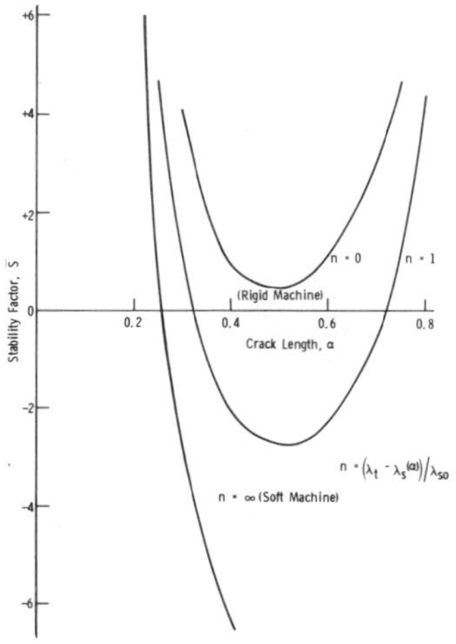

Figure 7 Influence of Machine Stiffness on Stability Factor

Fracture 1977, Volume 3, ICF4, Waterloo, Canada, June 19 - 24, 1977

ASSESSMENT OF FAILURES BEYOND THE LINEAR ELASTIC REGIME

I. Milne*

INTRODUCTION

The failure of a flawed structure has been shown to be bounded by two limits, the linear elastic one and the plastic collapse one [1]. Both of these regimes are now well understood for most geometries. The most difficult regime to assess is the intervening regime, where fast brittle fracture follows plasticity. This situation is met where sections are relatively thin and the material is relatively ductile, or where stress are elevated locally beyond yield due to geometric constraint. Here the adoption of either of the two limits can lead to an overestimate of the defect tolerance of a structure so that resort has to be made to some form of "post-yield" assessment. The following is an attempt to define and validate one of these post yield routes, and to show that even after appreciable plasticity, brittle fracture can still be described by the linear elastic failure parameter, K_{1C}.

THE MODEL

The model chosen is based upon the Bilby Cottrell Swinden [2] theory of yielding ahead of a crack. This is a crack opening displacement approach to fracture which can be reinterpreted in terms of K_{1C} following the suggestions of Heald Spink and Worthington [3]. Because of its theoretical basis this approach can be treated analytically [4] making it very versatile. In this way it has distinct advantages over empirical crack opening displacement approaches which are not universally applicable.

The basic equation can be generalised in terms of the failure stress, σ_f

$$\sigma_f = \frac{2}{\pi}\sigma_1 \cos^{-1} \exp - \frac{\pi^2 K_{1C}^{\ 2}}{8Y^2 a \sigma_1^{\ 2}} \tag{1}$$

where Y is the linear elastic compliance of the cracked body, and takes crack shape into account, and σ_1 is the collapse stress of the cracked body. The value of σ_1 must take into account the geometric constraint local to the crack. It can be obtained from conventional limit analysis, from slip line field theory, from finite element analysis or from the testing of scale models, whichever is the most appropriate. This has advantages over the J integral approach in that small scale tests can be used to confirm the predictions and avoid over-reliance on finite element techniques.

Following Harrison Loosemore and Milne [6] equation (1) can be plotted as a universal function in terms of the ratios $K_r = K_1/K_{1C}$ and $S_r = \sigma_f/\sigma_1$, Figure 1, where K_1 is calculated elastically at σ_f. It should be noted

*Central Electricity Research Laboratories, CEGB, Leatherhead, England.

that the stress ratio, S_r could equally be written in terms of loads or pressures, and need not be converted to stresses.

VALIDATION

For the model to be generally applicable it must be capable of describing failure in any geometry. Over the previous 10 years a whole fund of test data has become available on structural geometries, such as cylinders and spheres. Much of this data suffers from the lack of relevant materials data, particularly fracture toughness and so a direct comparison with the assessment line in Figure 1 is not possible. However, using this assessment line it is possible to obtain the ratio K_r and hence predict a K_{1C}. This should be constant with varying geometry and compare favourably with the expected values of K_{1C}.

Cylinder tests of Nichols Irvine Quirk and Bevitt [7]

These tests were performed on a variety of steel cylinders of varying diameters. Collapse can be expected in these geometries when $M\sigma = \bar{\sigma}$, where $\bar{\sigma}$ is a flow stress and M is the stress magnification factor due to bulging [8]. There is some speculation as to the actual value of the flow stress; the value adopted here is $\frac{1}{2}(\sigma_y+\sigma_u)$.

Table 1 groups the results obtained from these tests on .36 C steel, in order of temperature. It is apparent that failure occurred mainly at stresses well below the collapse limit. The predicted values for K_{1C} are reasonably constant at a given temperature and increase with temperature.

HSST 3 inch vessels of Derby [9]

Data on these tests are again not very complete. Failure was after considerable plastic bulging at the higher temperatures; thus the pressure at these temperatures was taken as the limit pressure. Consequently three of the tests failed with $S_r < 1$, as indicated in Table 2. (σ_u was assumed independent of temperature over this temperature range). Values for K_1 were difficult to predict, since there are no standard solutions for external part penetrating longitudinal cracks in pipe geometries, but they were obtained by applying a bulging factor to the solutions of Merkle et al [10]. The predicted values for K_{1C} are within the limits expected.

HSST large test vessels

For these vessels σ_1 was taken from the solutions of Duffy et al [8] for part penetrating defects, i.e.

$$\sigma_1 = \bar{\sigma}\left[\frac{t/a - 1}{t/a - 1/M}\right],$$

where t is the thickness of the vessel.

$\bar{\sigma}$ was again taken as $\frac{1}{2}(\sigma_y+\sigma_u)$, and K_1 was obtained using the ASME XI procedures. The predicted values of K_{1C} are within the range expected, Table 3.

Spherical vessel tests of Lebey and Roche [11]

Here different sized vessels made of different thicknesses of AMMO steel were tested with varying crack lengths. σ_1 was taken as 0.8 σ_u, since this is approximately the limit stress for the shorter cracks. Table 4 lists the predictions for these tests using the initiation of crack growth as the failure point. For a given sphere the variation in the predicted values of K_{IC} are within the range normally experienced in steels of this nature.

3-point bend specimens of Lubahn and Yukawa [12]

Here the specimen size was varied, so that the collapse stress could be taken as the failure stress in the smallest specimen. The predicted values for K_{1C} were very constant and of the value expected of a Ni-Mo-V steel, Table 5.

CKS specimen tests of Begley and Landes [13]

In this case σ_1 was obtained by the curve fitting technique of Chell and Milne [14]. This is important at high values of S_r (for the smaller series of tests $S_r \to 1$) in order to obtain precision in the region where the curve becomes asymptotic to $S_r = 1$, and the predictions were obtained using equation (1) rather than Figure 1. Nevertheless these predicted values of K_{1C} are in full agreement with those of Begley and Landes and the data obtained from large scale tests (Table 6).

DISCUSSION AND CONCLUSIONS

In the simplified form of Figure 1 equation (1) has been shown to predict consistent values for K_{1C} in both the linear elastic and the large scale yielding regimes. Hence it is proposed as a means for assessing the integrity of a structure regardless of the operational stress level. It is not evident from the foregoing, however, that an infinite value for K_{1C} is predicted at $S_r = 1$. This is consistent with the known behaviour that plastic collapse is independent of K_{1C}. Thus an increasing uncertainty occurs in the predicted K_{1C} as the Assessment Line in Figure 1 becomes asymptotic to $S_r = 1$. In this region σ_1 needs to be known very accurately for adequate predictions of K_{1C}, as was required for the CKS specimen tests. This is no disadvantage however, especially for assessment purposes where it is prudent to ensure pessimism by using upper bound criteria for S_r and K_r.

The advantages of the above approach using the assessment line of Figure 1 are manifold:

1) Any appropriate analytical technique can be used to obtain the parameters K_1 and σ_1. Thus the approach is as versatile and sophisticated as our knowledge of stress analysis.
2) Where an analysis is suspect, or inadequately defined, resort can be made to model testing for the evaluation of σ_1.
3) The effect of secondary stresses can be easily studied; e.g. if it can be demonstrated that a secondary stress influences only the linear elastic regime it is easy to allow for this without being unduly pessimistic.
4) The influence of various factors of safety on the final assessment can be readily explored; e.g. the effect of using factors in σ, or mat-

erials data, or in defect size etc.
5) The most likely regime of failure is immediately apparent.

ACKNOWLEDGMENT

The work was carried out at the Central Electricity Research Laboratories and is published by permission of the Central Electricity Generating Board.

REFERENCES

1. DOWLING, A. R. and TOWNLEY, C. H. A., Int. J. of Pressure Vessels and Piping, 3, 1975, 77.
2. BILBY, B. A., COTTRELL, A. H. and SWINDEN, K. H., Proc. Roy. Soc. A272, 1963, 304.
3. HEALD, P. T., SPINK, G. M. and WORTHINGTON, P. J., Mats. Sci and Eng. 10, 1972, 129.
4. CHELL, G. G., Int. J. of Fracture, 12, 1976, 135.
5. CHELL, G. G., Int. J. of Pressure Vessels and Piping, 1976.
6. HARRISON, R. P., LOOSEMORE, K. and MILNE, I., CEGB Research Department Report, 1976, R/H/R6.
7. NICHOLS, R. W., IRVINE, W. H., QUIRK, A. and BEVITT, E., Proc. 1st Int. Conf. Fracture, Sendai, 1966, p.1673.
8. KEIFFNER, J. F., MAXEY, W. A., EIBER, R. J. and DUFFEY, A. R., ASTM STP 536, 1973, 461.
9. DERBY, R. W., 1st Int. Conf. on Structural Mechanics in Reactor Technology, Berlin, 1971.
10. MERKLE, J. G., WHITMAN, G. D. and BRYAN, R. H., ORNL-TM-5090, 1975.
11. LEBEY, J. and ROCHE, R., 3rd Int. Conf. on Structural Mechanics in Reactor Technology, London, 1975.
12. LUBAHN, J. D. and YUKAWA, S., Proc. ASTM 58, 1958, 661.
13. BEGLEY, J. A. and LANDES, J. D., ASTM STP 514, 1972, 1.
14. CHELL, G. G. and MILNE, I., Mats. Sci. and Eng. 22, 1976, 249.

Table 1

Test	Temp. (^oC)	2a (mm)	σ_f	K_1	S_r	K_r	K_{1C} Predicted
V7T1	1	628.6	66.4	166.3	.47	.96	173
V13T1	10	304.8	95.75	115.5	.43	.97	119
V1T2	12	152.4	190	105.6	.62	.92	115
V8T2	13	304.8	123.5	128.7	.52	.95	135
V14T2	17	304.8	139	124.3	.46	.97	128
V2T1	29	304.8	130	134.2	.55	.94	143
V5XT1	45	152.4	222	124.3	.707	.88	141
V12T3	50	304.8	120.4	146.3	.55	.94	156
V4T4	51	304.8	145	150.7	.62	.92	164
V4T1	62	152.4	227	126.5	.73	.86	147
V3T1	62	609.6	88	215.6	.57	.94	229
V6T1	77	304.8	177.6	182.6	.77	.84	217
V12T1	79	304.8	161.4	195.8	.73	.86	227
V14T1	80	304.8	187	166.1	.62	.92	180
V5T1	84	304.8	183.8	190.3	.77	.84	226
V3XT1	88	609.6	105	108.9	.407	.97	112

Table 2

Temp. (^oC)	p_f	K_1	S_r	K_r	K_{1C} Predicted
-45	20.7	40.7	.885	.75	54
-18	20.7	40.7	.885	.75	54
-3	21.4	42.1	.914	.72	58.5
+16	23.4		1.0		
+54	23.4		1.0		

Table 3

Vessel	Temp. (^oC)	a (mm)	2C (mm)	σ_f	K_1	S_r	K_r	K_{1C} Predicted
1	54	65	209.5	365.4	194.7	.686	.89	219
2	0	64.25	210.8	355.8	171.6	.61	.92	186
3	54	53.6	215.9	397.1	183.7	.74	.86	213
4	24	76.2	209.5	337.8	176.0	.706	.88	200
6	88	47.5	133.4	406.8	215.6	.74	.86	279

Table 4

Sphere No.	2a (mm)	σ_f	K_1	S_r	K_r	K_{1C} predicted
9	35	330	116	.77	.84	138
	70	205	177	.48	.95	186
	105	132	203.6	.31	.98	207
10	40	318	127.5	.74	.86	151
	59	200	101	.46	.95	106
13	35	336	126	.78	.82	153
	50	254	139	.59	.92	151
	65	217	161.5	.5	.94	171
	72	180	153	.42	.96	159
	80	166	163.5	.39	.97	168
	95	132	173	.31	.98	177
	102	125	197	.29	.984	199
	110	115	186	.27	.99	188
	125	100	190.4	.23	.99	192
15	15	368	63.25	.86	.77	82
	25	368	94.5	.86	.77	123
10	56	203	152	.47	.95	160
	61	204	132.5	.47	.95	139
	75	185	165	.43	.96	172
	99	113	142	.26	.99	143

Table 5

Specimen size (mm)	σ_f	K_1	S_r	K_r	K_{1C} Predicted
5	1379	74.8	.93	.69	108.
10	1206	93.5	.814	.81	115.5
18	1103	112.2	.744	.855	127.6
40	724	112.2	.49	.96	117.7
100	414	101.2	.28	.99	102.3
240	276	103.4	.19	.995	103.4

Table 6

Specimen size (mm)	$a/_w$	σ_f	K_1	S_r	K_r	K_{1C} predicted	K_{1C} predicted from Begley & Landes (13)
25.4	.604	31.7	98.5	.99	.54	182	194
25.4	.573	37.2	102.5	.995	.49	208	
25.4	.547	41.4	104	.995	.49	212	
50.8	.576	41.7	140.8	.91	.72	195	202
50.8	.552	39.8	144.1	.9	.73	198	
50.8	.526	44.8	148.7	.9	.73	204	
50.8	.5	49.1	150.5	.86	.77	196	

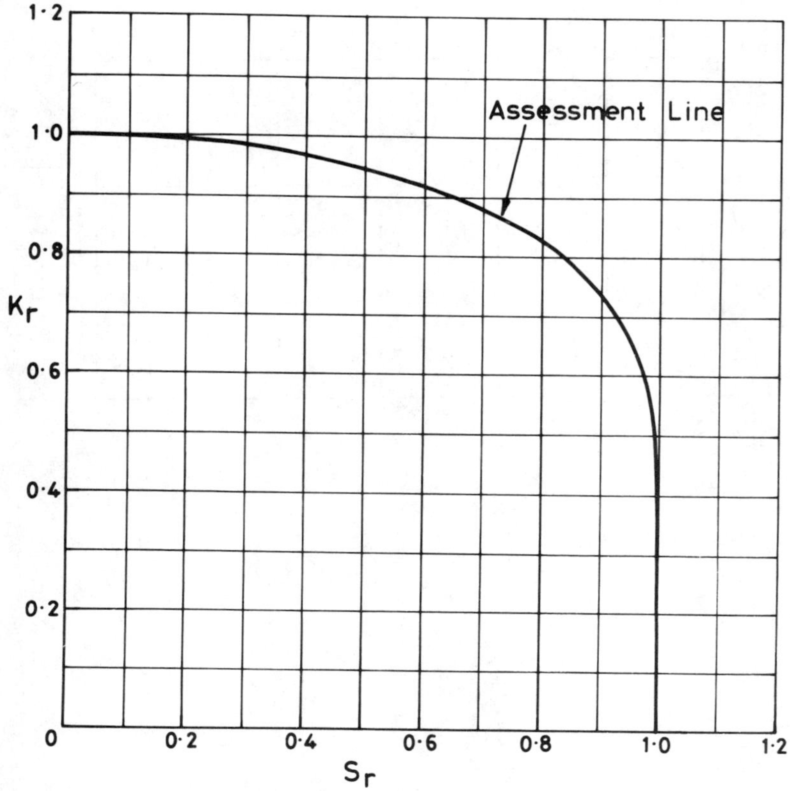

Figure 1

Fracture 1977, Volume 3, ICF4, Waterloo, Canada, June 19 - 24, 1977

ELASTIC CONTACT PROBLEMS IN FRACTURE MECHANICS

B. Billy E. Fredriksson*

INTRODUCTION

During a load cycle an existing crack may partially close causing the fractured surfaces to make contact. This would of course influence the stress field. The frictional forces arising at the contact surface could be expected to influence the direction of crack propagation. Contact problems in fracture mechanics have been studied by some authors. Aksogan [1] presents a solution to the elastic problem of a closing Griffith crack. Paris and Tada [2] studies a closing elastic single edge crack loaded in mode I. Newman [3] studies the effects of closing cracks in fatigue crack propagation. Erdogan and Gupta [4] studies contact and crack problems of elastic wedges. All of these solutions are restricted to specific types of geometries and loadings and do not include friction. In the present paper an attempt is made to do a unified approach to elastic contact problems taking frictional effects into account.

In order to obtain a solution which takes the frictional effects into account a general slip criterion with associated slip rule is introduced. As a special case of the general one a Coulomb type of slip criterion is used in the numerical calculations. The incremental governing equations for elastostatic contact problems with friction are solved by means of the finite element method. Attention is focused on an existing crack that may partially close during a load cycle. Considering a virtual crack growth the crack extension work is derived by applying the principle of virtual work. The energy dissipation due to friction at the contacting surfaces is obtained.

A finite element computer programme for two-dimensional elastic, plane and axisymmetric, problems has been developed. Stress intensity factors are calculated and the effect of crack closure is shown. Crack extension work for different virtual propagation directions is calculated and the effects of crack closure and frictional properties are shown.

STATEMENT OF THE PROBLEM

Consider a body containing a crack which might have closed due to the loading (Figure 1). The problem is studied in the orthogonal cartesian coordinate system x_1, x_2, x_3. In order to facilitate the study of oblique or curved cracks, a local coordinate system η_1, η_2, η_3 is introduced. η_1, η_2 defines the tangent plane to the crack surface. We indicate the material on one side of the crack with A and the other with B (Figure 1). η_3 is then defined as the outward normal vector from B. Temperature effects and dynamic terms are omitted. The body is assumed to be loaded by surface loads q_i on Γ_q and volume loads X_i in V. The displacements are assumed to

* Department of Mechanical Engineering, Institute of Technology, Linköping University, 581 83 Linköping, Sweden

be prescribed on Γ_u. The crack surfaces are assumed to be in contact on Γ_c. On Γ_{cr} the cracked surfaces are not in contact and are unloaded.

THE CONTACT PROBLEM

Bodies generally contact each other through small irregularities in their surfaces. The applied normal load forces the summits of irregularities to flow plastically and/or to crush down until their cross sections are sufficient to support the applied load. It is easy to understand that the type of contact is different at different summits. At some there may be cold welding and micro-seizure and at other summits we have merely elastic contact. When shear forces arise the joint surfaces are displaced. The displacement ceases when the micro-seizure points within the real contact area have reached sufficient numbers to be able to offset the applied tangential load. When a stable condition is attained some of the summits are in adhesion or a welded state and others are in a state of elastic contact. It may be reasonable to assume that elastic and small plastic deformations appear at the adhesion points and that relative displacements appear at the elastic contact points. As a consequence of these assumptions it can be understood that a micro-slip appears even though the applied tangential load is smaller than the sliding force as determined by using the macroscopic coefficient of friction. Based upon these ideas a general contact constitutive relation will be introduced. The contact surface is assumed to be ideal and free from the above mentioned irregularities. The constitutive relation relating the contact stresses and the slip will however be derived in order to satisfy the real case. The derivation is analogous to the derivation of the flow rule in the theory of plasticity.

Consider the cartesian coordinate system η_1, η_2, η_3 at a contact point given on Γ_c. The contact stress increment vector is written

$$dp_i = (dp_1,\ dp_2,\ dp_3) \text{ on } \Gamma_c \ . \tag{1}$$

In the following, when the indices α, β, γ, δ occur, they are assumed to range from 1 to 2 and refer to the local coordinate system η_1, η_2, η_3. A detailed derivation of the slip rule is given by Fredriksson [5]. The derivation of the slip rule is based upon two *basic assumptions*.

1) The slip increment dv_α is linearly dependent on the contact stress increment. That is,

$$dv_\alpha = du_\alpha^A - du_\alpha^B = h_{\alpha\beta}\, dp_\beta \tag{2}$$

2) There exists a *slip surface* $g(p_i) = 0$ in the contact stress space on which slip will occur. At each state of the slip no further slip will occur unless

$$\frac{\partial g}{\partial p_i}\, dp_i > 0 \ . \tag{3}$$

The first assumption implies that the slip has the same direction as the outward normal vector to the locus generated from the intersection between the surface and the plane p_3 = constant.

For *slip hardening* we obtain

$$dv_\alpha = \frac{1}{L}\frac{(\partial g/\partial p_\alpha)(\partial g/\partial p_\beta)}{(\partial g/\partial p_\delta)(\partial g/\partial p_\delta)}\, dp_\beta \text{ on } \Gamma_c^s \tag{4}$$

when

$$g = 0,\ \frac{\partial g}{\partial p_i}\, dp_i > 0,\ p_3 < 0\ .$$

Γ_c^s is the part of Γ_c, where the slip criterion is satisfied. In the part Γ_c^a of Γ_c where the slip criterion is not satisfied there is no slip increment and the displacement increment must satisfy

$$dv_\alpha = du_\alpha^A - du_\alpha^B = 0 \text{ on } \Gamma_c^a\ . \tag{5}$$

If we assume *ideal slip* (4) is replaced by the slip rule

$$dv_\alpha = \lambda\,\frac{\partial g}{\partial p_\alpha} \text{ on } \Gamma_c^s \tag{6}$$

$$\lambda \geq 0 \text{ when } g = 0 \text{ and } \frac{\partial g}{\partial p_i}\, dp_i = 0,\ p_3 < 0$$

$$\lambda = 0 \text{ when } g < 0 \text{ or, } \frac{\partial g}{\partial p_i}\, dp_i < 0,\ p_3 < 0$$

$\frac{\partial g}{\partial p_i}\, dp_i > 0$ does not exist in ideal slip. In ideal slip the slip surface is fixed.

The functions g and L depend on properties of the contact surface, for instance type of material and surface roughness. In the case of ideal slip the parameter λ is indeterminate.

Assuming *Coulomb isotropic slip criterion* [5] we obtain

$$g(p_i) = \frac{1}{\mu}\,(p_\alpha p_\alpha)^{1/2} + p_3,\ p_3 < 0 \tag{7}$$

and the associated slip rule for hardening slip

$$dv_\alpha = \frac{1}{L}\,\frac{p_\alpha p_\beta}{p_\delta p_\delta}\, dp_\beta \tag{8}$$

μ is the coefficient of friction.

For ideal slip we obtain

$$dv_\alpha = \lambda\,\frac{p_\alpha dp_\alpha}{\mu(p_\delta p_\delta)^{1/2}}\ . \tag{9}$$

Introducing the *effective contact stress*

$$p_e = (p_\alpha p_\alpha)^{1/2} \tag{10}$$

and the *effective slip*

$$v_e = \int dv_e; \quad dv_e = (dv_\alpha \, dv_\alpha)^{1/2} \tag{11}$$

it can be shown [5] that for slip hardening

$$L = -p_e \frac{d\mu(v_e)}{dv_e} . \tag{12}$$

Thus, the single curve $\mu = \mu(v_e)$ yields both the shear stress necessary to obtain slip and the function L, which might be called the *slip modulus*. For ideal slip L is zero and equation (9) has to be used. The parameter λ is indeterminate and the stiffness properties of the contacting bodies must be used to obtain the slip increment.

Furthermore, the displacement increment perpendicular to the contact surface must satisfy the kinematical condition

$$du_3^A - du_3^B = 0 \text{ on } \Gamma_c . \tag{13}$$

CRACK EXTENSION WORK IN CRACK CLOSURE PROBLEMS

Consider a virtual quasistatic crack growth. It is assumed that macroscopic (continuum mechanics) theory is applicable [6]. The virtual quasistatic crack growth generates a new crack surface Γ_{cr}^+ with a corresponding fracture area ΔS. The work done on the fracture process zone per unit of fracture area [7] during this virtual growth from state 1 to state 2 is the *crack extension work* G [7]

$$G = - \lim_{\Delta S \to 0} \frac{1}{\Delta S} \int_{\Gamma cr^+} \left(\int_1^2 q_i \, du_i \right) d\Omega . \tag{14}$$

$-q_i$ is, by definition, the stress vector acting from the continuum on the fracture process zone.

By applying the principle of virtual work the crack extension work can alternatively be expressed in global terms. Assume an infinitesimal virtual crack growth with corresponding displacements du_i. Applying the principle of virtual work to the total stress field we obtain

$$\int_V \sigma_{ij} \, d\varepsilon_{ij} \, dV + \int_{\Gamma_c} p_\alpha \, dv_\alpha \, d\Gamma = \int_V X_i \, du_i \, dV +$$

$$+ \int_{\Gamma q} q_i \, du_i \, d\Gamma + \int_{\Gamma_{cr}^+} q_i \, du_i \, d\Gamma \tag{15}$$

where it is assumed that $du_i = 0$ on Γ_u. The strain increment $d\varepsilon_{ij}$ and the slip increment dv_α are both compatible with the displacement increment.

Integrating equation (15) from state 1 to state 2 and introducing the total potential energy increment

$$\Delta\Pi = \int_V \left(\int_1^2 \sigma_{ij}\, d\varepsilon_{ij} - \int_1^2 X_i\, du_i \right) dV - \int_{\Gamma_q} \left(\int_1^2 q_i\, du_i \right) d\Gamma \tag{16}$$

we obtain

$$- \Delta\Pi - \int_{\Gamma_c} \left(\int_1^2 p_\alpha\, dv_\alpha \right) d = - \int_{\Gamma^+_{cr}} \left(\int_1^2 q_i\, du_i \right) d\Gamma \ . \tag{17}$$

Assume that the process is described using the fracture area S as a parameter [7]. Dividing equation (17) by the finite increment ΔS we obtain in the limit

$$- \Pi' - C' = G \tag{18}$$

where

$$\Pi' = \lim_{\Delta S \to 0} \frac{\Delta\Pi}{\Delta S}$$

is the change in total potential energy per unit of fracture area and

$$C' = \lim_{\Delta S \to 0} \frac{1}{\Delta S} \int_{\Gamma^c} \left(\int_1^2 p_\alpha\, dv_\alpha \right) d\Gamma \tag{19}$$

is the dissipated energy due to friction at the contact surface. Thus the sum of G and C' expresses the total energy dissipation.

From equation (18) it can immediately be concluded that the frictional properties influence the crack extension work G. When there is no friction present C vanishes. Although C vanishes the closure of the crack still influences G since the stress field is influenced and thereby the potential Π.

APPLICATIONS

The incremental governing equations for the contact problem are solved by means of the finite element method [8]. In the computer programme Coulomb slip criterion with associated ideal and hardening slip is included. The surfaces of the existing crack (or cracks) are defined as contact surfaces and the nodes in the finite element model are defined as contact nodes. The contact nodes at the crack tip may be allowed for cohesional forces. The external nodes are next applied and the contact nodes are checked for closure. When closure occurs iterations are performed until the slip rule is satisfied and convergence is achieved. By releasing the pair of contact nodes at the crack tip the crack extension work for a finite crack growth may be calculated. Relaxation must generally be performed incrementally because of the nonlinearity at the contact surface. This method of relaxation was first suggested by Andersson [9] and is also used by Hellan [7].

Stress Intensity Factor Calculations

In terms of the stress intensity factors the crack extension work for coplanar extension is written

$$G = \frac{1+\nu}{E}\left[\frac{\kappa+1}{4}\left(K_I^2 + K_{II}^2\right) + K_{III}^2\right] . \tag{16}$$

K_I, K_{II} and K_{III} are the stress intensity factors in mode I, II and III respectively [7]. For plane strain $\kappa = 3 - 4\nu$ and for plane stress $\kappa = (3-\nu)/(1+\nu)$.

The present method has been tested on a plate of unit thickness in a state of plane strain with a single edge crack subjected to tension and moment loads in mode I. In Figure 2a the stress intensity factor in pure tension is shown as function of the crack length and compared with the solution by Gross [10] for an infinite strip with a single edge crack. Next a moment load was applied and the stress intensity factor calculated. The result is shown in Figure 2b. As the moment is applied the stress intensity factor decreases linearly until the crack starts to close. The linear decrease in the stress intensity factor then ceases. When the crack is closing the problem becomes nonlinear. The stress intensity factor is almost constant after crack closure. This is in agreement with the result presented by Paris and Tada (a/W = 0.55). Due to conditions of symmetry the frictional properties do not influence the result.

Crack Extension Work in Crack Closure Problems

The crack extension work for the single edge crack previously studied was computed for different types of loading (Figure 3a). Different virtual crack propagation directions ϕ were studied and the crack extension work was calculated. Applying the criterion of maximum crack extension work [7] the crack propagation direction may be predicted.

The plate was first assumed to be loaded in pure tension σ_o and in pure tension plus antisymmetric shear F, M. The normalized crack extension work as a function of ϕ is plotted in Figure 3a. At this loading no contact forces arise. Computations were done for nine virtual propagation angles ϕ from -90° to +90°. G_{max} is the maximum crack extension work in tension plus antisymmetric shear loading. From the pure tension curve it is seen that G has a maximum at $\phi = 0$ and that coplanar extension is predicted. This result is in agreement with previous findings. Some discretization errors are observed. In the case of tension plus antisymmetric shear the maximum G appears at $\phi \simeq -40°$. If the critical G was reached the angle of propagation is predicted to -40°, that is, the crack tends to propagate downwards in a combined mode.

The plate was next simultaneously loaded in tension, in antisymmetric shear and in compression. The crack then partially closes. The influence of the frictional properties on the crack extension work G was studied. G was calculated for five different directions of virtual crack extension, from 0 to -90°. In view of the first example it is evident that the G has maximum for ϕ between 0° and -90°. In Figure 3b the normalized crack extension work for the frictionless case is compared with the case of friction. An ideal Coulomb model with $\mu = 1$ was assumed. G_{max} is the maximum crack extension work for frictionless case. From these results it could be concluded that the crack extension force is decreasing when taking frictional effects into account. Energy is dissipating at the surface of contact. This effect of course decreases the risk for crack propagation and gives a less dangerous situation. As the present loading conditions the crack surfaces starts to make contact at the left and the contact developes inwards. When the antisymmetric shear is applied slip takes place over the whole surface of contact. When the crack is virtually extended downwards the contact continues to develop inwards and the slip is increasing.

CONCLUDING REMARKS

A method of taking contact and frictional effects in crack closure problems into account was presented. Stress intensity factors was calculated from the crack extension work and compared with known solutions. It was also shown how the stress intensity factor is affected by a partial crack closure. By studying the crack extension work for virtual crack extensions at different angles the crack propagation direction was predicted. When studying cracks in combined modes it was shown how the present method could be used to study effects of frictional properties when the crack is closing.

ACKNOWLEDGEMENT

The present work is part of a research project supported financially by the Swedish Board for Technical Development and SKF/European Research Center. The author wishes to express his gratitude to Professor B. G. Allan Persson for valuable discussions and comments on the manuscript. The author also wishes to express his gratitude to Mr. Göran Rydholm for valuable coding work and to Mr. Jaroslav Mackerle for preparing input data and figure drawings.

REFERENCES

1. ASKOGAN, O., Int. J. of Frac. Mech., 11, 1975, 659.
2. PARIS, P. C. and TADA, H., Int. J. of Frac., 11, 1975, 1070.
3. NEWMAN, J. C., Eighth Nat. Symposium on Fracture Mechanics, Providence, Rhode Island, August 26 - 28, 1974.
4. ERDOGAN, J., Elasticity, 6, 1976, 57.
5. FREDRIKSSON, B., Linköping Studies in Science and Technology, Diss. No. 6, Linköping, Sweden, 1976.
6. CARLSSON, J., Fracture Mechanics, Div. of Solid Mech., Royal Inst. of Technology, Stockholm, Sweden, 1974.
7. HELLAN, K., Publ. No. 73:6, Norwegian Inst. of Technology, Norway, 1973.
8. FREDRIKSSON, B., An Int. J. of Com. and Struc., 6, 1976, 281.
9. ANDERSSON, H., J. Mech. Phys. Solids, 21, 1973, 337.
10. GROSS, B. et al, NASA TN D-2603, 1965.
11. FREDRIKSSON, B., Report LiTH-IKP-R-034, Linköping Inst. of Technology, Linköping, Sweden, 1975.

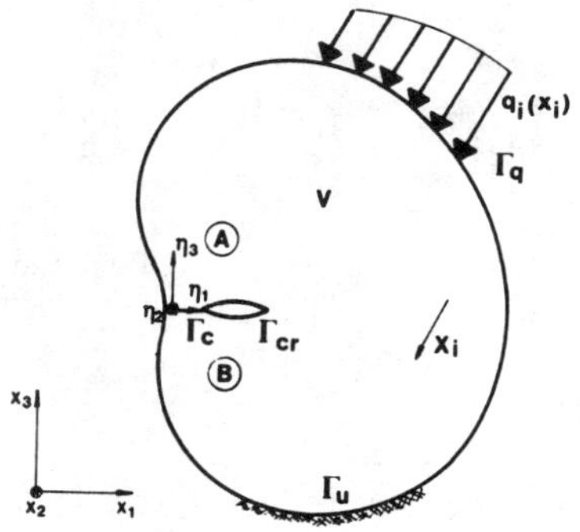

Figure 1 An Elastic Body Containing a Partially Closed Crack

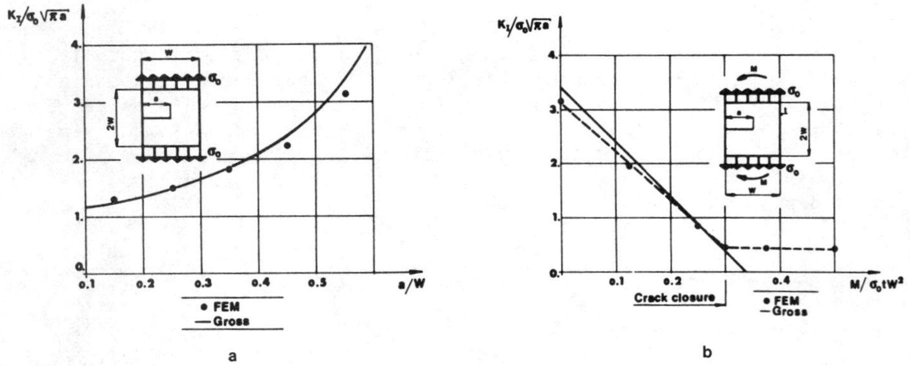

Figure 2 Single Edge Crack. FEM-Model: 178 Constant Strain Elements, 268 Degrees of Freedom. Stress Intensity Factors in

a) Pure Tension
b) Tension and Bending

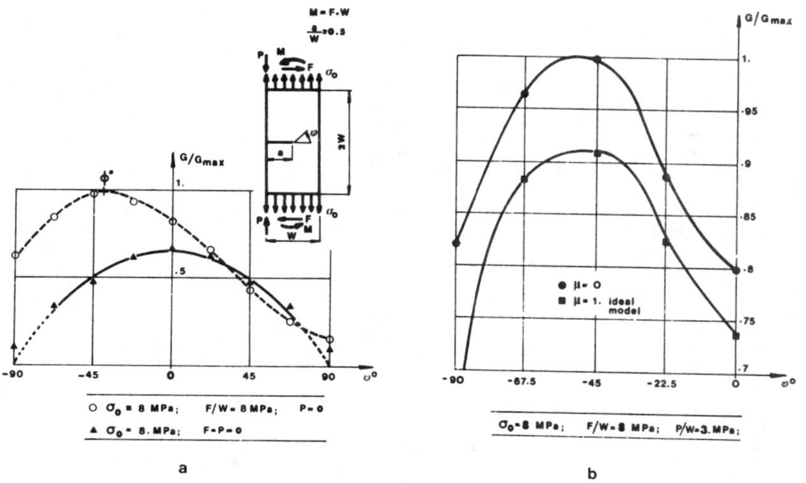

Figure 3 Single Edge Crack. Normalized Crack Extension Work in

a) Tension and Antisymmetric Shear
b) Tension, Compression and Antisymmetric Shear

Fracture 1977, Volume 3, ICF4, Waterloo, Canada, June 19 - 24, 1977

UNSTABLE FRACTURE CRITERIA UNDER LARGE PLASTIC DEFORMATION

Tatsu Fujita, Atsuo Mizuta and Osamu Tsuda*

INTRODUCTION

So far the ductile fracture of metals has been discussed mainly on the following three aspects: (a) the effect of some defects contained in the material on ductility of material through a continuum mechanics approach [1, 2]; (b) the development of linear fracture mechanics in reference to the stress and strain distribution near a crack tip after yielding [3, 4, 5]; (c) the generalization of ductile fracture criteria by studying failures in different kinds of material tests [6, 7, 8]. It seems, however, that ductile fracture has not yet been sufficiently studied in the aspect of the relation between a critical strain and a crack length, which might well be the most fruitful approach to the interpretation of the ductile fracture. The description of the relation is also very useful to resolve the various failure problems of engineering materials under large plastic deformation such as that in metal working, and could implicate the K_c concept in linear fracture mechanics in its extreme case.

UNSTABLE DUCTILE FRACTURE

If fracture occurs between the yield point and the load predicted by the ultimate tensile strength, then the linear fracture mechanics is no longer applicable, nor any other conventional fracture criteria. The present investigation is concerned with unstable ductile fracture criteria, which would complete a whole fracture concept along with linear fracture mechanics. In this study a crack length is limited to be (a) not so large as the plastic region does not cover entirely the test piece, and (b) not so small as a local necking takes place, providing a unfirom uniaxial plastic condition. The model for investigating the unstable fracture criterion is proposed and schematically shown in Figure 1. A curve OABC means the load-displacement relation of the specimen with an initial crack length a, and ODEF with an initial crack length a+Δa, respectively. The path C → F indicates that the state of C goes to a certain state F when a crack grows by Δa.

The following assumptions, designated KOBE-model, are postulated. (a) The state of F is independent of path. That is, the point F is also on the load-displacement curve of the specimen with initial crack length a+Δa. (b) The ratio of a load P_1 on the curve OABC to a load P_2 on the curve ODEF at any displacement is constant. Then, the load P-displacement λ relationship for the specimen with crack length a can be given by

$$P = F(a) \cdot f(\lambda) \tag{1}$$

* Central Research Laboratory, Kobe Steel, Ltd., Kobe, Japan

and the function F(a) is the same as that in the elastic range where it is well known. The validity of (b) has been assured by the finite element analyses as mentioned in the later section.

According to the general theory conservation law, the following equation is written at the onset of crack extension:

$$\dot{W} = \dot{U} + \dot{\Gamma} + \dot{K} \, , \tag{2}$$

where W is the external work, U the internal strain energy, Γ the effective fracture surface energy, K the kinetic energy, and the dot denotes differentiation with respect to time. From equation (2) and the first assumption (a), a criterion for crack extension is given by

$$\frac{\partial \Gamma}{\partial a} = \frac{\partial (W-U)}{\partial a} \, . \tag{3}$$

Recently, the J integral [9] and the COD (crack opening displacement) are used for the estimation of the fracture toughness of notched specimens, but it is difficult to find their exact values by experiments or by calculations. Equation (3) is more useful than other procedures if the loading displacement relation is given analytically as equation (1). So equation (3) is applied to derive the unstable ductile fracture criterion in our study.

THEORETICAL ANALYSIS

Though the criterion could be induced from equation (3) for any load-displacement relationship, here we study a typical one described by a power strain hardening law such as

$$\sigma = k \cdot \varepsilon^{n} \, , \tag{4}$$

where σ, ε, n and k are the true stress, the true strain, the strain hardening exponent and the material constant, respectively. According to the KOBE-model, the load P-displacement λ relationship for the specimen of such material with crack length a is represented by

$$P = F(a) \cdot [\ln(1+\lambda/l_0)]^{n}/(1+\lambda/l_0) \tag{5}$$

and from the linear fracture mechanics and the equation (1)

$$F(a) = A_0 \cdot k \cdot (1-2\pi a^2/w_0 l_0) \, , \tag{6}$$

where A_0 is the initial area of cross-section of the specimen, l_0 the initial gauge length and w_0 the initial width.

The internal strain energy U can be given from (5) and (6) by

$$U = \int_0^{\lambda} P \cdot d\lambda = \frac{A_0 l_0 k}{1+n} \cdot (1-2\pi a^2/w_0 l_0) \cdot [\ln(1+\lambda/l_0)]^{1+n} \, . \tag{7}$$

When a crack grows by Δa, the change of the strain energy ΔU becomes

$$\Delta U = \frac{\partial U}{\partial a}\Delta a + \frac{\partial U}{\partial \lambda} \cdot \Delta\lambda = \frac{\partial U}{\partial a} \cdot \Delta a + P \cdot \Delta\lambda \ . \tag{8}$$

And in that process, the external work is also given by $\Delta W = P \cdot \Delta\lambda$.

Putting (7), (8) and ΔW into (3), the criterion equation can be obtained:

$$\frac{\partial \Gamma}{\partial a} = \frac{2\pi a k}{1+n} \cdot [\ln(1+\lambda/l_0)]^{1+n} \ . \tag{9}$$

A crack growth occurs when the left-hand term in equation (9) reaches a constant value G_c, which is defined as the critical strain energy release rate. The unstable ductile fracture criterion is finally reduced to the form

$$\varepsilon_f \cdot a^{\frac{1}{1+n}} = \left[\frac{(1+n) \cdot G_c}{2 \cdot \pi \cdot k}\right]^{\frac{1}{1+n}} = \text{constant} \ , \tag{10}$$

where $\varepsilon = \ln(1+\lambda/l_0)$, and ε_f denotes a uniform strain enough away from notches at failure. The relationship (10) implicates the well known relations in linear fracture mechanics as its particular case, $n = 1$,

$$\varepsilon_f \cdot a^{1/2} = \left(\frac{G_c}{E \cdot \pi}\right)^{1/2} = \text{constant} \tag{11a}$$

or,

$$\sigma_f (\pi a)^{1/2} = K_c = \text{constant} \ , \tag{11b}$$

where E is the Young's modulus.

EVALUATION OF THE THEORY BY NUMERICAL ANALYSIS

The purposes of the numerical analysis are to examine the assumption (b) in the previous section and to calculate the values of the J integral and the COD. The finite element method based on the infinitesimal incremental theory by Y. Yamada [10] was used for this analysis. The specimen is divided into about 300 finite elements and the ratio of crack length to width varied from 0.01 to 0.1. Calculations were conducted for the double edge notched plates under plane stress condition. The maximum increment at each step was limited below 0.2% for strain increment or 0.1 times yield stress for stress increment. A uniform displacement was sequentially applied on both ends of the plate. The stress-strain behaviour of material was assumed to be $\sigma = E \cdot \varepsilon$ below the yield stress σ_y, and $\sigma = k \cdot \varepsilon^n$ over σ_y.

The values of $-(d \ln\varepsilon_f / d \ln a)$ obtained by these analyses on $n = 0.2635$ material are compared, that is, a theoretical prediction $1/1+n = 0.7915$, while 0.86 by the J integral, 0.85 by the COD criterion and 0.75 by the G criterion [11]. The theoretical value falls among three calculated values and seems reasonable taking account of cumulative errors of numerical analysis. It is also confirmed in course of calculation that the slope of the $\ln\varepsilon_f$-$\ln a$ curve is almost equal to -1/2 in the elastic range, which corresponds to the theoretical value of linear fracture mechanics.

EXPERIMENTATION AND EXPERIMENTAL RESULTS

The uniaxial tensile tests of 0.2 mm thick saw cut notched specimens of normalized 0.80% C steel were conducted. The test pieces were 46 mm wide, 7 mm thick and 100 mm wide, 6 mm thick, and the notch depths at both edges of the specimens were 0.2 mm to 32 mm. The variation of hardness throughout the thickness of a specimen was within Hv 10. The mechanical properties of non-notched specimen were as follows; ultimate tensile strength 91.9 kg/mm^2, elongation 12.9%, yield stress 44.5 kg/mm^2 and reduction of area 20.7%. The Young's modulus of the material is 20500 kg/mm^2 and the strain hardening exponent n is about 0.31 on true stress-true strain basis. Three quantities were measured in the test; first one is a uniform uniaxial strain away from notches by a plastic strain gauge, second one, an elongation of the gauge length 220 mm and third one, an applied load.

The experimental results are shown in Figure 2. Below about 0.2%, i.e., before overall yielding, the slope of $\ln\varepsilon_f$-ln a relationship is -1/2, while the slope is about -1.0 over 0.2% strain possibly due to unstable excess yield strain and the effect of the ratio of crack length to specimen width. On the other hand, in the range beyond 5% uniform strain, the slope tends to be flat due to the local necking of specimen. Between 0.2 mm and 1.0 mm in crack length, stable uniform strain conditions are satisfied, where the $\ln\varepsilon_f$-ln a curve has the slope of about -0.76 predicted by the present theory for n = 0.31 of 0.80% C steel. The agreement indicates that the fracture criterion represented by equation (10) is appreciably reasonable in case of unstable ductile fracture, so long as uniformity is kept and materials obey a power strain hardening law.

DISCUSSION AND IMPLICATION

(1) In order to get a better fit for various engineering materials, other expression of a stress-strain relation will be applied:

$$\sigma = Y + H \cdot \varepsilon_p^{n'}, \tag{12}$$

where Y and n' are material constants and ε_p is the plastic strain. In this case, by the same procedures as in the former, the following representations are obtained:

$$P = A_0(1-2\pi a^2/w_0 l_0)(Y+H\varepsilon^{n'})(1-Hn'\varepsilon^{n'-1}/E)/(1+\lambda/l_0) \tag{13}$$

$$a\cdot[\varepsilon_f(Y + \frac{H}{1+n'}\varepsilon_f^{n'}) - \frac{H}{E}\varepsilon_f^{n'}(Y + \frac{H}{2}\varepsilon_f^{n'})] = \text{constant} \tag{14}$$

or approximately

$$\varepsilon_f \cdot a^{\frac{1}{1+n'}} = \text{constant.} \tag{15}$$

Putting the value of each parameter into (14) and (15), the relationship ε_f-a is obtained for 0.80% C steel as shown in Figure 3. The former theoretical result and the experimental result are also shown. There is little difference among them and there comes a more simple conclusion that since the strain hardening exponents of most metals are usually between 0.25 and 0.35, the product of a critical strain and the 0.75 ~ 0.8th power of a crack length is almost constant, whichever representation is used for the stress-strain relationship.

(2) The present theory has the close relation with the Griffith criterion, and also throws light on the physical meaning of reduction of area in conventional tensile test, the historical basic ductility measure. If a material shows an ideal power strain hardening characteristics and a specimen has infinite width, an ideal $\ln\varepsilon_f$-ln a diagram will be drawn as Figure 4. Obviously, the unstable ductile fracture criterion is rewritten with reference to the conventional K_c value,

$$\varepsilon_f \cdot a^{\frac{1}{1+n}} = \frac{\sigma_y}{E}\left[\frac{K_{Ic}{}^2}{\pi \cdot \sigma_y{}^2}\right]^{\frac{1}{1+n}} . \tag{16}$$

While the true fracture strain ε_n derived from reduction of area in conventional tensile testing is plausibly given by

$$\varepsilon_n \cdot a_i^{\frac{1}{1+n}} = \text{constant}, \tag{17}$$

where a_i is the effective inclusion size of a particular material. Though the $\ln\varepsilon_f$-ln a relation over maximum uniform strain ε_u is somewhat ambiguous due to necking of specimen, for 0.80% C steel ε_n is about 0.19 as shown in Figure 2 and corresponding a_i is estimated about 0.04 mm which is reasonable value as the size of inclusions in the steel. So a whole physical interpretation is obtained throughout ductility by tensile test, i.e., unstable ductile fracture and brittle fracture. Another engineering application to estimate the K_c value from ε_n of tensile test is available on these lines. The relationship should be

$$K_c = (\pi \cdot a_i \cdot E^{1+n} \cdot \varepsilon_n{}^{1+n} \cdot \sigma_y{}^{1-n})^{1/2} . \tag{18}$$

CONCLUSION

1) A fracture criterion for unstable fracture under uniaxial tension of notched plates is proposed. The criterion for a power strain hardening material is represented by:

$$\varepsilon_f \cdot a^{\frac{1}{1+n}} = \text{constant}.$$

2) The theory has been approved by means of two procedures; the numerical analysis by the finite element method and the experiments with notched plates.

3) The present criterion is identical with that of linear fracture mechanics for n = 1, elastic body. In view of this theory, one can have better understanding about the reduction of area in conventional tensile test.

REFERENCES

1. McCLINTOCK, F. A., J. Appl. Mech., 35, 1968, 363.
2. THOMASON, P. F., J. Inst. Metals, 96, 1968, 360.
3. GERBERICH, W. W., J. Mat. Sci., 5, 1970, 283.
4. KE, J. S. and LIU, H. W., Eng. Frac. Mechanics, 5, 1973, 187.
5. HEALD, P. T., Mater. Sci. Eng., 10, 1972, 129.

6. COCKCROFT, M. G. and LATHAM, D. J., J. Inst. Metals, 96, 1968, 33.
7. LEE, P. W. and KUHN, H. A., Met. Trans., 4, 1973, 969.
8. OYANE, M., J. of the J.S.M.E., 75, 1972, 596.
9. RICE, J. R., J. Appl. Mech., 35, 1968, 379.
10. YAMADA, Y., Seisan-kenkyu, Monthly Journal of Institute of Industrial Science, University of Tokyo, 19, 1967, 75.
11. LIEBOWITZ, H., JONES, D. L. and POULOSE, P. K., The 1974 Symposium on Mechanical Behaviour of Materials, Kyoto, 1974, A(1).

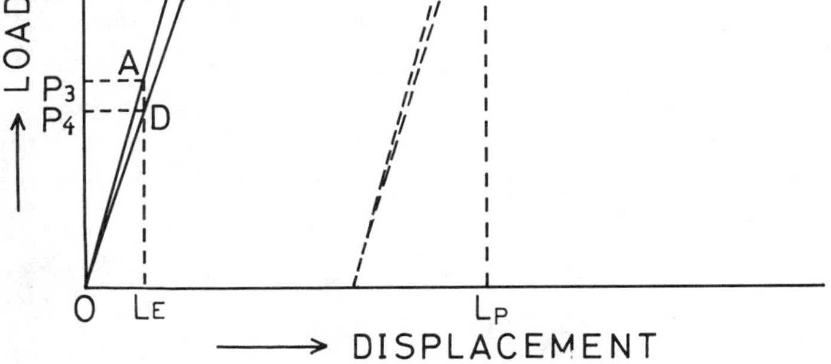

Figure 1 Schematic Diagram of the Model for the Load-Displacement Relation

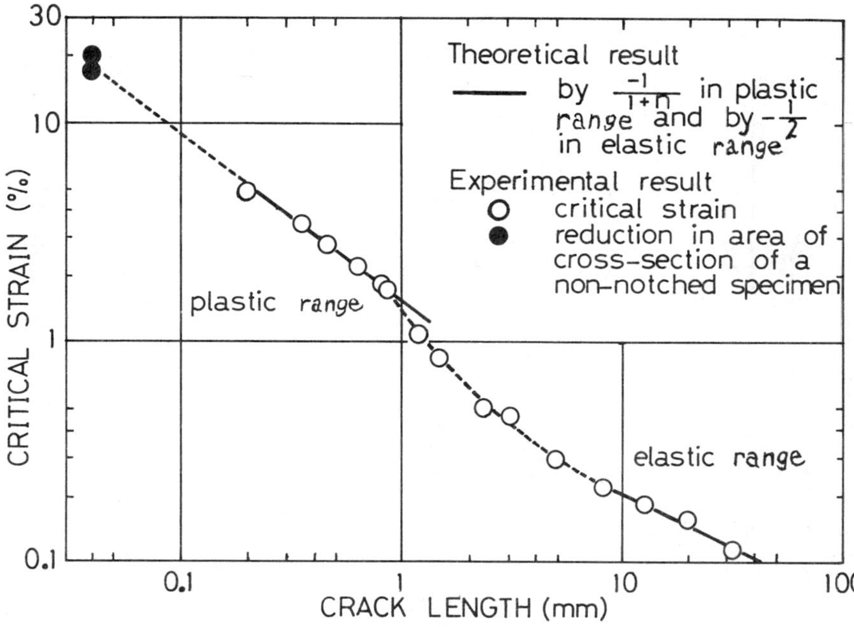

Figure 2 Theoretical and Experimental Relationship Between Strain to Fracture and Crack Length for .80%C Steel

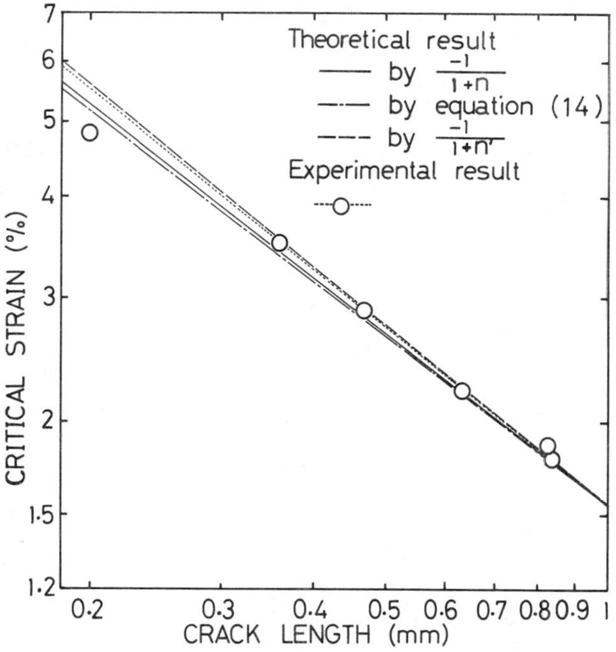

Figure 3 Theoretical and Experimental Relationship Between Strain to Fracture and Crack Length for .80%C Steel

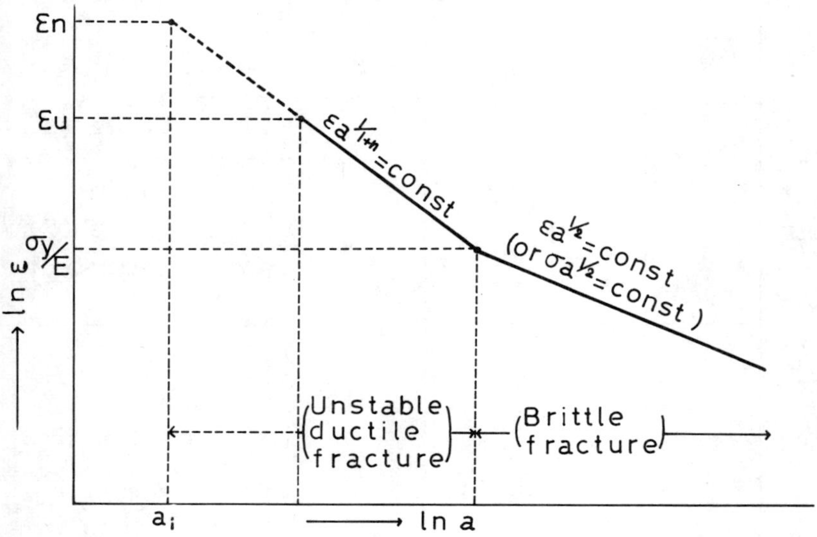

Figure 4 An Ideal Crack Length-Fracture Strain Relationship

Fracture 1977, Volume 3, ICF4, Waterloo, Canada, June 19 - 24, 1977

ON THE APPLICATION OF CONTINUUM THEORY OF DISLOCATIONS TO THE MECHANICAL BEHAVIOURS OF MULTI-PHASE MATERIALS

Hiroshi Miyamoto* and Masanori Kikuchi*

1. INTRODUCTION

The phase-boundary behaviour of multi-phase materials is an important problem. For example, pile-ups of dislocations occur at grain boundaries of metal polycrystals, and the internal stress field by these performs an important role for the mechanical behaviours of polycrystals. These internal stress fields are considered, from the view point of the continuum theory, due to the incompatibilities at the phase-boundary. So, it is useful to discuss the compatibility conditions. For this purpose, the method of continuum theory of dislocation (CTD) is introduced. The basic idea of this theory is established firstly by Eshelby [1], and Kröner [2] and Mura [3] extended his ideas to many problems.

In this method of CTD, a plastic deformation is considered to occur due to emissions and motions of dislocations. So, an elasto-plastic problem becomes to find out the equilibrium dislocation density, α_{ij}, which satisfies the boundary conditions. This is carried out by using three dimensional elasticity theory. Therefore, an elasto-plastic analysis becomes no other than an elastic analysis. In this paper, analyses are carried out by using finite element method (FEM). If α_{ij} is determined, the internal stress field of these dislocations is determined and the role of the incompatibilities is able to be evaluated.

On the other hand, many studies are carried out for the mechanical behaviours of multi-phase materials by using FEM. In general, a (D^P) matrix method, used for elasto-plastic analyses by FEM, assumes that compatibility conditions are satisfied, so this method is not suitable for the discussion of incompatibilities.

In this paper, two examples of multi-phase materials are analyzed. One is bi-crystal of metal and the other is spherical cast iron. Results are compared with those by a [D^P] matrix method.

2. NUMERICAL PROCEDURE

Assumption is needed for the creation and motion of dislocations which correspond to the plastic deformation. This is made by reference of dislocation theory. Once this is made, distribution of α_{ij} is determined corresponding to a stress field given by elastic analyses by FEM.

*Department of Precision Machinery Engineering, Faculty of Engineering, University of Tokyo.

The determination of equilibrium distribution of α_{ij} is carried out as follows:

(i) By Kröner, relationship between α_{ij} and β^*_{ij}, the plastic distortion tensor, is given as

$$\alpha_{ij} = \varepsilon_{ik\ell}\, \beta^*_{\ell j,k} \ , \tag{1}$$

where, ε_{ijk} is the unit permutation tensor.

(ii) The plastic strain, ε^*_{ij}, is given from plastic distortion tensor as

$$\varepsilon^*_{ij} = (\beta^*_{ij} + \beta^*_{ij})/2 \ . \tag{2}$$

(iii) Owen [4] obtained the internal stress assuming that ε^*_{ij} is equivalent to pre-strain in elastic domain. By FEM, this is carried out by substituting ε^*_{ij} into the following equation.

$$\{F_{in}\} = - \int [B]^T [D]\{\varepsilon^*\} d(vol) \tag{3}$$

where $\{F_{in}\}$ is equivalent nodal force. By $\{F_{in}\}$, internal stress and strain are determined. This consistent to the method of the self-consistent model by Eshelby.

(iv) Examination is performed whether the internal stress is in equilibrium to the dislocation density or not. If it is not, α_{ij} is redetermined by the sum of applied stress and the internal stress and process (iii) is repeated until it reaches the converging value.

(v) In the next stage, load is increased and repeated from procedure (iii).

3. APPLICATION TO THE MECHANICAL BEHAVIOURS OF F.C.C. METAL CRYSTALS

3.1 Assumption

(i) Emission of dislocations occurs when τ_a, the applied shear stress, reaches τ_c, the critical shear stress, until the back stress by emitted dislocations becomes equal to $\tau_a - \tau_c$. Therefore, τ_c on the slipped plane increases to τ_a.

(ii) Relations between the increment of stress, strain and number of dislocations are given by Seeger [5]:

$$d\varepsilon/dn = 9b/16x \ , \tag{4}$$

$$d\sigma = Gbdn/2\Pi L \ , \tag{5}$$

where $d\sigma$, $d\varepsilon$, dn denote the incremental value of stress, strain and number of dislocations, respectively. L is slip line length and x is slip line spacing.

3.2 Single crystal analyses

Figure 1 shows equivalent nodal force obtained by equation (3). Tensile axis is parallel to z-axis. At the free surface, nodal force becomes zero as the free surface condition $\sigma_{ij}n_j = 0$ must be satisfied. Evidently, surface integral of F_i, nodal force, becomes zero.

By the existence of this internal force, the crystal rotates and the deformation of single crystal is not equal all over the area. Figure 2 shows strain distriubtion of single crystals. By $[D^P]$ matrix method, this effect of rotation is not described clearly.

On the other hand, in early stages of deformation of f.c.c. single crystal, many experiments show that only one slip system becomes active. But in this analyses three or four slip systems become active in early stage of deformation. This means that the internal stress of dislocations on primary slip system performs an important role to prevent the activation of other slip systems. Perhaps this occurs at the tip of pile-ups of slip lines as shown in Figure 3(a). But in this analysis, dislocations are assumed to distribute uniformly as in Figure 3(b), therefore, it is difficult to describe these microstructures by this method.

3.3 Bi-crystal analysis

In analyses of bi-crystal, an additional assumption is added that only one ship system among 12 ones is able to be active on which the shear stress (which is the sum of the applied shear stress and the internal shear stress) is the maximum.

Three bi-crystals are analyzed and the tensile axes of which are shown in Figure 4. Figure 5 shows the strain distribution near grain boundaries obtained by a $[D^P]$ matrix method for type C. Strain near grain boundaries varies continuously and smoothly. But the results by CTD varies from Figure 5. Figure 6(a)-(c) show the results obtained by this method. Strain distribution is not smooth near grain boundary. A common feature of the three examples is that the strain parallel to the tensile axis decreases near boundary. In experiments, the decrease of strain near boundary is sometimes observed and these results agree with them, while the result by a $[D^P]$ matrix method is not able to describe this phenomenon.

Moreover, for every type A, B and C the active slip system near boundary is difficult from that far from boundary where the primary slip system is active. At grain boundaries, the phenomena that second slip system becomes active at first is observed. By Chalmers [6] the internal stress field by dislocation pile-ups in one side of bi-crystal provokes slip activations in adjacent crystal. The slip system activated is the one on which sum of the applies stress and the internal stress is the maximum. It is obvious that the method of CTD is able to treat these proceudres analytically, which, by a $[D^P]$ matrix method, is difficult. The state of the variation of shear stresses on first and second slip systems are shown in Table 1 for type C.

4. APPLICATION TO THE MECHANICAL BEHAVIOURS OF SPHERICAL CAST IRON

4.1 Assumption

Emission of dislocations occurs on the plane of the maximum shear stress when τ_a, the maximum shear stress, exceeds τ_c. The number of dislocations emitted is estimated from the solution by BCS model [7] as

$$n = \frac{\Pi(1 - \nu)\ell_o}{\mu b} \tau_{eff} . \tag{6}$$

The plastic strain corresponding to the emitted dislocations is determined as follows,

$$\gamma = \frac{1}{2} nb , \tag{7}$$

where ℓ_o is the average diameter of crystals and τ_{eff} is $\tau_a - \tau_c$.

Two dimensional analyses are carried out. Figure 7 shows the mesh division for analysis. The black part shows graphite. Volume ratio of graphite (V_g) is 17.1%. Table 2 shows the material constants. The yield stress is employed three kinds of combination. n_R, the ratio of the two is determined as τ_c^g/τ_c^{Fe}, where τ_c^g and τ_c^{Fe} denote the yield stress of graphite and ferrite, respectively.

4.2 Results and discussion

Figure 8(a)-(c) show the propagation states of plastic regions. Black and white region means the plastic region of ferrite and graphite, respectively, and serial numbers of figures express the load stages. The final figurations of plastic zones in ferrite matrix is the same for three n_R values, while they are different distinctly in graphite.

Miyamoto and Oda [8] analyzed the same problem for $n_R = 0.04$ using a $[D^P]$ matrix method. The result is that yield does not occur in graphite. In this analysis, the internal stress of dislocations piled up at the phase boundary is added to the applied stress. Therefore the plastic region of graphite is mainly due to the internal stress fields by the incompatibilities at the phase-boundary. In addition, the direction of propagation of the plastic region in graphite is along the phase-boundary. This corresponds to the result of SEM observation of the fracture surface.

Barnby [9] proposes a model that the fracture of the carbides in an austenitic stainless steel occurs due to the dislocation pile-up at the phase-boundary. The number of dislocations is obtained by one dimensional analysis. In this method, the same considerations are able to be performed. The result for $n_R = 0.081$ is shown in Figure 9. The region of oblique lines shows the region in which dislocations are emitted. Inclinations of lines indicate the slip planes. Dislocation density becomes maximum value ($2.8 \times 10^6/cm^2$) at point a and becomes minimum ($2.0 \times 10^4/cm^2$) at point b. In Figure 10(a) and (b), the equivalent nodal force is represented for $n_R = 0.04$ and 0.081. Comparing the two cases, it is noticed that the equivalent nodal forces for $n_R = 0.04$ are smaller than those for $n_R = 0.081$. It is reasonable that for $n_R = 0.04$ yielding occurs in graphite, while for $n_R = 0.081$ it is not observed until final stage of analyses. That is, for $n_R = 0.04$, stresses of dislocations are released at phase-boundary.

4. DISCUSSION

By Ashby [10], dislocations in non-homogeneous materials are divided into two parts, one is statistically-stored dislocations, ρ_s, which corresponds to a general, uniform deformation and the other is geometrically-stored dislocations, ρ_g, which corresponds to a local, non-uniform deformation. In this paper, ρ_g is mainly considered, and the results show that the method of CTD is useful to discuss the effect of ρ_g. But about ρ_s, it is difficult to describe the actual physical phenomenon accurately (see equations (4) and (6)). The difficulties occurred in single crystal analyses are due to this fact. So, it is necessary to develop this method combining with the experimental studies.

5. SUMMARY

(i) The method of CTD is intorudced to discuss the effect of incompatibilities at the phase-boundary. By this method an elasto-plastic analysis becomes an elastic analysis.

(ii) Mechanical behaviours of f.c.c. metal crystals are analysed by CTD. The effect of dislocation pile-ups is able to be evaluated, while it is difficult by a $[D^P]$ matrix method.

(iii) Spherical cast iron is also analyzed. Considerations similar to that of Barnby is carried out two-dimensionally.

(iv) It becomes obvious that this method is very useful to know about ρ_g, but about ρ_s, many experimental studies are needed.

ACKNOWLEDGEMENT

Authors with to thank Mr. Yoshito Mitani, post graduate student of the University of Tokyo, for his valuable comments and discussions.

REFERENCES

1. ESHELBY, J. D., Phil. Trans. A, 244, 1957, 87.
2. KRÖNER, E., Kontinuums Theorie der Versetzungen und Eigenspannungen, Springer Verlag, Berlin, 1958.
3. MURA, T., Inealstic Behaviours of Solids, 1970, 211.
4. OWEN, D. R. J., Journal for Numerical Methods in Engineering, Vol. 9, 1973.
5. SEEGER, A., KRÖNMULLER, H., MADER, S. and TRAUBLE, H., Phil. Mag. 6, 1961, 639.
6. CHALMERS, B. and LIVINGSTON, J. D., Acta. Met, 5, 1957, 322.
7. BILBY, B. A., COTTRELL, A. H. and SWINDEN, K. H., Proc. Roy. Soc. A, 272, 1964, 304.
8. MIYAMOTO, H. and ODA, J., Preprint of JSME 760-2.
9. BARNBY, J. T., Acta. Met. 15, 1967, 903.
10. ASHBY, M. F., Phil. Mag., 19, 1970, 399.

Table 1 Comparison of shear stress for type C

SHEAR STRESS (MPa)		
LOAD STAGE	PRIMARY SLIP SYSTEM $(\bar{1}\bar{1}1)$ $[\bar{1}10]$	SECONDARY SLIP SYSTEM $(\bar{1}11)$ $[01\bar{1}]$
1	-2.26	2.19
2	-2.49	2.42
3	-2.56	2.54
4	-2.58	2.63
5	-2.59	2.71
6	-2.60	2.80

Table 2 Material constants

	FERITE	GRAPHITE
E (MPa)	205800.0	4900.0
ν	0.29	0.16
τ_c (MPa)		
n_R = 0.04	490.0	19.6
n_R = 0.06	490.0	29.4
n_R = 0.081	362.6	29.4

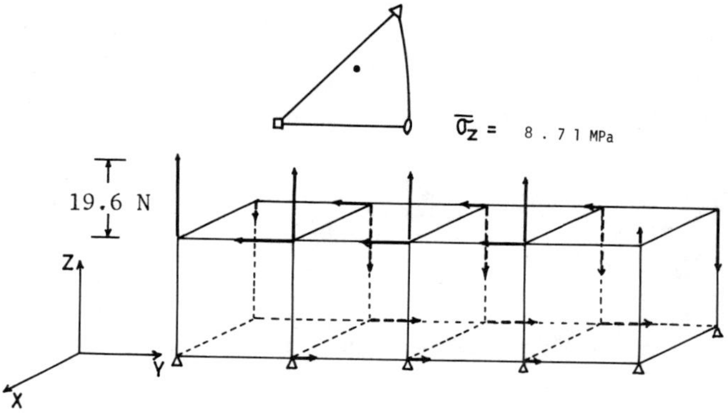

Figure 1

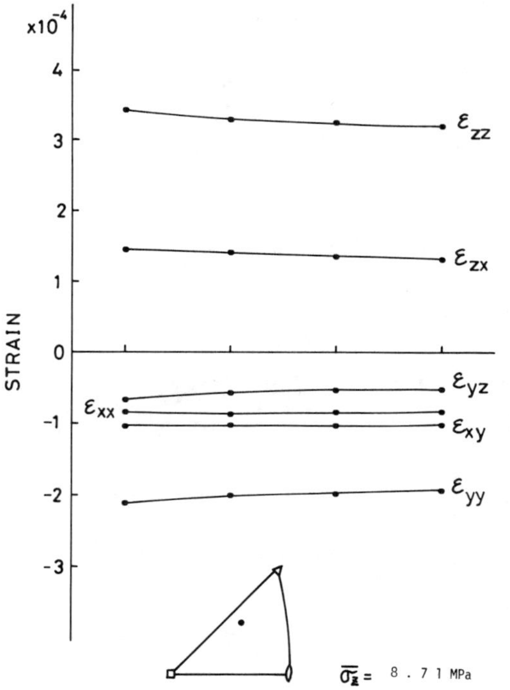

Figure 2 Strain distribution of single crystal

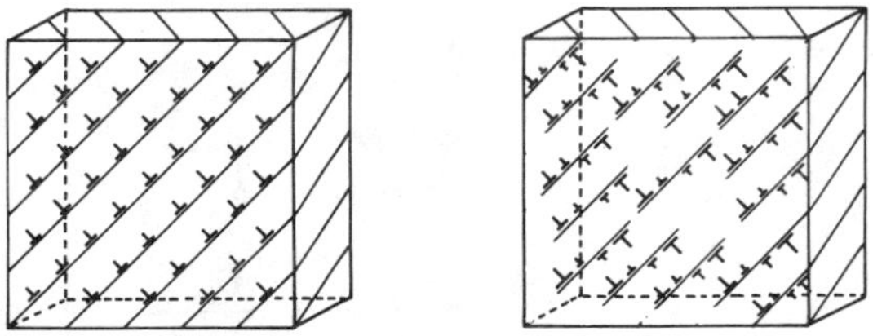

Figure 3

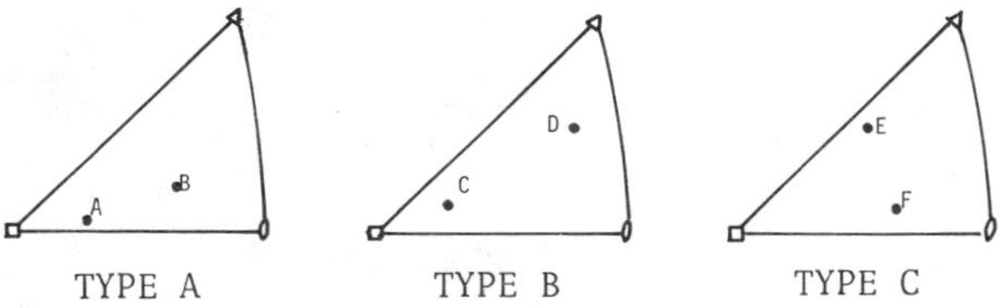

Figure 4 Tensile axes of three kinds of bi-crystals

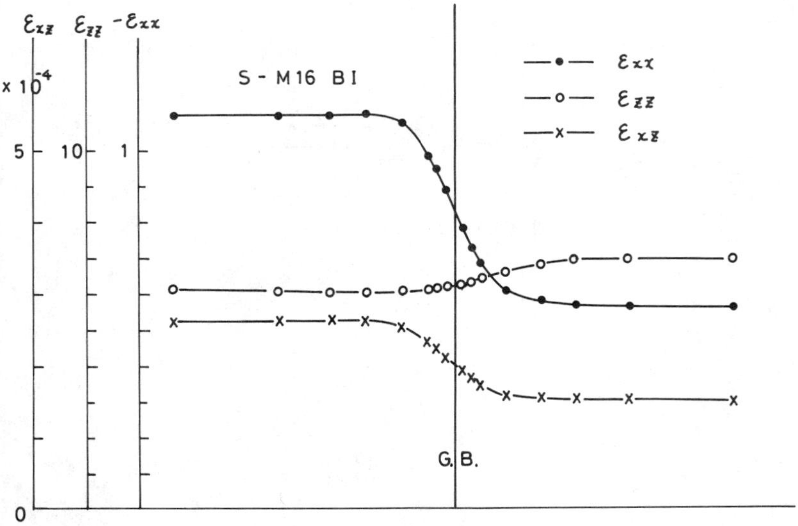

Figure 5 Strain distribution of bi-crystal by a $[D^P]$ matrix method

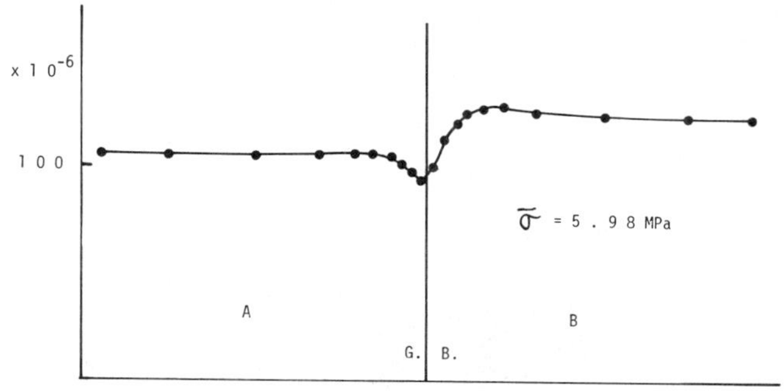

Fig.6 (a) For type A

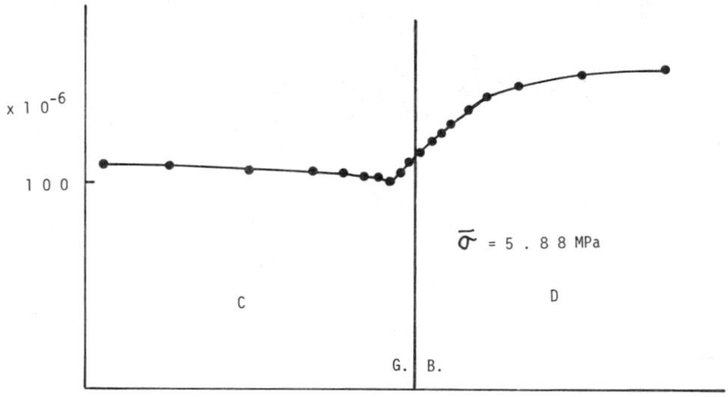

Fig.6 (b) For type B

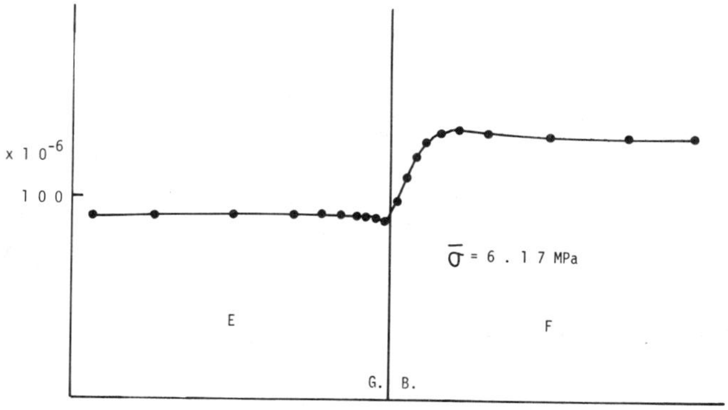

Fig.6 (c) For type C

Figure 6 Strain distribution near grain boundary

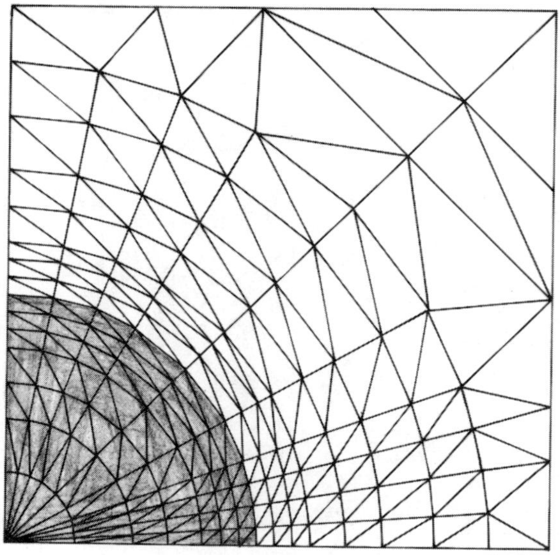

Figure 7   Mesh division of spheroidal cast iron

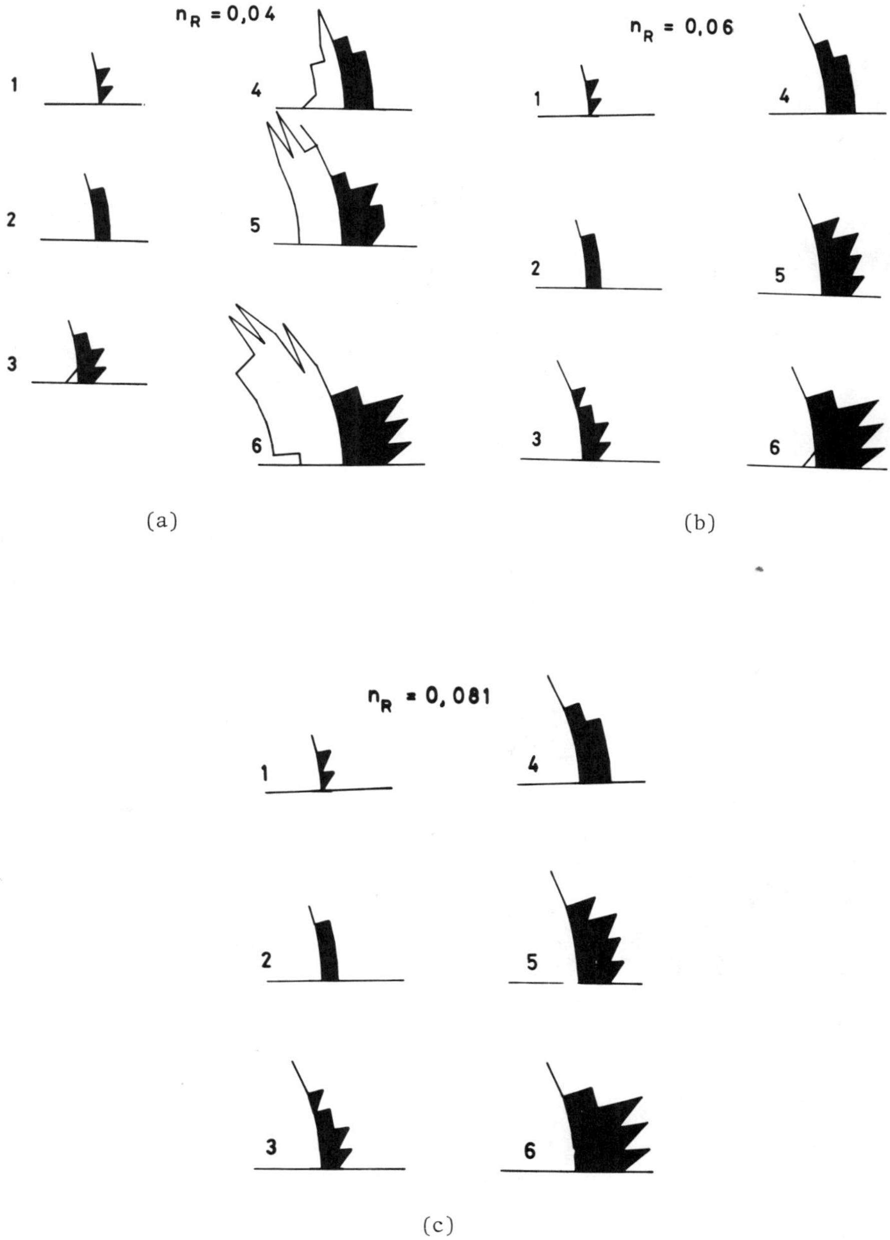

Figure 8 Propagation states of plastic region for three n_R

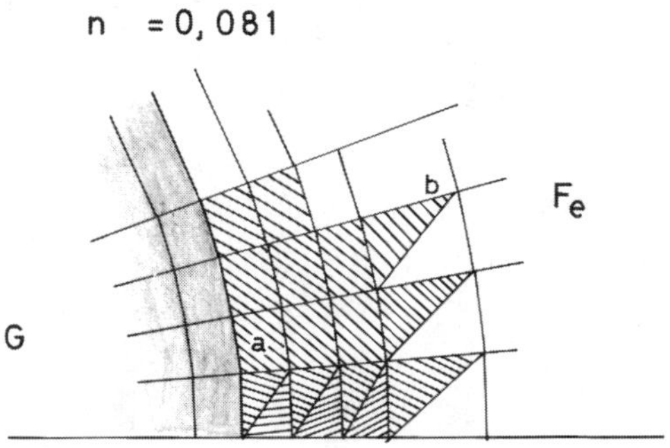

Figure 9

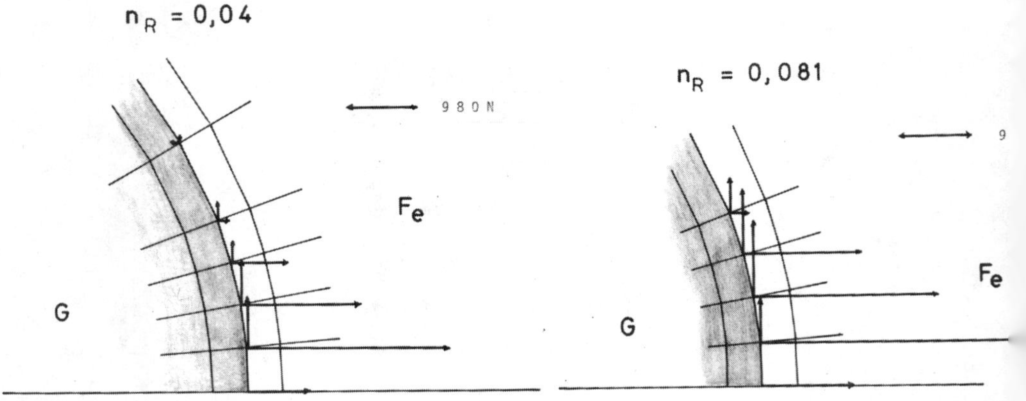

(a)

(b)

Figure 10

Fracture 1977, Volume 3, ICF4, Waterloo, Canada, June 19 - 24, 1977

STRESS ANALYSIS OF CRACKS IN ELASTO-PLASTIC RANGE

S. N. Atluri and M. Nakagaki*

INTRODUCTION

The structure of the dominant singularity in strains and stresses near a crack-tip in plane problems, for hardening materials ($\varepsilon \sim \sigma^n$), has been studied by Hutchinson [1] and Rice and Rosengren [2]. The amplitude of the above singularity for pure mode problems, within the limitation of a small-strain, J_2-deformation plasticity theory, is related to the well known J-integral. Begley and Landes [3, 4] have used the J-integral as an elastic-plastic fracture criterion in the presence of large-scale yielding, with considerable success, in a recent series of experiments. Attention has also been focused recently on other concepts such as the crack-tip opening displacement, nonlinear energy release rate, etc., to deal with the problems of nonlinear fracture mechanics. In the present paper we present a rigorous computational (finite element) method to analyse general in-service situations of ductile-fracture. As an example, analysis of an actual fracture test-specimen is presented, and particular attention is paid to the details of crack-tip stress-strain field, J-integral, COD, and their correlation.

METHOD OF APPROACH

A finite deformation, embedded singularity, elasto-plastic, finite element incremental method [5, 6] using J_2-flow plasticity theory has been developed. Strain and stress singularities of the type given in [1, 2] are embedded in elements near the crack-tip, and finite geometry changes near the crack-tip have been accounted for. A hybrid-displacement finite element model [7] is employed to enforce the conditions of displacement-continuity and traction-equilibrium between the near-tip elements and the far field elements. The incremental finite element method employed is of the "tangent-modulus" type (wherein, the stiffness matrix is up-dated at each step to account for finite-deformation effects as well as plastic flow effects) with a Newton-Raphson type corrective iteration in each step.

The example problem discussed here, is of a 3-pt bend fracture test specimen, of inplane dimensions shown in Figure 3, for which elastic-plastic experimental data for thicknesses of 10 and 20 mm, and uniaxial stress-strain data for the material is reported in [3]. For the present computational purposes, this data was fitted to a Ramberg-Osgood form: $\varepsilon = (\sigma/E) + (\sigma/B)^n$ where $E = 20.7 \times 10^{10}$ Pa, $\nu = 0.3$, $B = 1.26 \times 10^9$ Pa, and $n = 22.2$.

*School of Engineering Science and Mechanics, Georgia Institute of Technology, Atlanta, Georgia 30332, U.S.A.

MS166

CRACK-TIP STRESS AND STRAIN VARIATIONS

The θ-variation of the dominant singularity was determined in [1] from a numerical solution of a nonlinear, fourth order, ordinary differential equation for each value of hardening exponent, n. In the present computations, the r-dependence of the singular solution (i.e., $r^{-n/n+1}$ in strains) is correctly built into the near-tip elements. However, θ-variation is approximated in the usual sense of the finite-element method, by using sector-shaped near-tip elements and assuming a polynomial θ-variation of singular solution in each sector. The computed θ-variation of stresses at a loading stage when the near-tip elements have yielded, is shown in Figure 1. These variations for n = 22.2 are in excellent qualitative agreement with those reported in [1] for n = 13. The r-dependence of the computed strains and stresses is shown in Figure 2, and it can be seen that in the elastic range both the stress and strain singularities are of $1/\sqrt{r}$ type, whereas under fully-developed plasticity the strain singularity is of form $r^{-0.96}$ and the stress singularity is of form $r^{-0.045}$.

COMPLIANCE DATA AND J-INTEGRAL RESULTS

The computed load versus load-point displacement and crack-mouth-opening displacement curves are shown in Figure 3. At the non-dimensional load of 0.28, almost the entire net section ligament was found to have yielded. The linear portion of the P vs. U_L curve has the slope, $P/U_L = 1.33 \times 10^5$ N/mm, (corresponding to $K_I/P = 3137$ $m^{-3/2}$) both of which are found to agree excellently with the independent linear-elastic results of [8]. The J-integral (based on the definition as applicable to the case of finite deformations, given by Knowles and Sternberg [9]) is computed on four different paths (see insert, Figure 3). The variation of the value of J between different paths was ±1.4%. The average J is plotted as a function of load-point displacement, and as a function of crack-mouth-opening in Figure 4. The experimentally determined variation of J with load point displacement from [3] is also shown in Figure 4, for identical specimens of two different thicknesses. The excellent correlation of the experimental results for J with the presently computed values based on a stationary crack model, for all load-point displacement levels (including the critical level at fracture, as determined in the experiments) suggests that there may have been no appreciable stable crack growth prior to fracture in the particular experiments. However it has been found that for some cases [6] involving compact-tension, and centre-cracked specimens, the computed critical J was higher than the experimental value, thus pointing to the effect of stable crack growth prior to fracture in small test specimens. Since J-criterion is strictly valid to define only crack growth initiation, it appears that more experimental as well as analytical studies are necessary, firstly to precisely define the J_{IC} measurement point in the experiments, and secondly to analyze the continuum mechanics problem of stable crack growth. The authors are currently completing work on finite element modelling of crack growth, involving translation of the singularity field, to obtain information on energy balances during crack growth, and, more interestingly, during the terminal loss of stability.

A detailed comparison of the present finite-deformation analysis results, for the load-point displacement, stresses and strains near the crack-tip, plastic-yield-zones near the crack-tip, and crack-surface deformation profiles, at various load levels, with those obtained using only an infinitesimal deformation theory, is presented in [5]. One interesting observation in this comparison in [5] is that, at load levels corresponding to large-scale plastic yielding, and when there is noticeable crack-tip blunting, path-independence of the computed J (with ±1.5%) was noticed even for paths closest to the crack-tip, when the appropriate definition of J [9] applicable to the finite deformations is used. However, at the same load levels, when the well-known Rice's definition of J, applicable only in the case of infinitesimal deformations, was used, the computed J for the path closest to the crack-tip was substantially lower (by about 16%) than that computed from paths in the far-field.

CORRELATION OF J WITH COD

In the literature, there appears to be no precise definition of COD as essentially a near-tip quantity. However, for test specimens such as the present, calibration relations between COD and CGD (clip gauge or crack-mouth-opening displacement) have been given. For instance, Wells [10] gives a nonlinear relation between COD and CGD in the small scale yielding range, and a linear relation in the large scale yielding range, in agreement with theoretical considerations. Using this definition for COD, the relation between COD and J in the present computations, is shown in Figure 5, which shows that $J \sim 1.44 \cdot COD \cdot \sigma_y$. This result appears to be in agreement with the slip-line theory analysis of Rice and Johnson [11], who also account for finite geometry changes. A similar relation was found to hold for other test specimens also [6]. However, efforts to correlate the above calibration values for COD with essentially near-tip-geometrical quantities, such as the diameter of a circle inscribed near the crack-tip, etc. were not successful.

CLOSURE

A computational procedure to analyze elastic-plastic fracture situations is presented. From a computational viewpoint, the J-criteria appears to be the most attractive. However, further analytical framework to analyze stable crack growth is necessary to enhance the scope of computational methods to predict not only ductile fracture initiation, but also terminal loss of stability of crack growth.

ACKNOWLEDGEMENTS

This research was supported in parts by U.S. AFOSR Grant 74-2667, and U.S. NSF Grant ENG-74-21346.

REFERENCES

1. HUTCHINSON, J. W., Journal of Mechanics and Physics of Solids, 16, 1968, 13.
2. RICE, J. R. and ROSENGREN, G. F., Ibid., 16, 1968, 1.
3. BEGLEY, J. A. and LANDES, J. D., ASTM, STP 514, 1972, 1.
4. LANDES, J. D. and BEGLEY, J. A., ASTM, STP 560, 1974, 170.
5. ATLURI, S. N. and NAKAGAKI, M., ASTM, STP (in review) 1976 (Paper presented at ASTM Tenth National Symposium on Fracture, Philadelphia, August 1976.
6. ATLURI, S. N., NAKAGAKI, M. and CHEN, W. H., Journal of Pressure Vessel Technology (in review) and to be presented at Third International Congress of Pressure Vessel Technology, ASME, 1976.

7. ATLURI, S. N., KOBAYASHI, A. S. and NAKAGAKI, M., International Journal of Fracture, 11, 1975, 257.
8. BROWN, W. E. and SRAWLEY, J. E., ASTM STP 410, 1966, 13.
9. KNOWLES, J. K. and STERNBERG, E., California Institute of Technology, Div. of Eng. and Appl. Sci. TR 27, 1973.
10. WELLS, A. A., Proceedings of 3rd Canadian Congress of Applied Mechanics, 1971.
11. RICE, J. R. and JOHNSON, M. A., "Inelastic Behavior of Solids", McGraw Hill, N.Y. 1970, 641.

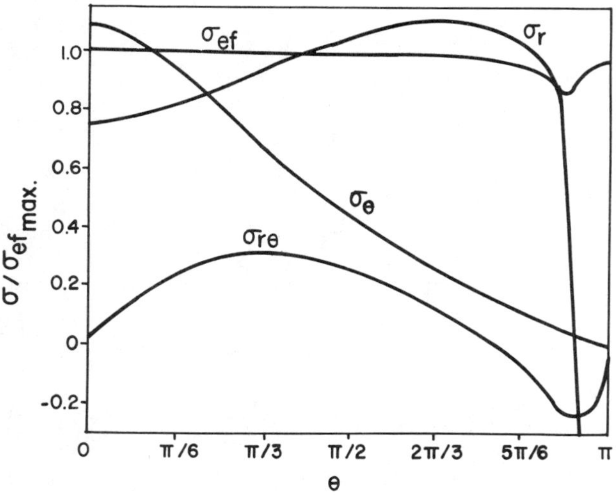

Figure 1 θ-Variation of Crack Tip Stresses in a Strain-Hardening Elastic-Plastic Material (n = 22.2)

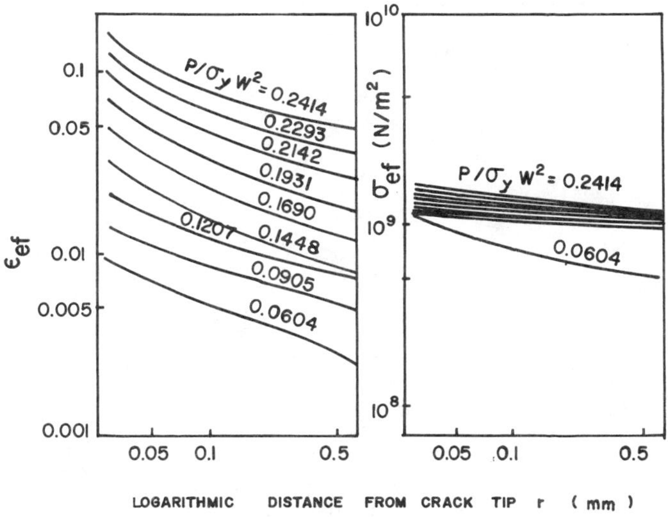

Figure 2 Strain and Stress Singularities Near the Crack-Tip in a 3-Point Bend Bar

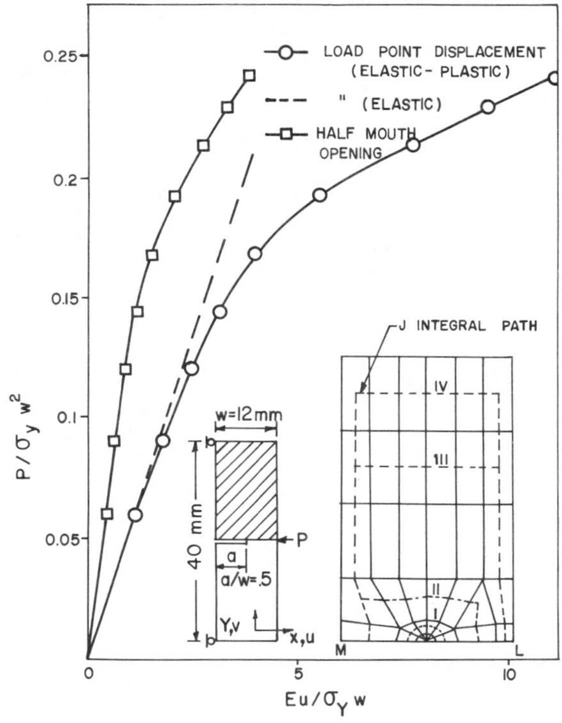

Figure 3 Load versus Load-Point and Crack-Mouth Opening Displacement Variations for a 3-Point Bend Bar

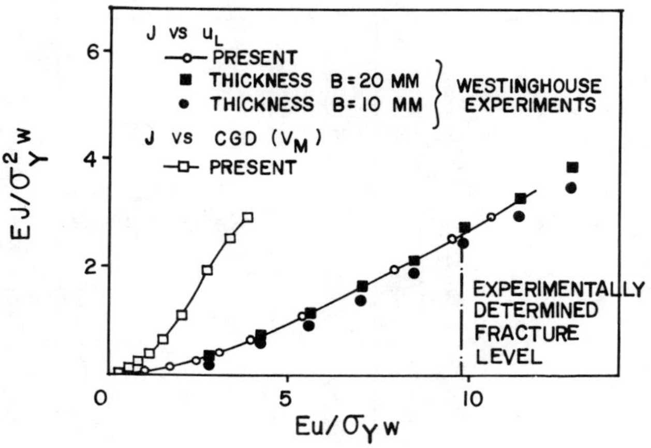

Figure 4 J-Integral Variation with Load-Point and Crack-Mouth Opening Displacement

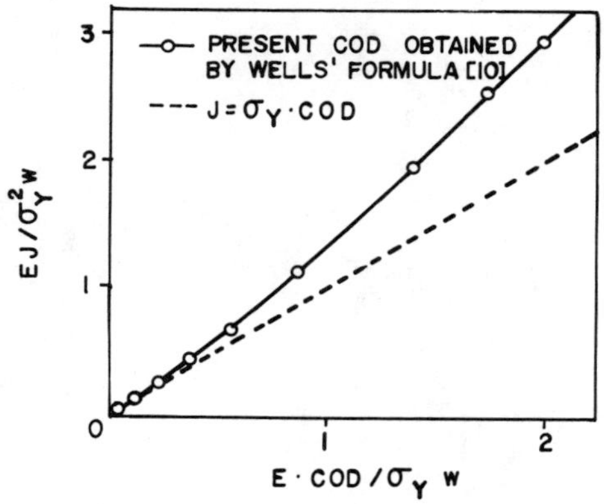

Figure 5 Correlation of J-Integral with Crack-Tip-Opening

Fracture 1977, Volume 3, ICF4, Waterloo, Canada, June 19 - 24, 1977

FRACTURE MECHANICS OF VISCOELASTIC SYSTEMS

V. N. Poturaev* and V. I. Dyrda*

INTRODUCTION

Consideration is being given to viscoelastic systems of inherited type. These systems, under cyclic load, display notable dissipative warming-up and their mechanical behaviour varies with time and depends upon environmental conditions in the form of radiation flows or chemically active agents. Local fracture theory is applied to analyze the fracture processes of such systems. All the relations outlining this process have been obtained on the basis of methods of thermodynamics of irreversible processes. An experimental procedure is presented which makes it possible to appraise characteristic properties and life time prior to fracture, based on the available present level of information on the physical, chemical and mechanical states of the system.

A set of equations is under consideration for calculation of real viscoelastic constructions. This set of equations included: 1) state parameters, 2) deformation equilibrium and coincidence equations, 3) rheological relations, 4) heat equation with internal heat sources, 5) fracture criterion equations.

Deformation equilibrium and coincidence equations are adopted with allowance made for particular system geometrical shape and its loading conditions. Rheological relations are selected with regard to system mechanical behaviour. Valter operators with a Rabotnov - type (relaxation with rational) exponential nucleus have been used in this specific case.

FRACTURE CRITERIA

Two fracture criteria have been utilized in this research:

1. An entropy criterion postulating the following: system fracture takes place at the moment when entropy increment density reaches a certain critical level $\Delta S_f(T)$ i.e. material behaviour at a given temperature. After integration this criterion appears as [1]:

$$\int_{t_o}^{t_f} \dot{S}(t)dt = \Delta S^f(T) = S(t_f) - S(t_o) \ , \tag{1}$$

where $S(t)$ = entropy density; $\Delta S^f(T)$ = entropy density at a critical level of material behaviour, serving as a temperature function.

Let us consider fracture in a real viscoelastic system and take as a system a prismatic rubber element loaded in accordance with the law of harmonics

*Ukrainian Academy of Science, Institute of Geotechnical Mechanics, Dniepropetrovsk, 1, U.S.S.R.

$\varepsilon = \varepsilon_0 \sin\omega t$. We take $\{T^\rho_\varepsilon, T, G\}$ as a complete set of thermodynamic parameters applicable to a 'special' one which on the one hand is sufficiently small to be considered as a continuous medium point and on the other hand -- sufficiently large to exhibit all the properties of this medium. By applying the first law of thermodynamics $U = T_\sigma \cdot \dot{T}_\varepsilon \pm q$, the available energy density definition $f = U - TS$, and the condition $f = f(T^\ell_\varepsilon, T, G)$ we derive the following expression for the rate of change of entropy density:

$$\dot{S} = \frac{1}{T}\left[T_\sigma \cdot \dot{T}^\rho_\varepsilon - \left(S + \frac{\partial f}{\partial T}\right)\dot{T} + \left(T_\sigma - \frac{\partial f}{\partial T^\ell_\varepsilon}\right)\dot{T}^\ell_\varepsilon - \frac{\partial f}{\partial G}\dot{G} \pm \dot{q}\right] \tag{2}$$

where U = internal energy density, G = shear modulus, q = heat flow, $T\sigma$ = stress tensor, T^ρ_ε = strain tensor (irreversible part), T^ℓ_ε - strain tensor (reversible part).

Substituting (2) into (1) and taking into account equalities [2]:

$$\frac{\partial f}{\partial T} = -S \quad , \quad \frac{\partial f}{\partial T^\ell_\varepsilon} = T_\sigma \ ,$$

we derive:

$$\Delta S_f = \int_{t_o}^{t_f} \frac{1}{T}\left[T_\sigma \dot{T}^\rho_\varepsilon - \frac{\partial f}{\partial G}\dot{G} \pm q\right] dt \tag{3}$$

Disregarding material ageing ($\dot{G} = 0$), and assuming the temperature field to be steady, this expression may be written as:

$$T\,\Delta S_f(T) = \int_{t_o}^{t_f} T_\sigma T^\rho_\varepsilon \, dt - \int_{t_o}^{t_f} \dot{q}\, dt \tag{4}$$

A negative q indicates dissipation of heat in the system. The first integral in (7) represents strain energy density, irreversibly dispersed in $[t_o, t_f]$ time; the second thermal energy density, released in the sample at the same time. For a closed strain path:

$$\int_{t_o}^{t_f} T_\sigma \dot{T}^\rho_\varepsilon \, dt = \int_{t_o}^{t_f} T_\sigma \dot{T}_\varepsilon dt \ , \tag{5}$$

which results from the condition that: work done by stress in elastic deformation equals zero in this case. With regard to (5) equation (4) assumes the form:

$$T\Delta S_f = \int_{t_o}^{t_f} T_\sigma \dot{T}_\varepsilon dt - \int_{t_o}^{t_f} \dot{q}\, dt \tag{6}$$

Supposing that rubber adheres to viscoelasticity linear inherited theory and that its properites do not vary in the process of loading, we may specify stress-strain bond as:

$$T_\sigma(t) = 3G_o\left[T_\varepsilon(t) - x\int_{-\infty}^{t} k(t-t')\, T_\varepsilon\,(t')dt'\right] \tag{7}$$

where k(t-t') is a nucleus of relaxation expressed as a Rabotnov rational - exponential function, x = rheological parameter, G_o = instantaneous shear modulus. With regard to (7) expression (6) assumes a final form:

$$T\Delta S_f = \left(\frac{3\omega}{8\pi^2}\,\psi\,G_o\,\varepsilon_o^2 - q\right)\,t_f \qquad (8)$$

Here ψ = material dissipation coefficient; ΔS_f occurred in the course of experimental studies of sample endurance, the samples being in the form of double-sided blades and was independent when investigating endurance of full-scale rubber structures in the form of a prismatic shear element. The heat flow q for the sample's most loaded point has been determined when solving a heat conduction equation of thermal conduction with an internal heat source.

For the sandwich-type rubber-metallic element with the following parameters G_o = 1,76 MPa, ε_o = 0,1, T = 320 K, ω = 12Hz, ψ = 0,31, ΔS^f = 2,03 $10^9 J/m^2$ deg, q = 5,31.$10^4 J/m^3$ sec estimated with regard to (8) time t_f = 4800 hours, time resulting from the experiment for the lot of 36 full-scale articles constitutes (4 to 5).10^3 hours. Agreement, as seen, is quite satisfactory.

2. The range under consideration as the second criterion of fracture is the process of achieving by the system a critical level of damage, when the system converts into labile state. A criterion equation for the shear element loaded by the law of harmonics has been obtained as follows:

$$P_{kp} = 1 - \sqrt{\frac{0,4}{\gamma G}}\,\tau \quad , \qquad (9)$$

and time-period prior to 'special' volume fracture:

$$t_f = \frac{P_{kp}\,\exp\,(Q/kT)}{C\,k_o\,I_o\,(\gamma\tau/kT)} \qquad (10)$$

where P_{kp} = critical concentration of stored defects; k_o constant of action; c = material behaviour, depending upon initial components concentration and forms of the elementary reactions; γ = specific energy of defect formation; I_o = Bessel function.

Expressions (9) and (10) are obtained for the case where thermo-conductancy equation is represented as:

$$\frac{\partial T}{\partial t} = a\left(\frac{\partial^2 T}{\partial x^2} + \frac{\partial^2 T}{\partial x_2^2} + \frac{\partial^2 T}{\partial x_3^2}\right) + \frac{W}{C_1 p}$$

and the revolution equation of material damage has been recorded due to the fact that damage concentration rate change is related to the intensity of physical-chemical processes as:

$$\dot{p} =. C\,k_o\,\exp\left[-\frac{1}{kT}\left(Q - N^0 T - N^1\frac{\partial T}{\partial \ell} - N^2 : T_6\right)\right]$$

Here a = thermo-conductance coefficients; N^0, N^1, N^2 - tensors of the null, first and second valency correspondingly; c_1 = specific heat capacity; p = density.

A number of problems have been solved estimating a period of time prior to local fracture of viscoelastic systems in the form of silent-blocks and sandwich-type shear elements under cyclic load. If material constants have been determined correctly then (10) shows a satisfactory agreement with practice.

Material damage in the process of continuous loading may also be disclosed in the course of direct physical experiments. In the general case for rubber, concentration of damage obeys the relation:

$$\Delta p(t) = P_{kp} [1-\exp(-nt)], \tag{11}$$

where ΔP_{kp} = damage storage from the moment of system loading till its failure; n = constant. For CKH - 3 based rubber P_{kp} = 5,3 (in terms of per unit value) and n = 0,0016. This concept has been confirmed in independent experimental studies wherein material damage (concentration of rupture with time up to system failure) has been determined on models by infrared spectroscopy.

We may derive pre-fracture time from (11) if $\Delta p(t)$ is known over the wide range of stress and temperature variations. Promising results have been obtained for thin films, as well as good agreement between theory and experiment.

REFERENCES

1. TCHOUDNOVSKY, A. I., O razrushenii macrotel, sb.Issledovania po uprugosty i plastychnosty, No. 9, izd. LGU, 1973.
2. VACKOULENKO, A. A., O svyazyach mazhdu napryazhenyamy v neuprugich sredach. sb.Issledovania po uprugosty i plastychnosty, No. 1, izd. LGU, 1961.

Fracture 1977, Volume 3, ICF4, Waterloo, Canada, June 19 - 24, 1977

FINITE ELEMENT ANALYSIS OF CRACK PROBLEMS IN HIGHLY ELASTIC MATERIALS

Neng-Ming Wang* and Hilario L. Oh*

1. INTRODUCTION

It is well recognized [1,2] that the Rivlin-Thomas criterion [3] for the tearing of rubber vulcanizates can be stated in terms of the J integral [4]: A cut (or crack) in a rubber vulcanizate sheet will spread if J reaches a critical value of an energy characteristic of the material. Thus, using the above criterion and a plane stress finite element procedure, the critical load that causes a crack to grow in uniaxial stretching of nicked rubber vulcanizate sheets was calculated in [2] and found to agree closely with existing experimental data. In [5], the tearing energy of two rubber test pieces was computed by using the J integral in conjunction with finite element calculations. The calculated results were again shown to agree with experimental data. It was from the recognition of the equivalence of J with the tearing energy that an experimental technique for measuring the tearing energy of rubber was developed in [6].

The purpose of this paper is to extend the computational procedure used in [2] to calculate J for two other "plane" crack problems. These are the plane strain stretching of a thick strip with an edge cut by a stress σ_o (Figure 1a) and the generalized out-of-plane shear of the same geometry by a shear stress τ_o (Figure 1b). By generalized out-of-plane shear, we mean that while the dominating displacement component may be in the out-of-plane direction, the components in the plane need not be vanishingly small. This is therefore in variance with the so-called anti-plane shear for which the displacement components in the plane are identically zero.

The organization of the paper is as follows: In Section 2, we list the basic equations for a class of two-dimensional finite deformations which contain the plane strain and anti-plane shear as special cases. The materials are assumed to be highly elastic and incompressible. Based on a virtual velocity equation, a finite element procedure is developed. Since formulations of finite element procedures for nonlinear elastic materials have been well documented in the literature [7], only a brief discussion of the present procedure is given. In Section 3, an expression for the J integral pertinent to the problems considered in this paper is derived, while its path independent property is illustrated by direct evaluation. It should be noted that this result is essentially contained in [1,8,9]. In Section 4, we report numerical results obtained for the two crack problems for Mooney materials. The values of J have been computed and plotted in several figures which depict the relationship between J and the applied load (σ_o or τ_o). For the plane strain stretching problem, a comparison is made of the computed J values with those of the corresponding plane stress case. It is found that the J values in the plane strain case normalized by the strain energy density corresponding of σ_o (without

*Research Laboratories, General Motors Corporation, General Motors Technical Center, Warren, Michigan.

MS304

a crack) are almost the same as those in the plane stress case. For the generalized out-of-plane shear problem, the J values are compared to the corresponding anti-plane shear solutions.

2. BASIC EQUATIONS FOR COMBINED EXTENSION AND SHEAR DEFORMATIONS

Consider in a cartesian coordinate system (x^i, $i = 1,2,3$) a body whose cross section at any x^3 occupies a plane domain D which is the same for all x^3 (Figure 2). The class of deformation to be considered is such that the displacement components u_i ($i = 1,2,3$) satisfy

$$u_i = u_i(x^1, x^2) \; . \tag{2.1}$$

We call (2.1) the combined extension and shear deformations which contain obviously the plane strain ($u_3 = 0$) and the anti-plane shear ($u_1 = u_2 = 0$) as special cases. For deformations characterized by (2.1) the deformed metric tensors G_{ij} and the strain tensors ε_{ij} are defined as follows:

$$G_{ij} = \delta_{ij} + 2\,\varepsilon_{ij} \; , \tag{2.2}$$

and

$$\varepsilon_{\alpha\beta} = \frac{1}{2}(u_{\alpha,\beta} + u_{\beta,\alpha}) + \frac{1}{2} u^k_{,\alpha}\, u_{k,\beta} \; , \tag{2.3}$$

$$\varepsilon_{\alpha 3} = \frac{1}{2} u_{3,\alpha} \; , \quad \varepsilon_{33} = 0 \; ,$$

where δ_{ij} denotes the Kronecker delta and a comma preceding an index denotes partial differentiation. Here we have used the convention that Greek indices range from 1 to 2 and Roman indices, from 1 to 3. The contravariant metric tensors G^{ij} are defined by the usual relations:

$$G^{ij}\, G_{jk} = \delta^i_k \; . \tag{2.4}$$

Let S^{ij} be the Kirchhoff stress tensors on convected coordinates initially coincide with the cartesian coordinates (x^i) and F^i be the surface loadings. Then, the well known virtual velocity equation (e.g., [10]) in its time-derivative form may be written as

$$\int \dot{S}^{ij}\, \delta\dot{\varepsilon}_{ij}\, dV + \int S^{ij} \delta \left[\frac{1}{2}\dot{u}^k_{,i}\, \dot{u}_{k,j}\right] dV = \int \dot{F}^i \delta\, \dot{u}_i\, dS \; , \tag{2.5}$$

where the dot over a symbol denotes the rate of change versus "time", or increment. The volume and surface integrals are referred to the undeformed configuration. (For simplicity body forces were omitted in (2.5)). We assume that the material is incompressible and, hence, the variations of the displacements in (2.5) must be required to satisfy

$$G^{ij}\, \delta\dot{\varepsilon}_{ij} = 0 \; . \tag{2.6}$$

Expressions (2.5) and (2.6) are valid for arbitrary domains. For the configuration shown in Figure 2 and for deformations characterized by (2.1), these expressions can be written as

$$\int_D \left[\dot{S}^{\alpha\beta} \delta \dot{\varepsilon}_{\alpha\beta} + 2\dot{S}^{\alpha 3} \delta \dot{\varepsilon}_{\alpha 3} \right] dA + \int_D S^{\alpha\beta} \delta \left[\frac{1}{2} \dot{u}^k_{,\alpha} \dot{u}_{k,\beta} \right] dA = \int_{\partial D} \dot{F}^i \delta \dot{u}_i ds \qquad (2.7)$$

$$G^{\alpha\beta} \delta \dot{\varepsilon}_{\alpha\beta} + 2G^{\alpha 3} \delta \dot{\varepsilon}_{\alpha 3} = 0 \quad , \qquad (2.8)$$

where ∂D is the boundary of the plane domain D.

Finally, we assume that the material of the body is elastic and that there exists a strain energy function W such that the Kirchhoff stresses S^{ij} satisfy

$$S^{ij} = \frac{1}{2} \left[\frac{\partial W}{\partial \varepsilon_{ij}} + \frac{\partial W}{\partial \varepsilon_{ji}} \right] + pG^{ij} \quad , \qquad (2.9)$$

where p is a scalar quantity representing the hydrostatic pressure. The stress increments $\dot{S}^{ij}$ can be derived from (2.9) by simply differentiating with respect to "time" to give

$$\dot{S}^{ij} = \frac{1}{2} \frac{\partial}{\partial \varepsilon_{k\ell}} \left[\frac{\partial W}{\partial \varepsilon_{ij}} + \frac{\partial W}{\partial \varepsilon_{ji}} \right] \dot{\varepsilon}_{k\ell} + \dot{p}\, G^{ij} - 2\, p\, G^{ik} G^{j\ell} \dot{\varepsilon}_{k\ell} \, . \qquad (2.10)$$

A finite element procedure

Based on the virtual velocity equation (2.7), the incompressibility condition (2.8) and the stress-strain relations (2.9) and (2.10), a finite element procedure has been developed. The details of the procedure appeared elsewhere [11]. Since the general formulations of finite element procedures for large strain and large displacement are well established (e.g., [7,12]), only a brief outline of the present procedure will be given in Appendix A. In the present procedure, the elements are quadrilaterals in the plane domain D. The incompressibility condition (2.8) is satisfied only in an approximate way, namely the sum of the integrand in (2.8) over all integration points is zero. This approximation appears to have alleviated the difficulties encountered when triangular elements were used [13, 14]. Accordingly, elemental hydrostatic pressure and stresses are likewise represented by their average values. By doing so, the quadrilateral element is essentially a constant stress element.

The incremental element-stiffness matrix equations assume the following form:

$$\begin{bmatrix} K_e & C^T \\ C & 0 \end{bmatrix} \begin{Bmatrix} \dot{\underset{\sim}{u}} \\ \dot{p} \end{Bmatrix} = \begin{Bmatrix} \dot{\underset{\sim}{f}} \\ 0 \end{Bmatrix} \quad , \qquad (2.11)$$

where the matrices K_e, C and the vectors $\dot{\underset{\sim}{u}}$, $\dot{\underset{\sim}{p}}$ and $\dot{\underset{\sim}{f}}$ are defined in Appendix A. We note that the elemental stiffness matrix equations (2.11) are derived for the combined extension and shear deformations (2.1) for which $u_i \neq 0$, $(i = 1,2,3)$. For plane strain deformations ($u_3 = 0$) and for anti-plane shear deformations ($u_1 = u_2 = 0$), the dimensions of the matrix equations (2.11) will accordingly be reduced. In fact, for anti-plane shear deformations, the incompressibility condition (2.8) satisfies automatically so that the matrix C vanishes identically.

For specific problems, the elemental stiffness matrix as given by (2.11) is assembled to form a master stiffness matrix. The area integrals in (A.11) and (A.12) may be computed by using any appropriate numerical quadrature. In the present work we have used a four-point Gaussian quadrature with the four integration points located at $s = \pm 0.57735$ and $t = \pm 0.57735$ in the para-metric s-t domain. The master stiffness matrix equations are then solved by a standard Gaussian elimination technique. For a prescribed load increment, the solution of the equations gives the corresponding incremental displacements and elemental pressure. The master stiffness matrix has the same form as in (2.11), which is apparently not banded. In order to reduce computing time, a reordering of the components of the vectors $\dot{u}$ and $\dot{p}$ has been made so that the resulting master stiffness matrix is banded.

3. THE PATH-INDEPENDENT J INTEGRAL

We now proceed to derive an appropriate expression for the J integral for the combined extension and shear deformations as described in (2.1). We assume that the elastic body has a crack, the crack face being perpendicular to the x^2 direction (see Figure 3). The J integral was defined originally in [4] by

$$J = \int_{\Gamma} \left[W dx^2 - \underset{\sim}{T} \cdot \frac{\partial \underset{\sim}{u}}{\partial x^1} \, d s \right] , \tag{3.1}$$

where W denotes the strain energy density, $\underset{\sim}{T}$ and $\underset{\sim}{u}$ denote traction and displacement vectors, respectively. The integral assumes the same value for any path Γ which surrounds the tip of the crack. For deformations described by (2.1), the J integral may be written in the following form:

$$J = \int_{\Gamma} \left[W dx^2 - S^{\alpha k} (\delta^j_k + u^j_{,k}) \, \nu_\alpha \frac{\partial u_j}{\partial x^1} \, d s \right], \tag{3.2}$$

where ν_α denotes the exterior normal of the contour Γ defined in the undeformed geometry. The path-independent property of (3.2) is shown in Appendix B.

4. TWO CRACK PROBLEMS

In this section, we report the numerical results for the two crack problems shown in Figure 1. We assume that the materials are of the Monney kind and the strain energy function can be expressed by

$$W = C_1 \, [(I_1 - 3) + \alpha(I_2 - 3) \tag{4.1}$$

where C_1, α are material constants and I_1, I_2 are strain invariants defined by

$$I_1 = G_{11} + G_{22} + G_{33} \, ,$$

$$I_2 = G^{11} + G^{22} + G^{33} \, . \tag{4.2}$$

The finite element procedure discussed in Section 2 is employed to calculate the stresses and deformation in the strip. Expression (3.2) is used to calculate the J integral.

The finite element grid which is used in both crack problems is shown in Figure 4, where by symmetry only half of the region needs to be analyzed.

The crack geometry is specified by c/b = 0.1. To test the accuracy of the grid, a linear elastic plane strain calculation is first made for a uniform tension in the x^2 direction. The J integral is evaluated along the contour Γ which consists of line segments (shown dotted in Figure 4) that join the mid points of two prescribed sides of an element. Using appropriate values of Young's modulus E and Poisson's ratio ν, the stress intensity factor K is determined from the computed J value and the relation

$$K = [EJ/(1 - \nu^2)]^{1/2} \quad ,$$

and found to be 2.05 $\sigma_o\sqrt{c}$. This compares well with the known numerical value of 2.15 $\sigma_o\sqrt{c}$ reported in [15].

We now proceed to discuss our numerical results for the plane-strain stretching problem and for the generalized out-of-plane shear problem.

(a) Plain-strain stretching: For α = 0.0, 0.5 and 1.0, incremental calculations for the stresses and deformation in the strip as functions of the applied load σ_o have been made. As in the case of plane-stress uniaxial stretching stretching [2], it is found that an increment of 0.2 for σ_o/C_1 is satisfactory. Figure 5 shows the computed J values plotted against the nominal extension ratio λ. For comparison purposes, the corresponding plane stress results from [2] have also been plotted in Figure 5. We see that for the same amount of stretching, the J values in plane strain are larger than those in plane stress.

When the J values in Figure 5 are normalized by the quantity $2W_oc$, all curves coincide into one for λ > 1.1 (see Figure 6). Here W_o is the elastic energy density corresponding to σ_o (without the crack) given by

$$W_o = \begin{cases} C_1\ [(\lambda^2 + \frac{2}{\lambda} - 3) + \alpha(\frac{1}{\lambda^2} + 2\lambda - 3)], & \text{for plane stress} \\ C_1\ (1 + \alpha)\ (\lambda^2 + \frac{1}{\lambda^2} - 2)\ , & \text{for plane strain.} \end{cases} \tag{4.3}$$

That the quantity $(J/2W_oc)$, or equivalently (tearing energy/$2W_oc$), is dependent mainly on stretch ratio λ was postulated by Rivlin and Thomas [3] for the stretching of rubber sheets with an edge cut (plane stress). This was verified experimentally by Greensmith [16] who showed that it holds true for c/b < 0.2. Thus for plane stress, results shown in Figure 6 are but a confirmation of results in [3,16]. However, that the quantity $(J/2W_oc)$ in plane strain should also be independent of the material constant α and that it coincides with values in plane stress for λ > 1.1 are new results. Although results shown in Figures 5 and 6 are obtained for a specific crack geometry with ratio c/b = 0.1, they are expected to be valid also for nearby ratios, e.g., 0 < c/b < 0.2, based on similar arguments advanced in [3].

(b) Generalized out-of-plane shear: For α = 0.0, 0.5 and 1.0, the computed J values are plotted against τ_o/C_1 in Figure 7. The dotted curve represents the results for the corresponding anti-plane shear problem ($u_1 = u_2 = 0$). For the latter problem, the governing differential equations are linear (Green and Adkins [10], p. 86) and, hence, the J integral can be explicitly expressed as

$$J = \frac{\pi c\ \ \tau_o^{\ 2}}{4C_1\ (1+\alpha)} \left[\frac{2b}{\pi c}\ \tan\left(\frac{\pi c}{2b}\right)\right] , \tag{4.4}$$

by using the elastic results in [17].

For $\alpha = 0$, the calculated displacements u_1 and u_2 are found to be identically zero, which means that the assumption underlying the anti-plane shear deformation has precisely been met. Moreover, the calculated pressure is found to remain constant for all stages of incremental loading. This is in complete agreement with the analytical relation

$$[p + 2(1+2\alpha)]/C_1 = -2\alpha\left[\left(\frac{\partial u_3}{\partial x^1}\right)^2 + \left(\frac{\partial u_3}{\partial x^2}\right)^2\right], \qquad (4.5)$$

derived in [10] for anti-plane shear deformations. For $\alpha \neq 0$, u_1 and u_2 do not vanish in general. The anti-plane shear assumption is therefore no longer valid as indicated by the J results in Figure 7. To elucidate this further, we have computed the quantity $[p + 2(1+2\alpha)]/C_1$ for $\alpha = 1$ using the following procedures:

(i) the present finite element procedure without assuming $u_1 = u_2 = 0$;

(ii) the same finite element procedure assuming $u_1 = u_2 = 0$; and

(iii) the analytical solution corresponding to anti-plane shear (using $c/b = \sim 0$ for simplicity).

These results have been plotted in Figure 8 for $\tau_o/C_1 = 4.2$. It can be seen that the results of (ii) agree well with those of (iii) as expected, but are significantly different from those of (i).

Finally, we compute the quantity $J/2W_oc$ where W_o denotes the strain energy density in the strip (without the crack) caused by the shear τ_o at $x^2 = \pm b$. Let λ be the extension ratio of an initial length in the x^2 direction, then the strain energy density W_o can be written as

$$W_o = C_1 (1+\alpha) (\lambda^2 - 1) . \qquad (4.6)$$

The computed values of $(J/2W_oc)$ for $\alpha = 0$, 0.5 and 1.0 are plotted in Figure 9. The dotted line corresponds to the anti-plane shear result which assumes a constant value of 1.58. From this figure it is seen that the dependence of the quantity $J/2W_oc$ on both the material parameter α and the extension ratio λ is rather small. Hence, for practical purposes, it will be sufficient to use the exact anti-plane shear solution (4.4) or, equivalently,

$$\left(\frac{J}{2W_oc}\right) = \frac{b}{c} \tan\left(\frac{\pi c}{2b}\right) \qquad (4.7)$$

to calculate J.

REFERENCES

1. CHANG, S. J., J. Appl. Math. Phys., 23, 1972, 149.
2. WANG, N. M., Int. J. Solids Struct., 9, 1973, 1211.
3. RIVLIN, R. S. and THOMAS, A. G., J. Polymer Sci., 10, 1953, 291.
4. RICE, J. R., J. Appl. Mech., 34, 1968, 379.
5. OH, H. L. and WANG, N. M., Proceedings of the SAE Int. Conf. Vehicle Struct. Mech., Detroit, Michigan, 1974, 89.

6. OH, H. L., Proceedings of the 8th National Symposium on Fracture Mechanics, Providence, R.I., August, 1974.
7. ODEN, J. T., Finite Elements of Nonlinear Continua, McGraw-Hill, 1972.
8. ESHELBY, J. D., The Energy Momentum Tensor in Continuum Mechanics, Inelastic Behaviour of Solids, Kanninen, M. R., et al (ed.), McGraw-Hill, New York, 1970.
9. KNOWLES, J. K. and STERNBERG, E., Arch. Rat. Mech. Anal., 44, 1973, 187.
10. GREEN, A. E. and ADKINS, J. E., Large Elastic Deformations and Non-Linear Continuum Mechanics, Oxford University Press, 1960.
11. WANG, N. M., Finite Element Procedure for a Class of Two-Dimensional Deformations of Highly Elastic Materials, General Motors Research Laboratories Research Publication GMR-1435, August 14, 1973.
12. HIBBITT, H. D., MARCAL, P. V. and RICE, J. R., Int. J. Solids Struct., 6, 1970, 1069.
13. ODEN, J. T., Aeron. Quart., 19, 1968, 254.
14. ODEN, J. T. AND KEY, J. E., Proceedings of the INTAM Colloq. on High Speed Computing Elastic Struct., Liege, 1971.
15. SRAWLEY, J. E., Plane Strain Fracture Toughness. Fracture, Liebowitz, H. (ed.), Vol. IV, Academic Press, New York, 1969.
16. GREENSMITH, H. W., J. Appl. Polymer Sci., 1, 1963, 993.
17. RICE, J. R., Mathematical Analysis in the Mechanics of Fracture. Fracture, Liebowitz, H. (ed.), Vol. II, Academic Press, New York, 1968.

APPENDIX A

<u>A finite element procedure</u>

Let the domain D shown in Figure 2 be divided into an assemblage of quadrilateral elements. For each element the initial cartesian coordinates (x^1, x^2) and the displacements $\dot{u}$ are mapped bilinearly onto a square s-t domain for $s \varepsilon [-1,1]$ and $t \varepsilon [-1,1]$ by

$$\begin{bmatrix} x^\alpha \\ \dot{u}_i \end{bmatrix} = \sum_n \begin{bmatrix} x^{\alpha(n)} \\ \dot{u}_i^{(n)} \end{bmatrix} q_n\ (s,t), \tag{A.1}$$

where

$$\begin{pmatrix} q_1 \\ q_2 \end{pmatrix} = (1 \mp s)\ (1-t)/4,$$

$$\begin{pmatrix} q_3 \\ q_4 \end{pmatrix} = (1 \pm s)\ (1+t)/4, \tag{A.2}$$

and the index n (n = 1,2,3,4) refers to the node number of the four nodes of the quadrilateral.

Denoting

$$\dot{\underset{\sim}{u}}^T = (\dot{u}_1^{(1)}, \dot{u}_2^{(1)}, \dot{u}_3^{(1)}, \dot{u}_1^{(2)}, \ldots, \dot{u}_3^{(4)}) ,$$

$$\dot{\underset{\sim}{E}}^T = (\dot{E}_{11}, \dot{E}_{22}, 2\dot{E}_{12}, 2\dot{E}_{13}, 2\dot{E}_{23}) ,$$

and making use of the strain-displacement relations (2.3), we obtain a matrix representation for the incremental strain-displacement relations,

$$\dot{\underset{\sim}{E}} = (H + \psi B)\, \dot{\underset{\sim}{u}} \quad . \tag{A.3}$$

The matrices H, ψ and B in (A.3) are defined as follows:

$$B = \begin{bmatrix} b_1 & 0 & 0 & b_2 & 0 & 0 & b_5 & 0 & 0 & b_6 & 0 & 0 \\ b_3 & 0 & 0 & b_4 & 0 & 0 & b_7 & 0 & 0 & b_8 & 0 & 0 \\ 0 & b_1 & 0 & 0 & b_2 & 0 & 0 & b_5 & 0 & 0 & b_6 & 0 \\ 0 & b_3 & 0 & 0 & b_4 & 0 & 0 & b_7 & 0 & 0 & b_8 & 0 \\ 0 & 0 & b_1 & 0 & 0 & b_2 & 0 & 0 & b_5 & 0 & 0 & b_6 \\ 0 & 0 & b_3 & 0 & 0 & b_4 & 0 & 0 & b_7 & 0 & 0 & b_8 \end{bmatrix} , \tag{A.4}$$

$$H = \begin{bmatrix} 1 & 0 & 0 & 0 & 0 & 0 \\ 0 & 0 & 0 & 1 & 0 & 0 \\ 0 & 1 & 1 & 0 & 0 & 0 \\ 0 & 0 & 0 & 0 & 1 & 0 \\ 0 & 0 & 0 & 0 & 0 & 1 \end{bmatrix} , \tag{A.5}$$

and

$$\psi = \begin{bmatrix} \frac{\partial u}{\partial x} & 0 & \frac{\partial v}{\partial x} & 0 & \frac{\partial w}{\partial x} & 0 \\ 0 & \frac{\partial u}{\partial y} & 0 & \frac{\partial v}{\partial y} & 0 & \frac{\partial w}{\partial y} \\ \frac{\partial u}{\partial y} & \frac{\partial u}{\partial x} & \frac{\partial v}{\partial y} & \frac{\partial v}{\partial x} & \frac{\partial w}{\partial y} & \frac{\partial w}{\partial x} \\ 0 & 0 & 0 & 0 & 0 & 0 \\ 0 & 0 & 0 & 0 & 0 & 0 \end{bmatrix} . \tag{A.6}$$

Here the coefficients b_i are given by

$$\begin{pmatrix} b_1 \\ b_2 \end{pmatrix} = [\mp\, s_x(1-t) - t_x(1\mp s)]/4,$$

$$\begin{pmatrix} b_3 \\ b_4 \end{pmatrix} = [\mp s_y(1-t) - t_y(1\mp s)]/4,$$

$$\begin{pmatrix} b_5 \\ b_6 \end{pmatrix} = [\pm s_x(1+t) + t_x(1\pm s)]/4,$$

$$\begin{pmatrix} b_7 \\ b_8 \end{pmatrix} = [\pm s_y(1+t) + t_y(1\pm s)]/4,$$

with

$$s_x = \frac{\partial y}{\partial t}/J, \quad s_y = -\frac{\partial x}{\partial t}/J, \quad t_x = -\frac{\partial y}{\partial s}/J, \quad t_y = \frac{\partial x}{\partial s}/J$$

and

$$J = \frac{\partial x}{\partial s}\frac{\partial y}{\partial t} - \frac{\partial x}{\partial t}\frac{\partial y}{\partial s}.$$

In the above expressions, notation (x,y) stands for (x^1,x^2). Similarly, notation (u,v,w) stands for (u_1,u_2,u_3).

Making use of the variational equation (2.7) and letting

$$\underset{\sim}{G}^T = (G^{11}, G^{22}, G^{12}, G^{13}, G^{23}),$$

$$\underset{\sim}{S}^T = (S^{11}, S^{22}, S^{12}, S^{13}, S^{23}),$$

$$\Psi = \begin{bmatrix} \gamma & 0 & 0 \\ 0 & \gamma & 0 \\ 0 & 0 & \gamma \end{bmatrix}$$

where

$$\gamma = \begin{bmatrix} S^{11} & S^{12} \\ S^{12} & S^{22} \end{bmatrix},$$

we obtain the following elemental equilibrium equations:

$$\int (H + \psi B)^T \dot{\underset{\sim}{S}}\, dA + \int B^T \Psi B \dot{\underset{\sim}{u}}\, dA = \dot{\underset{\sim}{f}}, \tag{A.7}$$

where the integral is defined over the initial elemental area and $\dot{\underset{\sim}{f}}$ denotes the vector of incremental nodal forces corresponding to $\dot{\underset{\sim}{u}}$. The displacements $\dot{\underset{\sim}{u}}$ must also satisfy the incompressibility condition (2.8) which is equivalent to

$$\int G^T (H + \psi B) \dot{\underset{\sim}{u}}\, dA = 0 \tag{A.8}$$

by assuming a uniform p inside the element.

We now write the incremental stress-strain relations (2.10) in the matrix form

$$\dot{\underset{\sim}{S}} = Q \dot{\underset{\sim}{E}} + \dot{p} \underset{\sim}{G}, \tag{A.9}$$

where the dimension of the matrix Q is 5 x 5. Substituting (A.9) into

(A.7) and combining the resulting equations with (A.8) gives the elemental stiffness equations:

$$\begin{bmatrix} K_e & C^T \\ C & 0 \end{bmatrix} \begin{bmatrix} \dot{\underset{\sim}{u}} \\ \dot{p} \end{bmatrix} = \begin{bmatrix} \dot{\underset{\sim}{f}} \\ 0 \end{bmatrix} , \tag{A.10}$$

where

$$K_e = \int[(H + \psi B)^T Q (H + \psi B) + B^T \Psi B]dA , \tag{A.11}$$

$$C = \int G^T (H + \psi B)dA . \tag{A.12}$$

Matrix Q for Mooney Materials

For Mooney materials (4.1) and for deformations characterized by (2.1) the stress-strain relations (2.9) are

$$\begin{aligned} S^{11} &= C_1[2 + 4\alpha(1+E_{22}) + p\, G^{11}] , \\ S^{22} &= C_1[2 + 4\alpha(1+E_{11}) + p\, G^{22}] , \\ S^{12} &= C_1[-4\alpha\, E_{12} + p\, G^{12}] , \\ S^{13} &= C_1[-4\alpha\, E_{13} + p\, G^{13}] , \\ S^{23} &= C_1[-4\alpha\, E_{23} + p\, G^{23}] . \end{aligned} \tag{A.13}$$

Using (2.10), the matrix Q in (A.9) is then given by

$$Q = C_1\alpha \begin{bmatrix} 0 & 4 & 0 & 0 & 0 \\ & 0 & 0 & 0 & 0 \\ & & -2 & 0 & 0 \\ & & & -2 & 0 \\ \text{symmetrical} & & & & -2 \end{bmatrix}$$

$$-C_1 p \begin{bmatrix} 2(G^{11})^2 & 2(G^{12})^2 & 2G^{11}G^{12} & 2G^{11}G^{13} & 2G^{12}G^{13} \\ & 2(G^{22})^2 & 2G^{12}G^{22} & 2G^{12}G^{23} & 2G^{22}G^{23} \\ & & (G^{11}G^{22}+G^{12}G^{12}) & (G^{11}G^{23}+G^{12}G^{13}) & (G^{12}G^{23}+G^{13}G^{22}) \\ & & & (G^{11}G^{23}+G^{12}G^{13}) & (G^{12}G^{33}+G^{13}G^{23}) \\ \text{symmetrical} & & & & (G^{22}G^{33}+G^{23}G^{23}) \end{bmatrix} . \tag{A.14}$$

APPENDIX B

To show that (3.2) is path independent, we begin with the volume integral (over the undeformed geometry)

$$\iiint_V \frac{\partial W}{\partial x^1}\, dx^1dx^2dx^3 = \iiint_V \frac{1}{2}\left[\frac{\partial W}{\partial \varepsilon_{ik}} + \frac{\partial W}{\partial \varepsilon_{ki}}\right] \frac{\partial \varepsilon_{ik}}{\partial x^1}\, dx^1dx^2dx^3 \tag{B.1}$$

where V is an arbitrary volume in the undeformed state.

Adding $\iiint p\, G^{ik} \frac{\partial \varepsilon_{ik}}{\partial x^1}\, dx^1dx^2dx^3$

which is identically zero for incompressible materials, to the right hand side of (B.1) and making use of (2.3) and (2.9) results in

$$\text{R.H.S. of (B.1)} = \iiint_V \left(S^{ik}(\delta^j_k + u^j_{,k}) \frac{\partial u_j}{\partial x^1}\right)_{,i} dx^1dx^2dx^3$$

$$= \iiint_V \frac{\partial}{\partial x^i}\left[S^{ik}(\delta^j_k + u^j_{,k}) \frac{\partial u_j}{\partial x}\right] dx^1dx^2dx^3$$

$$- \iiint_V \left[S^{ik}(\delta^j_k + u^j_{,k}\right]_{,i} \frac{\partial u_j}{\partial x^1}\, dx^1dx^2dx^3 \ .$$

The last term in the above expression vanishes because

$$[S^{ik}\ (\delta^j_k + u^j_{,k})]_{,i} = 0$$

are the equilibrium equations for finite deformation. Applying the divergence theorem to the remaining term and equating to the left hand side of (B.1) gives

$$\iiint_V \frac{\partial W}{\partial x^1}\, dx^1dx^2dx^3 = \iint_{\partial V} S^{ik}(\delta^j_k + u^j_{,k})\ \nu_i \frac{\partial u_j}{\partial x^1}\, dx^1dx^2dx^3\ , \tag{B.2}$$

where ∂V denotes the surface which encloses V. The path independent property of (3.2) follows immediately by integrating (B.2) over a volume which is of unit length in the x^3 direction.

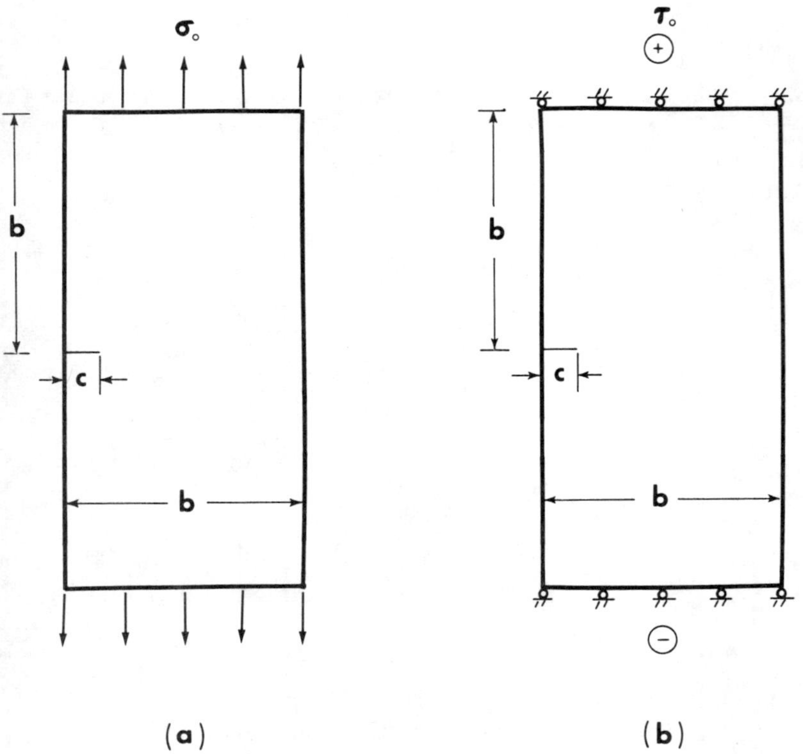

Figure 1 (a) Plane strain stretching and (b) generalized out-of-plane shear of a thick strip with a crack.

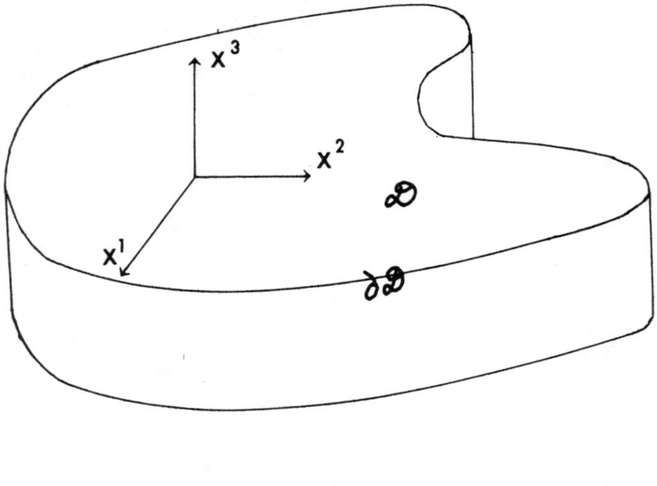

Figure 2 Domain and coordinate system.

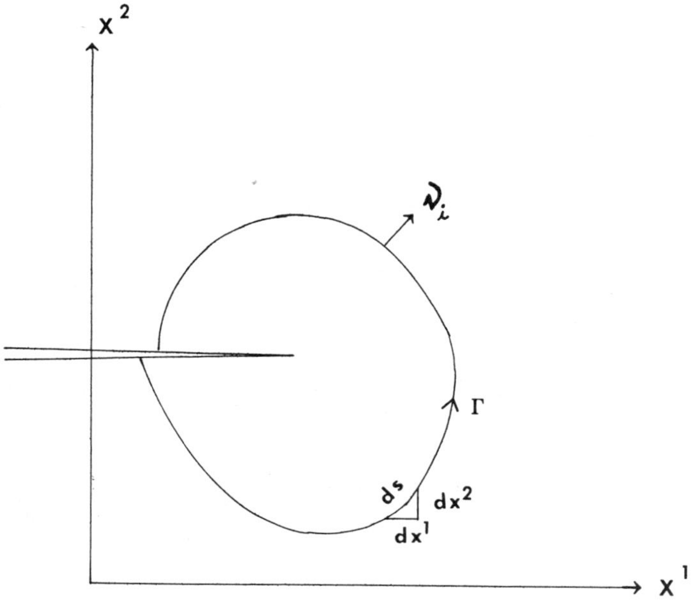

Figure 3 Notation for defining the J integral

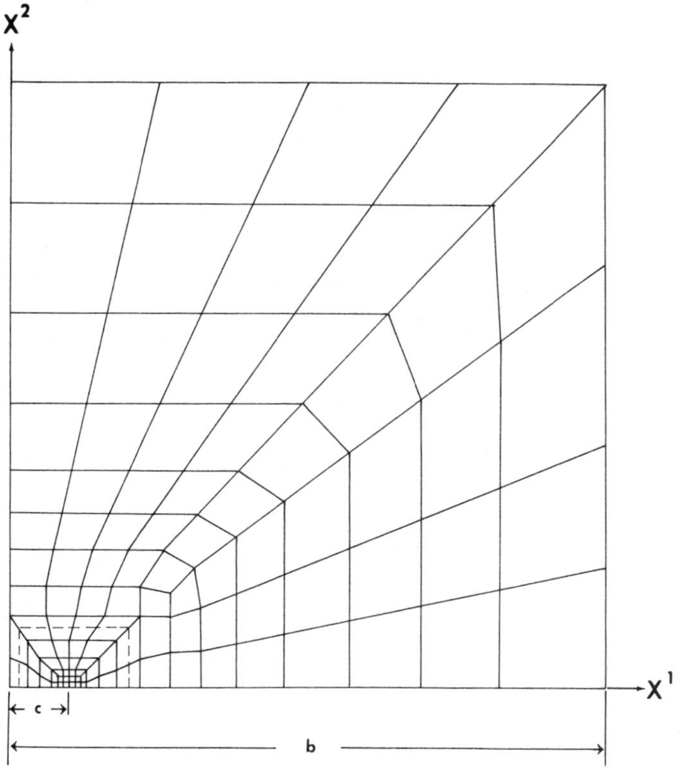

Figure 4 Finite element grid for half of strip for c/b = 0.1. Dotted contour indicates the path for calculating the J integral.

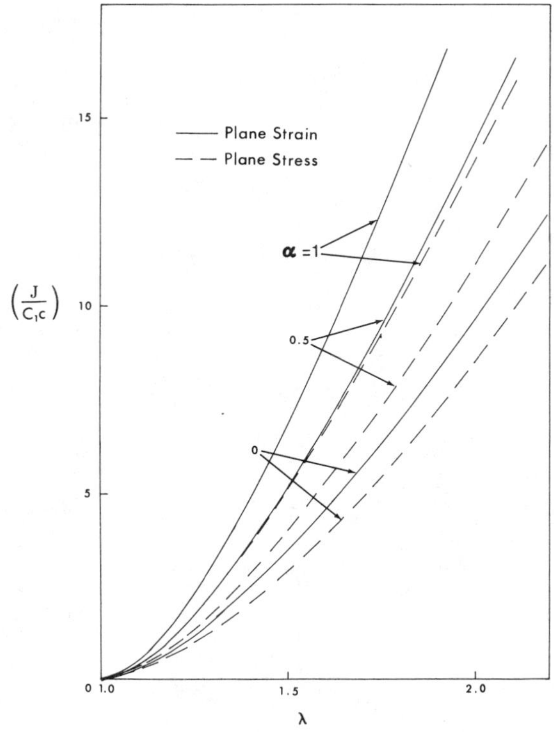

Figure 5 Calculated J values vs. nominal extension ratio λ for plane strain and plane stress stretching.

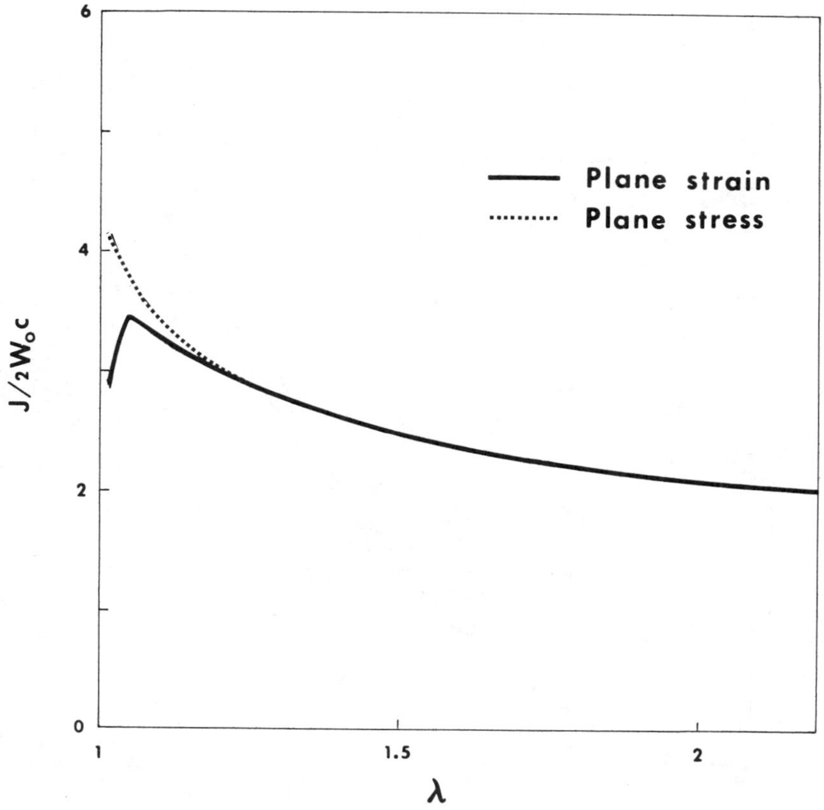

Figure 6 Calculated $J/2W_0c$ values vs. nominal extension ratio λ for plane strain and plane stress stretching.

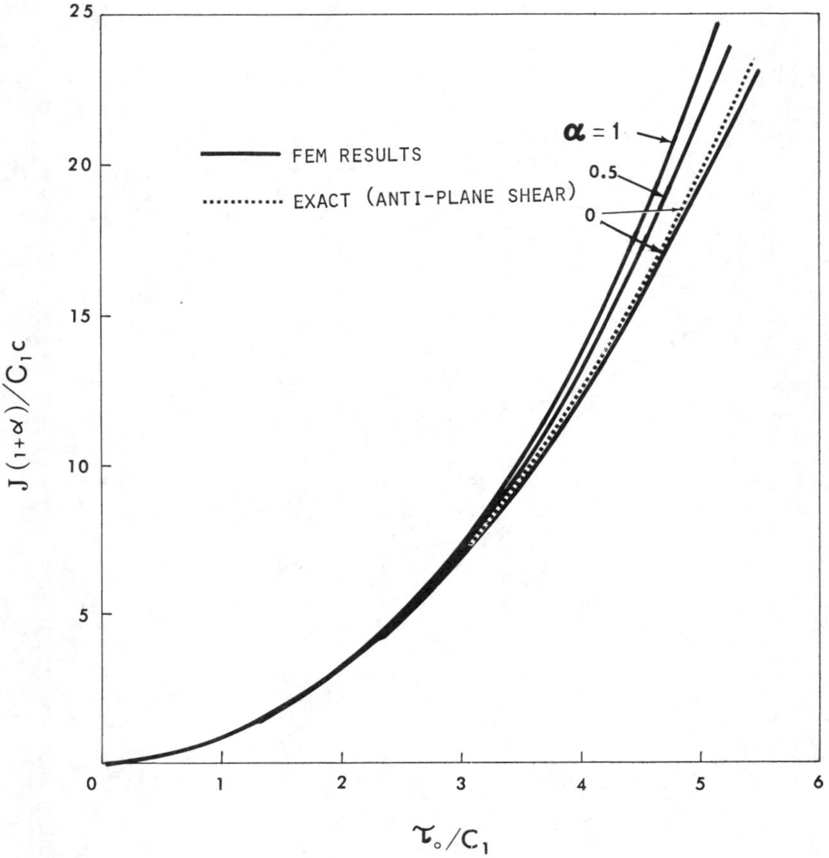

Figure 7 Calculated J values vs. shear stress τ_0 for generalized out-of-plane shear.

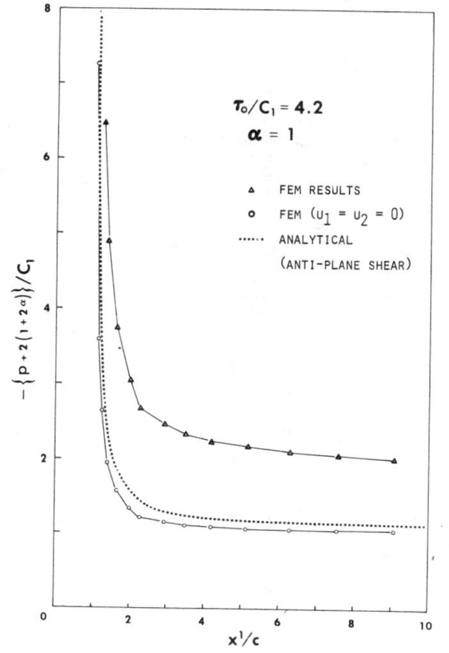

Figure 8 Distributions of normal stresses in the x^3 direction along $x^2 = 0$ calculated by different procedures.

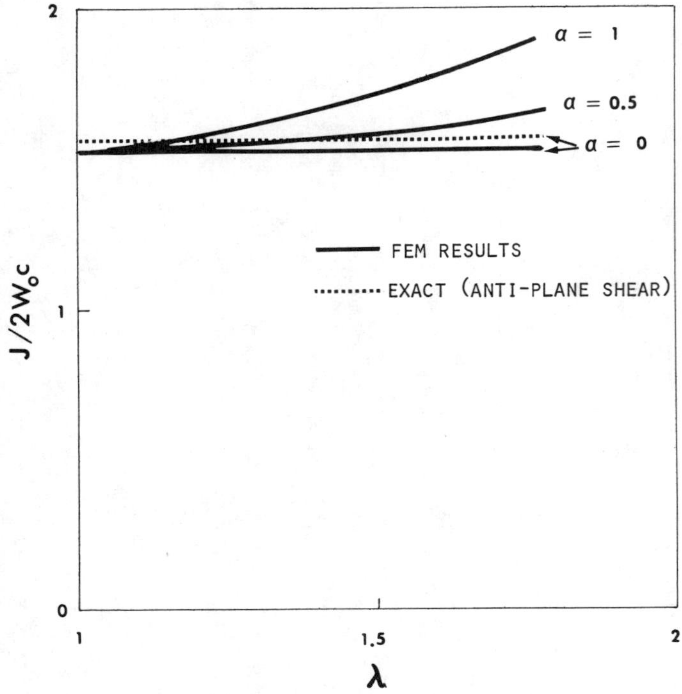

Figure 9 Calculated $J/2W_0c$ vs. nominal extension ratio λ for generalized out-of-plane shear.

Fracture 1977, Volume 3, ICF4, Waterloo, Canada, June 19 - 24, 1977

A FINITE ELEMENT METHOD ANALYSIS OF AN ELASTIC-PLASTIC SOLID CONTAINING HOLES IN THE VICINITY OF A CRACK TIP

Kjell Eriksson*

INTRODUCTION

Inclusions can sometimes reduce fracture toughness strongly i.e., in alloys where fracture is accompanied by extremely limited plastic flow. This feature is characteristic of modern high strength steels with, often unintentionally, a high inclusion content of sulphides and oxides. The reduction of fracture toughness due to inclusions must be a result of their influence upon the stress and strain distributions in a body, especially in the vicinity of the crack tip. The most accurate description of these distributions in an elastic-plastic body is today obtained by means of the finite element method.

The path independent J-integral is a function of stress and strain along a path enclosing the crack tip which characterizes the state at the crack tip and which can be used as a fracture criterion for both elastic and elastic-plastic conditions, as suggested by Rice [1]. In this work an attempt is made to estimate the reduction of fracture toughness due to inclusions by calculating the J-integral for equivalent paths around domains with and without inclusions.

According to a discussion in a previous paper [2] inclusions such as oxides and sulphides shall be considered as holes already from zero load. The results of this work is compared to experimental results given in the above previous paper.

THE FINITE ELEMENT PROGRAMME

The finite element programme used in this work was originally developed by Härkegård and Larsson [3]. They adopted a principle according to Yamada et al [4], based on the von Mises yield criterion and the Prandtl-Reuss equations, permitting an incremental treatment of elastic-plastic problems. The programme has subsequently been extended by Markström to include routines for calculations of path-independent integrals. Markström and Carlsson [5] have later studied, among other things, the effect of element size on the value of the J-integral and also the effect of different orientations of the integration path. One important result is that if the integration path intersects plasticized domains, then path independence is preserved only if the calculation includes the total elastic-plastic work. With their results in mind, it has been possible to choose an optimal element size and permissible integration paths.

The present problem is attacked by using the 'boundary layer' approach, that is, by assuming that the boundary stresses σ_{ij} of an elastic-plastic

* Department of Strength of Materials and Solid Mechanics, The Royal Institute of Technology, Stockholm, Sweden

domain around the crack tip are given by the singular term in the elastic stress solution:

$$\sigma_{ij} = \frac{K}{\sqrt{2\pi R}} f_{ij} (\theta) , \qquad (1)$$

where K is the stress intensity factor, R the radius of the domain considered and $f_{ij}(\theta)$ are given by the elastic solution.

The domain around the crack tip is divided into approximately 300 elements. Triangular constant strain elements are used. The element size at the crack tip is approximately 0.03 times the crack length. The load is applied at the boundary nodal points and the load distribution is determined according to equation (1) and to the principle of virtual work. For each load increment at most one element is plasticized.

Three geometries containing holes are studied. The first one comprises one hole, symmetrically situated in the elastic domain between the two branches of the plane strain plastic zone, and the second one also one hole asymmetrically situated with its centre on a line through the crack tip and the point of maximum extent of the plastic zone. The third one is a combination of the former two geometries, and is shown in Figure 1. As a reference a geometry without holes is used. The proportions of the geometry at the crack tip are chosen according to observations of the fracture surfaces of a real material.

COMPUTATION OF THE J-INTEGRAL

A necessary condition for path independence of the J-integral is that the domain inside the integration path is simply connected. In order to check if the presence of holes there implies any restriction of the path independence, the J-intregral was computed for paths either including or excluding the hole(s) in all geometries. All such integration paths coincide except in a neighbourhood of the hole(s), where for a given geometry one path is traced so as to include the hole(s) and the other so as to exclude it (them). For any pair of integration paths the corresponding difference in J is smaller than 1% for any value of J. Thus the presence of holes in the domain inside an integration path does not imply any restriction of the path-independence of the J-integral.

RESULTS AND DISCUSSION

The values of the J-integral of the geometries containing holes are given as functions of that of a massive material, in Figure 2 for 'small loads', that is, up to loads for which the crack tip plastic zone encloses all holes, and in Figure 3 for the whole load range investigated, where i = 1 refers to the two-hole model, i = 2 to one hole assymmetrically situated and i = 3 to one hole symmetrically situated.

In the whole load range the presence of holes implies increased stress and strain in every element of the hole-geometries. It may be concluded that the larger J-value of the hole-geometries correspond not only to a larger volume of plasticized material, but also, especially at the crack tip, to a more intense stress and strain distribution. Therefore, considering equally tough matrix materials, the larger value of the J-integral of the hole-geometries compared to that of a massive material is likely to correspond to a lower apparent fracture toughness of an inclusion material.

If this assumption is correct the results from all geometries indicate that the reduction of fracture toughness due to inclusions increases as the fracture toughness of the matrix material increases.

In Figure 4 the reduction of fracture toughness according to the model is compared to experimental results. In spite of its simplicity the model yields reliable results. Under plane strain conditions the model predicts accurately the reduction of fracture toughness and yields conservative, safe, values when extrapolated to mixed plane stress-plane strain conditions.

REFERENCES

1. RICE, J. R., "A Path Independent Integral and the Approximate Analysis of Strain Concentration by Notches and Cracks", J. Appl. Mech., 35, 1968, 379-386.
2. ERIKSSON, K., "Influence of Inclusions on the Fracture Toughness of a SIS 2140 Type Steel", Scand. J. Metallurgy, 4, 1975, 131-139.
3. HÄRKEGÅRD, G. and LARSSON, S. G., "On the Finite Element Analysis of Elastic-Plastic Structures under Plane Strain Conditions", Computers and Structures, 4, 1974, 293-305.
4. YAMADA, Y., YOSHIMURA, N. and SAKURAI, T., "Plastic Stress-Strain Matrix and Its Application for the Solution of Elastic-Plastic Problems by the Finite Element Method", Int. J. Mech. Sci., 10, 1968, 343-354.
5. MARKSTRÖM, K. M. and CARLSSON, A. J., "FEM-Solutions of Elastic-Plastic Crack Problems: Influence of Element Size and Specimen Geometry", Div. Strength Mat., Royal Institute of Technology, Stockholm, Publ. No. 197, 1973. Summary also in Int. J. Fracture, 9, 1973, 315-316.

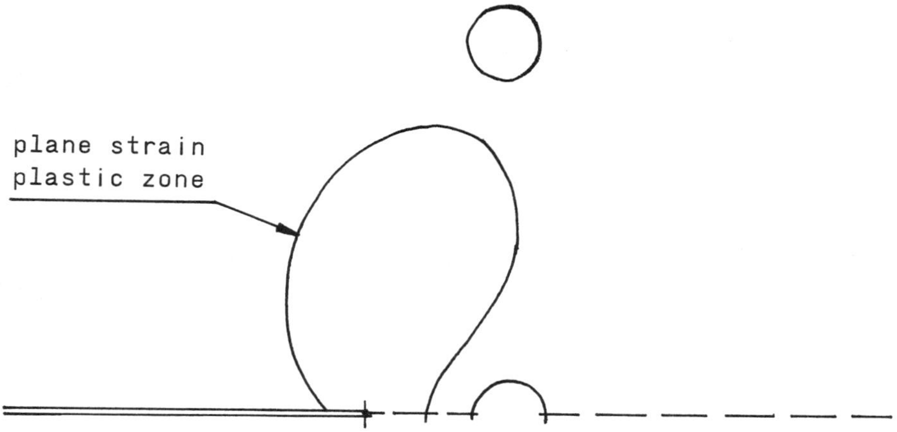

Figure 1 The 'Two Hole Model'

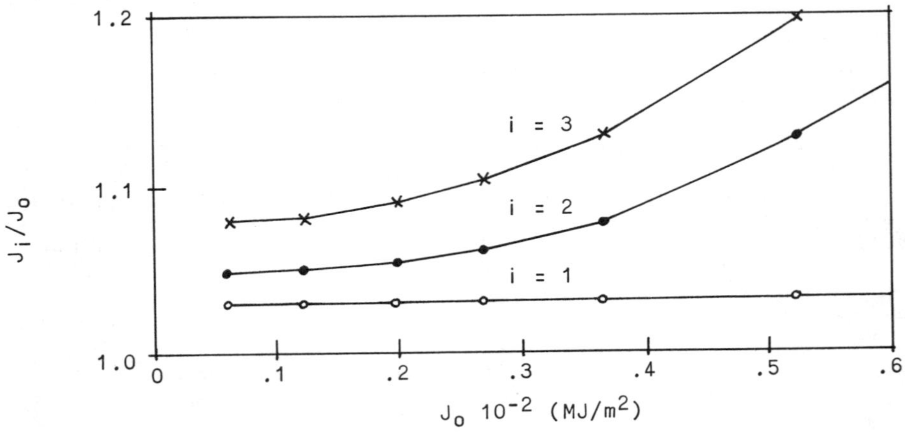

Figure 2 Effect of Holes Upon the J-Integral, 'Small Load' Range

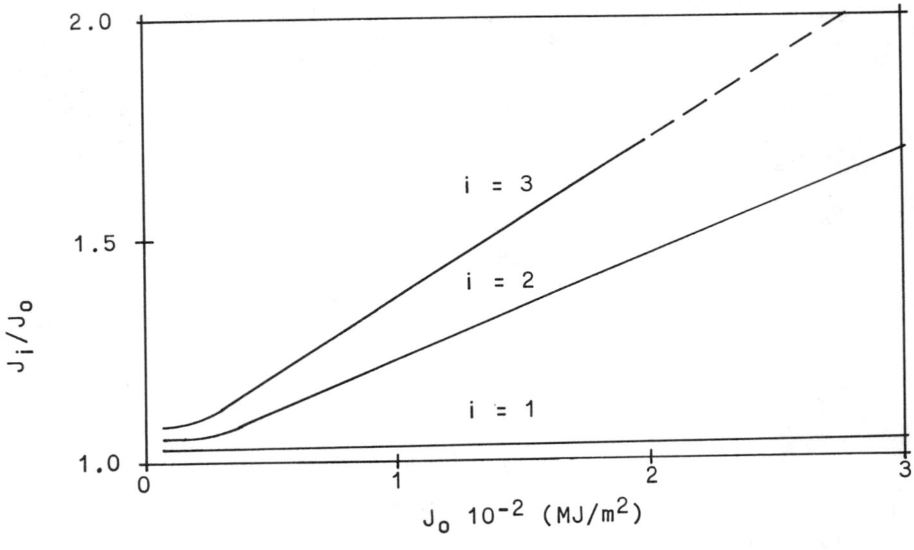

Figure 3 Effect of Holes Upon the J-Integral

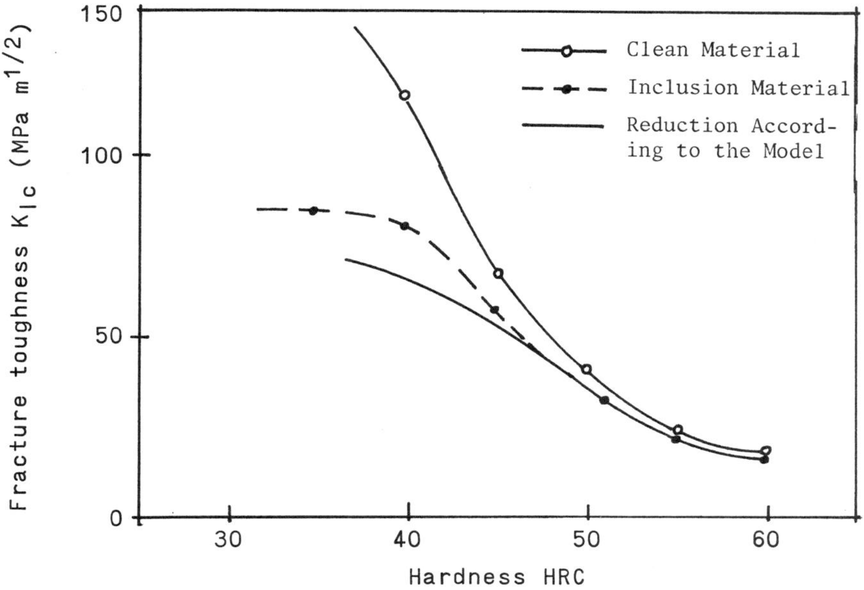

Figure 4 Comparison of Experimental and Theoretical Results

Fracture 1977, Volume 3, ICF4, Waterloo, Canada, June 19 - 24, 1977

FINITE ELEMENT ANALYSIS OF CRACK PROPAGATION UNDER COMPRESSION

H. Miyamoto*, S. Fukuda** and K. Kageyama*

INTRODUCTION

The behaviour of brittle materials under compression has been studied by, e.g., Hoek and Bieniawski [1] and Brace and Bombolakis [2]. Their experimental results can be summarized as follows:

1) Under compression, cracks propagate stably and further propagation of crack requires an increase of the applied stress.
2) Branching cracks emanate from the initial crack, deviate from the initial direction and gradually become aligned with the axis of the major compressive load.

As the stress at final catastrophic fracture is much greater than that at fracture initiation, as stated in (1), analysis of the propagation stage is quite important for the prediction of fracture under compression. Therefore, the authors paid their chief attention to the successive analysis of the change of stress states with crack propagation. Since the shape of cracks and stress states are quite complicated in the case of crack propagation under compression, they can be analysed only by finite element methods, and these must be more accurate finite elements than are conventionally used. Therefore, a 10 node, 20 degrees of freedom, triangular element, which makes it possible to adopt coarser meshes without the loss of high accuracy, was used in this analysis. A new finite element programme was developed, which can calculate the elastic contact stresses of crack surfaces at closure, in view of the fact that cracks might close under compression. The calculated results were closely comparable with the experimental results, and it was made clear that the fracture strength of brittle materials under compression cannot be evaluated without due consideration of the process of stable crack propagation.

THE 10 NODE 20 DEGREES OF FREEDOM TRIANGULAR FINITE ELEMENT

A triangular element of which the shape function is a complete cubic polynomial was used. This element possesses 10 nodes which correspond to the undetermined coefficients of the polynomial. The advantage of the use of a higher order shape function is that coarser mesh divisioning is possible without loss of high accuracy. Another advantage is that the value of stress and strain can be given by the node value.

---

* Department of Precision Machinery Engineering, Faculty of Engineering, University of Tokyo, Tokyo, Japan
** Department of Precision Machinery Engineering, Faculty of Engineering, University of Tokyo, (presently with Welding Research Institute, Osaka University, Osaka, Japan)

CONTACT PROBLEM OF CRACK SURFACES AT CLOSURE

Miyamoto and Shiratori [3] calculated contact stresses of crack surfaces at closure in order to study the opening and closure behaviour of fatigue cracks, but their analysis was limited to the case where the boundary condition of stress will be satisfied automatically due to symmetry if the boundary condition of displacement is satisfied on the contact surfaces. In the case of inclined cracks and their branching cracks which are to be studied in this analysis, the mere satisfaction of the boundary condition of displacement does not necessarily satisfy the boundary condition of stress on the closed crack surfaces.

Our newly developed finite element programme can deal with the contact problem on the closed crack surfaces where the state of contact cannot be assumed geometrically. In this finite element analysis, the boundary conditions of stress and displacement on the crack surfaces are replaced by those of the nodal force and nodal displacement on the crack surfaces. Let the node on the upper crack surface be i and the node on the lower surface be j, then the boundary conditions of closed crack surfaces are:

$$f_{Yi} + f_{Yj} = 0 \tag{1}$$

$$Y_i - Y_j + d_{Yi} - d_{Yj} = 0 \tag{2}$$

where f_{Yi}, d_{Yi} are respectively the force on and the displacement of node i in the Y direction. When there is no friction between the crack surfaces, the nodal forces in the X direction are equal to zero:

$$f_{Xi} = f_{Xj} = 0 \ . \tag{3}$$

The effect of friction is ignored in the following analysis for simplicity. As the area of contact of crack surfaces is generally not self-evident, the contact of crack surfaces is determined by the following conditons.

(1) If node i and node j contact each other, then $f_{Yi} > 0$.

(2) If node i and node j do not contact, $Y_i + d_{Yi} > Y_j + d_{Yj}$.

The correct solution satisfying the boundary condition of crack surfaces is obtained by repeating the same procedure, correcting successively the error of the boundary condition. This procedure is automatically carried out by the computer.

FRACTURE CRITERIA

As the stress state near the tip of a propagating crack is in the mixed mode I - II condition, the strain energy density criterion proposed by Sih [4] and the maximum stress criterion proposed by Erdogan and Sih [5], which are applicable to mixed mode fracture, were adopted in this analysis. The strain energy density criterion is based on the local density of the energy field in the crack tip region. For two-dimensional problems the strain energy density factor S is given by the quadratic form:

$$S = a_{11}K_I^2 + 2a_{12}K_IK_{II} + a_{22}K_{II}^2 , \tag{4}$$

where K_I and K_{II} are the stress intensity factors of mode I and mode II, respectively, and a_{ij} are the coefficients which are functions of Young's modulus, Poisson's ratio and the polar angle θ. The fracture criterion can be expressed mathematically for two dimensional problems in the following simple forms:

$$\partial S/\partial\theta = 0, \quad \partial^2 S/\partial\theta^2 > 0, \quad \theta = \theta_o \tag{5}$$

$$S_{min} = S(\theta_o) = S_{cr} \tag{6}$$

where θ_o is the fracture angle.

The maximum stress criterion postulates that the crack will open up in the plane normal to the direction of maximum stress and that crack propagation will occur when the maximum $\sqrt{2\pi r}\sigma_\theta$ value reaches K_{IC}. These conditions can be expressed as

$$K_I\sin\theta_o + K_{II}\ (3\cos\theta_o - 1) = 0 \tag{7}$$

$$\frac{1}{2}\cos\frac{\theta_o}{2}\left[K_I(1 + \cos\theta_o) - 3K_{II}\sin\theta_o\right] = K_{IC} . \tag{8}$$

SIMULATION PROCEDURE

The procedure for simulating crack propagation under compression was as follows. At each stage of propagation, the direction of the crack and the applied stress were determined by the above fracture criteria, using the stress intensity factor values calculated by the finite element method. The increments of crack growth were taken to be about 1/10 of the initial crack length, considering the experimental results [1], [2]. The width of the propagating crack was assumed to be infinitesimally small. The initial crack analyzed was of an elliptical form, but the crack growth increment Δa was about one hundred times greater than the radius ρ of the crack tip. Thus, it was considered that there was no effect of the finiteness of radius ρ on the direction and the stress of fracture initiation. Analysis was made in plane strain and the stress intensity factors were calculated by the displacement method. The simulation was carried out until complete failure of the whole specimen.

TEST SPECIMEN ANALYZED

The geometry and the mechanical properties of the test specimen analyzed were chosen so as to be the same as those of the glass specimen used by Hoek and Bieniawski [1] (see Figure 1). The mechanical properties of the glass specimen are given in Table 1.

SIMULATION RESULTS

1. Crack Path

The branching paths obtained are shown in Figure 2. The path shown in the upper right is obtained assuming the strain energy density criterion and the path shown in the lower left is obtained assuming the maximum stress criterion. If Δa is taken to be a/10, cracks propagate without increase of the applied stress after D1 and D2. In order to find the critical point Δa is taken to be a/20 after D1 and D2, then cracks propagate stably to F1 and E2. As the branching cracks become aligned with the axis of the compressive load, the cracks propagate in a zigzag manner. On the whole the crack path thus predicted agrees quite well with the experimentally obtained crack path. Meshes near the branching crack are shown in Figures 6 and 7.

2. Change of Stress Intensity Factors with Crack Propagation

The values of K_I and K_{II} at the tip of initial crack, calculated by the finite element method are:

$$K_I/\sqrt{\pi a}\sigma = 0.228 \quad ; \quad K_{II}/\sqrt{\pi a}\sigma = 0.397 \ .$$

Although the finite element values are respectively, 8.8% and 8.3% smaller than the theoretical values of K_I and K_{II}, the accuracy of the prediction of the crack path is expected to be high because the error in K_{II}/K_I, which determines the direction of the crack path, is only 0.53%. The initial crack closes when the applied stress, P_c, reaches 6400 MP_a. This is about 200 times greater than the crack initiation stress, and the crack does not close at crack initiation. The relationship between the branching crack length and normalized stress intensity factors is shown in Figure 3. It should be noted that, even under compression, K_I is positive, and the extending crack does not close during the stable process of propagation. No appreciable difference between the strain energy density criterion and the maximum stress criterion is observed.

3. Change of the Compressive Stress P_{cr} with Crack Propagation

The change in the compressive stress P_{cr} required for further crack extension with crack propagation is shown in Figure 4, together with the experimental data of Hoek and Bieniawski [1]. In the case of brittle materials, cracks, once initiated, will immediately lead to catastrophic failure if the applied stress is tensile. Therefore, the criterion for initiation can be regarded as the fracture criterion for total failure under tensile loading. Under compression cracks propagate stably and the "fracture hardening" phenomenon, which means that the applied stress must be increased in proportion to further propagate cracks, is observed; in the case of the strain energy density criterion, cracks do not propagate catastrophically until the L/2a value reaches 0.4. Therefore, quantitatively good agreement between experiment and simulation is found in the relation of P_{cr} to normalized crack length. Moreover, there is no appreciable difference between the maximum stress criterion and the strain energy density criterion. A great difference is observed, however, between simulation and experiment in terms of the fracture stress, but if we note the ratio of fracture stress/crack initiation stress, good agreement is found between simulation and experiment.

DISCUSSION

In this paper, the effect of friction was ignored for simplicity. This assumption can be regarded as appropriate, at least for this simulation, since the cracks did not close during the stable process of propagation. When the roughness of the crack surfaces becomes greater than the opening width of the crack, the crack surfaces are expected to contact. As the frictional force decreases the K_{II} value, the fracture stress will be increased if friction exists, and it is thought that this is one reason why the agreement between predicted and observed ultimate failure loads is so far off.

Several fracture criteria have been proposed for mixed mode problems, but they are originally derived for crack initiation and not for crack propagation. As cracks propagate stably under compression, a fracture criterion for crack propagation is required for analysis. In this paper, the strain energy density criterion and the maximum stress criterion are extended to crack propagation. From a continuum mechanics point of view, the ideal crack path can be mathematically expressed as a smooth differentiable curve. Because, when a finite change of angle occurs with an infinitesimal increment of crack growth, it is probable that the crack path depends on the crack growth increment Δa. Therefore, the fracture criterion for crack propagation should be compatible with the differentiability of the crack path curve. Many experimental results [4], [5] show that cracks change their direction in a zigzag manner if K_{II} is not zero. It follows from this that the fracture criterion for crack propagation must satisfy the condition that the K_{II} value be zero. Otherwise the differentiability of the crack path curve is not satisfied. Kitagawa and Yuuki [8] also suggested a "$K_{II} = 0$" criterion.

Nuismer [9] derived the following stress intensity factors for deflecting cracks when the crack growth increment converges to zero.

$$\overline{K}_I = \frac{1}{2} \cos \frac{\theta}{2} \left[K_I(1+\cos\theta)-3K_{II}\sin\theta\right] \tag{9}$$

$$\overline{K}_{II} = \frac{1}{2} \cos \frac{\theta}{2} \left[K_I\sin\theta+K_{II}(3\cos\theta-1)\right] \tag{10}$$

where K_I and K_{II} are the stress intensity factors at point 0, and $\overline{K}_I$ and $\overline{K}_{II}$ are those at point $\overline{0}$ respectively. Figure 5 shows the comparison of Nuismer's results and Kitagawa and Yuuki's results. Here, β = arc tan (K_I/K_{II}), and θ is taken positive in the clockwise direction. Good agreement between the two results can be observed. If Nuismer's solution is correct, then the "$K_{II} = 0$" criterion is no other than the maximum stress criterion. But Kitagawa and Yuuki's results show that the effect of the crack growth increment being finite is not negligible, and that, in the case of a finite crack growth increment the "$K_{II} = 0$" criterion and the maximum stress criterion agree only approximately. Oscillating solutions of K_{II}, as seen in Figure 3, can be explained by this K_{II} decreasing effect, and the zigzag crack path in Figure 2 could be made smooth if we let the crack growth increment converge to zero.

CONCLUSIONS

(1) Fracture behaviour in compression can be well explained by a finite element simulation in which special attention is paid to crack propagation.

(2) Even under compression, K_I was positive, and cracks propagated in the opening mode in this simulation after the first crack growth increment.

(3) It was made clear that the fracture criterion for crack propagation must contain a "$K_{II} = 0$" condition, from the continuum mechanics point of view.

ACKNOWLEDGEMENT

Part of this work is financially supported by a government grant from the Ministry of Education, for which the authors are grateful.

REFERENCES

1. HOEK, E. and BIENIAWSKI, Z. T., Int. J. Fract. Mech., 1, 1965, 137.
2. PAUL, B., Fracture, 2, 1968, 426.
3. MIYAMOTO, H. et al, J. of Fac. of Eng., University of Tokyo (B), XXXI, 1971, 217.
4. SIH, G. C., Mechanics of Fracture, 1, 1973, XXI.
5. ERDOGAN, F. and SIH, G. C., Trans. ASME, Ser. D, 85, 1963, 525.
6. SIH, G. C., Proc. of the 1974 Symposium on Mechanical Behaviour of Materials, 1974, 21.
7. KAGEYAMA, K., Graduation Thesis, 1976, (in Japanese).
8. KITAGAWA, H. et al, Japan - U. S. Seminar, The Strength and Structure of Solid Materials, 1974, 148.

Table 1 The Mechanical Properties of the Glass Specimens

E (MPa)	ν	Kc ($MPa \cdot m^{1/2}$)	Scr ($MPa \cdot m$)
73550	0.25	1.47	2.91

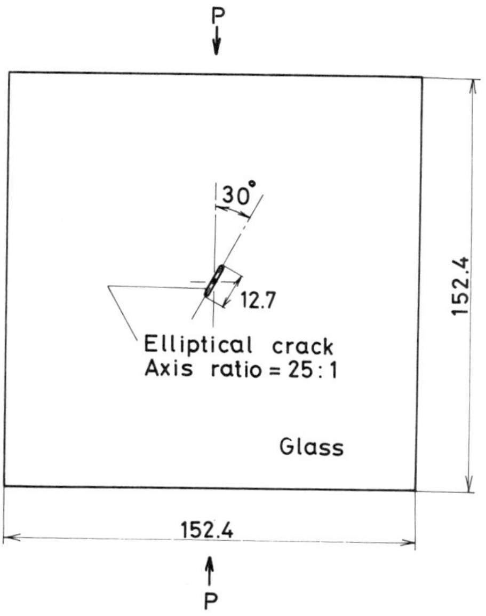

Figure 1 Test Specimen Analyzed

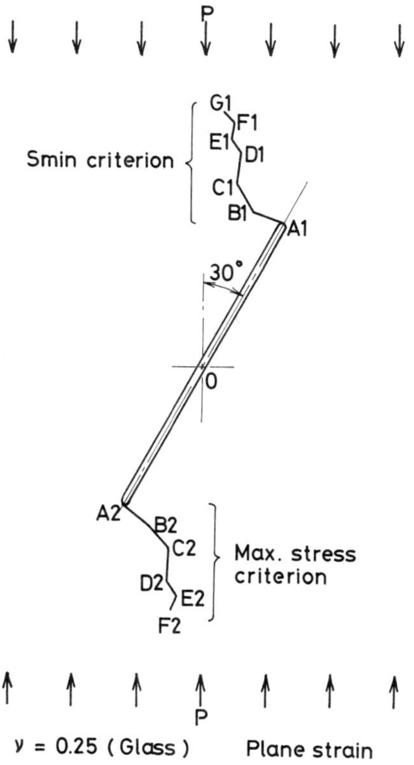

Figure 2 Crack Path

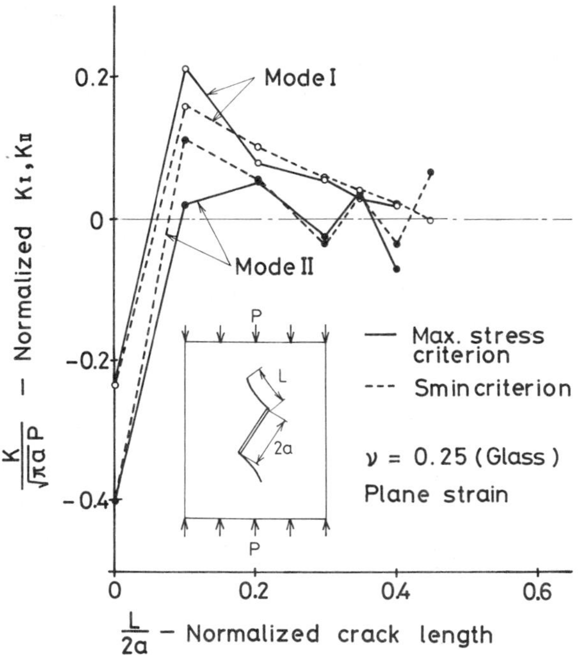

Figure 3 Change of Stress Intensity Factors with Crack Propagation

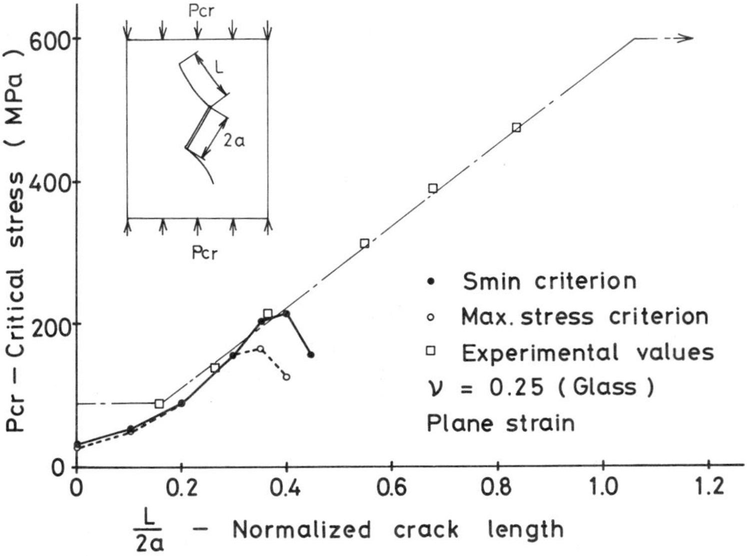

Figure 4 Change of the Compressive Stress with Crack Propagation

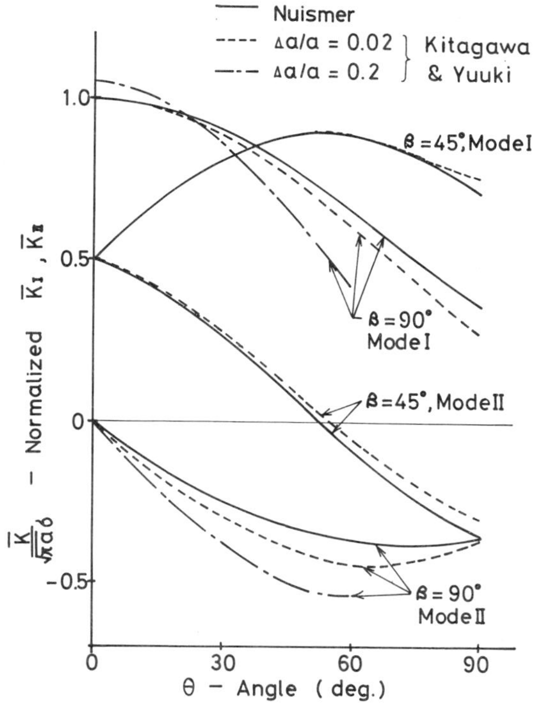

Figure 5 Stress Intensity Factors for Deflecting Crack

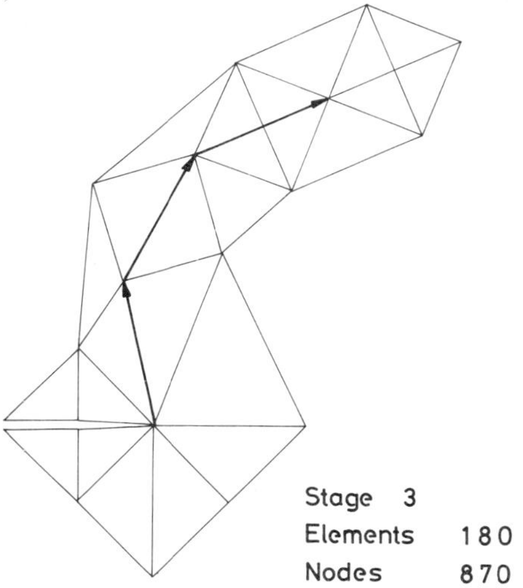

Figure 6 Breakdown of Branching Crack

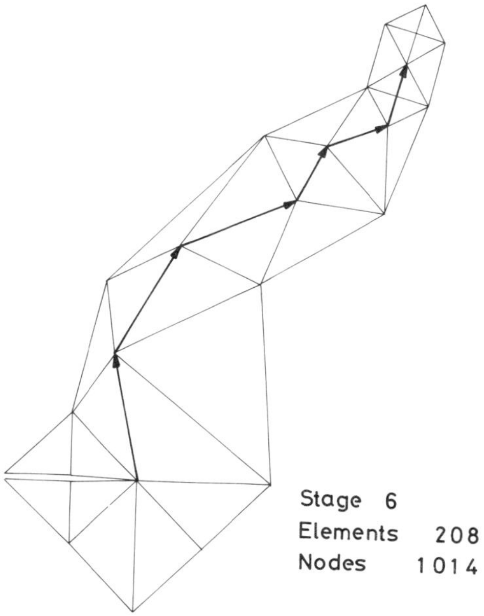

Figure 7 Breakdown of Branching Crack

Fracture 1977, Volume 3, ICF4, Waterloo, Canada, June 19 - 24, 1977

ON VIRTUAL CRACK EXTENSION METHODS FOR COMBINED TENSILE AND SHEAR LOADING

M. L. Vanderglas and R. J. Pick*

INTRODUCTION

The virtual crack extension methods described by Parks [1] and Hellen [2] have been shown to have advantage over other methods of applying finite element techniques of Linear Elastic Fracture Mechanics [3]. The methods of Parks and Hellen are both designed to compute the energy release rate (G) from (i) the displacement field before crack growth predicted by the finite element or other method, (ii) and the change of stiffness during growth, as described in equation (1),

$$G = -\frac{\partial V}{\partial a} = -\frac{1}{2}\{u\}^T \frac{\partial [K]}{\partial a}\{u\} + \{u\}^T \frac{\partial [f]}{\partial a} \tag{1}$$

where a is the crack length, u the displacements, [K], the finite element stiffness and f the loads. If the loads remain constant during crack extension then

$$G = -\frac{\partial V}{\partial a} = -\frac{1}{2}\{u\}^T \frac{\partial [K]}{\partial a}\{u\} . \tag{2}$$

Parks [1] computes this expression by summing the contributions to this equation from each element of a contour surrounding the crack. In contrast Hellen [2] bases his calculations on the assembled global stiffness matrix.

In both methods, stiffness derivative terms are approximated by

$$\frac{\partial [K]}{\partial a} \cong \frac{1}{\Delta a}([K]_{a+\Delta a} - [K]_a) = \frac{\Delta [K]}{\Delta a} . \tag{3}$$

Stiffness terms depend on the nodal coordinates and the difference $\Delta[K]$ is due to the alteration of the coordinates of some nodes by an amount Δa. A complication arises in combined tensile and shear loading because the calculated value of G depends on the direction (angle θ) in which the crack is assumed to extend. Most finite element techniques for the estimation of stress intensities (K_I, K_{II}) do not consider the variation of calculated values with the assumed (instanteous) propagation direction. It has been proposed that a crack will propagate in a direction favouring maximum energy release (G_{max}). If only to provide the most conservative value, the magnitude and direction of the maximum energy release (or equivalently, stress intensity) are desirable from an analysis point of view.

---

* Department of Mechanical Engineering, University of Waterloo, Waterloo, Ontario, Canada, N2L 3G1

VARIATION OF K_I, K_{II}, J_1, J_2 WITH θ

Defining W as the strain energy density, u_i the displacement vector, T_i the traction vector and s as the distance and n_k as the normal along a contour, Eshelby [4] has defined

$$J_K = \int \left\{ W n_K - T_i \frac{\partial u_i}{\partial x_K} \right\} ds \tag{4}$$

taken over any open contour starting at the lower crack face, surrounding the crack tip and ending at the upper face. He proves that J_K gives the energy release rate if the crack were to extend in the x_K direction (the crack is initially aligned with the x_K axis). Rice [5] shows further, that, by choosing a convenient contour the energy release rate can be calculated even when local crack-tip yielding is modelled.

Hellen and Blackburn [6] have shown that in two dimensional elasticity problems with combined tensile and shear loading, the stress intensity factors K_I, K_{II} are related to the J_1 and J_2 integrals by

$$J_1 = \frac{(1-\nu)(1+\kappa)}{4E} (K_I^2 + K_{II}^2) \tag{5}$$

$$J_2 = \frac{-(1+\nu)(1+\kappa)}{2E} K_I K_{II} \tag{6}$$

where K_{III} and J_3 are assumed to be zero and ν is Poisson's Ratio, E, Young's Modulus and $\kappa = (3-4\nu)$ for plane strain.

VARIATION OF G WITH J_1, J_2 AND θ

The value of G is simply J_1 for a tensile mode (I) of fracture however for tensile and shear loading (I, II) the calculated value of G depends on the value of θ. The virtual crack extension methods can be used to calculate

$$J_1 = -\left.\frac{\partial V}{\partial a}\right|_{\theta=0} \cong -\frac{1}{2}\{u\}^T \frac{\partial [K]}{\partial a}\{u\}\Big|_{\theta=0} , \tag{7}$$

$$J_2 = -\left.\frac{\partial V}{\partial a}\right|_{\theta=\pi/2} \cong -\frac{1}{2}\{u\}^T \frac{\partial [K]}{\partial a}\{u\}\Big|_{\theta=\frac{\pi}{2}} . \tag{8}$$

It is therefore desirable to have a relationship between J_1, J_2, θ and G. Hellen has predicted the trajectory of a two dimensional crack by calculating G for several value of θ and assuming growth in the direction of maximum G. Upon plotting the results of a test case, Hellen discovered a sinusodial variation of G with θ. Similar results can be obtained independent of the mesh size suggesting the relation between G and θ.

By studying the effect of Δa on the element stiffness matrix it can be shown that G at any angle θ is defined by

$$G(\theta) = J_1\cos\theta + J_2\sin\theta \; . \tag{9}$$

Furthermore it becomes apparent that this sinusodial behaviour of $G(\theta)$ is a consequence of the assumed linear material behaviour (strain energy is a quadratic form of the nodal displacements). It may be expected that formulations of the problem which include (nonlinear) plastic behaviour or based on plate theories will not exhibit this behaviour.

Considering relation (9) the maximum energy release rate occurs at

$$\theta = \text{arc tan}\left(-\frac{2K_I K_{II}}{K_I{}^2+K_{II}{}^2}\right) = \text{arc tan}\left(+\frac{J_2}{J_1}\right)$$

and has the value

$$G_{max} = \frac{(1+\nu)(1+x)}{4E}\left(K_I{}^4+6K_I{}^2K_{II}{}^2+K_{II}{}^4\right)^{1/2} = \left(J_1{}^2+J_2{}^2\right)^{1/2} \; .$$

The relationship between G and θ is shown graphically in Figure 1 and illustrates the following:

1) a polar plot of $G(\theta)$ versus θ gives a circle which intersects the origin,
2) if $K_I \neq 0$, $K_{II} = 0$ or if $K_I = 0$, $K_{II} \neq 0$, the circle is centered on the x-axis,
3) if $K_I \neq 0$ and $K_{II} \neq 0$, the centre of the circle will not lie on the x-axis,
4) the minimum and maximum values are the same in absolute value and are oppositely directed,
5) the maximum value of $G(\theta)$ is the same as the vector sum of J_1 and J_2.

In three dimensional applications, similar results are obtained except that, instead of a circle, a sphere is obtained:

$$G = G(\theta, \phi) = J_1\cos\theta\sin\phi + J_2\sin\theta\sin\phi + J_3\cos\phi$$

where θ, ϕ are angles shown in Figure 2 and G varies along the crack front.

If the crack lies on a plane of symmetry and boundary conditions are symmetrical with this plane, the component J_K normal to this plane must be zero. The parts of the contour contained in symmetric parts of the body are non-zero but opposite in sign. Thus, mixed mode crack problems cannot be analyzed using symmetry unless other information is available for the calculation of J_2.

EXAMPLE OF A CIRCULAR CRACK IN A HALF-SPACE

As an example, a circular crack in a half-space was analyzed under the action of 10,000 psi applied perpendicular to the crack face (Figure 3). Although this problem could be analyzed using an appropriate two-dimensional finite element formulation, a 90° segment was analyzed as an example of a crack problem giving rise to the two components J_1, J_2. A modified form of the stiffness derivative technique was used to predict the values of J_1 and J_2 for four points along the crack front. The displacement field was obtained using a variety of constant and linear isoparametric wedge and cube-like elements. Elements bordering the crack front were modified for singular behaviour.

The results of the analysis give the tabulated values of J_1, J_2 shown in Table 1. Because of symmetry the crack remains in its plane when growing and therefore J_3 is zero.

CONCLUSIONS

The method of virtual crack extension can be used to predict J_1, J_2, J_3 the energy release rates for crack growth in three mutually perpendicular directions. It has been shown that this can be related to the energy release rate $G(\theta, \phi)$ for crack growth in the directions described by θ and ϕ at any point along the crack front. $G(\theta, \phi)$ can therefore be considered a vector having both magnitude and direction. Prediction of the vector with maximum amplitude G_{max} is obtained from the vector sum of J_1, J_2, J_3. Since J_1, J_2, J_3 can be related to K_I, K_{II} and K_{III} which may be related to crack growth rate and direction, the value of G or G_{max} should also be indicative of crack growth rate and direction. There remains however the question as to the relation between G and the crack behaviour.

REFERENCES

1. PARKS, D. M., "A Stiffness Derivative Finite Element Technique for Determination of Elastic Crack Tip Stress Intensity Factors", Int. J. of Fracture, 10, No. 4, 1974, 487-502.
2. HELLEN, T. K., "On the Method of Virtual Crack Extensions", Int. J. for Numerical Methods in Engineering, 9, No. 1, 1975, 187-208.
3. VANDERGLAS, M. L., MUKERJEE, B. and PICK, R. J., "Stress Intensity Evaluation by the Finite Element Method", Symposium on Applications of Solid Mechanics, University of Toronto, 1976.
4. ESHELBY, J. D., "The Continuum Theory of Lattice Deflects", Solid State Physics, editors, F. Seitz and D. Turnbull, Volume 3, Academic Press, New York, 1956.
5. RICE, J. R., "A Path Independent Integral and the Approximate Analysis of Strain Concentration by Notches and Cracks", J. Appl. Mech., 35, 1968, 379-386.
6. HELLEN, T. K. and BLACKBURN, W. S., "The Calculation of Stress Intensity Factors for Combined Tensile and Shear Loading", Int. J. of Fracture, 11, No. 4, 1975, 605-617.
7. SNEDDON, I. N. and LOWENGRUB, M., "Crack Problems in the Classical Theory of Elasticity", John Wiley and Sons, New York, 1969.

Table 1 J_1 and J_2 for a Circular Crack in a Half-Space

ψ	J_1	J_2	$G_{max} = (J_1^2+J_2^2)^{1/2}$
11.25°	34.77	6.92	35.45
33.75°	29.48	19.70	35.46
56.25°	19.70	29.48	35.46
78.75°	6.92	34.77	35.35

Derivations by Sneddon [7] show that G_{max} has a constant value along the crack front of 34.76. The combination of the calculated values of J_1 and J_2 give a value of 35.46 within 2% of the theoretical value.

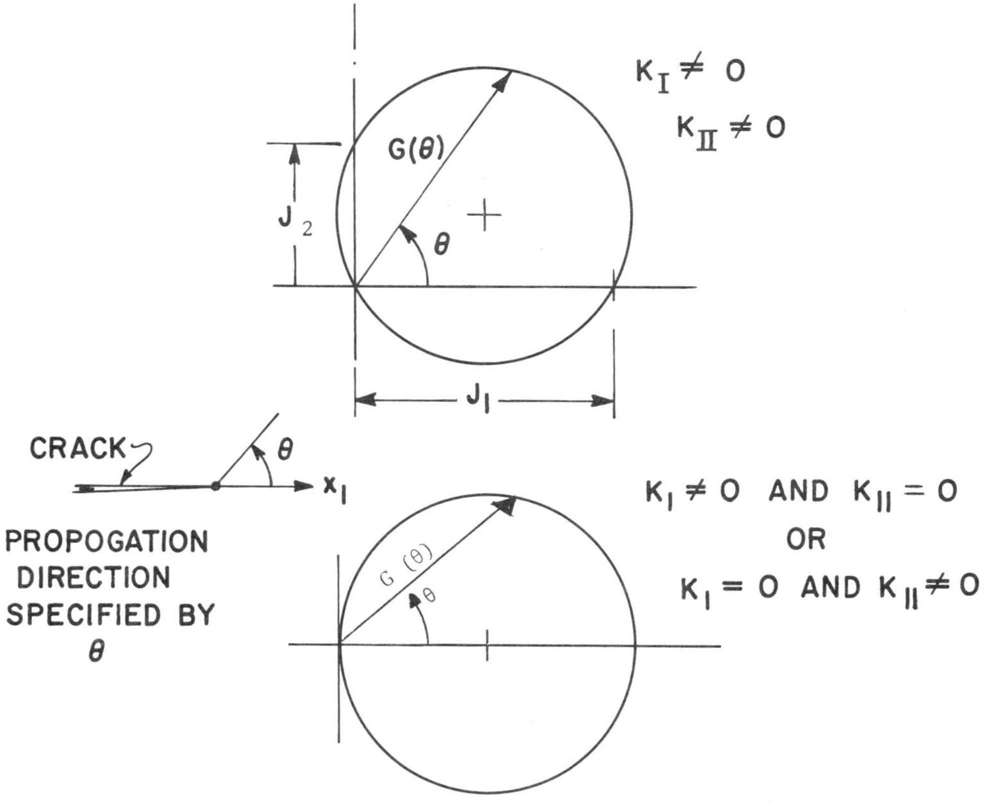

Figure 1

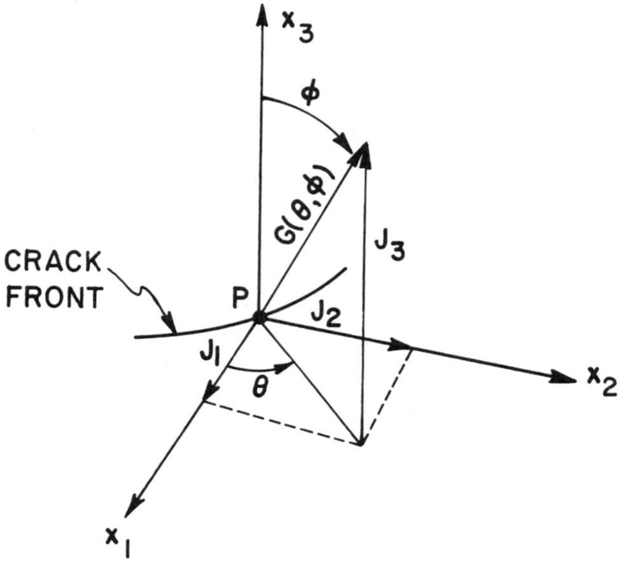

Figure 2

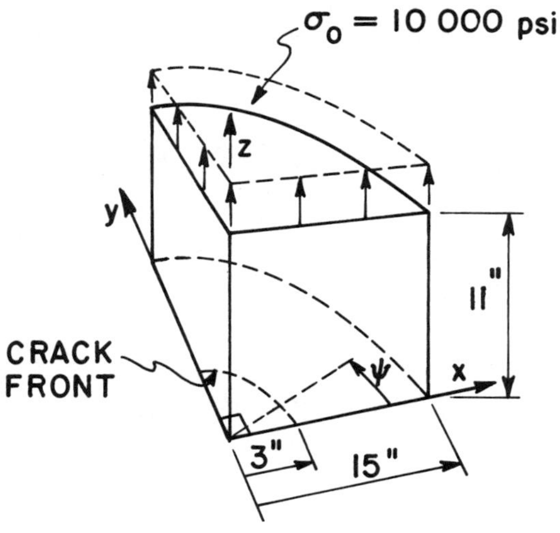

Figure 3

Fracture 1977, Volume 3, ICF4, Waterloo, Canada, June 19 - 24, 1977

A FINITE ELEMENT ANALYSIS OF A CIRCUMFERENTIALLY NOTCHED TENSILE SPECIMEN

D. K. Brown* and R. M. McMeeking**

INTRODUCTION

Over the past few years much research has been carried out at the University of Glasgow into ductile fracture. Much of the experimental work has involved the use of circumferentially notched tensile specimens [1]. By varying the ratio of minimum cross-section radius a_o to notch radius R, Figure 1, the constraint in the centre of the specimen can be varied. A series of tests on different materials using five different specimen geometries allowed failure curves to be drawn. The initial ratios (a_o/R) varied from 1 (A-notch) to 3 (D-notch). Forming the axes of such graphs are the parameters mean to effective stress ratio, $\sigma_m/\bar{\sigma}$, and strain to failure, $\bar{\varepsilon}_f$, both evaluated at the centre of the bar. In terms of principal stresses, $\sigma_m = (\sigma_1 + \sigma_2 + \sigma_3)/3$, $\bar{\sigma} = \bar{\sigma} = \{\frac{1}{2}[(\sigma_1 - \sigma_2)^2 + (\sigma_2 - \sigma_3)^2 + (\sigma_3 - \sigma_1)^2]\}^{1/2}$ and $\bar{\varepsilon}_f$ is defined in [1] as the average value of effective plastic strain across the minimum section at failure initiation, which is detected by a significant drop in load bearing capacity. Combining the failure curves with a material size parameter and the approximate expression [2], $\delta = 0.6K_I^2/E\sigma_y$ enabled prediction of C.O.D. and critical defect size. These latter parameters were determined using small scale circumferentially notched specimens, with obvious economic advantages.

In determining the stress ratio and strain from the tests use was made of the work of Bridgman [3], which dealt with naturally necking bars. Confirmation of the accuracy of applying Bridgman's analysis to pre-notched bars is obviously advisable and two approaches were made. The first approach [4] was experimental and involved the analysis of deformation in the mid-section of banded steel specimens. Three geometries were analysed, the results indicating much better agreement with Bridgman for the A-notch than for the D-notch. Comparison was also made to the results of Clausing [5] who also used banded specimens of different geometry. Clausing compared his results with a modified form of the Bridgman analysis.

The results from the second approach are presented below, they being derived from a finite element solution of the A-notch. A finite element solution was also used by Argon, Im and Needleman [6], but this was for a naturally necking bar.

PROGRAMME

The results were generated using an early version of the MARC finite element programme which was subsequently modified by Rice and Tracey [7]. The programme has a finite strain capability [8] and uses an incremental determination of the solution [7]. The elements used were isoparametric quadrilaterals over which uniform dilation was enforced [9].

*University of Glasgow, Scotland.
**Brown University, Providence, R.I., U.S.A.

Only a symmetrical quarter of the specimen cross-section need be considered and the finite element grid, Figure 1, was chosen to concentrate elements in areas of high stress and strain gradients. The shape of the elements on or near the section BC was chosen such that even after large deformations, element aspect ratios were kept to a minimum. The power law stress-strain curve used, can be defined by

$$\left(\frac{\bar{\sigma}}{\sigma_y}\right)^{1/n} - \frac{\bar{\sigma}}{\sigma_y} = \frac{3G\bar{\varepsilon}^p}{\sigma_y} \text{ for } \frac{\bar{\sigma}}{\sigma_y} \geq 1$$

where σ_y is the uniaxial yield stress. Two values of index n = 0.1, 0.2 were selected and the elastic perfectly plastic case was also solved. The grid was 'loaded' by application of uniform displacement along boundary AE, the displacement to cause initial yielding at C being denoted by 100%. 24 graded increments of uniform displacement produced a final 'load' of 990%.

RESULTS

The results presented here are for the case of n = 0.2 power hardening, this material response being closest to the ENIA low carbon steel used in the experiments of Earl and Brown [4]. Comparison is made between these latter results, the computer results, and the Bridgman approximation, based on the specimen shapes and computed by finite elements.

(1) *Profiles and Plastic Zone shapes.* Figure 2 illustrates the profile of the A notch, when elastic, and after deformation of 990%. In order to calculate the current radius of curvature, R, of the notch at C (Figure 1) the displacements of point 1, 2, 3 were considered. With the reduced cross-section radius, a, the ratio (a/R) can be determined and three values are shown on Figure 2. The initial value of ratio is 1.0 and, as plasticity develops, the value drops to 0.932 at 310% before returning to 1.0 when a significant proportion of the notch is flowing plastically. The increasing plastic zone is illustrated in Figure 3.

As the power hardening index drops there is a noticeable drop in plastic zone size, at the same boundary displacement of 990% but with a corresponding increase in plastic strain and deformation on the mid-section BC. The ratios (a/R) at 990% for n = 0.1 and the perfectly plastic case are 1.14 and 1.29 respectively.

(2) *Radial Displacement* U *on* BC. It is of interest to plot the ratio U/U_c (where U_c is the radial displacement of C) against radius across BC. Bridgman assumes a straight line and, as can be seen from Figure 4, the trend from the elastic to the plastic (990%) distribution is towards this straight line.

(3) *Distribution of Effective Plastic Strain.* Figure 5 shows the distribution of effective plastic strain $\bar{\varepsilon}^p$ along the boundaries ABCD (Figure 1) for 990% where $\bar{\varepsilon}^p$ can be defined, in principal strains, as

$$\bar{\varepsilon}^p = \int\left\{\frac{2}{9}\left[\left(d\varepsilon_1{}^p - d\varepsilon_2{}^p\right)^2 + \left(d\varepsilon_2{}^p - d\varepsilon_3{}^p\right)^2 + \left(d\varepsilon_3{}^p - d\varepsilon_1{}^p\right)^2\right]\right\}^{1/2}$$

Across the neck Bridgman assumed a uniform value of effective plastic strain, this best being considered an average value, which is given by

$$\bar{\varepsilon}^p_{ave} = 2 \ln\left(\frac{a_o}{a}\right)$$

At 990%, $\bar{\varepsilon}^p_{ave}$ is 21.1% and this value is shown on Figure 5. Values of $\bar{\varepsilon}^p_{ave}$ for n = 0.1 and the perfectly plastic case are 26.1% and 30.7% respectively for 990%, reflecting equally well the higher mean values of the $\bar{\varepsilon}^p$ distribution for the corresponding finite element solutions. As power hardening drops the peak value of $\bar{\varepsilon}^p$ moves from C to B. In Earl and Brown [4] the A4 specimen has $\bar{\varepsilon}^p_{ave}$ = 27.7% and cannot be compared exactly but the trend of $\bar{\varepsilon}^p$ is similar to Figure 5.

(4) *Stresses* σ_r σ_θ σ_z. On Figure 6 are plotted the normal stress distributions down ℄ AB and across the mid-section BC, for both the elastic (100%) and plastic (990%) conditions. From equilibrium, σ_r and σ_θ are always equal along AB and σ_r always drops to zero at C. However on BC at 990% σ_r, σ_θ follow almost a common curve down to zero at C a trend predicted by Bridgman. Quantitatively, however, for the A-notch, the distributions are not as accurate. They are calculated using assumed stress flow lines, from

$$\sigma_r = \sigma_\theta = \bar{\sigma} \ln\left[\frac{1}{2}\left(\frac{a}{R}\right)\left(1 - \left(\frac{r}{a}\right)^2\right) + 1\right]$$

Using the (a/R) ratio from Figure 2 the Bridgman approximation is drawn on Figure 5. Similarly the curve for σ_z is calculated from

$$\sigma_z = \bar{\sigma}\left\{1 + \ln\left[\frac{1}{2}\left(\frac{a}{R}\right)\left(1 - \left(\frac{r}{a}\right)^2\right) + 1\right]\right\}$$

The value for $\bar{\sigma}$ used above comes from the stress/strain curve at the value corresponding to $\bar{\varepsilon}^p_{ave}$.

At 990%, as the power hardening drops to the perfectly plastic case, the peak value of σ_r/σ_y, σ_θ/σ_y at B vary very little although the Bridgman value drops to 0.5. However the peak values of σ_z/σ_y drop to around 2.25 and the Bridgman values to around 1.5. Comparison with the Earl and Brown A4 specimen at $\bar{\varepsilon}^p_{ave}$ = 27.7% indicates peak values of σ_z/σ_y and σ_θ/σ_y, σ_r/σ_y at B of 2.2 and 0.5 respectively, the differences reflecting the experimental difficulty of determining mean stress.

(5) *Distribution of* $\sigma_m/\bar{\sigma}$. The distributions of this stress ratio, which indicates the severity of stress state are shown on Figure 7 for the elastic (100%) and plastic (990%) conditions. Little variation is detected with load and a favourable comparison can be made with the Bridgman approximation, which is calculated from

$$\frac{\sigma_m}{\bar{\sigma}} = \frac{1}{3} + \ln\left[\frac{1}{2}\left(\frac{a}{R}\right) + 1 - \frac{(r/R)^2}{2(a/R)}\right]$$

At 990% and lower work-hardening the Bridgman curves vary very little while the finite element solutions show the peak at B rising to over 1.5.

CONCLUSIONS

The results quoted are perforce a short summary of all the results generated and it is hoped in the near future to publish a more comprehensive comparison of the results of several notch shapes, such as the D-notch [1] and the notch used by Clausing [5]. From the comparisons made herein, it can be seen that the approximate Bridgman results are remarkably good for this pre-notched geometry and bear witness to his remarkable insight.

ACKNOWLEDGEMENTS

Thanks are due to the University of Rhode Island, Kingston, R.I., U.S.A. where the initial results were generated and to the University of Glasgow, Scotland where the final runs were completed. One of us (RMM) was supported by the U.S. Energy Research and Development Admiministration. Gratitude is expressed to Professor J. R. Rice of Brown University, Providence, R.I. for many helpful discussions.

REFERENCES

1. MACKENZIE, A. C., HANCOCK, J. W. and BROWN, D. K., To be published in Engng. Frac. Mechs.
2. RICE, J. R. and JOHNSON, M. A., "Inelastic Behaviour of Solids", (M. F. Kanninen et al. Eds.), McGraw Hill, 1970, 641.
3. BRIDGMAN, P. W., "Studies in Large Plastic Flow and Fracture", McGraw Hill, 1952.
4. EARL, J. C. and BROWN, D. K., To be published in Engng. Frac. Mechs.
5. CLAUSING, D. P., J. Materials, 4, 1969, 566.
6. ARGON, A. S., IM, J. and NEEDLEMAN, A., Met. Trans, 6A, 1975, 815.
7. RICE, J. R. and TRACEY, D. M., "Numerical and Computer Methods in Structural Mechanics", (S. J. Fenves et al. Eds.), Academic Press, 1973.
8. McMEEKING, R. M. and RICE, J. R., Int. J. Solids, Struc. 11, 1975, 601.
9. NAGTEGAAL, J. C., PARKS, D. M. and RICE, J. R., Comp. Meth. in Appl. Mech. and Engng., 4, 1974, 153.

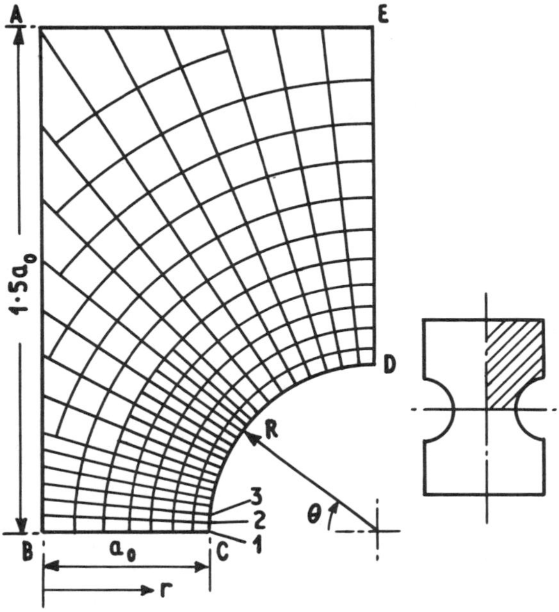

Figure 1

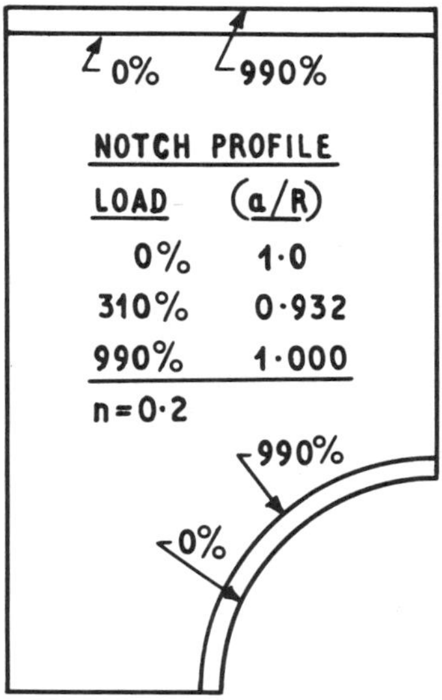

Figure 2

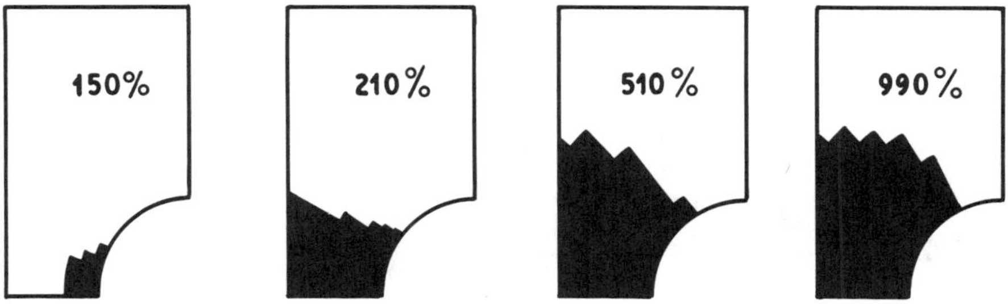

Figure 3 Development of Plastic Zone n = 0.2

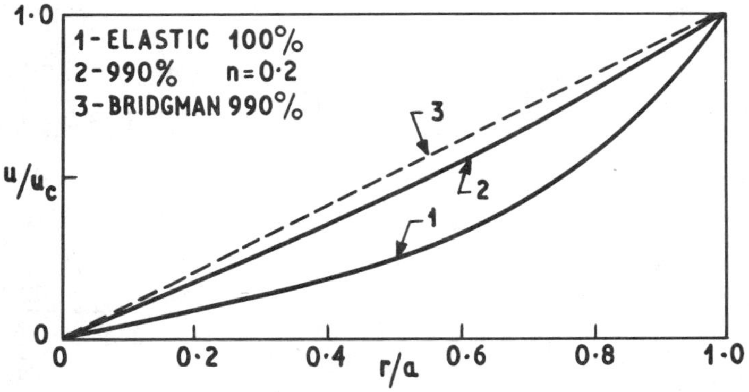

Figure 4 Radial Displacement at Neck

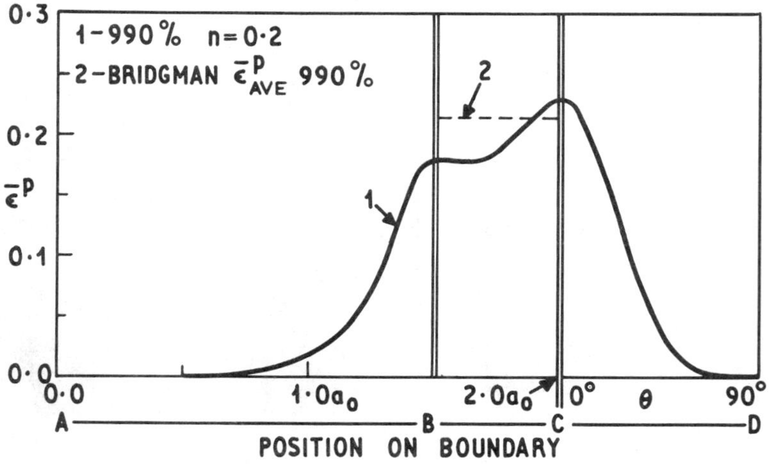

Figure 5 Distribution of Effective Plastic Strain

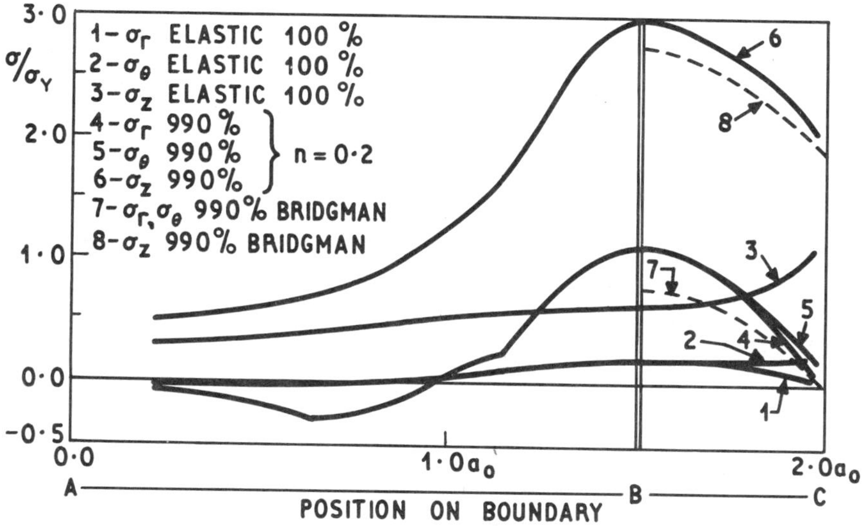

Figure 6 Stress Distributions

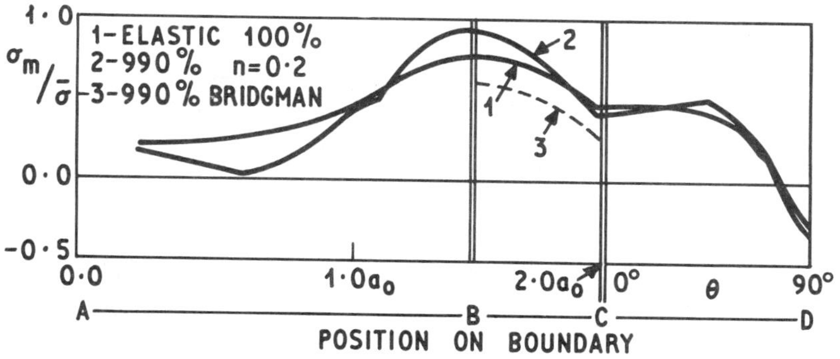

Figure 7 Distributions of $\sigma_m/\bar{\sigma}$

Fracture 1977, Volume 3, ICF4, Waterloo, Canada, June 19 - 24, 1977

A TRANSIENT FINITE ELEMENT ANALYSIS OF UNSTABLE CRACK PROPAGATION IN SOME 2-DIMENSIONAL GEOMETRIES

P. N. R. Keegstra*, J. L. Head** and C. E. Turner***

INTRODUCTION

Analytical solutions have been obtained to the problem of unstable crack propagation in an infinite plate by Freund [1] and in an infinite strip by Nilsson et al [2, 3, 4]. An extensive review of this work has been presented by Erdogan [5]. Recently, interest has extended to the problem of the analysis of crack propagation and arrest in finite geometries. There is interest in the analysis for standard 2-dimensional test piece geometries for which dynamic toughness data are known with some confidence. Also, there is interest in the modelling of crack propagation in real components, having more complex 2-dimensional, or, eventually, 3-dimensional geometries and for which the relevent toughness data are known. This paper describes, in outline, a 2-dimensional dynamic linear elastic finite element programme, based on triangular linear displacement plane strain elements, suitable for either purpose. Applications described here are restricted to test piece geometries.

It was thought convenient to use double cantilever beam (DCB) geometries, under fixed grip loading conditions, for the validation of the programme. For this geometry, there is a large volume of experimental data, assembled by Hahn et al [6, 7, 8]. The present paper gives the results of analyses for DCB geometry and a comparison with the published experimental results of Hahn et al. The paper also describes the application of the programme to finite strip geometries, also under fixed grip conditions. In conventional tests on metals, the problem is circuitous. Without a dynamic analysis of the test piece, the dynamic toughness (which is, in general, a function of crack speed) cannot be derived. On the other hand, the programme can be run only if the dynamic toughness data are input. The dynamic toughness can be measured however, without the use of a dynamic stress analysis, by the thermal wave technique [9], dynamic photoelasticity [10] or the shadow optical method [11].

The dynamic energy release rate, G_D, in a specimen under fixed-grip conditions, is [5]

$$G_D = -\left(\frac{dU}{da} + \frac{dS}{da}\right) \tag{1}$$

where S = kinetic energy

* Royal Netherlands Navy, seconded to Imperial College, London, England.
** Department of Mechanical Engineering, Imperial College of Science and Technology, Exhibition Road, London, SW7 2BX, London, England.
***Imperial College, London, seconded to National Physical Laboratory, Teddington, England.

(other symbols conform to the standard nomenclature list.) In the analyses described in the paper, it was assumed that energy is dissipated only at the crack front. There may be damping losses, either internally and/or at boundaries. The inclusion of these losses in the analysis would, in principle, present no difficulty, although the assessment of a reasonable magnitude would not be easy. The nature of energy dissipation at the crack front is not discussed in the paper, but dissipation is, by implication, taken into account by use of a generalised surface energy of the Irwin-Orowan type. The method by which this is included in the analysis is described in the paper. The energy balance equation is then

$$G_D = R(v) \tag{2}$$

where $R(v)$ = dynamic fracture toughness, which in general is a function of the crack speed v.

DYNAMIC ANALYSIS

The finite element discretisation of a continuum leads to the well-known matrix form of the equation of motion [12].

$$\underline{K}\,\underline{u} + \underline{C}\,\underline{\dot{u}} + \underline{M}\,\underline{\ddot{u}} = \underline{F}(t) \tag{3}$$

where
$\underline{K}$ = stiffness matrix
$\underline{C}$ = viscosity matrix
$\underline{M}$ = mass matrix
$\underline{u}$ = displacement vector
$\underline{F}(t)$ = force vector

and the dot indicates the derivative with respect to time. In the analyses described in the paper, the mass of each element was assumed to be concentrated at the nodes, thus diagonalising the mass matrix and reducing the computing time by up to 80% with an acceptable loss of accuracy (about 4%). Equation (3) is integrated stepwise in time in a manner similar to that described by Hitchins and Dance [13]. A second order Lagrangian polynomial is assumed to relate the accelerations at sequential time points, thus the displacements and velocities at time t + Δ are related to those at t and t - Δ by the following equation (see Keegstra [14]).

$$\begin{Bmatrix} \underline{u}_+ \\ \underline{\dot{u}}_+ \end{Bmatrix} = \begin{bmatrix} \underline{I} & \Delta\underline{I} \\ \underline{0} & \underline{I} \end{bmatrix} \begin{Bmatrix} \underline{u}_o \\ \underline{\dot{u}}_o \end{Bmatrix} + \begin{bmatrix} -\frac{\Delta^2}{24}\underline{I} & \frac{5\Delta^2}{12}\underline{I} & \frac{\Delta^2}{8}\underline{I} \\ -\frac{\Delta}{12}\underline{I} & \frac{2\Delta}{3}\underline{I} & \frac{5\Delta}{12}\underline{I} \end{bmatrix} \begin{Bmatrix} \underline{\ddot{u}}_- \\ \underline{\ddot{u}}_o \\ \underline{\ddot{u}}_+ \end{Bmatrix} \tag{4}$$

The subscripts +,o,- denote the values at times t + Δ, t, t - Δ respectively. $\underline{I}$ denotes unit matrix and Δ is the time interval. The predicted accelerations $\underline{\ddot{u}}_+$ are used with equation (4) to obtain a first estimate of the velocities $\underline{\dot{u}}_+$ and displacements $\underline{u}_+$, which are then substituted into equation (3) to obtain improved values of the accelerations $\ddot{u}_+$. The iteration is continued until adequate convergence is obtained. The solution is then advanced by a further time step. The length of the time step is taken to be

$$\Delta \leq s/5C_L \qquad (5)$$

where s = smallest finite element side length
C_L = speed of dilatational wave

CRACK EXTENSION CRITERION

At the present stage of the development of the programme, the crack path must be known. In DCB and strip geometries, the crack path may reasonably be assumed to be the plane of symmetry and only one half of the specimen is modelled. Figure 1 shows the DCB geometry for which the present analyses were made. The finite element mesh comprised 250 elements and 157 nodes. Nodes on the plane of symmetry are restrained against displacement normal to the crack plane until the crack front has passed, after which they are released. The forces on the restrained nodes are monitored.

It has been shown by Keegstra [14] that, for a given mesh, the force on the crack tip node is proportional to the stress intensity factor K and may, therefore, be used as a crack extension criterion. In the execution of the programme, the crack tip node is released when the restraining force reaches a prescribed value F_c, which depends on the mesh size and on the dynamic toughness K_D. The crack speed is calculated from the intervals between the release times of adjacent nodes. The crack is assumed to have arrested if the force on the crack tip nodes does not reach F_c within a reasonable period measured from the time at which the previous node is released. This time period is chosen, arbitrarily, to correspond to a crack speed of 0.01 C_L.

When the force on the crack tip node reaches F_c and the node is released, the force is not reduced instantaneously to zero but is reduced linearly with nodal displacement, according to the equation.

$$F_b = F_c(1 - \bar{u}/u_c) \qquad (6)$$

where

$$\bar{u} = \int_0^{t^*} \frac{u(t)}{t}\, dt \qquad (7)$$

and u_c = reference displacement
$u(t)$ = nodal displacement
t^* = current time

Thus $\bar{u}$ is a time-averaged displacement. The node therefore does work against the "holding-back" force F_b. This provides an energy sink which, by making an appropriate choice of the reference displacement u , represents the generalised surface energy. It has been shown [14] that the appropriate value of u_c is given by

$$u_c = F_c/K_n \tag{8}$$

where K_n is the stiffness of the node. This method of providing an energy sink is illustrated in Figure 2.

DCB ANALYSES

A force was applied to the node representing the loading pin, such that the initial stress intensity factor K_Q reached a value which exceeded K_{1C} by a factor chosen to characterise the bluntness of the initial crack. The dynamic calculation was initiated by releasing the crack tip node. Throughout the analysis, the displacement of the loading pin was held constant, equal to the initial value, modelling fixed-grip conditions.

To enable a comparison with the experimental results of Kobayashi and Mall (see [7]) and the analytical results of Hahn et al [7], the material properties used by those authors, which relate to Homalite-100, were used in the present analyses. Also, the same value of K_Q was used, although this was necessarily inferred from the quoted value of the initial strain energy. Figure 3 shows the assumed relationship between K_D/K_{1C} and crack speed v [7]. Figure 4 shows the relationship between the computed crack speed and crack length. For comparison, the figure also shows the analytical results of Hahn et al and the experimental results of Kobayashi and Mall. Figure 5 shows the variation with time and/or crack length of the various energy terms including the potential energy of the loading pin.

FINITE STRIP ANALYSES

The geometry for which these analyses were made is shown in Figure 6. The specimen was assumed to be of steel. As for the DCB analyses, fixed grip loading conditions were assumed. In each of these analyses, however, the crack tip nodes were released at prescribed time intervals, regardless of the magnitude of the force on the node. In other words, the specimens were assumed to be "sliced" at constant speed. The act of slicing at a given speed implies a certain ratio of static and dynamic energy release rates (G_{STAT}/G_D). The analyses covered a range of constant slicing speeds from 0.18 C_R to 1.8 C_R where C_R is the Rayleigh wave speed. Figure 7 shows the crack face displacement profile at a sequence of time values, for a single slicing speed (2917 m/s = 0.98 C_R). From the figure it may be seen that, for this slicing speed, the crack front propagates at a speed which is lower than the slicing speed. For slicing speeds below about 0.7 C_R, the crack front speed was found to be equal to the slicing speed. For slicing speeds above this value, the crack front speed was always lower than the slicing speed and approached assymptotically the Rayleigh wave speed as the slicing speed was increased. The relationship between the crack front speed and the slicing speed is shown in Figure 8.

For each value of slicing speed, G_D was calculated (using equation (1)) for a sequence of values of crack length. For this geometry, these values were nearly constant over the range of crack length. An average value of G_D was calculated for each slicing speed. The ratio $g(v) = G_D/G_{STAT}$ is shown plotted against the ratio v/C_R in Figure 9. The figure shows the analytical results of Nilsson [2] for an infinite strip. The figure also shows results of similar slicing runs on the DCB geometry shown in Figure 1 and the analytical results of Freund [1] for an infinite plate.

DISCUSSION AND CONCLUSIONS

The close agreement between the predicted and measured crack speeds in the DCB geometry, see Figure 4, gives some confidence in the validity of the computer programme and in the method of modelling the energy sink at the crack tip. Further evidence of the validity of the programme is provided by Figure 3, which shows, in addition to the input relationship between K_D/K_{1C} and v, values of K_D/K_{1C} obtained from the output values of G using the relationship

$$G_D = K_D^2 \, f(v)/E \qquad (9)$$

The function f(v), given by Nilsson [2], depends on the crack speed, shear wave speed, longitudinal wave speed and Poisson's ratio.

The potential energy of the loading pin, shown in Figure 5, exhibits fluctuations which are due to the arrival of stress waves at the loading pin. The propagation of stress waves, and reflections from the boundaries of the specimen, are clearly seen in various forms of graphical output which have been generated, also in a film which has been made from these outputs.

For the finite strip, Figure 9 shows the marked dependence of g(v) on the geometry of the strip. Additional results, obtained for longer strips, show that, as the strip length is increased, g(v) approaches Nilsson's result for an infinite strip. For the finite strip geometry, instantaneous values of G_D never exceeded G_{STAT} by more than a few percent. (Recall that Figure 9 shows the *average* value of g(v) for each crack speed). By contrast, in the DCB geometry, instantaneous values of G_D exceeded G_{STAT} by up to 50%. In other words, in a DCB specimen under fixed grip loading, kinetic energy plays a greater role in the mechanics of crack propagation.

The form of Figure 8 is not yet fully understood by the authors. One possible explanation of the non-linearity, for crack speeds above about 0.7 C_R, is the compressive stress which according to Baker [15] develops ahead of the crack at high crack speeds. Some evidence of this compressive stress was seen in the computer outputs, despite the coarse meshes which were necessarily used.

In conclusion, the results so far obtained give confidence in the validity of the programme. The results presented in the paper provide additional knowledge and understanding of crack propagation in DCB and finite strip geometries. The programme is now being used in the analysis of experimental results for these and other geometries.

REFERENCES

1. FREUND, L. B., J. Mech. Phys. Solids, 20, 1972, 129 and 141.
2. NILSSON, F., Royal Inst. of Tech., Stockholm, Report 8, 1975.
3. CARLSON, J., DAHLBERG, L. and NILSSON, F., Proc. Int. Conf. Dyn. Crack Prop. (ed. Sih, G. C.), 1972.
4. NILSSON, F., Eng. Fract. Mech., 6, 1974, 397.
5. ERDOGAN, "Fracture", (ed. H. Leibowitz), 2, 1968.
6. HAHN, G. T. et al, Battelle Report, BMI 1937, 1975.
7. HAHN, G. T. et al, Battelle Report, BMI 1959, 1976.
8. HAHN, G. T. et al, ASTM STP 601, 1976, 209.

9. WELLS, A. A., Welding Research, 7, 1953, 34.
10. KOBAYASHI, A. S. and MALL, S., Proc. Conf. Dyn. Fract. Toughness, London, 1976, 259.
11. KALTHOFF, J. F. et al, Proc. Symp. Fast Fracture and Crack Arrest, Chicago, 1976.
12. ZIENKIEWICZ, O. C., "The Finite Element Method", McGraw-Hill, London, 1971.
13. HITCHINS, D. and DANCE, S. H., J. Nucl. Eng. and Design, 29, 1974, 311.
14. KEEGSTRA, P. N. R., J. Inst. Nucl. Eng., 17, 89.
15. BAKER, B. R., J. Appl. Mech., 1962, 449.

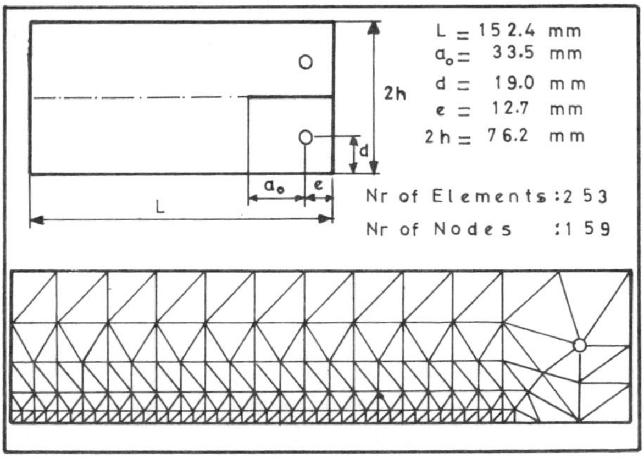

Figure 1 DCB Geometry and Mesh

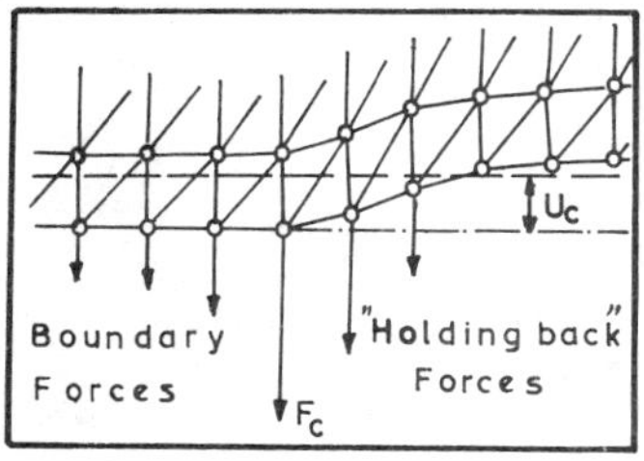

Figure 2 Fracture Energy Model

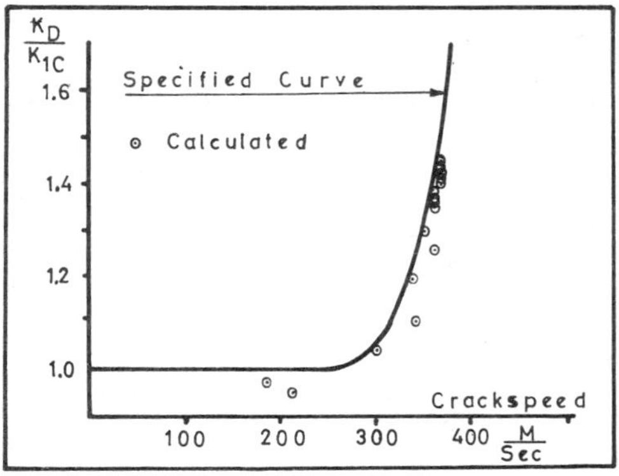

Figure 3 Assumed Relationship Between K_D and Crack Speed (Homalite-100

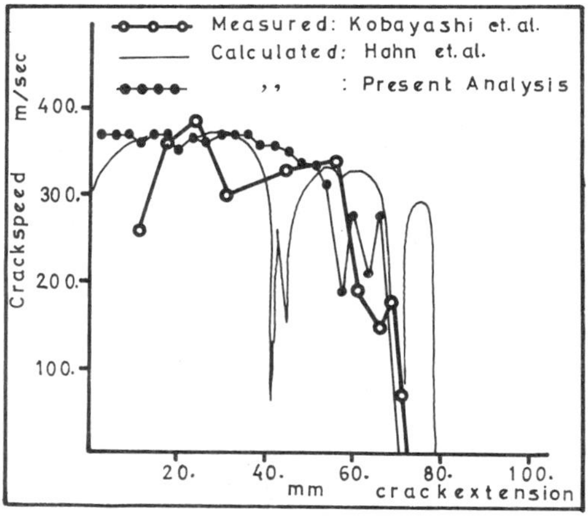

Figure 4 Calculated and Experimental Crack Speed Against Crack Length (DCB)

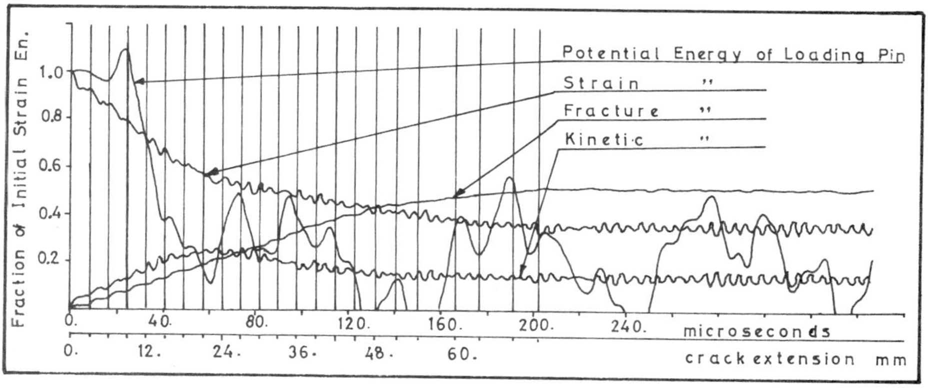

Figure 5 Calculated Energies Against Time and Crack Length (DCB)

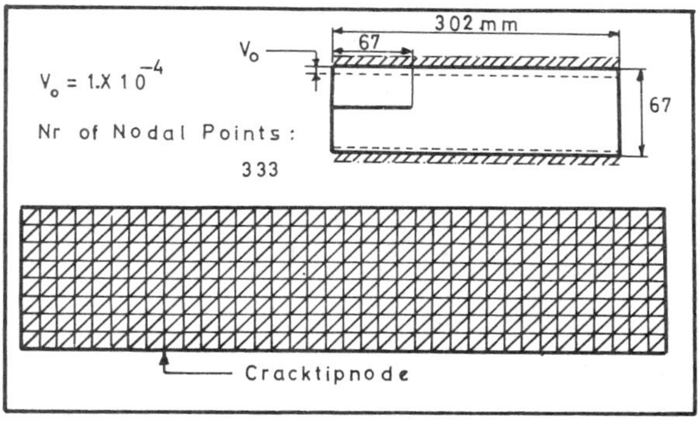

Figure 6 Finite Strip Geometry and Mesh

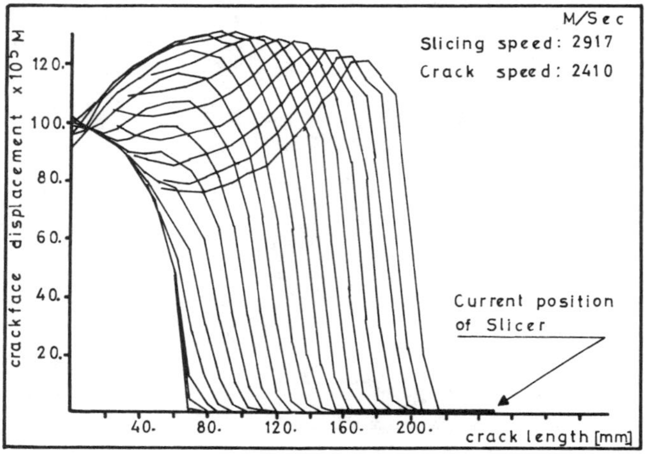

Figure 7 Crack Face Profiles (Finite Strip)

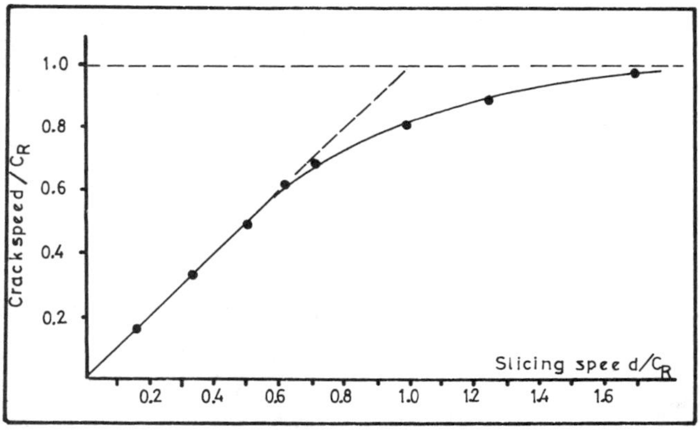

Figure 8 Crack Speed Against Slicing Speed (DCB and Finite Strip)

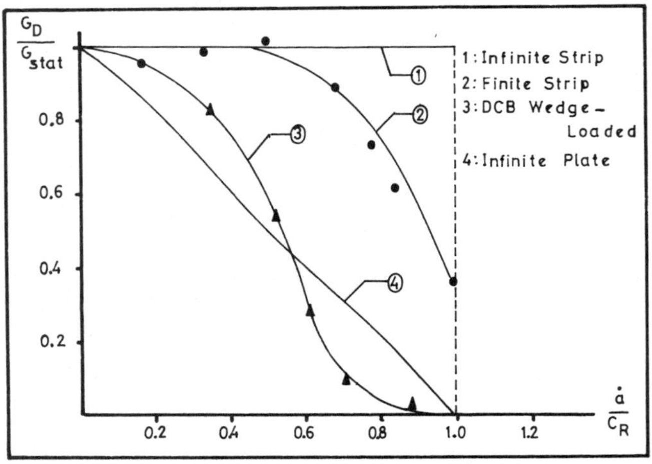

Figure 9 Ratios G_D/G_{STAT} Against Crack Speed

Author Index

The following provides an index of all the authors of plenary and workshop papers in the full proceedings. The first number indicates in which volume of the proceedings the paper is published and the second number the page in that volume.

D

E

F

G

L

M

Z